Fundamentals of Conservation Biology

Fundamentals of Conservation Biology

Second Edition

Malcolm L. Hunter, Jr.
Libra Professor of Conservation Biology
Department of Wildlife Ecology
University of Maine
Orono, Maine USA

Illustrated by
Andrea Sulzer

**Blackwell
Science**

Editorial Offices:

Commerce Place, 350 Main Street, Malden, Massachusetts 02148, USA

Osney Mead, Oxford OX2 0EL, England

25 John Street, London WC1N 2BS, England

23 Ainslie Place, Edinburgh EH3 6AJ, Scotland

54 University Street, Carlton, Victoria 3053, Australia

Other Editorial Offices:

Blackwell Wissenschafts-Verlag GmbH, Kurfürstendamm 57, 10707 Berlin, Germany

Blackwell Science KK, MG Kodenmacho Building, 7-10 Kodenmacho Nihombashi, Chuo-ku, Tokyo 104, Japan

Iowa State University Press, A Blackwell Science Company, 2121 S. State Avenue, Ames, Iowa 50014-8300, USA

Distributors:

The Americas
Blackwell Publishing
c/o AIDC
P.O. Box 20
50 Winter Sport Lane
Williston, VT 05495-0020
(Telephone orders: 800-216-2522; fax orders: 802-864-7626)

Australia
Blackwell Science Pty, Ltd.
54 University Street
Carlton, Victoria 3053
(Telephone orders: 03-9347-0300; fax orders: 03-9349-3016)

Outside The Americas and Australia
Blackwell Science, Ltd.
c/o Marston Book Services, Ltd.
P.O. Box 269
Abingdon
Oxon OX14 4YN
England
(Telephone orders: 44-01235-465500; fax orders: 44-01235-465555)

Acquisitions: Nancy Whilton
Development: Nancy Anastasi Duffy
Production: Shawn Girsberger
Manufacturing: Lisa Flanagan
Marketing Manager: Michael Rasmussen
Cover design by Gary Ragaglia
Cover photos by Malcolm Hunter
Interior design by Joyce C. Weston
Typeset by Graphic Composition, Inc.
Printed and bound by Edwards Brothers, Inc.
Printed in the United States of America
01 02 03 04 5 4 3 2 1

Library of Congress Cataloging-in-Publication Data

Hunter, Malcolm L.
    Fundamentals of conservation biology / Malcolm L. Hunter, Jr. ; [illustrated by Andrea Sulzer] — 2nd ed.
      p. cm.
    ISBN 0-86542-029-7 (hardcover)
    1. Conservation biology. I. Title.
QH75.H84 2001
333.95′16—dc21   2001001706

For Aram Calhoun, who inspires me daily with her delight in the natural world and dedication to conservation.

Contents

Preface

11:15 P.M. 20 June 1990   I'm not used to being this hot so late at night. I don't know the sounds coming through the window . . . crickets? . . . frogs? . . . a wheezing air-conditioning system? I don't know what to do.

I'm in a dorm at the University of Florida; the fourth meeting of the Society for Conservation Biology has just ended; I'm sifting through various conversations of the last 4 days. I wonder if I should postpone my plans to write a sequel to my book on managing forests for biodiversity, a sequel that would focus specifically on tropical forests. At the meeting I've discovered that professors are using my book for a much broader range of conservation courses than I ever anticipated and that tells me that there is a niche to be filled.

Apparently, various multiauthored books on conservation biology topics are not filling the need for a basic text. Perhaps I should add a brick to the foundation of the discipline before pursuing a more specific project. Now, if I can rough out an outline before I get too sleepy . . .

27 August 1993   Over 3 years later and I have just finished the first draft. Actually, the writing went reasonably quickly (I did not begin in earnest until May of 1992) because I chose a sort of stream-of-consciousness approach in which I wrote only what I knew or thought I knew. Now, I look forward to spending the next several months combing the literature and correcting, refining, and updating this draft. It might seem that this approach would make it easier to convey my original thinking about conservation biology as opposed to reporting on everyone else's thinking. Perhaps so, but I claim no truly original thoughts. I tend to think each person is no more than a unique melting pot for a vast community of ideas.

Unfortunately, I have already nearly reached my target for final length and thus keeping the book to a reasonable size and cost will be a challenge. Perhaps the best index of this is the fact that in *Wildlife, Forests, and Forestry,* I described managing forest ecosystems for biodiversity in 370 pages; in this draft the subject is covered in four pages. It has been particularly difficult trying to balance spanning the breadth of conservation biology with plumbing its depths. I have tended to err on the side of breadth on the assumption that most readers will use the book as part of a conservation biology class, and the instructor can easily focus on depth, for example, by describing applications of the principles outlined here.

24 August 1994   Sifting through the literature of conservation biology has been great fun, although it has entailed some difficult choices. If many of my readers will be North American, should I keep things familiar and easy by illustrating general principles with redwoods, bald eagles, and well-known foreign species like tigers? Or should I try to open some vistas by describing fynbos, huias, and thylacines? Many years of working abroad predispose me toward the latter approach, but I have curbed this temptation to some degree, partly to save the space it would take to describe the fynbos, and partly because I have tried to select literature that will be reasonably accessible.

As I enter the final stages of production, I often think about my readers and how they will use this book. My primary audience consists of students who have some background in biology and ecology, but who have not taken a previous conservation biology course. I also hope to reach some general-interest readers and have tried to keep the prose fairly lively so that they can manage at least half an hour of bedtime reading before dozing off.

This is an opportune place to explain some features of the book to my readers. First, you will note that there are no scientific names in the text; they are all in a separate list of scientific names, which also constitutes an index to all the species mentioned in the text. Unfortunately, there are almost no maps in this book—another decision to save space. Instead, I hope you will have an atlas on hand; after a dictionary it's the most important book in any personal library. Speaking of dictionaries, you will not find a glossary, but the Subject Index contains all the italicized words so you can easily find the first place I use and define a technical term. Finally, the Literature Cited section constitutes an index to authors, because after each citation the pages where it is cited are listed.

27 December 1994   Two more days before the book goes out to copyediting, and it is time to thank the scores of people who have helped. The first person to come to the fore is Andrea Sulzer, an exceptional artist, ecologist, and friend. We always enjoyed taking a time-out in the midst of a canoe or ski trip to go over some drawings, and we thank Zip and Aram for their patience during these episodes.

The longest list is all the people who critiqued the manuscript in its various drafts: Alan Cooperrider, Phillip deMaynadier, Ann Dieffenbacher-Krall, Alison Dibble, Carol Foss, Ed Grumbine, Vicki Ludden, Kimberly Peterson, Larry Alice, Drew Allen, Fred Allendorf, Mark Anderson, Doug Armstrong, Mike Baer, Steve Beissinger, Judy Blake, Kevin Boyle, Baird Callicott, Christopher Campbell, Jim Carlton, MaryEllen Chilelli, Tim Clark, John Craig, Eric Dinerstein, Dave Field, Jim Fraser, Tom Gavin, James Gibbs, Larry Harris, Leslie Hudson, David Jablonski, George Jacobson, Susan Jacobson, Steve Kellert, Roger King, Sharon Kinsman, Rick Knight, Irv Kornfield, Bill Krohn, Rich Langton, David Lindenmayer, John Litvaitis, Annarie Lyle, Mary Ann McGarry, Janet McMahon, Curt Meine, Laura Merrick, Ed Minot, Peter Moyle, Trinto Mugangu, Dara Newman, Dave Norton, Reed Noss, Miles Roberts, Kathy Saterson, Mark Shaffer, Michael Soulé, Bob Steneck, Kat Stewart, Stan Temple, Nat Wheelwright, Bob Wiese, Dave Wilcove, and E. O. Wilson. The first eight merit special mention for reading all, or virtually all, of the book. I could generate a list almost as long of people who have helped out in small, specific ways such as confirming a fact or locating a photo.

All the staff at Blackwell Science have been great, but I particularly wish to thank Jane Humphreys for her enthusiasm and support and Simon Rallison for encouraging me to sign on. Locally, the key people who helped with production were Bob Calhoun, copyediting; Julie Dodge, indexing and library work; Chris Halsted, computer graphics; and Shirley Moulton, typing—all did first-rate work.

The University of Maine and particularly the Department of Wildlife Ecology merit gratitude for providing a nurturing professional home for nearly 20 years. Recently, a chair endowed by Elizabeth Noyce through the Libra Foundation has made my work even more pleasant. A special thanks also goes to the University of Maine staff, alumni, and alumnae who have supported the International Biological Conservation Fund; all royalties from this book will go to this fund to support conservation students from developing countries. Finally, I am grateful to the University's Darling Marine Center for serving as a pleasant writing retreat for a few months.

The gratitude I feel toward Aram Calhoun—who has shared all but a month of our marriage with this book—is too profound for elaboration beyond the words of my dedication.

When I began writing this book, my goal was to fill a gaping hole, but now my colleagues have produced two other credible conservation biology textbooks (Primack 1993, Meffe and Carroll 1994), and more are in the pipeline. Still, I have absolutely no regrets about having embarked on this project for I have thoroughly enjoyed it, and if a small portion of my enthusiasm reaches my readers, it will be well worth the effort.

Preface to the Second Edition

January 26, 2001

Before undertaking this second edition I was rather dreading the prospect of replowing old ground, tearing apart my first edition, and putting it back together again. In hindsight, the last 9 months of sorting through the conservation biology literature have been rather enjoyable, especially after I realized that it was okay to be selective in my reading. With 651 new references there is a lot of fresh material to chew on here; most of it is very recent (my last trip to the library was this morning), although I have also added some older papers from the "classical period" of conservation biology (the 1980s). Some scepticism about the "authority" of information found on the World Wide Web has severely limited my use of these sources, but, on the other hand, I have provided many URLs to give readers a gateway to the organizations that make conservation biology happen.

Turning to acknowledgments, James Gibbs leads the list of reviewers because he read the entire draft with great speed and acumen. (Be sure to see Gibbs et al. (1998): "Problem-Solving in Conservation Biology and Wildlife Management" which is, in a sense, a lab manual for this text.) Many other people also helped with "quality control" on content, notably, Doug Armstrong, Kevin Boyle, Tim Clark, Richard Cowling, David Lindenmayer, Georgina Mace, Ed Minot, David Norton, Judith Rhymer, Bob Steneck, Eleanor Sterling, and Shelly Thomas. For production assistance Lincoln Hunt was my right-hand man locally, along with a small platoon from Blackwell Science led by Nancy Duffy and Shawn Girsberger. The most conspicuous changes in this edition will be the many new illustrations from Andrea Sulzer, my steadfast collaborator and friend. A glossary prepared by Sarah Lewis will make this edition a bit more user-friendly. Perhaps least conspicuous, but ultimately most important, is all the continued support and inspiration I receive from Aram Calhoun. Lastly, I want to thank all the readers of the first edition for various forms of support and encouragement; royalties from this book and other sources will be sufficient to endow a student fellowship in the foreseeable future.

Malcolm L. Hunter, Jr.

Part I

Sulger

Biodiversity and Its Importance

Think about our world and its wild things: a marsh splashed and flecked with the colors of flowers and dragonflies, the rhythmic roar and swoosh of waves punctuated by the strident calls of gulls, a dark forest pungent with the odors of unseen life teeming below a carpet of leaves and mosses. Imagine a future world utterly dominated by concrete and regimented rows of crops—a monotonous, ugly, and unhealthy home for us and the species we have chosen for domestication. This book is about hope in the face of forces that would degrade our world. This book is about the rich tapestry of life that shares our world now and about how we can maintain it.

Chapter 1
Conservation and Conservation Biology

What Is Conservation?

Since the beginning of humanity people have been concerned about their environment and especially its ability to provide them with food, water, and other resources. As our numbers have grown and our technology has developed, we have become increasingly concerned about the impact we are having on our environment. Newspapers herald the current issues:

- "Conservationists call for tighter fishing regulations."
- "Ecologists describe consequences of warmer climates."
- "Environmentalists request moratorium on toxin production."
- "Preservationists want more wilderness."

They also reveal an ambiguous terminology. Are we talking about conservation or preservation? Are the issues ecological or environmental? Students deciding which university to attend and which major to select are faced with a similarly bewildering array of choices—soil and water conservation, environmental studies, natural resource management, conservation biology, wild life ecology, human ecology, and more—that intertwine with one another and often cut across traditional departmental and disciplinary lines. In this chapter we will try to resolve these ambiguities by examining how they are rooted in human history and ethics. To start on common ground we will briefly examine some of the differences and similarities among conservationists, preservationists, environmentalists, and ecologists. In the second part of the chapter we will see where conservation biology fits into this picture.

A *conservationist* is someone who advocates or practices the sensible and careful use of natural resources. Foresters who prudently use trees and farmers who practice the wise use of soil and water are conservationists. Citizens who are concerned about the use of natural resources are also conservationists, and they often assert that the activities of foresters, farmers, and other natural resource managers are not prudent and wise. In theory arguments over who is, or is not, a conservationist should turn on the issue of what is sensible and careful. In practice, the foresters, farmers, and their colleagues have largely ceded the title "conservationist" to their critics. They have become reluctant to call themselves conservationists and instead use the word to describe the people they consider adversaries. The definition of conservation is further confused by the existence of a loose coalition of antienvironmental organizations that calls itself the Wise Use Movement.

A *preservationist* advocates allowing some land and some creatures to exist without significant human interference. Most people accept the idea that conservation encompasses setting a certain amount of land aside as parks and maintaining certain species without harvesting them. The divisive issues are how much land and which species. Many resource users believe that enough land has already been closed to economic use, and they use "preservationist" as a negative term for people they consider to be extremists. Because of this pejorative use relatively few people call themselves preservationists. People who find themselves labeled preservationists by others usually prefer to think of preservation as just one plank in their platform as conservationists.

An *environmentalist* is someone who is concerned about the impact of people on environmental quality. Air and water pollution are often the proximate concerns; human overpopulation and wasteful use of resources are the ultimate issues. There is enormous overlap between environmentalists and conservationists. Many environmentalists would say that environmentalism encompasses conservation, while many conservationists would say the reverse. The difference is a matter of emphasis. By focusing on air and water pollution and their root causes, environmentalists often emphasize urban and suburban situations where human-induced problems and human well-being are paramount. Because conservationists focus on natural resource use, they tend to emphasize rural areas and wildlands, as well as their associated ecosystems and organisms, including people.

Traditionally, an *ecologist* is a scientist who studies the relationships between organisms and their environments. However, in the 1970s the term developed a second meaning when the public failed to distinguish between environmentalists and the scientists (ecologists) who provided the scientific basis for the environmental movement. The confusion was understandable because most ecological scientists are also, politically speaking, environmentalists. Now "ecologist" is often used in the popular press as a synonym for "environmentalist." To make a hairsplitting distinction we can let the second definition of ecologist be a person who is concerned about the relationships between organisms (including people) and their environments.

A Brief History of Conservation

The primal roots of conservation are lost in prehistory (Fig. 1.1). No doubt there was a time when human reason, growing ever more sophisticated through the millennia, began to extend the idea of deferred gratification ("save this fruit to eat tomorrow rather than now") over much longer periods. "Leave these tubers so there will be more next year when we pass this place." "Take this calf home so that we can raise it and eat it next winter when it is bigger and we have little food." Certainly, such practices were simple, almost analogous to the food hoarding exhibited by many animals, but they represent conservation nevertheless. The roots of preservation are probably quite ancient too. With the development of spirituality and castes of priests and priestesses, some species were given special status as gods or totems that protected them from exploitation. Sometimes, large areas such

Figure 1.1. The roots of conservation can probably be found among the earliest *Homo sapiens.*

as sacred mountains were decreed off-limits or visited only on religious occasions.

Leaping forward, history records many examples of conservation throughout the ages and across cultures. For example, the biblical story of Noah's ark remains a popular metaphor for conservation. The Bible also contains the first-known game conservation law:

> If you come on a bird's nest, in any tree or on the ground, with fledglings or eggs, with the mother sitting on the fledglings or on the eggs, you shall not take the mother with the young. Let the mother go, taking only the young for yourself, in order that it may go well with you and you may live long. (Deuteronomy 22:6–7)
> (In other words, don't kill mother birds.)

A far broader law was promulgated by Asoka, emperor of India 274–232 B.C.:

> Twenty-six years after my coronation I declared that the following animals were not to be killed: parrots, mynahs, the aruna, ruddy geese, wild geese, the nandimukha, cranes, bats, queen ants, terrapins, boneless fish [shrimp] . . . tortoises, and porcupines, squirrels, twelve-antler deer, . . . household animals and vermin, rhinoceroses, white pigeons, domestic pigeons, and quadrupeds which are not useful or edible. . . . Forests must not be burned.

Many laws focused on regulating rather than prohibiting the exploitation of species. For example, Middle Eastern pharaohs issued waterfowl hunting licenses, and night hunting was banned in the city-states of ancient Greece (Alison 1981). Early regulations emphasized trees and birds, mammals, and fish caught for food, but all species and whole ecosystems benefitted from the popularity of declaring preserves. Starting at least 3000 years ago with Ikhnaton, king of Egypt, and continuing with the royalty of Assyria, China, India, and Europe, as well as with the Greeks, Romans, Mongols, Aztecs, and Incas, history has recorded many decrees setting aside land to protect its flora and fauna (Alison 1981).

Conservation was an issue during the period when European states were colonizing the rest of the world because colonization often led to disruption of traditional systems of natural resource use and rapid overexploitation, despite the protestations of some sensitive, farsighted people who argued for moderation. This was particularly true on some small, tropical islands such as Mauritius and Tobago, where the consequences of overexploitation became apparent very quickly (Grove 1992). Freedom from feudal game laws was a significant stimulus to colonization of North America. Imagine how attractive the promise of abundant, free game would seem to people who feared for their lives whenever their appetite for meat led them to poach one of the king's deer. The promoters of North American colonization knew this, and their claims became so exaggerated that one writer felt compelled to set the record straight:

> I will not tell you that you may smell the corn fields before you see the land; neither must men think that corn doth grow naturally (or on trees), nor will the deer come when they are called, or stand still and look on a man until he shoot him, not knowing a man from a beast; nor the fish leap into the kettle, nor on the dry land, neither are they so plentiful, that you may dip them up in baskets, nor take cod in nets to make a voyage, which is no truer than that the fowls will present themselves to you with spits through them. (Leven 1628, quoted from Cronon 1983)

Exaggerated or not, the bountiful resources of eastern North America succumbed quickly under the pressure of colonists, joined by Native Americans who were armed with guns and motivated to trade game and furs for manufactured goods (Cronon 1983). Soon the colonists found that they had to regulate themselves. As early as 1639 it was illegal to kill deer between May 1 and November 1 in Newport, Rhode Island (Trefethen 1964). This basic pattern—human populations growing, expanding into new areas, developing new technology, and then responding to overexploitation with an array of ever more restrictive conservation regulations—has been repeated across the globe and continues to this day.

With increasing human impacts, the abuse of resources other than trees and large animals also began to be recognized. For example, in the 1930s, when soil erosion in the midwestern United States generated clouds of dust that reached the East Coast (and the nation's capital), the Soil Erosion Service (now Natural Resources Conservation Service) was formed and the Taylor Grazing Act was passed. Society has been slower to recognize the importance of conserving resources that lack obvious economic value such as most invertebrates, small plants, amphibians, and reptiles. Aldo Leopold (1949) called for saving every species with his well-known admonition, "To keep every cog and wheel is the first precaution of intelligent tinkering," but it was not until the 1960s and 1970s that the idea of "endangered species" became a major issue for conservationists. During this period many nations passed laws (e.g., the United States Endangered Species Act) to form an umbrella under which all animal and plant species threatened with extinction can, in theory, benefit from conservation activities. In practice, however, smaller plants and animals still are not given equal treatment, and other biological entities such as microorganisms, genes, and ecosystems are usually not explicitly under the umbrella at all.

This brings us to the point of departure for conservation biology and this book, but first let us briefly return to preservation, environmentalism, and ecology to see where they fall in this history of conservation.

Although the early roots of preservation may lie in the proscriptions of religious leaders and royalty, many people would identify the establishment in 1872 of Yellowstone National Park, the world's first national park, as the beginning of a preservation policy. Here were 9018 square kilometers of evidence that society recognized the importance of removing some natural resources from traditional economic development. The national park movement has spread throughout the world and has been modified in many ways—some parks are preserved even from tourists, while some parks focus on human artifacts such as historic sites—but the underlying value system remains largely intact. This same preservationist value system has also ended the exploitation of many species. Not only have species on the brink of extinction become targets for preservation efforts, but societies also have decided to preserve some other species that could be harvested as a natural resource. Many countries, for example, have banned the harvesting of all songbirds even though some species could be harvested in a sustainable manner.

The first environmentalists were probably citizens of our earliest cities, more than 3000 years ago, who complained of water pollution and demanded the construction of sewer systems. The industrial revolution accelerated urbanization and brought its own problems such as coal burning and factory discharges into water bodies. Attempts to regulate such problems began centuries ago, but environmental issues coalesced into a broad-based political movement only after the first Earth Day in 1970, catalyzed significantly by Rachel Carson's 1962 treatise on pesticides, *Silent Spring*. By the 1990s, environmental issues were widely recognized as a global priority and were the focus of the largest gathering of world leaders in history at the 1992 Earth Summit (more formally, the United Nations Conference on Environment and Development).

As is true of most sciences, elements of ecology can be traced to Hippocrates, Aristotle, and other Greek philosophers, but it was not until 1869 that the word "ecology" was coined. Scientific societies of ecology and ecolog journals followed in the early 1900s, and ecology soon proved useful in developing a scientific basis for forestry and other areas of natural resource management. However, ecology did not move into the public eye until the advent of environmentalism in the 1960s and 1970s. The environmental movement spawned new government agencies, advocacy groups, and consulting firms, and universities educated large numbers of ecologists to fill these organizations. Schools at all levels began informing students about the relationships between organisms and their environment. Consequently, there are now many professional ecologists at work solving environmental problems and many more people who call themselves ecologists out of concern for these issues.

An Overview of Conservation Ethics

It is easy to describe the history of conservation in terms of political benchmarks such as the passage of laws, but these are only a manifestation of a more fundamental process: the evolution of human value systems or ethics. We will encounter conservation ethics in many chapters and will focus on the topic in Chapter 15, "Social Factors," but a brief preview here will complement our history of conservation and will provide a foundation for later chapters.

A milestone paper, "Whither conservation ethics?" by J. Baird Callicott (1990), placed conservation ethics in the United States into an historical context using the writings of three people—John Muir, Gifford Pinchot, and Aldo Leopold—to describe three ethics: the Romantic-Transcendental Preservation Ethic, the Resource Conservation Ethic, and the Evolutionary-Ecological Land Ethic, respectively (Fig. 1.2).

The Romantic-Transcendental Preservation Ethic found its American roots in the philosophy of Ralph Waldo Emerson and Henry David Thoreau. By the end of the 19th century this ethic had become the basis for political action in the hands of John Muir (1838–1914), the Californian writer and naturalist who founded the Sierra Club and who was best known in some circles for climbing trees during thunderstorms to experience nature at its fullest. Muir believed that communion with nature brings people closer to God and that visiting ancient forests and alpine meadows for this purpose is morally superior to using them to cut timber or graze livestock. In other words, nature is a temple that is sullied by the economic activities of people. Obviously, such an ethic puts a high premium on establishing parks and similar areas where nature is preserved reasonably intact.

At about the same time that Muir was calling for the preservation of extensive lands, Gifford Pinchot (1865–1946) was formulating a very different value system, the Resource Conservation Ethic. Pinchot was a wealthy, European-educated forester, the founder of the U.S. Forest Service, and a close ally of President Teddy Roosevelt. To Pinchot, nature consisted solely of natural resources and should be used to provide the greatest good for the greatest number of people for the longest time. This was not a call to plunder the land, but to use it in a way that distributes benefits fairly and efficiently among many people, rather than among a handful of lumber barons and cattle kings. It also advocated wise, judicious use of natural resources so that future generations would not be shortchanged. By recognizing aes-

Figure 1.2. John Muir, Gifford Pinchot, and Aldo Leopold (depicted clockwise) are credited with originating three conservation ethics: the Romantic-Transcendental Preservation Ethic, the Resource Conservation Ethic, and the Evolutionary-Ecological Land Ethic, respectively.

thetics as a resource, the Resource Conservation Ethic even found room for a modest amount of preservation to accommodate Transcendental philosophers and Romantic poets. Given these precepts and a history of overexploitation of the nation's natural resources, Pinchot believed that natural resources should be owned or regulated by the government. The Resource Conservation Ethic meshed well with the Progressive democratic ideals that were ascendant in the United States at that time and thus was a great success politically.

Although there was a profound gap between Muir and Pinchot's ethics, they both espoused an anthropocentric (people-centered) view of nature. They both wrote of nature's utility—its *instrumental value* in the terminology of philosophers. One promoted nature as a source of spiritual enlightenment, the other as a source of commodities, but neither claimed that nature had *intrinsic value,* value independent of its usefulness. However, in his journals, which were published posthumously,

Muir seemed to entertain the idea that nature may have intrinsic value as a work of God:

> Why should man value himself as more than a small part of the one great unit of creation? And what creature of all that the Lord has taken the pains to make is not essential to the completeness of that unit—the cosmos? The universe would be incomplete without man; but it would also be incomplete without the smallest transmicroscopic creature that dwells beyond our conceitful eyes and knowledge. (Muir 1916)

With the arrival of the science of ecology and the writings of Aldo Leopold (1886–1948)—founder of wild life conservation as a professional discipline, a man who began his career eradicating predators, but ended it as a strong advocate of wilderness—one finds a utilitarian perspective of species being questioned:

> Ecology is a new fusion point for all the sciences. . . . The emergence of ecology has placed the economic biologist in a peculiar dilemma: with one hand he points out the accumulated findings of his search for utility or lack of utility in this or that species; with the other he lifts the veil from a biota so complex, so conditioned by interwoven cooperations and competitions, that no man can say where utility begins or ends. (Leopold 1939)

Leopold was saying that because nature is a complex system rather than a random set of species with positive, negative, and neutral values, each species is important as a component of the whole. In other words, species have instrumental value because of their utility in an ecosystem. Furthermore, by this time in history members of the human species had been forced by Darwin to consider the proposition that, instead of having been created in the image of God, they might have evolved from other animals. If humans are just another product of evolution, then perhaps all species have intrinsic value. These were the key ideas that spawned the Evolutionary-Ecological Land Ethic. These were fundamentally different ideas that took Leopold's ethical vision beyond the choice of either preserving nature as inviolate or efficiently developing it. Muir wrote of the equality of species in religious terms; Leopold expressed equality in ecological terms. Pinchot (1947) stressed the dichotomy between people and nature ("there are just two things on this material earth—people and natural resources"); Leopold thought of people as citizen-members of the biotic system. Leopold's ideas gave people the right to use and manage nature *and* the responsibility of doing so in a manner that recognized the intrinsic value of other species and whole ecosystems.

All three of these ethics are still thriving. The Resource Conservation Ethic guides the actions of natural resource–based industries and their associated government agencies, although some would argue that the profit motive is too often the stronger guide. Many private conservation organizations are wedded to the Romantic-Transcendental Preservation Ethic, reflecting a membership that uses nature primarily for spiritual rejuvenation during brief forays out of the cities and suburbs. The Evolutionary-Ecological Land Ethic characterizes some conservation groups and government agencies (e.g., many park and wild life agencies) that try to balance the needs of people and wild life (Clark 1998). The idea that people have the rights and responsibility to manage nature carefully may be more common outside of North America, especially in Europe, where the hand of humanity is conspicuous on virtually every landscape, and, in developing countries, where the urgency of providing for the needs of poor, rural people is widely recognized.

In the conclusion to his essay, Callicott challenged conservationists with a provocative idea. If people are valid members of the biotic community as Leopold asserts, why do we turn to landscapes without people (at least without agricultural-industrial age people) to set benchmarks for what is natural? If beavers and reef-building corals can shape landscapes in positive ways, why can't people? Can people improve natural ecosystems? These are not simple issues, and we will return to them frequently in this book because this dynamic, often difficult, interface between people and nature is the crux of conservation and conservation biology.

What Is Conservation Biology?

Conservation biology is the applied science of maintaining the earth's biological diversity. A simpler, more obvious definition—biology as applied to conservation issues—would be rather misleading because conservation biology is both more and less than this. It is more than this because it reaches far beyond biology into disciplines such as philosophy, economics, and sociology that are concerned with the social environment in which we practice conservation, as well as into disciplines such as law and education that shape the ways we implement conservation (Jacobson 1990, Soulé 1985). On the other hand, it is less than this simple definition because there are many biological aspects of conservation, such as biological research on how to grow timber, improve water quality, or graze livestock, that are only tangentially related to conservation biology.

Thirty years ago maintaining biological diversity meant saving endangered species from extinction and was considered a small component of conservation, completely overshadowed by forestry, soil and water conservation, fish and game management, and related disciplines. Now with so many species at risk and the idea of biological diversity extending to genes, ecosystems, and other biological entities, conservation biology has moved into the spotlight as the crisis discipline focused on saving life on earth, which is perhaps the major issue of our time.

A Brief History of a Young Discipline

The origin of conservation biology is usually attributed to the First International Conference on Conservation Biology held in San Diego, California, in 1978, and to the book that followed, *Conservation Biology,* edited by Michael Soulé and Bruce Wilcox (1980). Eight years after this small beginning the Society for Conservation Biology was formed, and it launched a new journal, *Conservation Biology* in 1987. The society and its journal have flourished, and universities, foundations, private conservation groups, and government agencies have nurtured this growth with an array of conservation biology programs (Jacobson 1990).

Judging by their professional addresses, the founders of conservation biology had many more links to institutions of basic biological sciences (e.g., genetics, zoology, botany) than to natural resource management institutions. You might surmise that these were academic biologists who had begun their careers studying the earth's biota, later became aware of the extinction crisis, and then turned their talents to addressing this issue. In other words, curiosity led to awareness and awareness led to action.

By forming a new professional society dedicated to the maintenance of biological diversity, conservation biologists overlapped some of the domain of some older professional societies, especially The Wildlife Society. On the first page of the first issue of *The Journal of Wildlife Management,* management is described as "part of the greater movement for conservation of our entire native flora and fauna" (Bennitt et al. 1937), and the opening sentence of the second article in that issue refers to "the new and growing field of conservation biology . . ." (Errington and Hamerstrom 1937). Certainly, the extinction of the passenger pigeon and the near extinction of the American bison, wood duck, and other species were major catalysts for the emergence of wild life management as a professional discipline. Nevertheless, the dominant concern of wild life management in its early years was managing populations of mammals and birds for sport hunting. Over 60 years later this is still largely true. Recently, wild life management has placed a growing emphasis on endangered and nongame species, including reptiles, amphibians, and sometimes even invertebrates and plants, but the shift has been too slow and limited from some people's perspective. Perhaps, if more wild life managers had reached out to embrace all forms of life that are wild, not just the vertebrates, and to work with a constituency of all people who care about nature, not just hunters and anglers, then conservation biology might not have arisen as an separate discipline. (I prefer to define "wild life" as "all forms of life that are wild," but I recognize that many people use the word only for vertebrate animals, often excluding even fish, as in "United States Fish and Wildlife Service." To make it clear that this book uses a broad definition, the original, two-word spelling, "wild life," will be used. This follows the convention adopted by some Canadian conservationists.)

The overlap in their goals led to a bit of competitive friction between The Wildlife Society and The Society for Conservation Biology, primarily because some members of The Wildlife Society resented the claims of some conservation biologists that they were doing something new. Such claims did not give due credit to the work that members of The Wildlife Society had been undertaking for many decades (Jacobson 1990). Fortunately, in 1992 the two societies sponsored a joint conference, marking their realization that cooperation is better than competition.

A Model of Conservation Biology

Susan Jacobson (1990) devised a schematic model to illustrate the structure of conservation biology from an educational perspective (Fig. 1.3).

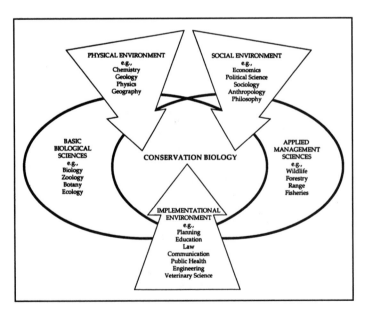

Figure 1.3. A schematic view of the relationship between conservation biology and other disciplines. (Reproduced by permission from Jacobson 1990.)

It shows how students wishing to become conservation biologists need to focus on courses in the basic biologic sciences and the applied sciences of natural resource management while acquiring some understanding of the subjects that shape the arena within which conservation operates. These include physical sciences such as geology and climatology, social sciences such as economics and politics, and subjects such as law, education, and communication that provide a vehicle for changing the structure of society.

We can also use this model to examine conservation biology itself. Conservation biology sits between basic biologic sciences and natural resource sciences because it originated largely with basic biologists who have created a new, applied natural resource science. It is different from traditional natural resource sciences because it places relatively greater emphasis on all forms of life and their intrinsic value, compared with other natural resource sciences that usually focus on a few economically valuable species (Soulé 1985). Like natural resource sciences, conservation biology is influenced by the physical sciences because it addresses issues with strong ecological and environmental linkages. Similarly, it is influenced by social sciences, law, education, and other disciplines because it operates in the world of human socio-economic-political institutions and seeks to change those institutions to allow people to coexist with the rest of the world's species.

Summary

People who care about nature and the natural resources we obtain from nature, such as clean air and clean water, come with many labels: conservationists and preservationists, environmentalists and ecologists. Although these people share many goals, their priorities can differ. For example, conservationists advocate the careful use of natural resources, whereas environmentalists often emphasize maintaining an uncontaminated environment.

The history of conservation has a recurring theme: people being forced to limit their use of natural resources more and more as human populations grow and technologic sophistication increases. Conservation history is marked by laws regulating our use of natural resources, but more fundamental is the evolution of our ethical attitudes toward nature and its intrinsic and instrumental values. Callicott (1990) has described three such ethical positions: 1) the Romantic-Transcendental Preservation Ethic (briefly, nature is best used for spiritual purposes); 2) the Resource Conservation Ethic (nature is natural resources to be carefully developed for human purposes); and 3) the Evolutionary-Ecological Land Ethic (people are part of nature and have both the right to change it and a responsibility for respecting the intrinsic value of other species and ecosystems in general).

Conservation biology is the applied science of maintaining the earth's biological diversity. It is a cross-disciplinary subject lying between basic biologic sciences and natural resource sciences. It differs from basic biologic sciences because it reaches out to economics, law, education, politics, philosophy, and other subjects that shape the human world within which conservation must operate. It differs from traditional natural resource sciences because it places relatively greater emphasis on all forms of life and their intrinsic value, compared with other natural resource sciences, which typically focus on a few economically valuable species.

FURTHER READING

A world history of conservation is not available, but there are good treatments for certain times and places such as Costa Rica (Evans 1999), India (Gadgil and Guha 1992), and colonial New England (Cronon 1983), plus a compilation of readings on U.S. conservation history (Nash 1990), a reference book (Davis 1983), and a journal, *Environmental History*. Related books include histories of ecology (Worster 1977), environmental ethics (Nash 1988), environmental policies (Andrews 1999), and human impacts on the environment (Ponting 1991). Articles by Soulé (1985), Callicott (1990), and Jacobson (1990) form a foundation for the latter parts of the chapter and merit further reading. For relevant websites, checkout the Society for Conservation Biology's website at www.conservationbiology.org and some of the major international conservation groups at: www.iucn.org, www.wwf.org, www.nature.org, and www.worldwildlife.org.

TOPICS FOR DISCUSSION

1. Do you think of yourself primarily as a conservationist, environmentalist, ecologist, or preservationist, or none of these? Why?
2. Can you draw some linkages between the history of conservation and conventional history, which emphasizes politics and economics?
3. Which of the three ethics described by Callicott do you think will be predominant 50 years from now? Why?
4. Name some organizations that exemplify each of the three ethics today. Have any of these organizations changed their philosophy?
5. Can you identify some specific examples of how each of the disciplines in Figure 1.3 has contributed to conservation biology?

Chapter 2
What Is Biodiversity?

Vestiges of tropical rain forests covered in soot and mud, elephant carcasses bereft of their ivory, sea turtles drowned in a shrimp trawl—these images have become well known, and for many people they all revolve around a central issue and a single word, "biodiversity." Some have argued that "biodiversity" is too vague and trendy to be a useful word, but it does succinctly imply a fundamental idea: life on earth is extraordinarily diverse and complex. This idea is not as well captured in other words such as "nature" or "wild life." Furthermore, "biodiversity" has entered the public vocabulary at a time when global concerns about the survival of life are at their zenith, and thus to many people the term carries a conviction to stem the loss of the planet's life-forms.

Definitions of biodiversity usually go one step beyond the obvious—the diversity of life—and define biodiversity as the diversity of life in all its forms and at all levels of organization. "In all its forms" reminds us that biodiversity includes plants, invertebrate animals, fungi, bacteria, and other microorganisms, as well as the vertebrates that garner most of the attention. "All levels of organization" indicates that biodiversity refers to the diversity of genes and ecosystems, as well as species diversity. The idea that biodiversity has levels of organization introduces a depth of complexity that we will explore in the next three chapters, "Species Diversity," "Ecosystem Diversity," and "Genetic Diversity," after an overview here.

Species, Genes, and Ecosystems

It is easiest to comprehend the idea of maintaining biodiversity in terms of species that are threatened with extinction. We know about blue whales, giant pandas, and whooping cranes, and we would experience a sense of loss if they were to disappear, even though most of us have never encountered them except in films and magazine articles. For most mosses, lichens, fungi, insects, and other small species that are unknown to the general public, it is much harder to elicit concern. Nevertheless, many people are prepared to extend some of the feelings they have for whales, pandas, and cranes to species they do not know, as an expression of their belief that all species have some intrinsic value.

Like tiny obscure species, genes are rather hard to understand and appreciate. These self-replicating pieces of DNA that shape the form and function of each individual organism are obviously important, but so are water, oxygen, and thousands of other molecules. It is not the genes themselves that conservation biologists value; it is the diversity that they impart to organisms that is so essential. If two individual strawberry plants have a different set of genes, one of them might be better adapted to fluctuations in water availability and thus would be more likely to survive a period of climate change. One of them might be less susceptible to damage from ozone and other types of air pollution. The fruit of one might be more resistant to rotting and therefore its progeny might prove useful to strawberry breeders and farmers. Perhaps the fruits simply taste different and thereby provide aesthetic diversity. The diversity of life begins with genetic differences and the processes of evolution.

Unlike genes, ecosystems are large and conspicuous, and thus anyone with the most rudimentary understanding of ecology appreciates the value of lakes, forests, wetlands, and so on. Nevertheless, ecosystems can be hard to define in practice. Where do you draw the boundary between a lake and the marsh that surrounds it when many organisms are moving back and forth between the two? This sort of problem can complicate the role of ecosystems in biodiversity conservation. Conservation biologists often advocate protecting examples of all the different types of ecosystems in a region, but how finely should differences be recognized? Is an oak-pine forest ecosystem that is 60% oak and 40% pine appreciably different from one that is 40% oak and 60% pine? If you look hard enough, every ecosystem will be unique. The rationale for protecting ecosystem diversity also differs. Some conservationists advocate protecting ecosystems as independent biological entities that are not just a loose assemblage of species, whereas others think of protecting ecosystems simply as an efficient way to protect the species that compose the ecosystem.

Structure and Function

The definition of biodiversity provided above emphasizes structure—forms of life and levels of organization—but sometimes ecological and evolutionary functions or processes are also included in a definition of biodiversity. For example, The Wildlife Society (1993) defines biodiversity as "the richness, abundance, and variability of plant and animal species and communities and the *ecological processes that link them with one another and with soil, air, and water*" (emphasis added).

The diversity of ecological functions is enormous. First, each of the earth's millions of species interacts with other species, often many other species, through ecological processes such as competition, predation, parasitism, mutualism, and others. Second, every species interacts with its physical environment through processes that exchange energy and elements between the living and nonliving world such as photosynthesis, biogeochemical cycling, and respiration. All of these functional interactions must total in the billions. The diversity of evolutionary functions is even more complex. It includes all these ecological processes because they are key elements of natural selection, in addition to processes such as genetic mutation that shape each species' genetic diversity.

Functional biodiversity is clearly important. For example, a management plan designed to keep a species from becoming extinct will almost certainly fail in the long run unless the processes of evolution, especially natural selection, continue, allowing the species to adapt to a changing environment. Sometimes, focusing on a functional characteristic, for example, the hydrological regime of a wetland (Turner et al. 1999), is the most efficient way to maintain biodiversity of an ecosystem. Nevertheless, conservation biologists usually focus on maintaining structural biodiversity rather than functional biodiversity for two reasons. First, maintaining structural biodiversity is usually more straightforward. In particular, it is easier to inventory species than their interactions with one another. Second, if structural diversity is successfully maintained, functional biodiversity will probably be maintained as well. If we can maintain a species of orchid and one of its insect pollinators together in the same ecosystem, then we will probably have a pollination interaction between the two. Similarly, if we can maintain the orchid's genetic diversity, we will probably have orchid evolution. The qualifier "probably" has been added here because one can imagine circumstances in which structural diversity is maintained without maintaining functional biodiversity in full. For example, natural selection may not have the opportunity to operate on the genetic diversity represented in the seeds that plant breeders store in a freezer to maintain the structural diversity of a species of crop plant. On the other hand, it is much easier to think of circumstances where some major ecological processes are maintained, but structural diversity is severely degraded; for example, a plantation of exotic trees that maintain normal rates of photosynthesis and biogeochemical cycling.

In short, both the structural and functional aspects of biodiversity are important; however, if genetic, species, and ecosystem diversity are successfully maintained, then ecological and evolutionary processes will probably be maintained as well.

Measuring Biodiversity

It is easy to provide a simple definition of biodiversity such as "the diversity of life in all its forms and at all levels of organization," but this is only a starting point. To monitor biodiversity and develop scientific management plans, we should have a quantitative definition that allows us to measure biodiversity at different times and places.

The first step in measuring biodiversity is to determine which elements of biodiversity are present in the area of interest. Ideally, we would have a complete inventory, including genes, species, and ecosystems. In practice, logistical constraints commonly limit us to a partial list of species, often listing only vertebrates and perhaps vascular plants. (Sometimes a list of ecosystems is compiled, although the basis for distinguishing among the different types is often unclear; we will focus on the species level of biodiversity here for simplicity.) Lists can be tallied to provide a crude index of biodiversity. In Table 2.1, for example, ecosystem A is easily recognized as more diverse than B or C because it has four species instead of three. This characteristic is called *species richness* or just richness, and it is a simple, commonly used measure of diversity.

Ecologists also recognize a second component of diversity called *evenness* that is based on the relative abundance of different species. In Table 2.2 ecosystem C is more diverse than B because in C the three species have similar levels of abundance, or high evenness. The concept of evenness is not as intuitively obvious as the idea of richness. It may help to think of a jury that has five women and five men versus one that has eight women and two men; the five plus five jury is more diverse because it is more even.

The ecological importance of species richness seems quite evident,

TABLE 2.1 Hypothetical lists of species for three ecosystems.		
Ecosystem A	Ecosystem B	Ecosystem C
Black oak	Black oak	Black oak
White pine	White pine	White pine
Red maple	Red maple	Red maple
Yellow birch		

TABLE 2.2 Abundance of species (number/hectare) in three ecosystems and measures of richness, evenness, and the Shannon diversity index (H).

Ecosystem	A	B	C
Black oak	40	120	80
White pine	30	60	60
Red maple	20	20	60
Yellow birch	10	—	—
Richness	4	3	3
Evenness	0.92	0.88	0.99
H	0.56	0.39	0.47

$H = -\Sigma\, p_i \log p_i$ where p_i is a measure of the importance of the ith species. Evenness $= H/H_{max}$ where H_{max} is the maximum possible value of H.

especially if you consider the loss of richness through extinction. Similarly, most conservation biologists would be concerned about any process that reduced evenness, because this would mean uncommon species are becoming less common, while common species are becoming more common. To return to our jury metaphor, this would be analogous to losing a man from the jury that only had two men.

Richness and evenness are often combined into a single index of diversity using mathematical formulas. Sometimes, they are both plotted together on a graph to represent diversity patterns visually. Devising ways to quantify diversity is a fertile field for quantitative ecologists (see Magurran 1988), but, surprisingly, it is not of prime interest to conservation biologists for two reasons. First, as mentioned, a total inventory of all elements of biodiversity is essentially impossible to obtain because of the difficulties in defining genes and ecosystems and inventorying genes, microbes, fungi, small invertebrates, and other small things. Second, as we will discover in the next section, mathematical formulas treat all elements of biodiversity as equally important, but both conservation biologists and the general public do not see things this way.

The Mismeasure of Biodiversity

Often, being precise and quantitative will reveal solutions to a difficult problem, but using quantitative indices of diversity can be misleading when maintaining biodiversity is the goal. Consider the following three lists of species, each one representing (in very abbreviated form) a sample of the species found in three different types of ecosystems (Fig. 2.1).

Forest	Marsh	Prairie
Black oak	Reed-grass	White prairie-clover
Shagbark hickory	Painted turtle	Horned lark
Gray squirrel	Red-winged blackbird	Black-footed ferret
White-tailed deer	Muskrat	
Raccoon		

If someone were asked which of these tracts is most important from the perspective of maintaining biodiversity, one measure of biodiversity—species richness—would suggest that the forest be chosen. However, conservation biologists would know that the black-footed ferret is one of the rarest animals in North America and that all the other species listed are very common. Therefore, in the absence of any additional information, they would select the prairie tract as most important for maintaining biodiversity. Why?

The issue of how to set priorities for maintaining biodiversity involves many factors that we will review in Chapter 14, "Setting Priorities," but one of the most critical criteria is the likelihood of a life-form becoming extinct. Black-footed ferret populations have been reduced to a few score individuals while raccoons number in the millions and are as common as ever, perhaps more so because throughout most of their range we have provided them with abundant food and eliminated the large carnivores that might prey on them. Therefore black-footed ferrets are more important than raccoons if maintaining biodiversity is the goal. From a purely quantitative perspective, the extinction of either raccoons or black-footed ferrets is equal—in both cases the number of species in the world would be reduced by one—but this is a narrow way to view the issue. As long as black-footed ferrets have a greater risk of extinction than raccoons, anyone who cares about the earth's biodiversity will make a tract of prairie harboring ferrets a higher priority than a forest with raccoons and other common species. To put it another way, this particular prairie tract is more essential to the world's black-footed ferrets than this forest is to the world's raccoons. (Incidentally, readers who suffer from math phobia should not be sighing with relief, because quantitative techniques are important tools for conservation biologists; for example, in Chapter 7, "Extinction Processes," we will explore how to estimate extinction probabilities.)

Biodiversity and Spatial Scales

Extinction usually refers to the disappearance of a species from the earth, but the term is also routinely used, with modifiers, to describe the

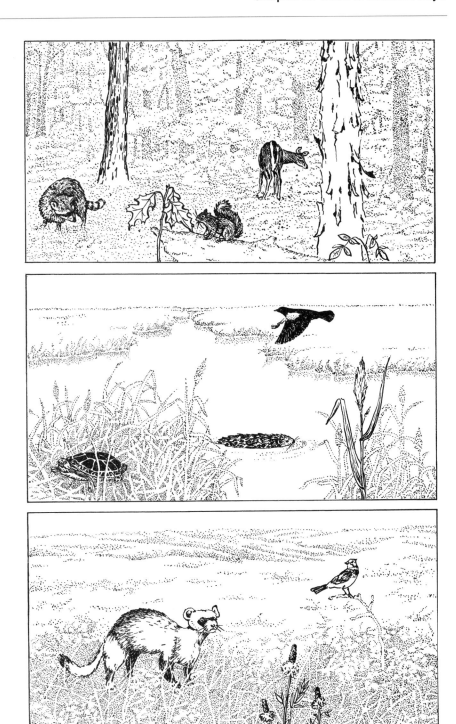

Figure 2.1. From a biodiversity perspective which of these three ecosystems would be most important to protect?

disappearance of a species from a smaller area. For example, when a species disappears from a small area, this is called a *local extinction*, even though the area may later be recolonized by immigrants. Beavers often become locally extinct in a small watershed and then return in a few years. Sometimes, a small-scale extinction is called *extirpation*. Although conservation biologists are most concerned about global extinctions, smaller-scale extinctions are also of some concern because they may foreshadow extinctions on a larger scale. Another key term is *endemic*, which refers to species found only in a defined geographic area; thus, koalas are endemic to Australia. If a species is found only in a small area (e.g., many inhabitants of the Galapagos and other isolated islands), it is called a *local endemic*.

The risks of extinction at different spatial scales are a key consideration when deciding which endangered species are a high priority. The larger the scale at which an extinction is likely to occur, the more important it is to try to prevent it. For example, if we had to choose between a plan that protected black-footed ferrets, a species facing global extinction, versus gray wolves, a species threatened with extinction in the continental United States but relatively secure in Canada and Alaska, we would favor the ferret plan. Similarly, there are likely to be a few locales in the Rocky Mountains where raccoons are rarer than gray wolves, but the gray wolves would be a higher priority than raccoons because wolves are more vulnerable at the regional scale.

The ecologist Robert Whittaker (1960) devised a simple system for classifying the scales at which diversity occurs; he described three scales of diversity as alpha, beta, gamma (A, B, C in Greek). *Alpha diversity* is the diversity that exists within an ecosystem. In Figure 2.2 two hypothetical lizard species, spotted lizards and long-tailed lizards, illustrate alpha diversity by coexisting in the same forest, living at different heights within the forest. A third species, banded lizards, illustrates *beta diversity* (among habitats diversity) by occurring in a nearby field. Finally, if you imagine spotted, long-tailed, and banded lizards living on one island, and a fourth species, speckled lizards, living a thousand kilometers away on another island, this would represent *gamma diversity*, or geographic-scale diversity.

We can use this hypothetical example to show how a narrow-scale perspective on maintaining biodiversity can lead would-be supporters of biodiversity astray. Some people might look at Figure 2.2 and think, "There are more lizard species in forests, so let's plant trees in the field." By doing so they might increase the alpha diversity of the field from one lizard to two (from banded lizards to spotted and long-tailed lizards), but they might also decrease the beta diversity of the island from three species to two because banded lizards would no longer have any suitable habitat. Similarly, they might think, "Let's bring some of the speckled lizards from the other island to our forest and have four species here." However, the

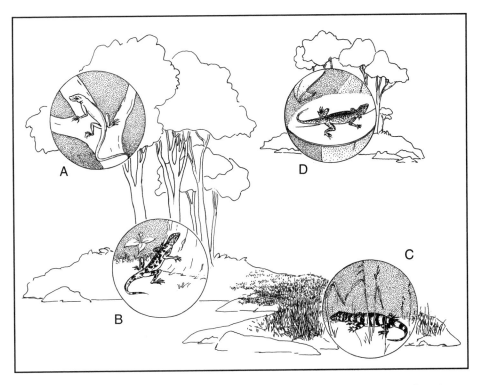

Figure 2.2. The distribution of four hypothetical lizard species showing alpha diversity (within an ecosystem, A plus B), beta diversity (among ecosystems, A/B plus C), and gamma diversity (geographic scale, A/B/C plus D). See text.

speckled lizards might outcompete and replace one of the local lizards or introduce a disease. The whole archipelago could end up with only three, two, or one lizard species instead of four and thus decreased gamma diversity.

The idea of spatial scale is so fundamental to maintaining biodiversity that a mnemonic phrase is worth remembering: "Scale is the tail that *w-a-g-s* biodiversity" (*w*-within ecosystem diversity, *a*-among ecosystem diversity, *g-s*-geographic-scale diversity).

Perspicacious readers may think that some intuitively obvious ideas are being belabored here, but these ideas are frequently overlooked in the real world of natural resource management. For example, natural resource managers responsible for the management of large tracts of contiguous forest often claim that they can increase the biodiversity of their forest by cutting moderate-sized patches in their forest (Hunter 1990). This claim is usually true, and the quantitative indices of diversity described above can be used to prove it. Cutting patches in a forest provides habitat for many species associated with early successional ecosystems and with the edges between early and late successional ecosystems. Moreover, most of the species associated with a forest ecosystem will persist despite limited cutting; therefore the total diversity of a forest will likely increase after cutting. On the other hand, what about the few forest species that may not persist after cutting? Some birds associated with forest interiors are vulnerable to nest predators that may be more common in a forest with openings (Robinson et al. 1995). Some plant species are vulnerable to being consumed by deer, and deer populations often increase after cutting (Miller et al. 1992). Global populations of some of these forest interior species are probably declining as large tracts of unbroken forest become scarcer. If they become extinct, then gamma diversity at the global scale will have been reduced, while the beta diversity of some forested landscapes was being increased.

This perspective also returns us to a question posed in the preceding chapter: Can people improve natural ecosystems? Some conservationists have answered this question by relating an anecdote from Gary Nabhan (1982), a botanist who studies traditional agriculture of Native Americans. One year, Nabhan made three trips to a pair of oases, fifty kilometers apart, on back-to-back days. One oasis was in the U S. Organ Pipe Cactus National Monument and has been protected from human activity since 1957, when farming by Papago Indians was stopped. At the second oasis, across the border in Mexico, traditional Papago farming persists. Nabhan and his colleagues counted 32 species of birds during their visits to the protected oasis and 65 species at the farmed oasis.

So, is this an example of people improving an ecosystem? Maybe yes, maybe no; we cannot decide from the available data because the numbers of bird species do not tell the whole story. What about beetles, lizards, and cacti? What if the sanctuary had a large population of masked bobwhite, a globally endangered bird, and the farming oasis had no bobwhites but a large population of starlings, an abundant exotic pest? Many conservation biologists would argue that manipulating the biodiversity of a natural ecosystem should always be avoided because both adding and removing species compromise an ecosystem's integrity (Angermeier 1994).

In sum, whenever we manipulate diversity at a local scale, we should consider the consequences at a larger scale and not rely on simple measurements of local biodiversity to judge the outcome. We will return to this issue in a case study.

Biodiversity Verbs

People change, manipulate, and manage the world and, consequently, affect biodiversity. Most of our activities have a negative impact on biodiversity; conservation biologists promote positive actions and use a variety of verbs to describe these activities. The verb *maintain* is dominant in this book because I believe that the major goal of conservation biology should be to keep all the elements of biodiversity on earth, despite human-induced changes that tend to diminish biodiversity. In this section we will evaluate some alternative verbs that are often encountered in the conservation biology literature. This may seem like a pedantic exercise, but some verbs carry implications that are not always consistent with the goal of maintaining biodiversity. For example, to *maximize* biodiversity implies manipulations such as increasing alpha diversity of an ecosystem, even importing exotic species, without considering the big picture. What is the natural level of biodiversity in that type of ecosystem? What will be the consequences for biodiversity at a larger scale? Manipulating the lizard populations in Figure 2.2 was a good example of this. To *increase* or to *enhance* biodiversity may imply the same shortsightedness, unless we are referring to an ecosystem in which biodiversity has been diminished by previous human activity and the goal is to return it to its previous state. If this is the case, it is probably best to refer to *restoring* biodiversity. *Protecting* biodiversity is similar to maintaining biodiversity but with a heavier emphasis on the negative impact of most human activities. To *preserve* biodiversity carries a connotation comparable with "to protect," but it may also imply that the only way to maintain biodiversity is to isolate it from human influence as much as possible; this is not always feasible or desirable. To *benefit* or *optimize* biodiversity are rather vague terms, sometimes used by people who have unusual ideas about what is beneficial or optimal. Finally, to *conserve* biodiversity implies using it carefully in a manner that will not diminish it in the long term. This is a reasonable goal, but it tends to overlook the idea that many elements of biodiversity have little or no instrumental value for people.

Related Concepts

"Biodiversity" is only one of several concepts that have been competing for the attention of natural resource managers in recent years; it has been joined by "sustainability," "ecosystem integrity," "biotic integrity," and others. In this section we will attempt to clarify the linkages and differences between these terms and biodiversity with a distillation of two syntheses, Callicott et al. (1999) and, primarily, Hunter (1999) (Fig. 2.3).

Biotic Integrity

Biotic or biological integrity refers to the completeness or wholeness of a biological system, including the presence of all the elements at appropriate densities and the occurrence of all the processes at appropriate rates

Figure 2.3. What is the state of this intertidal zone? From a biodiversity perspective we would focus primarily on having a complete set of the native species (especially any that might be in danger of disappearing from the system), as well as genetic and ecological attributes. A biotic integrity perspective would be similar, but would put more emphasis on having an appropriate density of each species and the appropriate rate of ecological processes. In terms of ecosystem integrity, the emphasis would be on the ecological processes driving this system. A focus on sustainability would center on the prospects for maintaining this system in the future.

(Angermeier and Karr 1994), and thus it is quite similar to the concept of biodiversity. The difference is mainly a matter of emphasis. Biotic integrity emphasizes the overall balance and completeness of biological systems, while biodiversity emphasizes that all the biotic elements are present. Furthermore, biotic integrity gives almost equal weight to functions and structure, whereas biodiversity usually emphasizes structure. Consequently, a person who was judging the biotic integrity of an ecosystem would be likely to focus on the ecosystem's key species and processes and might overlook the disappearance of a rare species. The well-being of rare things—species, ecosystems, and sometimes genes—is always in the spotlight from a biodiversity perspective. A biotic integrity perspective does avoid some of the misunderstandings about biodiversity described earlier in this chapter. This is accomplished primarily by focusing on the condition of an ecosystem with respect to a reference condition, usually what the ecosystem would be like if it were in a natural state (Hunter 1996, Angermeier 2000, Cam et al. 2000). For example, no one could ever claim that they had increased biotic integrity by adding an exotic species of lizard to an island.

Ecosystem Integrity

Ecosystem health and ecosystem integrity (ecological health and integrity) are effectively synonymous. I prefer "ecosystem integrity" because the inevitable analogy between ecosystem health and human health can be misleading (to take just one example, an ecosystem that is profoundly affected by a native pathogen is not necessarily unhealthy) (Suter 1993, Rapport 1998, De Leo and Levins 1997). In some ways, ecosystem integrity is broader than both biotic integrity and biodiversity because it encompasses the physical environment; for example, soil erosion and sedimentation are key aspects of ecosystem integrity, but they are primarily physical processes. On the other hand, biotic integrity and biodiversity are broader because they include genes and evolutionary processes, at least conceptually. In practice, the biotic integrity concept is usually applied to ecosystems and only rarely to genetic systems. Because ecosystem ecologists often focus on overall processes, ecosystem integrity is usually evaluated in terms of ecosystem functions, rather than the suite of species that constitute the biological portion of an ecosystem (Callicott et al. 1999).

Sustainability

"Sustainability" is simply the ability to maintain something over time without diminishing it. In a natural resource management context, sustaining the resources that are most directly used by people—timber, water, recreational opportunities, and so on—usually comes first (Lélé and Norgaard 1996). The key idea here is "intergenerational equity" or, in plainer language, not messing things up for our children and grandchildren. Obviously, sustaining biodiversity is inextricably tied to sustaining natural resources. However, sustaining biodiversity can also be driven by concern for the intrinsic value of life. Some conservationists are reluctant to focus on sustainability because it implies that the status quo is satisfactory; they would argue that we should be restoring ecosystems and populations in many cases. Furthermore, a focus on sustainability can be misleading if we fail to recognize that most ecosystems are highly dynamic (Chapin et al. 1996).

Values

People's values are clearly reflected in their choices of what should be sustained. It is also true, but less obvious, that the ways we judge biotic integrity and ecosystem integrity are also shaped by values (Lackey 1995, Lélé and Norgaard 1996, De Leo and Levins 1997). Proponents of the biotic integrity concept are quite explicit that their ideas about "all appropriate elements and occurrence of all processes at appropriate rates" are based on using natural systems as benchmarks, that is, those with little or no human influence (Angermeier and Karr 1994, Hunter 1996). For example, they would decide whether a particular species of lizard belongs on a given island by whether it would be there without human intervention. Many biologists would share this standard, but there is nothing sacred about using a natural system as the basis for comparison. For example, Robert Lackey (1995) has argued that "An undiscovered tundra lake and an artificial lake at Disneyland can be equally healthy." For him the key question is whether the lake is in a desired state; i.e., is it satisfying human expectations? The bottom line is that to use any of these concepts, including biodiversity, requires some kind of benchmark, and the selection of benchmarks inevitably reflects human values.

CASE STUDY   ### Clear Lake

In the northeastern corner of California lies Clear Lake, a large body of water (17,760 ha) that is shallow, warm, and productive; thus it supports a great abundance of fish. Originally, Clear Lake was home to 12 native kinds of fish, at least 3 of which were endemic to the lake: the Clear Lake splittail, Clear Lake hitch, and Clear Lake tule perch (Moyle 1976a, personal communication) (Fig. 2.4). Two of the native species, Pacific lamprey and

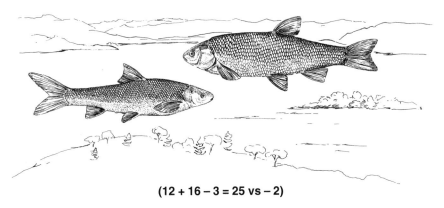

(12 + 16 – 3 = 25 vs – 2)

Figure 2.4. Clear Lake in northern California used to be inhabited by 12 native species of fish until fisheries managers began introducing new fish species, 16 in all. These introductions decimated the native fish populations, but still produced a net increase in alpha diversity of 13 species. This increase came at the expense of global diversity because two of the original species, the Clear Lake splittail and the thicktail chub, are now globally extinct.

rainbow trout, migrated between tributaries of the lake and the sea and practically disappeared from the lake when a dam was built on the lake's outlet. Other species were decimated largely because of human attempts to increase the fish diversity of the lake by importing exotic species. By 1894 carp and two species of catfishes had been introduced to Clear Lake, and they flourished there. During the 1900s 13 additional species were introduced, primarily members of the Centrarchidae family (sunfishes and basses) native to the eastern United States. One species introduced in 1967, the inland silversides, soon became the most abundant species in the lake. In the face of this competition, the native species have declined dramatically, and only four native species remain common in the lake. Worse still, two of the native species that have disappeared from the lake (the Clear Lake splittail and the thicktail chub) are globally extinct. The net scorecard: misguided attempts to enrich the fish fauna of Clear Lake have increased the number of fish species there from 12 to 25 by adding 16 exotic species, but these introductions have decimated the lake's native fish fauna, eliminating two elements of biodiversity from the entire planet and reducing gamma diversity. This was not a very good trade.

Summary

Biodiversity is the diversity of life in all its forms (plants, animals, fungi, bacteria, and other microorganisms) and at all levels of organization (genes, species, and ecosystems). Biodiversity includes these structural components, as well as functional components; that is, the ecological and evolutionary processes through which genes, species, and ecosystems interact with one another and with their environment. Conservation biologists focus on maintaining structural biodiversity because, if genetic, species, and ecosystem diversity are successfully maintained, then the diversity of ecological and evolutionary processes will probably be maintained as well.

Some elements of biodiversity can be measured with quantitative indices of diversity based on richness, the number of elements of biodiversity (usually number of species), and evenness (their relative abundance). However, these indices can be misleading because a higher biodiversity index is not always desirable if the goal is maintaining biodiversity. It is more important to assess the risk of extinction of different species and emphasize those that are most endangered. The risk of extinction needs to be evaluated at different scales, and emphasis needs to be placed on those species most at risk at the global scale because they are irreplaceable. The biodiversity and scale issue can also be addressed by thinking of diversity at three scales (alpha—within an ecosystem; beta—among ecosystems; and gamma—geographic scale) and by always assessing the large-scale consequences whenever one manipulates biodiversity at a small scale. Thinking about biodiversity at large scales will often reveal that it is inappropriate to advocate maximizing biodiversity. Instead, the goals should be to maintain natural levels of biodiversity or to restore biodiversity in ecosystems degraded by human activity. The goal of maintaining biodiversity is closely related to some other goals such as maintaining ecosystem or biotic integrity and ensuring sustainability of natural resource management.

FURTHER READING

Wilson (1992) and Heywood and Watson (1995) provide good introductions to the concept of biodiversity, and Angermeier (1994, 2000), Povilitis (1994), and Hunter (1996) discuss some of the difficulties in moving from a conceptual definition to action. DeLong (1996) reviews definitions of biodiversity. The two major biodiversity journals are *Conservation Biology* and *Biological Conservation*, but there are many other journals also worth perusing for conservation biology topics: *Biodiversity and Conservation, Bioscience, Ecological Applications, Oryx,* and *Pacific Conservation Biology,* to name just five among dozens. There is one electronic journal of relevance, *Conservation Ecology,* www.consecol.org/Journal/.

TOPICS FOR DISCUSSION

1. Genes, species, and ecosystems are just three of many levels of biological organization. What are some other important levels?
2. Can you think of some circumstances under which it might be preferable to focus on the functional aspects of biodiversity rather than on the structural aspects?
3. Is it desirable to increase alpha- and beta-scale diversity if it can be done without apparently decreasing gamma-scale diversity?
4. If you were managing a forested stream valley, would you consider putting a small dam on the stream to add a pond ecosystem to the valley? What if the pond would be inhabited by a globally endangered species of turtle?
5. Think of some place in which you have observed change over time. How did these changes affect biodiversity? Make a list of some places in which the changes have been positive and some in which the changes have been negative. Are there any common themes?

Chapter 3
Species Diversity

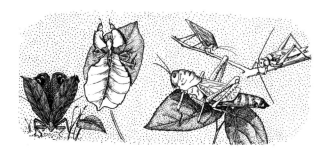

Imagine flocks of parrots flashing green and gold over the piedmont forests of Virginia, a raft of penguin-like birds paddling up a Norwegian fjord, or a marsupial wolf coursing kangaroos through the eucalypt woodlands of Australia. These sights will never be seen again because the Carolina parakeet, great auk, and thylacine are gone forever. And they are not alone. Over a thousand species are known to have been driven into extinction by people just since 1600 (Hanski et al. 1995), and we can only guess at the total number of species that have disappeared because of human activities. Nothing highlights the need for maintaining biodiversity like the fate of these species and the many more that still survive but are sliding toward extinction. Keeping the wave of species extinctions from becoming a flood is the core of conservation biology.

In this chapter we will first address two fundamental questions: 1) What is a species? 2) How many species are there? Then we will explore the importance of species diversity in terms of both intrinsic and instrumental values.

What Is a Species?

When we try to classify the natural world, it seems relatively easy to recognize different species—peregrines and redwoods are much easier to distinguish from closely related entities than are ecosystems and genes. Nevertheless, the question "What is a species?" is more complex than most people realize. One widely used definition is based on reproductive isolation: "Species are groups of actually or potentially interbreeding natural populations, which are reproductively isolated from other such groups"

(Mayr 1942). For example, mammalogists classify brown bears in Eurasia and North America as the same species, even though they have been separated by the Bering Strait for about 10,000 years, because they would interbreed given the opportunity. On the other hand, American black bears and brown bears are considered separate species because they do not interbreed. Occasionally, interbreeding does occur between two apparently distinct species, and the offspring are considered hybrids. Here some difficult questions arise. How much hybridization can occur before you decide that the two parent species are really just one species? Are gray wolves and coyotes, which occasionally interbreed, separate species? And what if the hybrid offspring form self-perpetuating populations? Some biologists believe that coyote-gray wolf hybridization produced the red wolf, an endangered canine that used to be widely distributed in the southeastern United States (Wayne and Gittleman 1995, Roy et al. 1996, Nowak and Federoff 1998).

These questions are more familiar to botanists than to zoologists. Look through any comprehensive list of plant species, and you will find many listings such as *Typha angustifolia X latifolia*, indicating that hybrids of the narrow-leaved cattail *(angustifolia)* and the broad-leaved cattail *(latifolia)* occur routinely. However, this is only the tip of the iceberg; it has been estimated that 70% of angiosperms (flowering plants) owe their origins to hybridization (Grant 1981, Whitham et al. 1991, Arnold 1992). Plant species are also harder to define in terms of reproductive isolation than animal species because they are more likely to exhibit asexual reproduction, self-fertilization, polyploidy (multiple sets of chromosomes), and other variants of what we usually consider "normal" reproduction. Similarly, most microorganisms reproduce asexually, thus confounding the idea of reproductive isolation. Their extremely rapid reproduction adds another complexity: Is the flu virus that embarks on a transoceanic voyage with a ship's crew the same species when it returns to shore a week later?

Evolutionary biologists and taxonomists are wrestling with these issues and have proposed many other species definitions—evolutionary, phylogenetic, ecological, cladistic, morphological, and more. (See Otte and Endler 1989 and Claridge et al. 1997 for reviews.) Different definitions serve different purposes, and no one of them is "best" or "correct." The differences among definitions would be an academic issue except that species identified by different definitions do not always correspond to one another. For example, there may be from one to 30 species of *Drimys*, a kind of tree, in New Guinea, depending on the definition you use (Stevens 1989), and this would be an important issue for a conservation biologist trying to protect *Drimys* diversity. Another example comes from Mexico where two quite different conservation plans based on the geographic distribution of endemic birds have been proposed: one based on species defined by reproductive isolation and one based on phylogenetic species (Peterson and

BOX 3.1

Defining species
Judith Rhymer[1]

Defining a group of organisms as a species, subspecies, or distinct population is often difficult and controversial because of the lack of clear criteria for classification, and even systematists working on the same taxonomic group often disagree. In addition, variation considered subspecific in one taxonomic group may be considered worthy of species recognition in another. Because existing taxonomy may not reflect underlying genetic diversity, Ryder (1986) introduced the term "Evolutionary Significant Unit" (ESU) to provide a rational basis for delineating conservable units of biological diversity. An ESU refers to a group of organisms that has been isolated long enough to have undergone significant genetic divergence from other groups of the same species (Fig. 3.1). In a similar vein, Management Units (MU) are local populations that, because they have so little dispersal among them, have evolved some genetic differences (Moritz et al. 1995), for example, commercial fisheries stocks.

Problems arise in trying to define how much genetic differentiation should be considered significant or sufficient for groups of organisms to be considered an ESU or MU (Paetkau 1999, Avise 2000). Moritz (1994) suggested operational genetic criteria to delineate these categories, but given that it could take many generations and tens of thousands of years for genetic differences to accrue, particularly for species with long generation times, common sense in appraising levels of intraspecific and interspecific genetic variation among groups must prevail.

Ultimately, there are two separate issues: Are groups of organisms distinct based on some scientific criteria, and if so, are they worthy of protection? Determining if groups are worthy of protection involves policy decisions, in addition to scientific evidence. For example, the U.S. Endangered Species Act (ESA) defines species to include any distinct population segment of any species of vertebrate that interbreeds when mature. The term Distinct Population Segment (DPS) has been applied in two ways. First, it is similar to the concept of an ESU or MU, but can include differences in one or more of morphological, behavioral, physiological, or ecological criteria, in addition to molecular genetic differences. Second, the concept of a DPS can also be used based only on political boundaries. For example, woodland caribou living in the Rocky Mountains in the United States are considered a DPS even though they are only separated from Canada by a political boundary (Karl and Bowen 1999 call these "geopolitical species"). These may be appropriate for management, but have no sound scientific basis. Note that the ESA does not recognize DPS for plants and invertebrates; this, too, is a political, not a scientific, decision (Clegg et al. 1995).

---

[1] Department of Wildlife Ecology, University of Maine, Orono, Maine

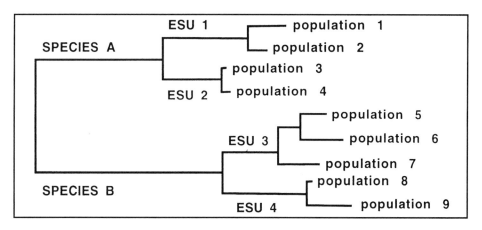

Figure 3.1. Hypothetical example illustrating the relationships between species, Evolutionary Significant Units (ESU), and populations. In this example all populations could be considered separate Management Units (MU) except populations 3 and 4, which are too closely related to be managed separately. The length of the lines joining species, ESUs, and populations are generally equivalent to the genetic distances among them.

Navarro-Sigüenza 1999). A third example comes from the fact that the U.S. Endangered Species Act is based on a reproductive isolation definition of species, and therefore it may or may not protect hybrids; this has made arguments over the taxonomy of the red wolf controversial and critical (Gittleman and Pimm 1991, O'Brien and Mayr 1991, Whitham et al. 1991, Ranker and Arft 1994).

Conservation biologists need to be aware of the debate over species definitions, but they cannot allow themselves to be paralyzed by it (Rojas 1992). It is better to use a fallback definition such as a species is "what a competent taxonomist says it is" (Stevens 1990), rather than do nothing for lack of definitive information. Fortunately, uncertainty over species definitions actually bolsters the overall goal of maintaining biodiversity because it highlights the critical importance of maintaining diversity below the species level, namely, genetic diversity. This means that conservation biologists can often sidestep the definition of species and use a term such as "evolutionarily significant units," or more succinctly "taxa," to refer to both species and subspecific groups such as subspecies, races, varieties, or even populations (see Box 3.1 and Vogler and Desalle 1994). As we will see in Chapter 5, Genetic Diversity, all of these merit some attention from conservationists.

How Many Species Are There?

Carolus Linnaeus, the Swedish biologist who founded modern taxonomy, described about 12,000 species in his 1758 opus *Systema Naturae*, but

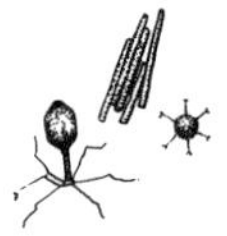

	Described Species	Estimated Species Richness
Viruses	4,000	400,000
Bacteria	4,000	1,000,000
Fungi	72,000	1,500,000
Protozoa	40,000	200,000
Algae	40,000	400,000
Plants	270,000	320,000
Arthropods	1,065,000	8,900,000
Other animals	255,000	900,000

Figure 3.2. Roughly 1.7 million species have been described by scientists; arthropods, primarily insects, constitute almost half this number. The estimated number of species is far greater, especially for smaller life-forms. (The data presented here are summarized from Table 3.1–2 of Heywood and Watson 1995. Redrawn from Hunter 1999.)

must have been well aware that this list was incomplete because in the 18th century much of the world remained unexplored by scientists. It is interesting to speculate how many species he might have estimated to exist. Today, roughly two centuries later, scientists have described about 1.7 million species using Linnaeus's system (Hammond 1995), but we can still only guess how many species there might be (Fig. 3.2). Hammond (1995) describes a range of estimates from 3.6 to 111.7 million species and suggests 13.6 million as a reasonable "working figure."

Attempts to make a systematic estimate of the number of species have often revolved around insects. We have known for quite some time that insects represent a substantial portion of the world's species. In just one order, the beetles, roughly 400,000 species have been described, far more than the number of known species of vascular plants. Biologists like to make this point with an anecdote about J. B. S. Haldane, a 19th-century biologist

(Gould 1993). When asked by a group of theologians what he had learned about God from having spent a lifetime studying His creations, Haldane is said to have replied, "He seems to have an inordinate fondness for beetles."

The scope for describing new beetles and other insects remains enormous, although some of the attention is shifting toward even smaller creatures. Consider a study undertaken by a group of Norwegian microbiologists (Torsvik et al. 1990a, 1990b). First, they collected two tiny soil samples: 1 gram of Norwegian forest soil and 1 gram of sediment from off the coast of Norway. Next, they extracted the bacteria from the samples and then extracted the DNA from all the bacteria. They then estimated the diversity of DNA strands, made a conservative assumption that bacteria are different species if less than 70% of their DNA is identical, and arrived at a rough estimate that each sample contained over 4000 species of bacteria, with little or no overlap in species between the two samples. Over 4000 bacteria species in a pinch of Norwegian soil is doubly impressive when you realize that only about 4000 species of bacteria have ever been described from all environments in the whole world. Other large, unexplored lodes of species diversity exist among mites, nematodes (Hammond 1992), fungi (Hawksworth 1991), parasites (Embley et al. 1994), and organisms living on the deep-sea floor (Grassle and Maciolek 1992). Finally, the number of species may be bolstered by the existence of *sibling species* or *cryptic species*, species that scientists cannot readily distinguish based on morphology but that are genetically isolated. Consider the red crossbill, a well-known North American bird, which is believed to be eight separate cryptic species (Benkman 1993, Adkisson 1996).

Do we really need to know how many species there are? From a conservation perspective we do not even have the resources to address adequately the problems of a few hundred well-known vertebrates and plants that are slipping toward extinction. Does it matter whether there are ten million or a hundred million other species that we ignore? The number of species may not matter strategically, but these estimates do convey two fundamental ideas. First, the number of species that may ultimately be at risk is enormous; in other words, we have a lot to lose. Second, we have a great deal to learn about the world (Blackmore 1996, Oliver and Beattie 1996).

The Intrinsic Value of Species and Their Conservation Status

Many conservationists believe that every species has intrinsic value. Its value is independent of its usefulness to people. Strictly speaking, its value is even independent of its usefulness to other species or within an ecosystem. In other words, every species has value without reference to anything but its own existence (Fig. 3.3). The idea of things having value

Figure 3.3. A species' intrinsic value is independent of its relationship with any other species, whereas its instrumental value depends on its importance to other species, including people. Its uniqueness value is based on how different it is from other species that might provide similar values.

without reference to humans is hard for many philosophers to accept (Hargrove 1989), but it does appeal to many conservationists because of its simplicity and equity. If you accept the idea of species having intrinsic value, it is relatively straightforward to decide which species merit more attention from conservation biologists: they are those species most threatened with extinction, the ones whose continued existence is jeopardized by people. In the task of assigning conservation status to various species, the probability of extinction is the primary consideration, as illustrated in Box 3.2.

The World Conservation Monitoring Centre (WCMC) publishes books, commonly called Red Data Books, and maintains a website that lists the species that fall into these categories. These resources provide the primary international standard for the conservation status of various species, but there are others. For example, the Convention on International Trade in Endangered Species of Wild Fauna and Flora (CITES) lists species in various appendices depending on how endangered they are: Appendix I includes the most-threatened species and confers the highest degree of protection, and so on. At a more local level, many national and state governments also maintain lists of species that are threatened within their borders. Sometimes, WCMC categories are used at these local levels (Palmer et al. 1997), but more often different sets of criteria are used. Most of these organizations also maintain lists of species that are not yet endangered but that are declining and need to be monitored. These are often called "species of special concern" or "species to watch."

For all of these organizations, the decisions on which species to list and in which category have historically been based on the best judgment of biologists rather than specific, quantifiable criteria. With a better understanding of the process of extinction and better data about species (e.g., population size, rate of decline), we are moving toward a process for making systematic decisions (Mace and Lande 1991). In the future, statements like, "Blue whales are endangered because they are very likely to become extinct," will be replaced by statements like, "This species is endangered because it has a greater than 10% probability of extinction within 100 years unless causal factors change." See Box 3.3.

The decision to list or not list a species as endangered is likely to be controversial when it is made by a government agency because most governments give listed species some degree of protection. At a minimum, harvesting of the species is likely to be prohibited. In some cases any activity that could harm the species or its habitat, directly or indirectly, may be limited. This means that the decision to list a species can run head-on into substantial economic interests. In the United States this kind of controversy has forced the U.S. Fish and Wildlife Service to be very careful about its listing decisions. This fact and an underfunded, overworked staff has produced a backlog of hundreds of taxa, waiting to be listed or at least evaluated for possible listing, and no doubt this list is incomplete for many groups of smaller organisms.

The phrase "rare and endangered" has become a bit like "assault and battery"; most people use it without really understanding what it means. You might be surprised to know that many species that are quite rare are not highly endangered with extinction, and conversely, that some endangered species are not particularly rare. For example, the African elephant has a total population over 500,000, but was listed by the IUCN as Vulnerable in 1989 because its numbers were declining precipitously as a result of ivory poaching. On the other hand, in the fynbos and succulent karoo ecosystems of southwestern South Africa there are hundreds of plant species with very small population sizes that live in fairly pristine environments and show no evidence of population decline (Cowling 1992). In other words, rarity seems to be their natural state. Should such species be listed as "endangered" even in the absence of any threat? The IUCN used to have a separate category, "Rare," for such species. However, it is moving toward listing these species as "endangered," "vulnerable," etcetera in recognition of the fact that extreme rarity always puts a species in some jeopardy. The standards for listing a species that is rare, but not currently in decline, are tighter. For

BOX 3.2

Categories of the IUCN Red List

The following categories are used by the World Conservation Monitoring Centre[1] to classify species for the IUCN Red List, a global compilation of species of concern to conservationists. Their relationship is depicted in Figure 3.4.

EXTINCT (EX)
A taxon is Extinct when there is no reasonable doubt that the last individual has died. The great auk, Carolina parakeet, thylacine, and over 1000 other species have become extinct since 1600.

EXTINCT IN THE WILD (EW)
A taxon is Extinct in the Wild when it is known only to survive in cultivation, in captivity, or as a naturalized population. Dozens of species are currently found only in captivity (e.g., the Guam rail and several tree snails) or used to be Extinct in the Wild until they were successfully reintroduced (e.g., the wisent and nene goose).

Species that fall in the next three categories are collectively called Threatened; see Figure 3.4. Note that the U.S. Fish and Wildlife Service uses "threatened" as a category of jeopardy one step below "endangered."

CRITICALLY ENDANGERED (CR)
A taxon is Critically Endangered when available scientific evidence indicates that it meets any of the criteria A to E in Box 3.3, and it is therefore considered to be facing an extremely high risk of extinction in the wild. Well-known examples include the Sumatran, Javan, and black rhinoceroses, the Philippine eagle, California condor, and hawksbill turtle.

ENDANGERED (EN)
A taxon is Endangered when available scientific evidence indicates that it meets any of the criteria A to E in Box 3.3, and it is therefore considered to be facing a very high risk of extinction in the wild. Many high-profile endangered species fall in this group: for example, giant pandas, tigers, snow leopards, gorillas, chimpanzees, Asian and African elephants, blue and fin whales, whooping and Siberian cranes, and loggerhead and green turtles.

VULNERABLE (VU)
A taxon is Vulnerable when available scientific evidence indicates that it meets any of the criteria A to E in Box 3.3, and it is therefore considered to be

---

[1] A note on organizational taxonomy: The World Conservation Monitoring Centre in Cambridge, UK, is an office of the United Nations Environmental Program (UNEP) that was originally created as a joint venture of the UNEP, the World Conservation Union (which is still widely known as the IUCN, the initials of its former name), and the World Wide Fund for Nature (WWF, which is the new name of the World Wildlife Fund everywhere except in the United States, where the old name has been retained). Unfortunately, conservation biology is not free of the alphabet-soup syndrome. The wording used here follows that of the WCMC with minor differences. For the exact and latest wording see the WCMC website at: www.unep-wcmc.org/

BOX 3.2 *Continued*

facing a high risk of extinction in the wild. Most threatened species are listed as Vulnerable; examples include the cheetah, orangutan, humpback whale, and snail darter.

NEAR THREATENED (NT)

A taxon is Near Threatened when it has been assessed against the criteria and does not qualify for Critically Endangered, Endangered, or Vulnerable now, but is close to qualifying for, or is likely to qualify for, a threatened category in the near future. Also included here are taxa that are the focus of a conservation program, the cessation of which would result in the taxon qualifying for one of the threatened categories. Jaguars, maned wolves, white-tailed eagles, and Atlantic sturgeon are listed as Near Threatened because their status is of some concern, but they do not meet any of the criteria listed below. Polar bears, giraffes, and white rhinos are listed as Near Threatened species because their survival depends on conservation programs.

LEAST CONCERN (LC)

A taxon is Least Concern when it has been evaluated against the criteria and does not qualify for Critically Endangered, Endangered, Vulnerable, or Near Threatened. Widespread and abundant taxa are included in this category.

DATA DEFICIENT (DD)

A taxon is Data Deficient when there is inadequate information to make a direct, or indirect, assessment of its risk of extinction based on its distribution and/or population status. Listing of taxa in this category indicates that more information is required and acknowledges the possibility that future research will show that threatened classification is appropriate. Many molluscs, fishes, and nocturnal birds and mammals have been evaluated, but could not be listed as Threatened because there was not enough information.

NOT EVALUATED (NE)

A taxon is Not Evaluated when it is has not yet been assessed against the criteria. Most of the world's species, notably all the invertebrates and other small life-forms, fall into this category.

---

[2] The examples used here are all animals because the WCMC has yet to produce lists of other groups using this terminology. For example, the plant list uses a 1990 set of terminology with very different categories such as "rare" and "out of danger."

[3] In some cases a species is assigned to one category overall, while various subspecies or populations may be designated differently. For example, tigers are Endangered as a species but the Amur, Sumatran, and South China subspecies are all considered Critical. Blue whales are Endangered overall, while the Atlantic population is only listed as Vulnerable.

[4] The specificity of these criteria may seem rather naive given the uncertainty that often surrounds these kinds of data; see Akçakaya et al. (2000) for a system for dealing with this uncertainty.

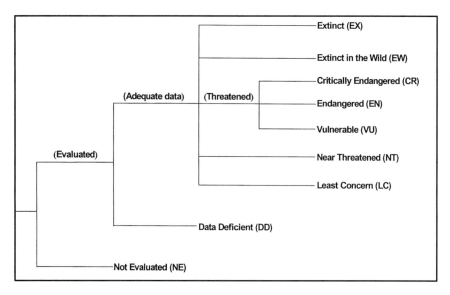

Figure 3.4. The relationships of different categories used in the IUCN classification system are depicted here.

example, to be listed as "critically endangered" a population that is in decline needs to have fewer than 250 individuals, but a population that is stable has to have fewer than 50 individuals.

The idea that rarity can be a natural state is easier to understand if we go beyond simply equating rarity with having a small total population. Deborah Rabinowitz (1981) described rarity on the basis of three separate characteristics:

1. *Geographic range:* Some species are rare because they are found only in a small geographic area such as a single island or lake; in other words, they are local endemics.
2. *Habitat specificity:* Species that occur only in specific, uncommon types of habitat such as caves or desert springs are likely to be rare.
3. *Local population size:* Some species occur at low population densities wherever they are found.

Some species are rare in more than one respect. The alpine lily occurs only in the Snowdonia mountains of Wales (geographically restricted) where it grows only in vertical fissures on cliffs (habitat restricted), often as solitary individuals, and never commonly (small local populations). (See Arita et al. 1990 and Goerck 1997 for two other examples of evaluating rarity. We will return to Rabinowitz's classification in Chapter 7, Extinction Processes.)

Although rare species may not be immediately threatened with extinction, especially if their habitat is secure from human intervention, they need to be monitored carefully because they can quickly shift from secure to endangered.

BOX 3.3   Quantitative criteria for assessing threatened status

The IUCN red lists of threatened species now require that a species meet at least one of five quantitative criteria. Shown below are the five criteria for Critically Endangered. The primary differences for Endangered and Vulnerable are certain key numbers; these are shown in parentheses as values for EN and VU.

A taxon is Critically Endangered when the best available evidence indicates that it meets any of the following criteria (A to E), and it is therefore considered to be facing an extremely high risk of extinction in the wild:

A. Reduction in population size based on any of the following:
  1. An observed, estimated, inferred, or suspected population size reduction of ≥ 90% (EN 70%; VU 50%) over the last 10 years or three generations, whichever is the longer, where the causes of the reduction are clearly reversible AND understood AND ceased, based on any of the following:
     a. Direct observation
     b. An index of abundance appropriate for the taxon
     c. A decline in area of occupancy, extent of occurrence, and/or quality of habitat
     d. Actual or potential levels of exploitation
     e. The effects of introduced taxa, hybridization, pathogens, pollutants, competitors or parasites.
  2. An observed, estimated, inferred, or suspected population size reduction of ≥ 80% (EN 50%; VU 30%) over the last 10 years or three generations, whichever is the longer, where the reduction or its causes may not have ceased OR be understood OR be reversible, based on any of (a) to (e) under A1.
  3. A population size reduction of at least 80% (EN 50%; VU 30%), projected or suspected to be met within the next 10 years or three generations, whichever is the longer (up to a maximum of 100 years), based on any of (b) to (e) under A1.
  4. An observed, estimated, inferred, projected, or suspected population size reduction of ≥ 80% (EN 50%; VU 30%) over any 10-year or three-generation period, whichever is longer (up to a maximum of 100 years), where the time period includes both the past and the future, and where the reduction or its causes have not ceased, based on any of (a) to (e) under A1.
B. Geographic range in the form of either B1 (extent of occurrence) OR B2 (area of occupancy) OR both:
  1. Extent of occurrence estimated to be less than 100 km^2 (EN 5000; VU 20,000) and estimates indicating at least two of a–c:
     a. Severely fragmented or known to exist at only a single (EN 5; VU 10) location
     b. Continuing decline, observed, inferred, or projected, in any of the following:
       (i) Extent of occurrence
       (ii) Area of occupancy
       (iii) Area, extent, and/or quality of habitat

BOX 3.3 *Continued*

      (iv) Number of locations or subpopulations

      (v) Number of mature individuals.

  c. Extreme fluctuations in any of the following:

      (i) Extent of occurrence

      (ii) Area of occupancy

      (iii) Number of locations or subpopulations

      (iv) Number of mature individuals.

2. Area of occupancy estimated to be less than 10 km^2 (EN 500; VU 2000), and estimates indicating at least two of a–c:

  a. Severely fragmented or known to exist at only a single (EN 5; VU 10) location

  b. Continuing decline, observed, inferred, or projected, in any of the following:

      (i) Extent of occurrence

      (ii) Area of occupancy

      (iii) Area, extent, and/or quality of habitat

      (iv) Number of locations or subpopulations

      (v) Number of mature individuals.

  c. Extreme fluctuations in any of the following:

      (i) Extent of occurrence

      (ii) Area of occupancy

      (iii) Number of locations or subpopulations

      (iv) Number of mature individuals.

C. Population size estimated to number less than 250 (EN 2500; VU 10,000) mature individuals and either:

1. An estimated continuing decline of at least 25% (EN 20%; VU 10%) within 3 years (EN 5; VU 10) or one generation (EN 2; VU 3), whichever is longer, OR

2. A continuing decline, observed, projected, or inferred, in numbers of mature individuals AND at least one of the following (a–b):

  a. Population structure in the form of one of the following:

      (i) No subpopulation estimated to contain more than 50 (EN 250; VU 1000) mature individuals, OR

      (ii) At least 90% (EN 95%; VU 100%) of mature individuals are in one subpopulation.

  b. Extreme fluctuations in number of mature individuals.

D. Population size estimated to number less than 50 (EN 250; VU 1000) mature individuals.

E. Quantitative analysis showing the probability of extinction in the wild is at least 50% (EN 20%; VU 10%) within 10 years (EN 20; VU 100) or three generations (EN 5), whichever is the longer (up to a maximum of 100 years).

Note: Criterion D can also be met for Vulnerable by the following: Population with a very restricted area of occupancy (typically less than 20 km^2) or number of locations (typically five or fewer) such that it is prone to the effects of human activities or stochastic events within a very short period in an uncertain future, and is thus capable of becoming Critically Endangered or even Extinct in a very short time.

The Instrumental Values of Species

When we think about the instrumental value of a species, we are likely to go straight to the basics: Can I eat it? Can I make it into clothing or shelter, or burn it to keep me warm? Or, in the market-based economies in which most of us live: Can I sell it? Materialistic uses of a species may be the core of instrumental values, but this is not the whole story. People also value species for purely aesthetic or spiritual reasons; species have instrumental value as members of ecosystems and as models for science and education; and conservation biologists use certain species to expedite their larger goal of maintaining biodiversity.

Economic Values

Food: Except for salt and few other additives, everything we eat started out as an organism, an element of biodiversity. Often, we do not even recognize all the organisms involved, for example, the array of microorganisms that are essential in the production of cheese, bread, and alcoholic beverages. Despite their fundamental role, the instrumental value of species as food is usually considered an issue for agricultural scientists rather than conservation biologists, because the vast bulk of our food comes from a relatively small number of domesticated species (Prescott-Allen and Prescott-Allen 1990). Maintaining the genetic diversity of domestic species is a component of conservation biology as we will see in future chapters, but it is not in the mainstream of conventional conservation biology, which tends to focus on wild species. Nevertheless, there are at least three ways in which conservation biologists who work with wild species are involved with the issue of species as food for people.

First, most domesticated species are closely related to species that are still wild, and these wild relatives are a critical source of genetic material, *germplasm,* for agricultural breeders who are trying to improve domesticated species. Indeed, in many cases (e.g., pigs, coconuts, and carrots) there are both wild and domesticated populations of the same species. Maintaining viable populations of the wild relatives of crop plants and livestock falls squarely within the purview of mainstream conservation biology, especially if the wild relatives are threatened with extinction. For example, yaks and water buffalos are important livestock in parts of Asia, and the wild populations of both species are in danger of extinction. We lost the wild version of the domestic cow, the auroch, back in 1627 (Szafer 1968). A well-known example of the potential role of wild relatives is found in the perennial teosintes, wild relatives of corn (or maize) that were thought to be extinct until rediscovered in southern Jalisco, Mexico, in 1978 (Iltis et al.

1979, Dorweiler et al. 1993). The perennial habit of these teosintes suggests that some of their genetic material, if transferred to corn, could increase its resistance to some diseases and, perhaps, could even enable it to regrow annually without the expense of tilling and sowing. The rediscovery of perennial teosintes also led to the establishment of a major reserve in the area where they were found.

Second, wild species may be a source of new domesticates in the future. Domestication is almost as old as humanity, and it is still practiced. In fact, concern about world food supplies, especially shortages of protein, has kindled a new interest in domestication (Janick 1996). For example, the National Research Council of the United States (1991) produced a book, *Microlivestock: Little-known small animals with a promising economic future*, that describes a sample of the many wild mammals, birds, and reptiles suitable for farming. Domesticating wild species is sometimes stimulated by a concern for the fate of wild populations, as well as a need for new food supplies. In Latin America green iguanas (a large lizard with tasty meat and eggs) are such prized sources of food that they have been overhunted in many areas. This was a significant catalyst to initiating experimental attempts to farm them. Some of the food items that we associate with wild species are already produced primarily in captivity. Most of the venison sold in markets comes from captive herds of deer. Deer (mostly red deer) are the third most common type of livestock in New Zealand after sheep and cattle. Potential domesticates that will thrive in lands that are marginal for mainstream agriculture are also of great interest. For example, the buffalo gourd from semiarid portions of southwestern North America is rich in protein and oil and has abundant carbohydrate reserves; it is being domesticated for use as a crop plant in semiarid regions (DeVeaux and Shultz 1985).

Third, wild plant and animal populations are still major food sources for people (Fig 3.5). It is well known that many rural people rely heavily on wild plants and animals for food, and it is easy to attribute this to the poverty that pervades much of the world. For example, it has been estimated that 85% of the animal protein consumed by people in the Ucayli region of Peru comes from wild animals (Prescott-Allen and Prescott-Allen 1982). However, it would be wrong to assume that people consume wild species only out of necessity. In many countries plants and animals gathered from the wild abound at marketplaces and command a good price. In Nigeria giant rats sell for much more than mutton, and in Liberia monkey meat costs about the same as beef (Ajayi 1979). Of course, you do not have to visit the marketplaces of Ibadan or Monrovia to find wild species for sale; most of the fish and shellfish (more correctly molluscs and crustaceans) sold in the world—whether dipped from a bucket on the streets of Calcutta or wrapped in plastic at a New York supermarket—come from wild popu-

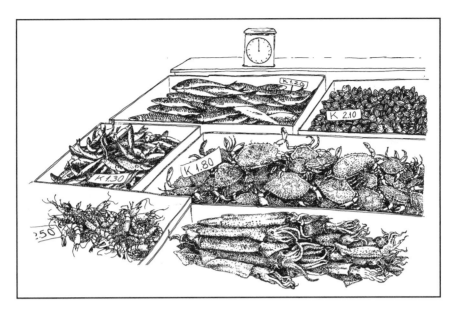

Figure 3.5. Although most of our food comes from domestic species, a wide variety of wild species are consumed.

lations. Brazil nuts are harvested exclusively from the wild because the trees cannot be grown in orchards, which lack the necessary pollinating insects. It is particularly important for conservation biologists to be concerned with the harvesting of wild species because 1) populations are often overexploited (Chapter 9, Overexploitation); 2) conserving populations requires regulating harvests (Chapter 12, Managing Populations); and 3) maintaining biodiversity requires special consideration for the well-being of rural people because they share most of the habitat of wild plant and animal populations (Chapters 15, 16, and 17, Social Factors, Economics, and Politics and Action).

Medicine: There was a time when essentially all of our medicines, like all of our foods, came directly from other organisms. Today, traditional medicines, predominantly herbal medicines, remain a conspicuous and valuable legacy of this past, not just a quaint relic. An estimated 80% of the people in developing countries, where most of the world's population resides, rely on traditional medicines (Farnsworth 1988), and even in industrialized countries herbal medicines are worth billions of dollars per year (Tyler 1986). A less obvious legacy persists in modern pharmacology (Fig. 3.6). A significant portion of our modern medicines contains active ingredients that are directly obtained from organisms (Akerele et al. 1991); in the United States the figure

Figure 3.6. Two chemicals produced by the rosy periwinkle are used to treat cancer.

is about 41%: 25% from plants, 13% from microorganisms, and 3% from animals (Oldfield 1984). Additionally, many other medicines such as aspirin and various anesthetics contain chemicals that were originally isolated and identified in an organism and then later synthesized by chemists. It is nearly impossible to attach a monetary figure to all these values, but it is certainly in the hundreds of billions of dollars per year, perhaps trillions (Principe 1996).

Plants are a primary source of medicinal chemicals, largely because they have developed a wide diversity of complex organic chemicals (often known as secondary compounds) for deterring plant-eating animals and for other purposes. One of the best-known medicinal plants is the rosy periwinkle from Madagascar, which produces two chemicals that cure most victims of two virulent forms of cancer, Hodgkin's disease and acute lymphocytic leukemia. Another example comes from the Pacific Northwest of the United States where the Pacific yew was transformed from a "trash tree" that simply was in the way of forestry operations into an important medicinal plant (Joyce 1993, Walsh and Goodman 1999). The Pacific yew contains high concentrations of taxol, a chemical that has proven very effective in the treatment of ovarian cancer and breast cancer. Medicines derived from microorganisms include penicillin, tetracycline, and virtually all other antibiotics, as well as a variety of vaccines, hormones, and antibodies (Madigan et al. 1997).

Although animals are the source of some medicines—for example, chemicals used to prevent blood clots have been isolated from the saliva of two blood-sucking animals, leeches and vampire bats—they are generally more widely used in medical science as biological systems to be studied. The role of mice, rats, and primates as surrogates for people in medical research is well known, but animals' contribution to science goes far beyond this. Research on the metabolism of black bears during their winter dormancy has given insight to researchers concerned with kidney function and bone metabolism, specifically, ways to avoid osteoporosis, a disease that afflicts 20 million people in the United States alone (Floyd and Nelson 1990). Researchers studying leprosy have found armadillos to be the best experimental surrogates for humans (Maugh 1982). Animals may also be useful in making the search for medicinally active plants less of a needle-in-the-haystack exercise. Medicinal surveys of plants have long been expedited by consulting with local people about their use of local plants, a field known as *ethnobotany* (Prance et al. 1994), and, more recently, researchers have discovered that some mammals, especially primates, may serve a similar role (Dossaji et al. 1989, Newton 1991).

The role of different species in medicine is of particular interest to conservation biologists because it so clearly highlights the need to maintain biodiversity. From a biochemical perspective every species is unique and thus potentially could be the source of a major scientific breakthrough. If we lose a species, we may have lost an invaluable opportunity. We have only begun to screen organisms for their biochemical properties, and it promises to be an endless task because by the time we have completed one round of screening, medical technology will likely have advanced to the stage where another search could be productive.

Clothing, Shelter, Tools, and Trinkets: Plastics, metals, glass, and concrete may constitute the bulk of materials people use today, but more traditional materials such as wood, cotton, thatch, sisal, wool, silk, leather, fur, and others remain very important to us. In industrialized nations natural materials often command a premium price because people prefer to walk on hardwood floors rather than linoleum and to sit on leather upholstery rather than plastic. In places that are far from industrial centers, or where a subsistence economy prevails over a cash economy, natural materials may still be dominant.

A conservation biology perspective on the use of organisms for materials parallels our earlier discussion about using organisms for food—wild relatives of domestic populations, wild species that might be domesticated, and direct use of wild species. One issue stands out. The overexploitation of wild populations for materials seems particularly unacceptable when they are used to produce nonessential items: trinkets and toys for wealthy adults such as spotted cat fur coats, ivory knickknacks, rhino-horn dagger handles, or Brazilian rosewood guitars.

Fuel: The single biggest use we make of other living creatures, as measured in tons, is burning them as biomass fuel. Trees provide most of this material, about 3.2 billion cubic meters per year (United Nations Development Programme et al. 2000); agricultural residues are another significant portion. Of course, all forms of life are full of carbon and will burn given sufficient heat and oxygen. Therefore the prospect of harnessing the sun's energy by taking advantage of photosynthesis does not provide a particularly compelling argument for maintaining species diversity. In other words, biomass fuel is pretty much biomass fuel whether it comes from ubiquitous weeds or rare herbs. There are some exceptions to this generalization. People interested in producing biomass fuel as efficiently as possible favor certain fast-growing plants such as cattails, willows, and aspen (Cook et al. 1991).

Closely related to fuel are various oils and waxes used for lubricants, chemical feedstocks, and other specialized uses. Some of these substances are unique to certain species. For example, sperm whale oil has special properties as a lubricant, properties so valuable that sperm whale populations have been grossly overexploited. Fortunately, scientists have discovered that a plant, the jojoba, which can easily be cultivated, produces an oil with qualities very similar to sperm whale oil.

Recreation. A person's requirements for food, clothing, shelter, tools, and fuel are fundamental, but we also have emotional needs that drive our search for pleasure. Virtually all of us find pleasure in interacting with other people, and most of us also seek enjoyment from our interactions with other living creatures. Enjoying another species does not necessarily require economic activity, but, in practice, our attraction to other species involves large sums of money (Fig. 3.7). Keeping pets and growing ornamental plants are the basis for enormous businesses. Dogs, cats, and roses may be a large part of this trade, but thousands of species from ants to zinnias are involved, and most of them are not domesticated. The selling of encounters with wild plants and animals is the basis for a substantial enterprise that has become known as ecotourism (Whelan 1991, Cater and Lowman 1994). People pay to travel long distances for the privilege of seeing redwoods, coral reefs, whales, lions, and many other species. Most ecotourists carry expensive cameras and binoculars; some of them carry guns or fishing rods. Overall, hunters and anglers pay the highest sums to pursue their recreation, sometimes thousands of dollars per person per day. Closer to home, backyard interactions with wild creatures are the basis for large sales: bird seed, bird feeders, birdhouses, and birdbaths tally over $2.5 billion per year in the United States alone (USFWS 1993). In the home, hobbyists assemble collections of butterflies and mollusc shells, as well as books, paintings, sculptures, and stamps with flora and fauna themes.

Diversity is the spice of life, and species diversity is a key element in the recreational value of organisms. Many gardeners, exotic pet fanciers, and shell and butterfly collectors want to own species that their friends do not have, and they will pay for the privilege. Similarly, birders, botanizers, hunters, and anglers covet experiences with species they have not encountered before.

Services. Most of the economic values described above involve species that serve as goods—physical objects that people can use—but there are some

Figure 3.7. People enjoy the diversity of nature in many ways.

exceptions. When wild relatives of domestic species provide genetic information for plant and animal breeders, or when wild species give enjoyment
to outdoor recreationists, they are providing services rather than goods.
Other examples include the pollination services rendered to farmers by
bees and other species, the degradation of oil spills by bacteria, the aeration
of soils and decomposition of organic matter by earthworms and many
other organisms, and the removal of pollutants from air and water by plants
and other organisms. Many of these services are not routinely purchased
and could be described as ecological values, which we will address below.
On the other hand, the absence of these services often has direct, easily

measured economic costs—for example, farmers having to rent beehives because the wild pollinators have been decimated by insecticides.

Spiritual Values

People love life, a phenomenon called "biophilia" by E. O. Wilson (1984). We delight in the beauty of a calypso orchid. We are inspired by the majesty of a golden eagle. We find spiritual comfort in the transformation of a caterpillar into a monarch butterfly. It is easy to find evidence of our aesthetic, spiritual, and emotional affinity for other species. This linkage is revealed in the symbols we choose for our governments, religions, and athletic teams: the sugar maple leaf emblem of Canada, the raven totem of the Vikings and several Native American tribes, the banana slug mascot of the University of California at Santa Cruz. We show it in the motifs we use to decorate our clothing, jewelry, and dwellings and in the places we select to visit in our leisure time. Our language—busy as a bee, an eager beaver—reveals the depth of this linkage (Lawrence 1993).

Sometimes, our feelings for other species are revealed in the way we spend our money; sometimes, they are not. Imagine a woman who lives her whole life in landlocked Hungary who will never see a living blue whale, but who derives pleasure from simply knowing that they exist. Her love for whales is real and valuable, but costs her nothing. It is hard for society to account for feelings like hers when making policy decisions because economic issues are usually paramount, and her feelings are not easily expressed in monetary units. But this does not make her feelings unimportant. It also does not diminish the political impact of her feelings. For example, the decision to curtail exploiting Newfoundland's baby harp seals for their fur was made because of the feelings of people who had no direct contact with harp seals and no economic stake in their survival. Economists are trying to devise methods for estimating the monetary value of blue whales and harp seals for people whose only relationship with them is knowing that they exist; we will discuss *existence values* further in Chapter 16, Economics.

Scientific and Educational Values

The world is a complex place, but our knowledge of it is increasing all the time, and some of the credit goes to our fellow inhabitants (Fig. 3.8). There are many examples. Birds offered both the inspiration to fly and a model from which to learn, and, similarly, the ability of bats to fly in the dark inspired the development of sonar and radar. Mendel's peas opened the door to genetics, and the convenience of working with *Drosophila* fruit flies has greatly facilitated genetic research. For Charles Darwin, the diversity of

Figure 3.8. Other organisms teach us about our world.

some Galapagos birds that now bear his name—Darwin's finches—was instrumental in his development of the theory of natural selection. Many anthropologists who seek insight into human social interactions study our nearest relatives, the roughly 180 other members of the primate order.

Of course, scientific inquiry is just an advanced form of the intellectual curiosity about the world that begins in infancy. Our education would suffer greatly without a diverse world to explore, without bean seeds to plant, without frog eggs to watch develop into tadpoles. Whether we want to learn about ourselves or the world we share with other species, we need models to observe.

Ecological Values

Every population of every species is part of an ecosystem of interacting populations and their environment and thus has an ecological role to play. There are producers, consumers, decomposers, and many variations of these roles and others—competitors, dispersers, pollinators, and more. In this sense, every species has ecological value; it is of instrumental use to other species that share the same ecosystem, including people. Although all species have ecological roles, not all roles are of equal importance. Some species are ecologically important simply because of their great abundance. Sometimes they are called *dominant species,* a term that usually implies that they constitute a large portion of the biomass of an ecosystem such as sugar maples in a sugar maple forest or various species of planktonic copepods in many marine ecosystems. Sometimes, they are called *controller species*, which implies that they have major roles in controlling the movement of energy and nutrients. This would include dominant species such as sugar maples and various copepods, as well as many species of bacteria and fungi that are important decomposers but too small to have a sizable biomass.

Some species play critical ecological roles that are of greater importance than we would predict from their abundance; these are called *keystone species* (Power et al. 1996). The classic example of a keystone species is the purple sea star, an intertidal predator that preys on several species of invertebrates, apparently allowing many species to coexist without any one species becoming dominant (Fig. 3.9). After these sea stars were experimentally removed from a rocky shore in the state of Washington, the population of one prey species, the California mussel, dominated the site, and

Figure 3.9. The purple sea star is a keystone species because its predatory activities allow many species to coexist.

the system shifted from 15 species of invertebrates and macroscopic algae to only eight species (Paine 1966). In some tropical forests many animal species depend on fruit for the bulk of their diet, but during certain seasons only a few tree species bear fruit (Terborgh 1986). These off-season fruit producers are keystone species. The endangered red-cockaded wood-pecker might play a keystone role in those southern United States pine forests where it persists; because it is the only woodpecker that routinely ex-cavates cavities in living trees, it provides habitat for a number of other cav-ity-dwelling species incapable of making their own cavities. The aquatic ecosystems created by their dams make beavers a keystone species.

As a rule conservation biologists tend to focus more on the population health of keystone species than dominant or controller species because many keystone species are uncommon, while, by definition, dominant and controller species are relatively abundant. Of course, being abundant does not mean that these species are secure from population crashes. Many is-land plants have gone from being ecological dominants to being quite rare following the introduction of exotic herbivores or competitors (Cuddihy and Stone 1990). Even continental species have plunged from dominance to rarity in a short period; such was the case for the American chestnut fol-lowing invasion of an exotic fungus disease.

When assessing the ecological roles of species, conservation biologists are typically conservative and assume every component of an ecosystem is critical until proven otherwise (Ehrlich and Mooney 1983, Berlow 1999). Our understanding of ecosystems is usually so limited that it would be unwise not to take this position. Furthermore, it is possible that one should look beyond the role of individual species because overall species richness of an ecosys-tem may be an important attribute. We will return to this issue in the next chapter, Ecosystem Diversity. Finally, it is important to realize that a species that is relatively unimportant now may become more important as an ecosys-tem changes through time. For example, during the last 12,000 years the east-ern white pine has varied from being quite rare to being an ecosystem dominant over large areas (Jacobson and Dieffenbacher-Krall 1995).

Incidentally, there are many ways in which ecological values interface with economic values. Most notably, the health and productivity of people have huge economic consequences, and these are directly dependent on ecological integrity. Similarly, each species we use directly for economic gain as food, medicine, materials, and so forth depends on ecosystems and the continuing existence of a whole suite of other species.

Strategic Values

With a large agenda and limited resources, conservation biologists have to be efficient strategists, and this often leads them to target certain species to advance their overall goal of maintaining biodiversity (Simberloff 1998, Caro and O'Doherty 1999). Best known are the *flagship species*, the charismatic species that have captured the public's heart and won their support for conservation. Some species have won converts to conservation across the globe; consider the cuddliness of the giant panda, the haunting songs of the humpback whale, and the grandeur of the tiger or gorilla. Some species have been rallying points for local action, engendering local pride and concern. In northeastern Peru, for example, conservationists built a program around the yellow-tailed woolly monkey, an endangered species endemic to the area, using special T-shirts, posters, and other means. Once the local people learned how special their monkey was, it was much easier to enlist their support for conservation of all the local biota.

Mammals, especially those with big brown eyes, are often the most successful flagships, but many other species have been successfully used too. In northern Maine an inconspicuous plant with an unprepossessing name, Furbish's lousewort, became a flagship species for the effort that stopped a dam that would have flooded 35,000 hectares of forest. This was a case where concern for an ecosystem pushed a species into the flagship role. A better-known example of the flagship process in reverse comes from the northwestern United States where concern for old-growth forests has made the spotted owl a flagship species. Conservation has also known some flagship individuals. The most famous is Digit, the gorilla whose death while defending his family from poachers unleashed a flood of support for gorilla conservation (Fossey 1983).

Many flagship species are relatively large animals, making them good candidates to be *umbrella species* as well. Umbrella species usually have large home ranges, and thus by protecting enough habitat for their populations, adequate habitat for many other species will also be protected. The most effective umbrella species are also found in a wide variety of ecosystems across a broad geographic range and can thereby provide an umbrella for a large set of species. Perhaps the single best example of an umbrella species is the tiger. With a geographic range reaching from the Russian Far East south to Indonesia and west to India (formerly to Turkey and Iran), the tiger ranges across a broad set of ecosystems—boreal forests, mangrove swamps, rain forests, dry deciduous woodlands, and more. Efforts to keep the tiger from going extinct have benefitted other wild creatures throughout much of Asia (Tilson and Seal 1987) (Fig. 3.10).

Some species are useful to conservation biologists because the health of their populations is an easy-to-monitor indication of environmental conditions or of the status of other species; these are called *indicator species*. They are the "miners' canaries" that can warn us about environmental degradation. The classic example comes from the impact of DDT on peregrines, brown pelicans, and some other birds. It was the catastrophic de-

Figure 3.10. Because tigers range over a broad region and many different types of ecosystems, efforts to save them have benefitted many other species, thus making tigers an umbrella species.

cline of these species that first alerted scientists to a subtle but pervasive and serious problem. Smaller species are often sensitive indicators (Kremen et al. 1993); lichens reveal air pollution problems, and aquatic invertebrates are monitored to track water pollution (Clark and Samways 1996). Some indicator species provide "easy access"; for example, monitoring colonial seabirds to assess the health of the marine realm is often easier than deploying oceanic survey vessels (Monaghan 1996). Indicator species may also reflect undisturbed ecosystems that are prime candidates for reserves. If, for example, you find an area with a sizable population of curassows, chachalacas, or guans (a family of large, delicious birds that are avidly sought by hunters throughout Latin America), you can be fairly confident that it is not heavily hunted and therefore might be a relatively easy place to establish a reserve (Silva and Strahl 1991).

Realized Values and Potential Values

When we assess the instrumental values of species, we generally focus on their usefulness here and now, but this is a shortsighted viewpoint as revealed in our discussion of medicinal research and biodiversity. Our rudi-mentary understanding of biology and ecology leaves an enormous gap be-tween the currently realized value of a species and its potential future value. This gap is particularly wide because we have only a vague idea of what our future lives will be like—technologically, culturally, and ecologi-cally. Consider the technology commonly known as DNA fingerprinting that has become so important in identifying criminals from the bits of pro-tein, usually blood or semen, they may have been left at the scene of a crime. Not long ago we could barely imagine such a technology. We still might not have the technology if scientists had not isolated an enzyme capable of sep-arating DNA strands and remaining functional at very high temperatures. They found this enzyme in a bacterium, *Thermus aquaticus,* that grows in the boiling hot springs of Yellowstone National Park (Gelfand 1989).

It may be even harder to guess at the potential ecological role that a species might assume in the future. It would certainly have taken a very pre-scient biologist to guess that the shrewlike mammals that shared the earth with dinosaurs would lead to the earth-dominating *Homo sapiens.* More-over, we do not need to think in evolutionary time scales to find examples of changes in ecological importance. It is likely that many of the species that currently dominate the backyards, farmlands, and small woodlots of east-ern North America were far less common a few centuries ago when forests covered most of the region. During a lifetime of exploring North America the 19th-century ornithologist, John James Audubon, only once encoun-tered a chestnut-sided warbler (Forbush 1929). This species has become very common, presumably because the early successional habitats that it prefers are more abundant now than they were in Audubon's time.

The core idea in this section is nicely captured in a phrase that could be a motto for conservation biology: keep options alive. We must take this approach because we know so little. We can never say of any species that it lacks value.

The Uniqueness Value of Species

Imagine a question on your vertebrate zoology final: What do the aardvark, ostrich, and bowfin have in common? "They are all vertebrate an-imals" might get you grudging partial credit. "They each are the only species in their respective orders: Tubulidentata, Struthioniformes, and Amiiformes" would earn you full credit and maybe a "Good!" penciled in the margin. These are three special species because they are unique at the taxonomic level of an order, a level of taxonomy that also encompasses such large groups as rodents (Rodentia, ca. 1700 species), songbirds (Passeri-formes, ca. 5300 species), and perches and their relatives (Perciformes, ca.

9000 species). We could argue about how artificial taxonomic classifications are, but in the end we would agree that a white-eyed vireo is much more similar to a red-eyed vireo than an aardvark is to one of its nearest living relatives, the African elephant.

The uniqueness of a species is a value that can amplify all the other values elaborated above. All other things being equal, a conservationist focusing on intrinsic value might give more importance to a spectacled bear (the only member of its genus) than a polar bear (one of three members of the genus *Ursus),* because the spectacled bear is more different from other bears than the polar bear is. The spectacled bear lineage split off from the main bear line over 10 million years ago, while polar bears evolved from brown bears only about 70,000 years ago and have even produced fertile hybrids with brown bears in zoo matings (O'Brien 1987). (See Vane-Wright et al. 1991 and subsequent articles by Erwin 1991, Crozier 1992, Faith 1992, and Purvis et al. 2000 for more on setting conservation priorities based on taxonomic uniqueness.)

In terms of instrumental values, a species that has close relatives is more likely to be replaced than a species without close relatives. Huckleberries may not taste exactly like blueberries, but they are not a bad substitute. In contrast, nothing tastes very much like a pineapple. There is some bad news lurking here. The process of replacing one species with another one that has similar economic values can spread the web of overexploitation. Whalers started with the species that were most profitable to catch, mainly the right whale (so named because it was the "right" whale to catch), and, as each species was depleted, they concentrated on the next one in line. If there were only one species of whale, would the whalers have realized their shortsighted folly and instituted conservation measures, or would they have driven that one species to extinction because they had no other species to turn to?

The instrumental values that are determined by a species' role in an ecosystem may also be influenced by a species' uniqueness. Although the exact ecological role or niche of each species in an ecosystem may be different from that of every other species, there is often considerable overlap or redundancy in the functional roles of species. For example, no other forest herb may fill the exact niche of a Canada mayflower, but there are other species that are fixing carbon, providing a substrate for soil fungi, providing food for pollinating insects and fruit-consuming small mammals, and so on. If these functional overlaps are sufficient, then it is likely that the ecosystem would not be profoundly changed by the disappearance of its Canada mayflower population. On the other hand, if a species is very distinctive, it is more likely that its disappearance from an ecosystem would cause significant changes because it is a keystone or dominant species. For example, the loss of African elephants from many grass and shrub ecosystems has profoundly changed the structure of these ecosystems by allowing them to become forested (Laws et al. 1975).

CASE STUDY   The Neem Tree

Wheat, corn, rice, potatoes—many species of plants have been profoundly important to the welfare of humanity. Indeed, some scholars have argued that one of the key defining events in western civilization was the hybridization, about 10,000 years ago in the Middle East, of two species to produce a form of wheat amenable to cultivation. From an historical perspective, at least one animal might rival these plants in its value: the horse, backbone of early transportation, exploration, and too often, war. When we consider species in terms of the diversity of their instrumental values, not many species equal the neem tree, a member of the mahogany family from southern Asia.

The most remarkable thing about the neem is the myriad ways it is used as a health product. People use neem products to treat boils, burns, cholera, constipation, diabetes, heat rash, indigestion, malaria, measles, nausea, parasites, pimples, rheumatism, scorpion stings, sleeplessness, snake bites, stomach aches, syphilis, tumors, and ulcers, and they drink neem tea as a general tonic. They clean their teeth with neem twigs and neem-derived toothpaste and make a disinfectant soap with the oil of neem seeds. Recent research suggests that neem products may provide the basis for a birth-control pill for men and as a spermicide.

These marvelous features may account for the spiritual importance of neem as well. It is considered sacred by many Hindus, and its leaves are hung in the doors of a house to ward off evil spirits and burnt as an incense to drive evil spirits out of anyone who inhales the smoke. Some Hindu holy men place neem twigs in their ears as a charm. The wood of the neem, attractive, strong, and durable, is one of few types used for carving idols. Returning to secular uses, neem wood is also used for fuel, furniture, and house building; neem foliage and seeds are used as livestock fodder; and neem seed oil is used as lamp fuel and to make lubricants and disinfectants. Neem trees grow well on marginal sites, making them appropriate for reforestation, and they produce a deep shade that is especially valued in hot climates. People place neem leaves in their cupboards, grain bins, beds, and books to repel insect pests. Various neem extracts are also effective as repellents and antifeedants for insects and nematodes that are agricultural pests.

The qualities of the neem are well known among millions of people in the Indian subcontinent: it is often called the "village pharmacy." It is being explored by foreigners as well. The breadth of interest is evidenced in three volumes (Vietmeyer 1992, Jacobson 1988, Schmutterer 1995) that provided the basis of this account.

Summary

There are many ways to define species, and decisions on what constitutes a species can have significant ramifications for conservation activities. It is generally desirable for conservationists to seek to maintain all distinguishable taxa, whether or not there is full agreement on definitions of "species," because they represent significant genetic diversity. Approximately 1.7 million species have been described by scientists, but the actual number of species that exists is certainly much greater because there are large numbers of undescribed species, notably, tropical forest insects, marine invertebrates on the deep-ocean floor, and microorganisms in all ecosystems. Although conservation biologists cannot hope to work with each species, it is useful to know the magnitude of what we might lose if environmental degradation continues.

One can argue that every species has intrinsic value; in other words, its importance is independent of its relationships with people and all other species. From this perspective, conservationists usually evaluate the importance of a species relative to how endangered it is. This is the basis for the lists of species jeopardized with extinction maintained by many organizations. Instrumental values, which are based on the usefulness of species, differ among species. Many species have economic value because they provide food, medicine, materials, fuel, recreation, and various services for people. Species also have aesthetic, spiritual, scientific, and educational values that go beyond economics. They have ecological importance to many other species because of their roles in ecosystems. They can be of strategic value to conservationists by serving as flagship, umbrella, or indicator species. Some of these instrumental values are currently realized; many of them are potential values because they have not yet been expressed. Finally, species vary in their taxonomic uniqueness, and species that have no closely related species are generally considered more important than species with many close relatives.

FURTHER READING

For a popular account of how species diversity arises, how many species may exist, and related issues, see Wilson (1992). More detailed accounts are available in Huston (1994) and Heywood and Watson (1995) and in the *Proceedings of the Royal Society of London, Series B, 1994,* Vol 256 (1345). The World Conservation Monitoring Centre has published many Red Data Books that list and describe endangered species, plus a broad compendium, *Global Biodiversity* (Groombridge 1992), that are essential references. For a review of the instrumental values of species see Oldfield (1984) and Prescott-Allen and Prescott-Allen (1982, 1986). The World Conservation Monitoring Centre website is www.unep.wcmc.org. Also see www.species2000.org for a global effort to list all the world's species, www.gbif.org for a global biodiversity site, www.natureserve.org for information on species in Canada and the United States, and phylogeny.arizona.edu/tree/phylogeny.html for information on the taxonomic relationships among species. Read Bisby (2000) and Edwards et al. (2000) for treatments of internet resources for biodiversity information.

TOPICS FOR DISCUSSION

1. Under what circumstances should conservationists attempt to maintain hybrids? Are there circumstances under which conservationists would wish to eliminate hybrids?
2. If you had a large budget for conservation biology research, say $200 million per year, what percentage of it would you allocate to 1) estimating the number of species in existence 2) surveying and classifying little-known groups of organisms, and 3) studying species and ecosystems known to be threatened? Defend your budget.
3. Should a species' instrumental value be evaluated when deciding whether to place it on a list of endangered species that will be a priority for conservation efforts? Why or why not?
4. What approaches would you use to estimate a species' potential instrumental value?
5. Should we seek to eradicate species such as the smallpox virus, or should we confine them to research laboratories where they cannot harm people?

Chapter 4
Ecosystem Diversity

Flying over the countryside in an airplane you see patterns: blue patches and ribbons that are lakes and rivers, dark green patches that are forests, brown patches that are tilled fields, and so on. These are the coarse manifestations of an enormously complicated web of ecological interactions, a myriad of species interacting with one another and their physical environment. Despite this complexity, all is not chaos. There are patterns; some are so obvious that they can be seen from far above the earth, and some are so subtle that we have little awareness or understanding of them. These patterns of interactions are the basis for ecosystems, and they are fundamental to the goal of maintaining biodiversity.

What Is an Ecosystem?

It is easy to define an ecosystem conceptually. It is a group of interacting organisms (usually called a *community*) and the physical environment they inhabit at a given point in time. It is much harder to delineate ecosystems in the real world—to decide where one ecosystem ends and another begins—because the web of interactions does not have clean breaks (Fig. 4.1). Most ecologists would say that a forest and an adjacent lake are different ecosystems because the assemblages of organisms inhabiting them are almost completely different and have relatively few direct interactions. This said, there are some interactions across the shoreline through the movements of frogs, insects, autumn leaves, water, and so on, and these interactions can be quite important. Conversely, many ecologists would say that a young oak forest and an adjacent old oak forest are the same ecosystem even though a fair number of their species would be different, as would some key

Figure 4.1. Deciding where one ecosystem begins and another ends is a complex task because the web of ecological interactions does not have clean breaks. In this example, distinguishing between the forest ecosystem and the lake ecosystem may be relatively easy, but is the young forest on the left a different ecosystem from the older forest on the right?

ecological processes such as succession and decomposition. Separating two adjacent ecosystems is particularly difficult when the edge between them, often called an *ecotone,* is a gradual transition zone. For example, on the side of a mountain, ecosystems change continuously in response to the climate gradient that parallels altitude, and it is probably arbitrary to draw lines among them.

Distinguishing ecosystems is also difficult because ecologists think about ecosystems at a variety of spatial scales. A pool of water that collects in the crotch of a large tree and is home to some algae and small invertebrates can be considered an ecosystem. At the other extreme, ecosystems are sometimes defined on the basis of the movements of wide-ranging animals. When biologists speak of the Greater Yellowstone Ecosystem, they are referring to an area of about 50,000 km^2 defined in large part by the habitat needs of a grizzly bear population (Clark et al. 1999). At the largest known scale, the earth's entire biosphere can be considered an ecosystem.

The key thing to understand is that the term "ecosystem" is a conceptual tool that makes it easier for us to organize our understanding about ecological interactions and to communicate that understanding to other people. For the purposes of this book we can think of ecosystems at a scale that is easy to detect from an airplane, typically from a fraction of a hectare to a few hundred hectares. We can draw the boundaries between adjacent ecosystems where they will separate significantly different sets of species. We will sidestep the question "what is significant?" because this may change depending on the circumstances.

A note about language ambiguities is necessary here. We have two different words for a type of organism and a particular example of that type; we call the former a species and the latter an individual. Conservationists focus on maintaining species rather than every individual of a species. We have no parallel words to distinguish between a given type of ecosystem (e.g., an alkaline eutrophic lake) and a particular example of an ecosystem (e.g., Smith Lake). The distinction is usually obvious from context but not always. If people say "Protecting Mojave Desert springs is a high priority for Californian conservationists," are they talking about a type of ecosystem labeled Mojave Desert springs for which there are many different examples? Or, are they talking about some desert springs that happen to occur in the Mojave Desert but that are not substantially different from desert springs in the Sonoran Desert? The use of a definite or indefinite article—*the* Mojave Desert Springs versus *a* Mojave Desert spring—will usually make the distinction clean. This is an important distinction because conservationists must give priority to maintaining *types* of ecosystems; we cannot realistically expect to protect every example of an ecosystem type.

Classifying Ecosystems

Just as it can be difficult to delineate particular ecosystems on the ground, it is also difficult to classify them into different types once they are delineated (Whittaker 1973). How similar must two different ecosystems be to be considered the same type of ecosystem? Although there are several quantitative methods for assessing similarity of community composition, there is no standard level of similarity used to decide whether two ecosystems are of the same type (Fig. 4.2 and Table 4.1).

It is often useful to approach ecosystem classification hierarchically. For example, at the highest level we could separate terrestrial and aquatic ecosystems; at a lower level freshwater, marine, and estuarine ecosystems; then freshwater ecosystems into lakes and rivers; and so on. However, there is no universally accepted system for doing this analogous to the kingdom-phylum-class-order-family-genus-species system.

Geography also needs to be considered when classifying ecosystems. Two alkaline eutrophic lakes that share a very similar biota and other at-

TABLE 4.1 Relative abundance of species (%) in three hypothetical ecosystems.

Ecosystem	A	B	C
Black oak	40	30	10
White pine	30	40	10
Red maple	20	10	10
Yellow birch	10	20	70
Similarity index	A vs B	0.96	
	B vs C	0.54	
	A vs C	0.40	

Figure 4.2 and Table 4.1. Based on the limited data presented, most ecologists would probably classify A and B as belonging to one type of ecosystem and C to a different type. Note that the similarity index (which has a range of 0 to 1) is much higher between A and B than between B and C or A and C. However, there is no standard level of similarity used to determine if two ecosystems are of the same type. (See Magurran 1988 for calculation of the Morisita's similarity index, used here, and others.)

tributes (e.g., chemical structure) would probably be considered the same type of ecosystem even if they are hundreds of kilometers apart and on either side of a mountain range. On the other hand, the mountain range might be a geographic barrier for many species, or perhaps the climate is significantly different on either side of the mountain because of a rain shadow effect. In this case the two alkaline eutrophic lakes might have quite different biotas, and we might decide that they are different types of ecosystems. Yet, they still would be more similar to one another than to an acidic lake.

How can we recognize both the basic similarity of the two alkaline eutrophic lakes and the differences that occur because of their geographic separation? One approach involves dividing the world into regions based on biologically meaningful patterns that shape the distribution and abundance of species such as climatic zones, mountain ranges, oceans that isolate terrestrial biota, or continents that isolate marine biota (Hunter 1991). There are many examples of such maps; they use a variety of criteria and names such as ecoclimatic zones, biogeographic provinces, and biophysical regions (e.g., Kuchler 1964, Udvardy 1975, Bailey et al. 1994, Bailey 1995; see Bailey 1996, Rowe 1996 for an overview). Figure 4.3 shows one example. By using such a map we could recognize the differences that exist between the two lakes because they are in different ecological regions, but we could still recognize their basic similarity by calling them both alkaline eutrophic lakes.

From a conservation perspective we could largely avoid the issue by

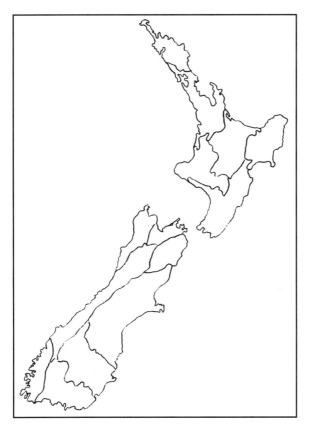

Figure 4.3. Map of the 14 ecological provinces of the main islands of New Zealand, which are based on land form, geology, climate, and species distributions; four more provinces are associated with smaller islands. New Zealand has been further subdivided into 85 ecological regions and 268 ecological districts. (Redrawn by permission from MacEwen 1987.)

organizing conservation efforts for each ecological region. However, conservation efforts are usually organized around political units—states, provinces, nations—and political boundaries do not usually coincide with ecological boundaries.

The Values of Ecosystems

Species cannot survive in isolation from other species; they are all part of some ecosystem. Therefore all ecosystems have value because the species they support have value. In other words, at a minimum the value of an ecosystem is the summation of the value of all its constituent organisms. This idea is simple enough, but it is not the end of the story. We must also consider the possibility that ecosystems have special attributes that make them valuable beyond the sum of species-specific values. Let's consider each of the major types of values that we evaluated in Chapter 3 from this perspective.

Intrinsic Value

Whether or not ecosystems have intrinsic value independent of the intrinsic value of their constituent species is an issue that hinges on a complex and controversial question. Are ecosystems tightly connected, synergistic systems built around a set of closely coevolved species, or are they based on a loose assemblage of species that happen to share similar habitat needs and end up interacting with one another to varying degrees because they are in the same place at the same time? To put it another way, are ecosystems analogous to supraorganisms in which different populations are closely connected, or are they just a collection of competing populations? This question has stimulated ecologists for decades (McIntosh 1980). The Gaia hypothesis—the idea that all life on earth might constitute a giant, well-organized, self-regulating organism (Lovelock 1979)—is one well-known manifestation of this debate. Undoubtedly, the truth lies somewhere between the poles presented here and varies somewhat from ecosystem to ecosystem, but for our purposes we do not have to worry too much about exactly where (Fig. 4.4). We will return to this question in other contexts, but for now it is sufficient to note that the closer ecosystems lie to the "tightly connected" pole of the spectrum, the easier it is to acknowledge that they have intrinsic value.

If ecosystems do have intrinsic value, then conservationists need to protect some examples of each different type of ecosystem, especially those that are in danger of disappearing. Some types of ecosystems are rare because they occur only in uncommon environments. For example, cool forests and alpine areas are rare in Africa because the continent has only a few,

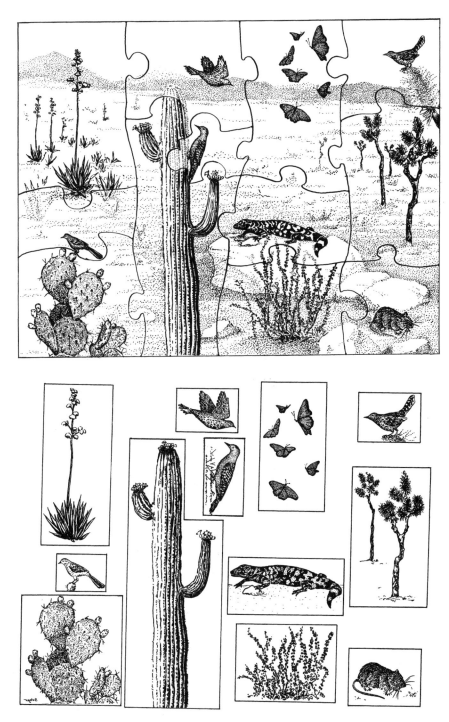

Figure 4.4. Are ecosystems tightly connected systems of closely coevolved species, or are they a loose assemblage of species that happen to share similar habitat needs and end up interacting with one another?

isolated mountains tall enough to support these ecosystems (Kingdon 1989). Other ecosystem types have become uncommon because of human activities. In particular, many types of forest and grassland ecosystems with fertile soils and benign climates have largely been converted to agricultural lands.

Conservationists recognize the importance of protecting a representative array of ecosystems, but they have not yet developed endangered ecosystem lists with legal status analogous to the endangered species lists of CITES, the U.S. Fish and Wildlife Service, and others (Noss et al. 1995). Political hurdles may be paramount, but there is also substantial difficulty in classifying ecosystems as discussed above. Are the spruce-fir forests that occur on a few summits in the southern Appalachians a different type of ecosystem from the spruce-fir forests that stretch across Canada? If so, they are a very rare ecosystem; if not, they are just a peripheral variation of one of the planet's most widespread ecosystems. Decisions like this are absolutely critical if you are trying to protect ecosystems for their intrinsic values, but they are not quite so important if your focus is on the instrumental values of ecosystems.

Instrumental Values

Economic Values:If we think of the economic values of ecosystems in terms of goods and services, the material goods provided by ecosystems can generally be accounted for by summing the goods provided by various species such as the lumber from tree species, the food from fish species, and so on. It is services rather than goods, however, that are of primary economic importance at the ecosystem level (Oldfield 1984). For example, wetlands are often used for tertiary treatment of municipal wastewater, a service that would be quite expensive to duplicate with a treatment plant. Dune and salt-marsh ecosystems provide an invaluable service during coastal storms by buffering upland areas. Coastal wetlands export nutrients and organic matter to adjacent estuaries where they support economically valuable fisheries (Fig. 4.5). Forests export high-quality water to aquatic ecosystems. This list could go on and on because for virtually every ecosystem we could identify services that would be very expensive to replace artificially. Access to the recreational services of ecosystems is the basis for an enormous array of commercial enterprises. These can be as simple as providing buses to take city dwellers out to a forest or field on a Saturday afternoon, or they can be as all-inclusive as completely catered tours to coral reefs, tropical forests, Antarctic islands, and so on (Whelan 1991, Cater and Lowman 1994).

The economic values of ecosystems for both goods and services have been compiled (Daily 1997) and a grand tally of their economic value has been estimated by multiplying a value-per-hectare figure for each major type of ecosystem times the total global area of that ecosystem (Costanza

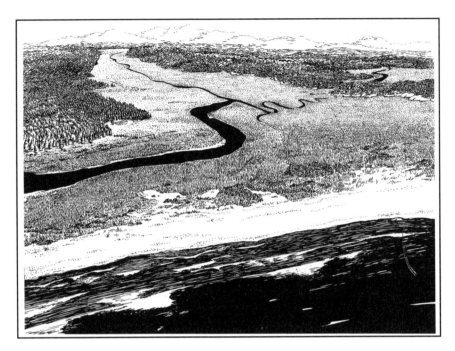

Figure 4.5. Relatively few species can tolerate the special conditions of salt marshes, but those that do create ecosystems of great importance. This is in part because salt marshes export large amounts of organic matter to adjacent estuaries, which constitutes a key component of the estuarine food web.

et al. 1997a). The mean estimate of $33 trillion per year was considered a minimum figure because of the nature of various uncertainties. To put that figure in perspective, the gross national products of all the world's nations total about $18 trillion.

Spiritual Values: The journeys people make to natural ecosystems, to places where the hand of humanity is hard to detect, are often too profoundly important to be reduced to dollars and cents. The forty days Moses spent in the desert, the walkabouts of Australian Aborigines, and perhaps the night you spent watching the tide ebb and flood are periods of spiritual re-creation and revitalization that many people find of immeasurable value. For some people, particularly those who are pantheistic (i.e., believe that God is nature and nature is God), ecosystems provide far more than an aesthetic setting for these experiences. The ecosystems themselves, with their depth and complexity, are a source of inspiration, a vehicle for feeling connected to something larger and more permanent than one's self.

Scientific and Educational Values: Ecology has become a very sophisticated science, but we still cannot hope to understand an ecosystem fully. This dilemma is apparent when you think of ecology as the apex of a pyramid with biology as the next layer below, earth sciences such as geology and climatology forming the third layer, chemistry the fourth, and physics the foundation. Of course, ecologists do not have to be intimately familiar with quantum physics to be effective, but they do have to have a basic understanding of thermodynamics, electromagnetic radiation, and many other aspects of physics. In contrast, a physicist can be successful and understand nothing about ecology. The fact that ecosystems integrate so many phenomena makes them a focal point for scientists trying to monitor how the earth is changing, particularly in response to human activities. This feature also means that ecosystems are fascinating models for researchers interested in complex systems. The computer models developed to predict global climate change are perhaps the most obvious example (Hoffman and Jackson 2000).

Ecosystems are also wonderful models for showing children and adults how everything can be connected to everything else. Drawing lines between boxes to represent the functional relationships of those boxes can become an extremely complex exercise, or it can be very simple. It can be as simple as drawing lines between the sun, a plant, and an animal to form a food chain and then adding more boxes and lines to create a food web. In short, we can all learn a great deal from ecosystems.

Ecological Values: The ecological interactions that are the basis of ecosystems are absolutely fundamental to life. Try to imagine a planet where dead things did not decompose, or where plants did not replenish oxygen. Consequently, it is not really profound or insightful to say that ecosystems have ecological value. Nevertheless, it is extraordinary how often some industrialists and politicians try to draw a line between the well-being of people and the well-being of the ecosystems on which our lives ultimately depend.

Do all ecosystems have equal ecological value? No. Obviously, a large salt marsh will usually provide more ecological values than a small salt marsh, and, similarly, a dominant type of ecosystem such as spruce-fir forests will have more total ecological value than an uncommon type of ecosystem such as caves. Certain types of ecosystems may have a significantly greater importance to other nearby ecosystems than we would predict based on their area. We can call these *keystone ecosystems*, analogous to calling species with disproportionately significant ecological roles keystone species (deMaynadier and Hunter 1997). For example, salt marshes can play a keystone role by providing critical resources—nutrients and organic matter—for an adjacent estuary (see Fig. 4.5). Keystone ecosystems can also shape disturbance regimes that affect large areas by either inhibiting or facilitating the spread of a disturbance. To take two examples: a river can inhibit the spread of a fire, while certain types of woodlands that burn easily can facilitate the spread of fires to other ecosystems.

Strategic Values: From the perspective of maintaining biodiversity at all levels—genes, species, and ecosystems—the single most essential value of ecosystems may be their strategic value. Conservation biologists have often proposed that by protecting a representative array of ecosystems, most species and their genetic diversity can be protected as well (Noss 1987b, Hunter 1991). This idea is often described using a metaphor of coarse filters and fine filters first proposed by The Nature Conservancy (1982) (Fig. 4.6).

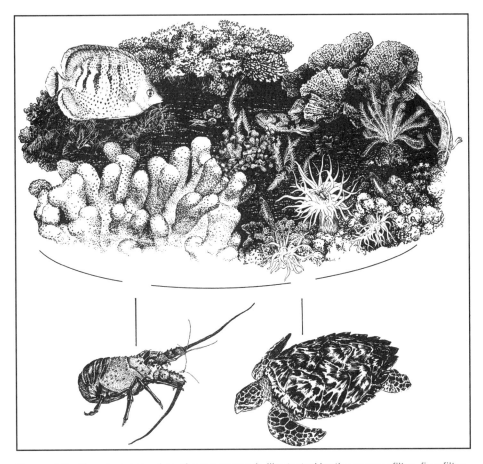

Figure 4.6. The strategic value of ecosystems is illustrated by the coarse-filter–fine-filter approach to conserving biodiversity. Protecting a representative array of ecosystems constitutes the coarse filter and may protect most species. However, a few species will fall through the pores of a coarse filter because of their specialized habitat requirements or because they are overexploited. These species will require individual management, the fine-filter approach. In this example, a coral reef ecosystem with all its constituent species is protected by the coarse-filter approach, but fine-filter management is still required for the hawksbill turtle and spiny lobster.

The coarse-filter approach to conserving biodiversity is appealing because it is efficient and provides broad protection. It is efficient because compared with the number of species in the world there are relatively few different types of ecosystems, perhaps numbering in the thousands, and protecting a representative array of these ecosystems in each ecological region may protect 85% to 90% of the species (The Nature Conservancy 1982). The coarse-filter approach is broad because it is likely to protect most unknown species, as well as known species, plus their genetic diversity to some degree.

Importantly, the coarse-filter approach can be an effective strategy regardless of whether ecosystems are tightly connected systems or loose assemblages of species. It is only necessary that the distribution of ecosystems correspond reasonably well with the distribution of species so that a complete array of ecosystems will harbor a nearly complete array of species (Hunter et al. 1988). We will return to this point and the coarse-filter approach in general in Chapter 11, Managing Ecosystems.

Uniqueness Values

The ambiguities of ecosystem classification cloud the issue of ecosystem uniqueness. If we define many different types of ecosystems, each type of ecosystem will not be very different from similar types. For example, if we consider young oak forests to be a distinct type of ecosystem, then they will not be very different from old oak forests. Alternatively, if we make coarse distinctions (e.g., all forests are one type of ecosystem), then each type of ecosystem will clearly be unique. Some types of ecosystems may seem unique under any classification, for example, caves and hot springs, but there is a danger of confusing uniqueness and rarity. In short, different ecosystems may have different uniqueness values, but these would be difficult to evaluate without an objective, universally accepted classification scheme.

Ecosystem Diversity and Species Diversity

The coarse-filter–fine-filter metaphor (see Fig. 4.6) captures the strategic value of protecting ecosystems as a vehicle for maintaining species diversity, but the relationship between ecosystem-level conservation and species-level conservation is more complex than this. Some of this complexity is captured in two related questions that have long intrigued ecologists. First, are species-rich ecosystems more stable than species-poor ecosystems? Second, why do some ecosystems have more species than other ecosystems?

Diversity and Stability

Conservation biologists have long been concerned that species extinctions could have dire consequences for the stability of entire ecosystems. This idea is captured in a well-known metaphor suggested by Anne and Paul Ehrlich (1981). Imagine you were flying in a plane, looked out the window, and saw a rivet fall out of the wing. You might not worry too much because there are thousands of rivets in a plane, and the loss of one rivet would not make it fall apart and crash. In fact, several rivets could probably fall out before the situation became dangerous, but, eventually, if enough rivets fell out, the plane would crash. By analogy, an ecosystem could survive the loss of some species, but if enough species were lost, the ecosystem would be severely degraded. Of course, all the parts of a plane are not of equal importance, and, as explained in Chapter 3's discussion of keystone, controller, and dominant species, not all species are of equal importance in an ecosystem (Ehrlich and Mooney 1983). Thus it is possible that even the loss of a single important species could start a cascade of extinctions that might dramatically change an entire ecosystem (Diamond 1989, Myers 1989a). A good illustration of this occurred after fur hunters eliminated sea otters from some Pacific kelp bed ecosystems: the kelp beds were practically obliterated too, because, in the absence of sea otter predation, sea urchin populations exploded and consumed most of the kelp and other macroalgae (Estes et al. 1989). The likelihood of such calamities is related to the "supraorganism" versus loose assemblage debate we discussed earlier (see Fig. 4.4); obviously, significant degradation is more likely if ecosystems are analogous to "supraorganisms."

Three mechanisms for higher diversity increasing ecosystem stability have been proposed by Chapin et al. (1997). First, if there are more species in an ecosystem, then its food web will be more complex, with greater redundancy among species in terms of their trophic roles. In other words, in a rich system if a species is lost, there is a good chance that other species will take over its function as prey, predator, producer, decomposer, or whatever. Second, diverse ecosystems may be less likely to be invaded by new species, notably exotics, that would disrupt the ecosystem's structure and function. Third, in a species-rich ecosystem, diseases may spread more slowly because most species will be relatively less abundant, thus increasing the average distance between individuals of the same species and hampering disease transmission among individuals.

Scientific evidence to illuminate these ideas has been slow in coming and many shadows remain (Johnson et al. 1996, Chapin et al. 1997, Tilman 1999). One of the first studies to provide data supporting a relationship between diversity and stability examined how grassland plants in Minnesota responded to a drought. Tilman and Downing (1994; also see Tilman 1996) used the ratio of aboveground biomass in 1988 (after 2 years of drought) to that in 1986 (predrought) in 207 plots as an index of ecosystem response to disruption by the drought. They compared these values with the number of plant species in each plot and discovered that the plots with a greater number of plant species experienced a less dramatic reduction in biomass (Fig. 4.7). Apparently, species-rich plots were likely to contain some drought-resistant plant species that grew better in drought years, compensating for the poor growth of less-tolerant species.

To put this result in more general terms, a species-rich ecosystem may be more stable because it is likely to have species with a wide array of responses to variable conditions such as droughts (McCann 2000). Furthermore, a species-rich ecosystem is more likely to have species with similar ecological functions, so that if a species is lost from an ecosystem, another species, probably a competitor, is likely to flourish and occupy its functional role (Tilman 2000). Both of these, variability in responses and functional re-

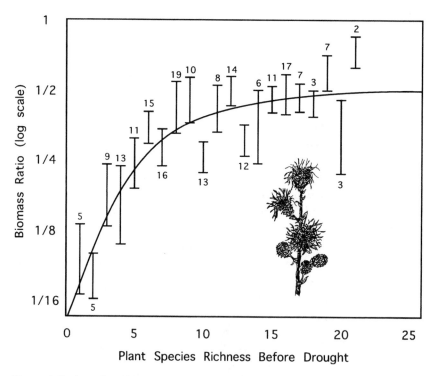

Figure 4.7. A study of Minnesota grassland plants found a significant relationship between the number of plant species in 0.3-m² plots and the ratio of aboveground biomass in those plots between 1988 (after 2 years of drought) and 1986 (predrought) (Tilman and Downing 1994). Plots with more than 10 species had about half as much biomass in 1988 as in 1986, whereas those with fewer than five species only produced roughly 1/8 as much biomass after a 2-year drought.

dundancy, could be thought of as insurance against perturbations (McCann 2000).

The Minnesota grassland research has been widely accepted as strong evidence for the diversity-stability theory; however, its findings have been questioned (Givnish 1994, Huston 1997, Kaiser 2000), and analogous studies on other ecosystems have not always found a positive relationship between diversity and stability (Johnson et al. 1996, Sankaran and Mc-Naughton 1999). Clearly, this is a complex issue that requires further field research with a broad spectrum of ecosystems and species; grassland plants and computer models (e.g., Kaufman et al. 1998, Naeem 1998) will only take us so far. In the end, despite insightful attempts to distill some general patterns (Tilman 1999), we may find it very difficult to reduce this topic to a simple, universal "truth."

The Species Richness of Ecosystems

Lying just below the diversity-stability question is a more fundamental issue: Why are some ecosystems more diverse than others? Even the most casual observer of nature realizes that a tropical coral reef is extraordinarily more diverse than an alpine pond, but why? What factors shape the rates at which species accumulate in an ecosystem (through colonization or speciation) versus disappear from an ecosystem (through local or global extinction)? Once again, there is no simple, universally accepted answer, but here is a brief overview of some of the ideas that have been proposed, distilled largely from Jablonski (1993), Ricklefs and Schluter (1993), Ricklefs (1995), Rosenzweig (1995), Cowling et al. (1996), and Gaston (2000).

Life flourishes in warm, moist places; think about tropical forests, or consider what would happen to a bowl of egg salad left on a picnic table for a couple of summer, versus winter, days. This simple observation has been supported in the scientific literature by many positive correlations between species richness and temperature, precipitation, energy flux, and complex metrics such as potential evapotranspiration. There are some exceptions to this general pattern (e.g., species richness is often greatest at intermediate levels of gross primary productivity), and, obviously, the availability of water is usually not an issue in aquatic ecosystems. Nevertheless, the overall pattern is clear, and it makes sense: more species should be able to evolve and persist in places with adequate water and energy where they can channel their resources into growth and reproduction rather than a struggle for survival. One can also make the converse argument: that relatively few species can survive in stressful ecosystems that are extremely cold, dark, or dry. The idea that stress may explain the low diversity of some ecosystems can be extended to additional circumstances such as very high or low pH levels, high rates of disturbance (e.g., the slope of an active volcano), extreme seasonality, and so on.

Interestingly, moderate levels of disturbance may actually promote species diversity. For example, a forest that is subject to occasional windstorms or ground fires may harbor more species than a forest that is rarely affected by disturbance. There are two possible explanations here. First, occasional disturbances are likely to prevent a few species from dominating the ecosystem. In this context, predation may also be considered a form of disturbance that limits dominance by a few species. Recall from Chapter 3 how predation by the purple sea star leads to richer intertidal communities (see Fig. 3.9). In other words, the absence of predators or disturbance may allow a few species to prosper with minimal competition, and, while such ecosystems may be highly productive (e.g., a salt marsh), they will be species-poor overall. Second, disturbances are usually patchy, and this will generate spatial heterogeneity that allows many species to coexist. For example, a forest that is a mosaic of different age patches created by small windthrow events would have all the species associated with different stages of ecological succession, while an undisturbed forest would have just those species associated with a late-successional stage. Of course, many ecosystems have heterogeneous environments with or without the patchiness of disturbances, and this is also an important source of niches for additional species. For example, an ecosystem that has an array of substrates ranging from clay to boulders will support more species than one that is covered by only clay. Similarly, the vertical dimension of forests and aquatic ecosystems is a form of spatial heterogeneity that adds opportunities for many species.

Perhaps the simplest explanation for variance in species richness among ecosystems is size. Not surprisingly, more species can fit into a large ecosystem than a small one. There are many reasons for this; we will discuss them in Chapter 8 in the section on fragmentation. That discussion will also cover isolation, another factor that limits species richness by curtailing colonization, especially on islands.

Finally, we need to recognize that species richness probably operates in a positive-feedback loop, a "snowballing effect" in more colloquial language, to further increase the diversity of species-rich ecosystems. Compare two ecosystems, one with 50 species of plants and the other with 200. The latter is likely to support a much wider spectrum of herbivores, pollinators, parasites, pathogens, and so on (Wright and Samways 1998). From this perspective one could argue that the primary driver of species richness is the physical environment, especially, how big, warm, and wet it is and how much it varies in space and time because of disturbances and other factors. Secondarily, the dominant species in the system (plants in terrestrial systems and a mixture of plants, algae, corals, and more in aquatic systems) shape diversity by enhancing spatial heterogeneity and providing the basis of a food web. Every species plays some role, if only as food for its suite of predators, parasites, and pathogens.

If you find the interweaving of ideas presented in this section rather complex, do not despair. Read it again, making a list of the key ideas, and recognize that no one really understands how they all interact to shape the patterns of species richness we see in different ecosystems. Moreover, new ideas are constantly being offered (e.g. Ritchie and Olff 1999).

An Important Postscript

This focus on the relative species richness of different ecosystems returns us to our earlier discussion about mismeasuring biodiversity by overemphasizing species richness (Chapter 2). Becoming fixated on species richness can lead conservation managers astray. For example, although maintaining the stability of ecosystems is an important argument for avoiding the loss of species, the converse of this argument does not hold: we should not seek to increase the stability of ecosystems by artificially augmenting the number of species, e.g., by planting additional tree species in a forest. Similarly, although sustaining species-rich ecosystems like tropical forests may be a somewhat higher priority than sustaining species-poor ecosystems, overemphasizing species-richness to the exclusion of species-poor ecosystems would be very short-sighted. Recall the discussion about salt marshes, home to a narrow range of species but a very important type of ecosystem because of their productivity (see Fig. 4.5). Finally, because each type of ecosystem harbors a unique suite of species, the coarse-filter approach requires protecting a complete array of ecosystems, even those that may have relatively few species (Fig. 4.8). In particular, many islands support a precious biota of endemic species, but are not very diverse overall; the Galapagos islands may be the best example of this.

Ecosystems and Landscapes

The mosaic of ecosystems we see from a plane is not just a random array. There are patterns to the spatial configurations of ecosystems. Lakes are drained by rivers and bordered by marshes, woodlots are patches embedded in a matrix of agricultural ecosystems, clearcuts are patches in a matrix of forests, and so on. Ecologists call these mosaics of ecosystems *landscapes*, and a subdiscipline called landscape ecology has developed to study ecological phenomena that exist at this scale (Forman 1995). For example, landscape ecologists are interested in ecosystems that occur as long, narrow strips such as rivers and their associated riparian (shore) ecosystems be-

Figure 4.8. The extreme cold and dryness of a high-latitude or high-altitude ecosystem are just two reasons why they support far fewer species than the coral reef depicted in Figure 4.6. Such ecosystems still merit conservation because of their unique biota and other attributes.

cause these ecosystems may serve as corridors that facilitate organisms moving among ecosystems. Also of interest to landscape ecologists are the edges between ecosystems. The interface between a forest and a field is one example: it will be avoided by some species and preferred by other species (Hunter 1990).

Conservation biologists are interested in landscape phenomena for a number of reasons that we will examine further in subsequent chapters. Two brief examples will suffice here. First, many endangered species are large animals that have large home ranges—tigers, wolves, elephants, etcetera—that encompass many ecosystems. If we wish to maintain habitat for these species, we must maintain entire landscapes that provide for all their needs. Second, human activities have left many natural ecosystems isolated in a matrix of human-altered ecosystems, and conservation biologists are concerned with what happens along the edges of these small, residual patches. Are they being degraded by factors that originate externally such as exotic species, pesticides, and changes in local climate?

These and similar issues have led conservation biologists to advocate maintaining biodiversity at the landscape scale (Noss 1983). This is a way of saying that it is not sufficient to protect a representative array of ecosystems. We must also ensure that these arrays occur in spatial configurations that maintain the natural relationships among ecosystems. In short, we must maintain natural, functioning landscapes.

Mangrove Swamps

Despite popular impressions, tropical shores are not all white-sand beaches lined by coconut palms. In many places the transition from the terrestrial to marine realms is marked by dense stands of trees and shrubs that form a type of ecosystem known as mangrove swamps or mangal. The seaward edge of mangal is usually quite sharply delineated, but moving inland mangal often grades into other types of swamps as the elevation rises and the water becomes

less saline. This gradation is one reason that the term "mangrove" is rather ambiguous. "Mangrove" is a quasi-taxonomic term that is routinely used for at least fifty species of woody plants from twenty families that inhabit tropical intertidal environments (Tomlinson 1986). Depending on the breadth of your definition, many more species could be added. Of course, on a global scale fifty species of plants is not very many—you could find that number of tree species in a fraction of a hectare of tropical rain forest, and in any given mangrove swamp only one or a few species of mangrove may occur. The biotic diversity of mangrove swamps is quite low because relatively few vascular plant species have evolved mechanisms such as salt secretion for living in saline environments.

Despite modest levels of species diversity, mangrove swamps are very important and interesting ecosystems (Christensen 1983, Osborn and Polsenberg 1996). First, they are extremely productive, capturing sunlight and collecting nutrients imported by the tides, and exporting huge amounts of organic matter to the adjacent aquatic ecosystems where they support aquatic food webs and economically valuable fisheries. For certain commercial fish species, mangrove swamps provide cover, as well as food, especially for young individuals. Consequently, it is common to refer to them as nurseries. They also provide a sort of cover for shoreline human communities by creating a buffer against the storm tides of hurricanes and typhoons. Conversely, they buffer coral-reef and sea-grass ecosystems from siltation stemming from inland erosion. Mangrove swamps also provide resources—timber and fuelwood—that can, unfortunately, lead to their over-exploitation (Walsh 1977). Limited wood harvest might be sustained, but it is often overdone, especially considering the risk to fisheries production. Worse than the threat of excessive timber harvesting is the wholesale destruction of mangal to make room for aquaculture, agriculture, and coastal development (ranging from garbage dumps to high-rise hotels). Because they occupy a narrow band between the land and the sea, mangrove swamps have never occupied a large total area, and this makes it doubly tragic that so many have been lost. In the Indo-Pacific region alone 1.2 million hectare (ha) of mangal had been converted to aquaculture ponds by 1977 (Saenger et al. 1983), and the economic incentives for this continue to be enormous (Janssen and Padilla 1999). Fortunately, the great ecological value of mangrove swamps is being recognized in some quarters. For example, in a court case involving restoration of an 8.1-ha mangrove swamp in Puerto Rico damaged by an oil spill, an oil tanker was initially fined over $6 million ($751,368 per hectare) (Lewis 1983). Mangrove restoration may be feasible (Field 1999), but it certainly would be preferable to avoid damaging them in the first place.

Summary

The conceptual definition of an ecosystem is straightforward—a group of interacting organisms and their physical environment—but deciding where one ecosystem ends and another begins can be difficult. Evaluating the differences and similarities among many ecosystems and classifying them into different ecosystem types is even more challenging. Despite these difficulties, recognizing and classifying ecosystems are useful exercises for organizing our understanding of the patterns of ecological interactions.

The value of an ecosystem, at a minimum, consists of the sum of all the values of the species that occupy the ecosystem. Beyond this, the instrumental values of ecosystems are primarily based on services; for example, exporting clean water and other economically valuable functions, providing complex models for research and education, and serving as sites for spiritual renewal. From a conservation biology standpoint, ecosystems have a critical strategic role because protecting a representative array of ecosystems will protect biodiversity at the species and genetic level to a significant extent. The idea of ecosystems having intrinsic value revolves around an unresolved controversy: To what extent are ecosystems loosely organized collections of species versus highly integrated systems of coevolved species? The closer they are to being highly integrated, the more likely it is that loss of species could lead to ecosystem degradation. Maintaining ecosystem diversity also requires maintaining the spatial arrangements in which ecosystems occur; in other words, natural landscapes require protection.

FURTHER READING

A 29-volume series, *Ecosystems of the World,* published by Elsevier of Amsterdam, is the single most comprehensive treatment available. For material on many of the world's most threatened ecosystems, read Groombridge (1992). For further reading on the issue of how tightly organized ecosystems are, see Lovelock (1979), Botkin (1990), Pimm (1991), and Schulze and Mooney (1993). The species-richness patterns of ecosystems are covered in Ricklefs and Schluter (1993), Huston (1994), and Rosenzweig (1995). See Forman (1995) for a landscape-scale perspective on ecosystems and Bailey (1996) to read about ecological regions. For web-based information about ecological communities in Canada and the United States, including a classification system, see www.natureserve.org

TOPICS FOR DISCUSSION

1. In the area where you live, which types of ecosystems are easiest to define? Which are hardest? Why?
2. Draw a map of the ecological region you inhabit. How did you distinguish it from surrounding regions?
3. What is the rarest type of ecosystem in your region? Have many examples of it been protected?
4. What services are provided by the major types of ecosystems in your region?
5. What evidence can you cite that supports the idea that ecosystems are just loose collections of species? What evidence refutes the idea? If you do not specifically know of such evidence, how would you design a research program to obtain it?

Chapter 5
Genetic Diversity

The process by which sequences of four simple chemicals—adenine, thymine, cytosine, and guanine—shape a molecule of DNA and, ultimately, all the organisms that comprise the earth's biota is an extraordinary story. It is a story about the foundations of biological diversity. It can be a rather complex story, and if your recollection of Hardy-Weinberg equilibria, phenotypes versus genotypes, alleles, diploidy, and so on has rusted a bit, you will find it helpful to review the genetics and evolution sections of a biology textbook before proceeding.

What Is Genetic Diversity?

One of the best places to appreciate genetic diversity is at a county fair. Peppers, squashes, chickens, horses, cattle, and most other domestic species come in an extraordinary array of colors, shapes, and sizes. Some of this phenotypic diversity was shaped by environmental conditions such as the soil in which the peppers were grown, but most of it is based on genotypic differences. In other words, you are seeing the expressions of genetic diversity based on differences in the types and distributions of the thousands of genes that occur within every individual.

It is useful to think of genetic diversity as occurring at four levels of organization: among species, among populations, within populations, and within individuals. Most conspicuous are the genetic differences that distinguish one species from another, horses from cows or peppers from squashes. We do not always think of the differences between cows and horses as manifestations of genetic diversity because we can usually distinguish species readily without knowing anything about their genes. Species that are an exception to this generalization are called cryptic or sibling species (see Chapter 3).

The genetic diversity among the populations of a single species can also be quite substantial. Someone who had never encountered the diversity of dogs would hardly believe that a St. Bernard and a Chihuahua represent genetic diversity within the same species. Also, most people who have eaten cabbage, cauliflower, broccoli, kale, kohlrabi, and brussels sprouts their whole lives do not realize that they are genetic variations of the same species, *Brassica oleracea*. Of course, these differences have been generated by artificial selection. Among wild populations, genetic diversity is usually not manifested in conspicuous characteristics unless perhaps the populations are widely separated geographically. Nevertheless, genetic diversity among populations (e.g., differences in tolerance to thermal stress) can be profoundly important.

Within populations of most wild species, different individuals look quite similar, but they are probably genetically distinct from all other individuals. Exceptions would include individuals that have identical siblings, because a single zygote split into two or more during its development, and individuals produced by asexual reproduction. We will discuss the importance of genetic differences within populations in some detail in this chapter.

Finally, genetic diversity exists within a single individual wherever there are two alleles for the same gene or, to state it more explicitly, different configurations of DNA occupying the same locus on a chromosome. Differences in the distributions of alleles are the foundation of measuring genetic diversity. (For details on how genetic diversity arises through processes such as mutation and natural selection, see Hartl and Clark 1997 and Hartl 2000.)

Measuring Genetic Diversity

There are six basic methods to determine which alleles are present at a given locus. An indirect technique, called protein electrophoresis, involves determining the rate at which enzymes move through a gel when subjected to an electrical field. Different alleles produce different variations of enzymes that move at different rates; enzymes that differ because of allelic differences are called *allozymes*. The most direct method, called DNA sequencing, involves directly determining the sequence of adenine, thymine, cytosine, and guanine for a given allele. Four intermediate methods break DNA into fragments and then separate and characterize these fragments using electrophoresis. Different alleles produce different fragment lengths. These methods are described briefly in Box 5.1 and in Hedrick and

Molecular analysis of genetic diversity
Judy Blake[1]

Electrophoresis is the basic tool for all molecular genetic techniques. It begins by placing purified proteins or DNA molecules obtained from different individuals in separate wells containing a gel (e.g., agarose) and running a strong electric current through the gel. Because the molecules differ in size and/or charge, they will migrate from the origin through the gel at different speeds in response to the current and will end up in different positions. The final positions of the tested molecules can be determined using dyes or radioactive probes.

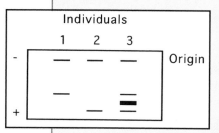

Protein Electrophoresis. Tissues are collected from many individuals, macerated, and separated by electrophoresis. They are then stained for specific enzymatic or protein activity. Comparisons of electrophoresis patterns are made among many individuals, and gels are scored for presence or absence of particular bands. Variation may represent functionally similar forms of enzymes (isozymes) or variants of enzymes that are different allelic forms of the same gene locus (allozymes). Detection of alleles permits study of variation among populations or similar taxonomic forms.

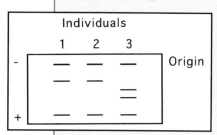

Restriction Fragment Length Polymorphism (RFLP). High-molecular-weight DNA is cut with a site-specific restriction enzyme (e.g., *HpaI*, which cuts double-stranded DNA at 5′... GTT ↑ AAC ... 3′ sites), electrophoresed, and stained for nucleic acids. Lengths of fragments revealed through electrophoresis distinguish sequences that have gained or lost a site specific to the restriction enzyme used. These are repeated for multiple restriction enzymes. Analysis of closely related taxa assumes that a particular size fragment reflects the same cleavage sites.

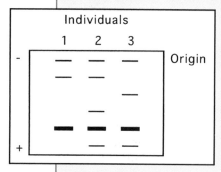

Random Amplification of Polymorphic DNA (RAPDs). Using purified DNA, random fragments are isolated using the polymerase chain reaction (PCR). Following electrophoresis, nucleic acids are stained and diagnostic bands are determined. Diagnostic bands between parental or outside groups form the basis for comparisons within and between populations. (A major advantage of PCR technology for conservation

[1] Jackson Laboratory, Bar Harbor, Maine

BOX 5.1   *Continued*

biology studies is that it allows analysis of very small samples such as a tiny piece of skin tissue.)

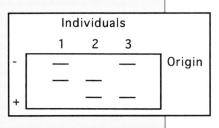

Microsatellites or Simple Sequence Repeat (SSR) Polymorphisms. In this technique the DNA banding patterns of individuals at a single locus are visualized by electrophoresis after using PCR. Different banding patterns result from the number of simple nucleotide repeats that an individual carries. For example, the sequence ATATATAT would produce a band at a different position on the electrophoresis gel than that produced by the sequence ATATATATATAT.

Amplified Fragment Length Polymorphisms (AFLP). This is a highly variable method of DNA fingerprinting of the nuclear genome that uses selective PCR amplification of restriction fragments. As with RAPDs, if a mutation occurs, the mutant form will not show up on the electrophoresis gel; these types of markers are called dominant markers. Usually, about 50 to 100 restriction fragments are amplified per individual.

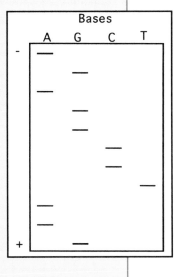

DNA Sequencing. In preparation for DNA sequencing, specific DNA (gene) fragments are isolated using PCR or cloning techniques. Sequences differing in length by one base pair are separated by size in an electrophoretic sequencing gel. In one technique, sequences are made radioactive and gels are exposed to x-ray film to expose the radioactive signal; another sequencing technique relies on fluorescent dyes. The DNA sequence in the example shown here is GAATC-CGGAGA, reading from the bottom. Sequences are aligned, and each site along the length is compared among different individuals. Differences in molecular sequences provide the information for computing measures of genetic relatedness and subsequent phylogenetic analysis.

Besides sorting out systematic relationships (species, subspecies, and evolutionary significant units) and assessing genetic diversity within and among populations, another practical application of this DNA technology is conservation forensics, which involves the identification of illegally collected species such as whales, salmon, deer, and turtles (e.g., Cipriano and Palumbi 1999).

Miller (1992), Amos and Hoelzel (1992), and Hartl (2000), and, more completely, in Hoelzel (1998).

Because these methods are quite laborious, it is not generally feasible to determine allelic distributions for the many thousands of genes found in most organisms; the human genome project is one notable exception. Usually, a sample of genes must be selected. Similarly, it is usually not possible to test all the individuals in a population; thus a sample of individuals is used. After the allelic distribution for a sample of genes from a sample of individuals has been determined, then an index to describe these distributions quantitatively can be calculated. Conservation biologists often use two indices—polymorphism and heterogeneity—to quantify genetic diversity and understand its role in population survival.

Polymorphism: Polymorphism (usually abbreviated P) is defined as the proportion or percentage of genes that are polymorphic. A gene is considered polymorphic if the frequency of the most common allele is less than some arbitrary threshold. This threshold is usually 95% (Hartl and Clark 1997), although with the advent of techniques that have much higher powers of resolving different alleles (Box 5.1), a threshold of 99% may be more appropriate (D. Hartl personal communication). This definition is easier to explain with numbers than with words; we will use data collected from five American bison sampled from the descendants of bison that were moved to Badlands National Park as part of a reintroduction program (McClenaghan et al. 1990). The allelic distributions for 24 different genes were determined using electrophoresis of blood samples, and only one gene was polymorphic. That is, for the other 23 genes sampled, a single allele accounted for at least 95% of the samples. The polymorphic gene was called malate dehydrogenase-1 (abbreviated MDH-1) for the enzyme it encoded. MDH-1 had two different alleles that we will call X and Y. Among the five bison, two individuals (A and B) were heterozygous (X/Y), two individuals (C and D) were homozygous for the Y allele (Y/Y), and one (E) was homozygous for the X allele (X/X) (Table 5.1). In this case Y was the most common allele; its frequency was 0.6 or 60% (i.e., 6 of the 10 alleles were Y), and the frequency of the X allele was 0.4. Because the frequency of the most common allele, Y, was less than 95%, the MDH-1 gene was considered polymorphic. Because out of the 24 genes sampled, only MDH-1 was polymorphic, the estimated polymorphism was 1 divided by 24 or 0.042 or 4.2%.

Although it is common for a single allele to comprise close to 100% of any gene that is not polymorphic, very few genes consist of absolutely 100% of a single allele. Thus, if you search a large enough sample of individuals, you are likely to find *rare alleles*. Rare alleles are defined as having a frequency of less than 0.005, 0.01, or 0.05, depending on the techniques em-

TABLE 5.1 Distribution of two alleles, MDH-1^X and MDH-1^Y, among five bison.

Bison	X allele	Y allele	Genotype
A	1	1	X/Y
B	1	1	X/Y
C	0	2	Y/Y
D	0	2	Y/Y
E	2	0	X/X
Total	4	6	
Gene frequency	0.4	0.6	

Sampled by McClenaghan et al. (1990).

ployed and how the information is being used (D. Hartl personal communication and Hartl and Clark 1997).

Finally, we need to emphasize that polymorphism is based on the distribution of alleles, not genotypes. This means that if you had a population without any heterozygotes, a gene could still be polymorphic; for example, a population of four homozygous Y/Y bison plus one homozygous X/X bison would be polymorphic at this locus, and P would still be 4.2%.

Heterozygosity: A second index, called *heterozygosity* (usually abbreviated H), is defined as the proportion or percentage of genes at which the average individual is heterozygous (Hartl and Clark 1997). In the bison example, two out of five individuals were heterozygous at the MDH-1 locus, so heterozygosity $2/5 = 0.4$ for this gene. We can calculate H by averaging the heterozygosity of each gene across all 24 genes. In this case,

$$\frac{0.4 \text{ (for MDH-1)} + 0_1 + 0_2 + \cdots + 0_{23} \text{ (for the other 23 genes)}}{24} = \frac{0.4}{24} = 0.017$$

Two uses of heterozygosity measurements merit description. First, geneticists often compare the heterozygosity that they measure—the observed H, or H_o—with the heterozygosity they would expect to find, H_e, given the relative frequency of alleles. The expected heterozygosity is calculated by using the middle component ($2pq$) of the Hardy-Weinberg equation, $p^2 + 2pq + q^2 = 1$. In this example, given a frequency of $p = 0.6$ for the Y allele and $q = 0.4$ for the X allele (see Table 5.1), the Hardy-Weinberg equation is

$$0.36 \ (Y/Y) + 0.48 \ (X/Y) + 0.16 \ (X/X) = 1$$

Figure 5.1. The relative distribution of genetic variation between and among populations of desert fishes differs substantially. Meffe and Vrijenhoek (1988) describe two models: the Death Valley Model (A) in which populations reside in isolated desert springs; and the Stream Hierarchy Model (B) in which populations are connected by a stream system and can exchange genes at a rate that will be affected by their proximity and by the permeability of the intervening habitat. D_{st} (variability among the populations) will probably be significantly higher in populations that fit the Death Valley Model.

Consequently, H_e for MDH-1 is 0.48 ($2 \times 0.6 \times 0.4 = 0.48$). The H_e based on all genes for these five bison is $0.48/24 = 0.02$, which is probably not significantly different from the H_o of 0.017.

Second, geneticists often use the heterozygosity index to estimate how much of a species' total genetic diversity (H_t) is due to genetic diversity within the populations that compose the species (H_s) versus how much is due to variability among the populations (D_{st}) (Nei and Kumar 2000). Mathematically, this can be expressed as $H_t = H_s + D_{st}$. (This concept is often expressed with different formulas, but the basic idea is the same: partitioning variability within and between populations.) If a species has a relatively high D_{st}, then it is necessary to maintain many different populations to maintain the species' genetic diversity. Alternatively, if most of the species' genetic diversity exists within each population (i.e., H_s is relatively high), then it is less critical to maintain many different populations (Fig. 5.1). This is often a key issue for people who manage populations of endangered species, and we will return to it in Chapter 12, Managing Populations.

Finally, how do estimates of polymorphism and heterozygosity compare? Figure 5.2 shows that P and H values vary widely among different taxonomic groups, but are closely correlated with one another. Despite this correlation, it is often useful to estimate both P and H (Hartl and Pucek 1994).

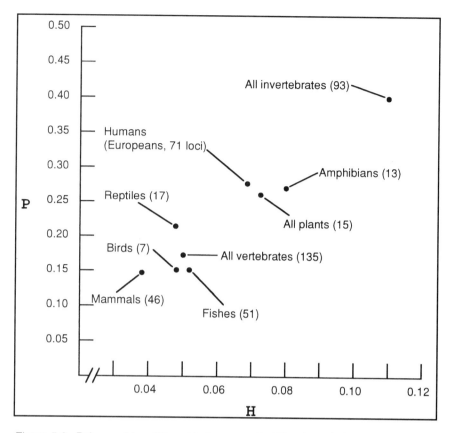

Figure 5.2. Polymorphism (P) and heterozygosity (H) values derived from allozyme studies of a wide array of organisms show a positive correlation. (Redrawn by permission from Hartl and Clark 1997.)

The Importance of Genetic Diversity

To assess the importance of genetic diversity, it is useful to think of genes as units of information rather than tangible things. As tiny amounts of carbon, hydrogen, oxygen, nitrogen, and some other common elements, genes have little value in and of themselves. As sources of information, however, genes are clearly essential; they shape the synthesis of the biochemicals that control cellular activity and, ultimately, all biological activity. The quantity of information encoded by genes is enormous; a typical mammal might have 100,000 genetic loci. E. O. Wilson (1988) estimated that the amount of genetic information encoded in a single mouse, if translated into letters, would fill the first 15 editions of the *Encyclopaedia Britannica.*

Of course, most of this wealth of genetic diversity is encapsulated in the diversity of species and their interspecific genetic differences. The key issue to address here is the distribution of alleles. Why is it important to maintain different versions of the same gene and, in many circumstances, to have them well distributed in a population dominated by heterozygotes rather than homozygotes? There are three basic answers: evolutionary potential, loss of fitness, and utilitarian values.

Evolutionary Potential

A key requisite for natural selection is genetic-based variability in the fitness of individuals; that is, some individuals must be more likely to survive and reproduce than others. If every individual were genetically identical and only chance determined which ones left progeny, then populations would change through time very slowly, if at all. If they are to persist, populations must change because the world is changing. The physical world changes as continents drift over the globe, mountains rise and erode, oceanic currents and jet streams shift paths, and the earth's orbit around the sun varies. The biological world changes as species evolve, become extinct, and shift their geographic ranges, coming into contact with new species that may be predators, prey pathogens, or competitors. Changes have been particularly dramatic during the last few decades as human populations and their technological capabilities have grown.

Species with greater genetic diversity are more likely to be able to evolve in response to a changing environment than those with less diversity. To put it another way, the rate of evolution is directly proportional to the amount of variability in a population. The classic example of this involves many species of moth, notably *Biston betularia*, that occur in two different forms: a light form that is hard to detect against a lichen-covered tree trunk and a dark form that is not cryptic among lichens (Fig. 5.3) (Kettlewell 1973). The light moths were much more common than the dark forms until the 19th century when air pollution killed lichens and covered trees with soot. On the darkened trees the light moths were more conspicuous to predators, and the dark form became dominant. More recently, air pollution has been curbed in some forests, and the light moths are increasing again. Without the genetic diversity expressed in two color forms, the species might not have survived these changes. A similar story could be told for many species of plants and fungi that have evolved a tolerance for the high concentrations of toxic metals often found at mine sites (Antonovics et al. 1971).

Environments change through space, as well as time, and a species with greater genetic diversity is more likely to colonize a wider range of environments than a species with limited genetic diversity. For example, a survey of the heterozygosity and polymorphism of 189 species of amphibians indicated that genetic diversity was greatest in amphibians that lived in the

Figure 5.3. Genetic diversity allows species to adapt to changing environments, as when dark forms of certain moths helped the species to survive after air pollution darkened the trees they inhabited.

most heterogeneous environments (e.g., forests) and least in homogeneous environments (e.g., aquatic ecosystems and underground) (Nevo and Beiles 1991). A similar pattern has been shown for plants (Gray 1996). This is one of those "Which came first, the chicken or the egg?" issues, because it is also likely that a species inhabiting a wide variety of environments would evolve and maintain a relatively high degree of genetic diversity.

Loss of Fitness

Populations that lack genetic diversity may also experience problems (e.g., low fertility and high mortality among offspring, etc.) even in environments that are not changing. A loss of fitness in genetically uniform populations is often called inbreeding depression because it usually develops from breeding between closely related individuals. It is a well-known phenomenon in zoos, where populations of captive animals are often small and

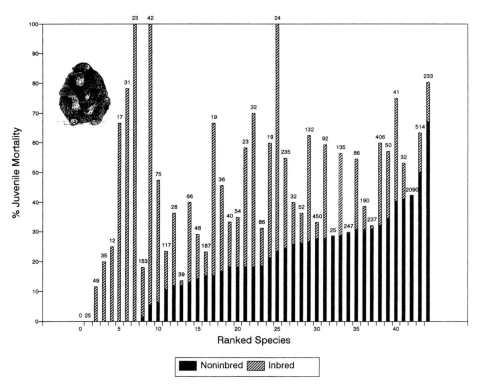

Figure 5.4. Juvenile mortality in 44 species of mammals (16 ungulates, 16 primates, 10 rodents, 1 marsupial, and an elephant shrew) bred in captivity. Open bars represent mortality rates with inbred parents; black bars represent mortality rates from matings between unrelated parents. Species are arranged from left to right by increasing mortality from unrelated parents. Numbers on the tops of the bars are the sample size. (Data from Ralls and Ballou 1983.)

individuals are often closely related (Ralls and Ballou 1983, Ralls et al. 1988) (Fig. 5.4). It is also a problem for plant and animal breeders who breed individuals that are genetically similar to one another to promote desirable characteristics that they share such as a preferred color or resistance to a certain disease.

There are three general explanations for relatively low fitness in genetically uniform populations (Packer 1979). First, there is more homozygosity in genetically uniform populations, and this may lead to the expression of recessive deleterious alleles that are suppressed in heterozygous individuals. Hip dysplasia in purebred dogs is a widely known current example; at one time, inbreeding within the royal families of Europe resulted in many family members having a split upper lip. Some alleles, called *lethal recessives,* are even fatal when they come together in a homozygous recessive individual. Second, heterozygous individuals may be more fit in terms of phenotypic characteristics than homozygous individuals, a phenomenon known as *heterosis*. For example, evidence suggests that heterozygous animals tend to be more resistant to disease, grow faster, and survive longer than homozygotes (Allendorf and Leary 1986); this effect seems to be present, but not as strong, among plants (Ledig 1986). The third reason is closely tied to the "evolutionary potential" issue discussed in the preceding section. In a population dominated by heterozygotes there will be more genetic variability among offspring (some heterozygotes, some homozygous dominants, and some homozygous recessives), and in an unpredictable environment perhaps at least some of the young will survive. In other words, it may be preferable not to put all your eggs in one basket or all your zygotes into one genotype (Williams 1966).

Documentation of inbreeding in wild animal populations is scarce, probably in large part because many animal species employ behavioral mechanisms, such as juveniles dispersing away from the place where they are born, to avoid breeding with close relatives (Ralls et al. 1986). Nevertheless, evidence for low fitness in some genetically uniform populations has slowly been accumulating for a small, but diverse suite of species: e.g., lions (Packer et al. 1991), song sparrows (Keller et al. 1994), adders (Madsen et al. 1996), and black-footed rock-wallabies (Eldridge et al. 1999). More importantly, a group of Finnish biologists has explicitly shown a link between low heterozygosity and an increased probability of population extinction in a study of 42 populations of a butterfly, the Glanville fritillary (Saccheri et al. 1998). During the study 35 populations persisted and seven went extinct and the degree of heterozygosity was a strong predictor of which populations survived even after ecological and demographic factors were taken into account. Populations with low heterozygosity had lower egg-hatching rates, lower larval survival, and shorter female life spans. Despite the slowly growing evidence, it is still quite possible that conventional wisdom about the negative consequences of genetic uniformity does not apply to all species (Thornhill 1993). In particular, many plant species seem to have evolved a greater tolerance for inbreeding depression than animals, presumably because they are less mobile, and in many species self-fertilization is common (Barrett and Kohn 1991). With self-fertilization, deleterious recessive alleles will often appear together in homozygous recessive individuals, and natural selection should soon remove them from a population, although there are exceptions (Ellstrand and Elam 1993). Some have argued that island populations may also be relatively tolerant of inbreeding (Craig 1994), but this point is disputed (Frankham 1998).

The limited evidence for loss of fitness in genetically uniform wild populations does not mean that the problem is of little interest to conservation biologists. First, few studies capable of revealing this phenomenon in wild populations have been undertaken, especially for very small populations. In other words, the problem may be far more widespread than we realize (Charlesworth and Charlesworth 1987). Second, the demonstrated importance of inbreeding depression in captive populations is of great concern to conservation biologists who work with captive populations of wild species that are endangered or extinct in the wild, and those who manage rare breeds of domestic species. We will return to loss of fitness in genetically uniform populations later in this chapter and in Chapter 13, Zoos and Gardens.

A loss of fitness can also occur when mating occurs between individuals that are too genetically dissimilar; this is called *outbreeding depression* (Templeton 1986). For example, when the ibex population of the Tartra Mountains of Slovakia was extirpated, conservationists replaced it with ibex from nearby Austria (*Capra ibex ibex*), and later added ibex from Turkey (*Capra ibex aegagrus*) and the Sinai (*Capra ibex nubiana*) (Turcek 1951, Greig 1979). The offspring of these subspecific crosses mated in the fall rather than the winter as the Austrian ibex had, and their young were born during the winter, rather than spring, and died (Fig. 5.5). The reintroduction failed. In this case, genetic diversity in the form of local adaptations to the seasonality of local environments was lost. Turcek (1951) also reported a more extreme case of outbreeding depression after the Siberian subspecies of the roe deer was introduced to Slovakia. When females of the European subspecies mated with the much larger Siberian males, they died during parturition because they were unable to deliver the large fawns. Although

Figure 5.5. Outbreeding among ibex translocated from Austria, Turkey, and the Sinai led to a population that produced offspring in the winter. The young perished, and the population disappeared.

outbreeding depression usually refers to intraspecific mating, botanists also use the term to refer to a loss of fitness that occurs when individuals of two closely related species interbreed, what zoologists would call hybridization. (Recall that botanists often do not use the reproductive-isolation definition of species described in Chapter 3.) Interspecific outbreeding depression or hybridization is a problem among some rare plants that may be exposed to large amounts of pollen from closely related common species (Ellstrand 1992, Levin et al. 1996).

In summary, inbreeding and outbreeding may lead to a loss of fitness because 1) with inbreeding, mating within a genetically uniform population means there are fewer heterozygotes and more homozygotes (some of which may express recessive deleterious alleles), and 2) with outbreeding, adaptive genetic differences among populations are lost through interbreeding. It must be emphasized that these are both generalizations that do not necessarily apply to all species and may even vary within different populations of the same species (Fenster and Galloway 2000). For a review of inbreeding and outbreeding, see Thornhill (1993).

Utilitarian Values

The surreal images of a St. Bernard sitting on someone's lap or a Chihuahua wading through alpine snows highlight the importance of the genetic diversity of domestic species. This diversity allows people to act as agents of selection and develop different forms of the same species for a variety of purposes: lap dogs and rescue dogs, corn for silage and corn for the dinner table, cherries to eat and cherry trees to admire, and so on (Fuccillo et al. 1997, Maxted et al. 1997, Virchow 1999). Just as important, it allows us to grow the same species in a variety of environments, each with a different climate and local suite of pathogens, predators, competitors, and so forth. Wheat thrives in the deserts of the Middle East (its original home) and in the northern prairies of Canada; cattle range from alpine meadows to tropical grasslands. Genetic diversity can also be exploited simply to satisfy our appetite for variety. Flower gardens are strong testament to the saying, "variety is the spice of life."

The genetic diversity of some wild populations is also important to plant and animal breeders because wild relatives of domestic species are a significant source of genetic material. For example, when scientists at the International Rice Research Institute in the Philippines set out to develop a variety of rice that would be resistant to a major disease, grassy stunt virus, they screened over 6000 varieties of rice and found only one variety that was resistant to the disease. That variety, a wild species of rice called *Oryza nivara*, was represented in their collection by only 30 kernels, of which only three showed resistance (Hoyt 1988). Returning to the area in north-central India where the rice sample had been collected, they could find no new material; the original collection site had been inundated by a dam. Fortunately, this story still had a happy ending because they were able to use the genetic information in these three kernels to develop a new variety of rice, IR36, that is resistant to this virus and is planted across millions of hectares in Asia.

Differences within a species can be of strategic value to conservation because they provide a clear justification for protecting a species across its entire geographic range, including all subspecies and major populations. This is particularly important if the species is a flagship or umbrella species because a wide variety of other biota may benefit. We will return to this issue in Chapter 12, Managing Populations.

Postscript: Careful readers may wonder why we have departed from the taxonomy of values used for species and ecosystems: intrinsic, instrumental, and uniqueness. We could squeeze genes into this classification, but it seems a bit contrived to talk about intrinsic value and uniqueness of molecules. The value of genes lies in what they do, rather than what they are, and in this sense all of their value is instrumental. The classification used here distinguishes between values that are important to the species itself (evolutionary potential and loss of fitness) versus those that are important to people and other species (utilitarian values).

Processes That Diminish Genetic Diversity

To better understand the relationship between reduction in genetic diversity and loss of fitness, we will now consider the processes that diminish genetic diversity, especially in small populations: genetic bottlenecks, random genetic drift, and inbreeding.

Bottlenecks and Drift

Some populations are quite large: thousands of individuals are loosely connected through a web of breeding that ensures genetic *flow* throughout the population. On the other hand, some populations are quite small, perhaps because they are confined to small, isolated patches of habitat and have limited dispersal abilities. In this section we are primarily concerned with what can happen to the genetic diversity of small populations, especially among species that usually live in large populations. Sometimes, large populations experience a catastrophe such as a hurricane and collapse to a few remnant individuals. Sometimes, a few individuals arrive in a new area and establish a new population that is inevitably small at first; this is called a *founder event*. When a population collapses or a new population is established, the genetic diversity of the original larger population is likely to be reduced because only a sample of the original gene pool will be retained. If you start with a population of 1000 bison with 2000 alleles for MDH-1 and reduce it to 50 bison, only 100 alleles will remain. Moreover, the remaining sample is not likely to be representative of the whole. This phenomenon is called a *genetic bottleneck*. Passing through a genetic bottleneck can create two problems: 1) a loss of certain alleles, especially rare alleles, and 2) a reduction in the amount of variation in genetically determined characteristics. For example, a population that ranged across a continuum from very dark individuals to very light individuals might, after a bottleneck, have only intermediate colored individuals or only dark or only light individuals (Frankel and Soulé 1981).

The proportion of genetic variation and number of alleles likely to be retained after a bottleneck can be estimated using the formulas presented in Table 5.2. From this table we can see that most of the genetic variation is retained even in a tight bottleneck, 95% with just 10 individuals. The situation is worse, however, for retention of uncommon alleles. In this example,

---

TABLE 5.2 Proportion of genetic variation remaining after a genetic bottleneck.*

Sample Size (N) After Bottleneck	Proportion of Heterozygosity Retained	Average Number of Alleles Retained from an Original Set (m) of 4	
		$p_1 = 0.70$, $p_2 = p_3 = p_4 = 0.10$	$p_1 = 0.94$, $p_2 = p_3 = p_4 = 0.02$
1	0.50	1.48	1.12
2	0.75	2.02	1.23
6	0.917	3.15	1.64
10	0.95**	3.63	2.00***
50	0.99	3.99	3.60
∞****	1.00	4.00	4.00

* Based on Tables 3.1 and 3.2 in Frankel and Soulé (1981).
** Retention of heterozygosity is approximately equal to $1 - 1/(2N)$, where N is the population size after the bottleneck. If a population crashed to 10 individuals, about $1 - 1/2(10) = 1 - .05 = 0.95$ of the genetic variation of the original population would remain.
*** The formula for estimating how many alleles would remain after a bottleneck is $E = m - \Sigma_j (1 - p_j)^{2N}$, where m is the number of alleles before the bottleneck, p is the frequency of the jth allele, and N is the population size after the bottleneck. From an original set of four alleles the remaining number would be
$4 - \Sigma (1 - 0.94)^{20} + (1 - 0.02)^{20} + (1 - 0.02)^{20} + (1 - 0.02)^{20} =$
$4 - \Sigma 0.06^{20} + 0.98^{20} + 0.98^{20} + 0.98^{20} =$
$4 - \Sigma \sim 0 + 0.666 + 0.666 + 0.666 = 2$
**** With a population of infinite size no genetic bottleneck occurs.

TABLE 5.3 The proportion of genetic variation retained in small populations of constant size after 1, 5, 10, and 100 generations is approximately $[1 - 1/(2N)]^t$, where N is the population size and t is the number of generations. For example, $0.95^5 = 0.77$.*

Population Size (N)	Generations			
	1	5	10	100
2	0.75	0.24	0.06	<<0.01
6	0.917	0.65	0.42	<<0.01
10	0.95	0.77	0.60	<0.01
20	0.975	0.88	0.78	0.08
50	0.99	0.95	0.90	0.36
100	0.995	0.975	0.95	0.60

*Based on Frankel and Soulé (1981).

10 individuals are likely to retain only two of four alleles if three of the alleles were uncommon (2% each of all the alleles). This figure improves to an estimate of 3.63 alleles retained if the alleles are more common, 10% of the total in this example. Genetic data from the whooping crane illustrate this phenomenon; six genotypes were detected in a sample of old museum specimens, but only one of these persists in the modern population after a 1938 bottleneck in which only 14 adults survived (Glenn et al. 1999).

Closely allied to genetic bottlenecks is a term called *random genetic drift*. Random genetic drift is the random change in gene frequencies, including loss of alleles, that is likely to occur in small populations because each generation retains just a portion of the gene pool of the previous generation, and that sample may not be representative (Frankel and Soulé 1981, Hartl and Clark 1997). Table 5.3 presents a formula for estimating the effect of random genetic drift on genetic diversity and some sample results. The formula is identical to the one for estimating the loss of genetic variation in a bottleneck, with an exponent added to represent the number of generations that a population has continued to remain small. In other words, random genetic drift is the same thing as passing through a genetic bottleneck except that the drift lasts multiple generations (compare column 2 in Table 5.3 with column 2 in Table 5.2). Although we typically use the term "random genetic drift" when a population remains small for many generations and "bottleneck" for a short phenomenon, it would not really be wrong to speak of bottlenecks lasting more than one generation or of drift occurring during one generation.

TABLE 5.4 **Expected number of alleles remaining after *t* generations for a population of six individuals with 2, 4, or 12 alleles for a gene, assuming equal frequency of each allele.***

Generations	Number of Alleles		
	$m = 2$	$m = 4$	$m = 12$
0	2.00	4.00	12.00
1	1.99	3.87	7.78
2	1.99	3.55	5.88
8	1.67	2.18	2.64
20	1.24	1.36	1.44
∞	1.00	1.00	1.00

*Based on Frankel and Soulé (1981).

We can see that although a population of 10 individuals may retain 95% of its genetic variation after one generation (or after one bottleneck), with random genetic drift for 10 generations only 60% of the variation is likely to be retained, and after 100 generations virtually all the original genetic variation would be lost. A similar pattern exists for the loss of alleles; after many generations of random genetic drift, small populations will usually retain only one allele for a given gene (Table 5.4). In the language of genetics, the gene will have been *fixed* for that allele. In sum, random genetic drift in a population that remains small for many generations is much more likely to lead to a loss of genetic diversity than is a single bottleneck from which a population recovers quickly.

Effective Population Size: To estimate the effects of bottlenecks and random genetic drift, as presented in Tables 5.2 and 5.3, it is necessary to make some simplifying assumptions that are often violated. These estimations assume that the organism is diploid, sexually reproducing, and has nonoverlapping generations; that the population is of constant size, has equal numbers of females and males, random mating, and no migration; that reproductive success of all individuals is the same; and that no mutation or natural selection occurs. In particular, these assumptions have allowed us to avoid a major complexity: the difference between total or census population size (the actual number of individuals in a population) and the *effective population size*. To take a very simple example of this idea, consider a population of 100 bison in which 25 are too young to breed and 15 adults are infertile; 60 is the number of breeding adults and therefore the effective size of the population. In practice, the issue is usually more complicated than this and involves considerations such as fluctuations in population size, unequal family size, and unequal numbers of females and males. We will begin with a definition and then show two examples of how to calculate effective population size. (See Lande and Barrowclough 1987 and Nunney and Elam 1994 for further details.)

The effective population size (N_e) of a population is the number of individuals in a theoretically ideal population (i.e., one that meets all the assumptions stated above) that would have the same magnitude of random genetic drift as the actual population.

Example 1. Population fluctuations. The effective size of a population that is fluctuating through time (as most do) is less than the actual population size. In this case, N_e is estimated to be the harmonic mean of the actual size of each generation (Hartl and Clark 1997). Mathematically,

$$\frac{1}{N_e} = \frac{1}{t}\left(\frac{1}{N_1} + \frac{1}{N_2} + \cdots + \frac{1}{N_t}\right)$$

In words, the harmonic mean is the reciprocal of the average of reciprocals of the population size for each of t generations. This method of estimating an effective population gives more weight to small N's. For example, the N_e for three generations ($t = 3$) in which $N_1 = 1000$, $N_2 = 10$, and $N_3 = 1000$, would be

$$\frac{1}{N_e} = \frac{1}{3}\left(\frac{1}{1000} + \frac{1}{10} + \frac{1}{1000}\right) = 0.034$$

$$N_e = \frac{1}{0.034} = 29.4$$

which is far less than 670, the arithmetic mean of 1000, 10, and 1000. (Also see Vucetich [2000] for the effect of population fluctuations.)

Example 2. Unequal numbers of females and males. If a population has an unbalanced sex ratio, the effective population size is less than the actual size and can be estimated

$$N_e = \frac{4N_f N_m}{N_f + N_m}$$

where N_f is the number of breeding females and N_m is the number of breeding males (Hartl and Clark 1997). For example, if 96 females mated with four males,

$$N_e = \frac{4 \times 96 \times 4}{96 + 4} = \frac{1536}{100} = 15.4$$

This kind of imbalance may be more common than is usually realized. Recent research using genetic techniques to determine the mother and father of offspring has indicated that in many species relatively few individuals, especially among males, are responsible for a disproportionate share of a population's reproduction (Parker and Waite 1997). Many apparently healthy adults do not leave any offspring. Such inequity is generally not a problem—it is the basis for natural selection—but it may lead to difficulties in very small populations because of its effect on genetic diversity. For more complex models of effective population size, see Reed et al. (1993); for a good example of using effective population size, see Grant and Grant (1992). The bottom line to remember is that the effective population size is often substantially less than the actual number of individuals in a population, often only 10% to 20% (Vucetich et al. 1997). Thus, if you want a population of bison with $N_e = 100$ (hopefully sufficient to retain 99.5% of its genetic variability through at least one generation; see Table 5.3), you actually need a field population of about 500 to 1000.

Inbreeding

Inbreeding refers to the mating between closely related individuals; closely related individuals are likely to share identical copies of some of their genes because they have ancestors in common. Quantitatively, the inbreeding coefficient, F, is the probability that two copies of the same allele are identical by descent—in other words, derived from a common ancestor (Templeton and Read 1994). For example, in our bison example, if both MDH-1 X alleles in the X/X homozygous individual were derived from its grandmother, those alleles would be considered identical by descent.

There are several methods to estimate F: one of the simplest involves counting links in the pedigree chain: $F = (1/2)^n$, where n is the number of individuals or links in the pedigree chain starting with one parent, going back to the common ancestor, and then going down the other branch to the other parent. Figure 5.6A shows the pedigree chain for the offspring (A) of a half-sister (B) mating with her half-brother (C) (i.e., B and C have the same mother, D, but different fathers). The inbreeding chain has three links—B, D, and C—and thus F is equal to $(1/2)^3 = 1/8 = 0.125$. If B and C were full siblings (i.e., they had both the same mother D and the same father E) (Fig. 5.6B), then there would be two chains, one for each common ancestor (B, D, and C for the mother plus B, E, and C for the father). In this case the F values for each chain would be added: $(1/2)^3 + (1/2)^3 = 1/4 = 0.25$.

As described in our discussion of inbreeding depression, inbreeding is known to lead to problems among captive populations and thus is of great concern to conservationists who propagate endangered species in

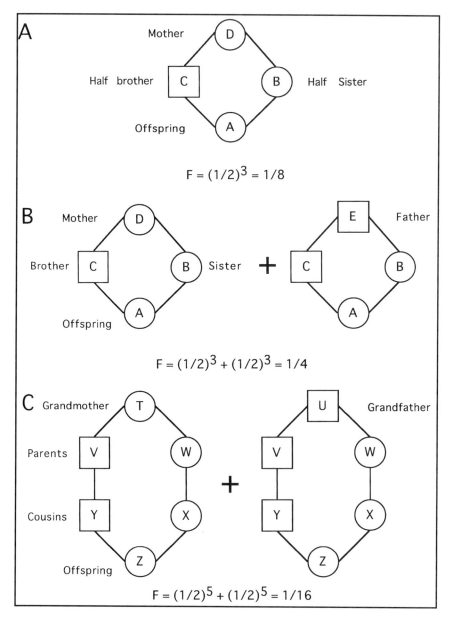

Figure 5.6. Inbreeding pedigrees for matings between: (A) a half-sister with her half-brother, (B) full sister and brother, and (C) full cousins (different parents but identical grandparents). See the text for an explanation of A and B. Circles represent females and squares represent males.

captivity. Documentation of inbreeding depression in wild populations is limited, but it could be a problem among very small populations.

An Important Caveat

It must be emphasized that the equations presented in this section provide only estimates of the likely effects of processes that diminish genetic diversity. Exceptions may be fairly common. For example, Indian rhinoceros appear to have retained a high level of genetic diversity despite having passed through a bottleneck, perhaps because of high mobility of some individuals and long generation times (Dinerstein and McCracken 1990). Similarly, an isolated population of pinyon pine retained its genetic diversity over several hundred years (Betancourt et al. 1991). Moreover, even if the predicted effects on genetic diversity occur, they may not have catastrophic consequences for a population. For example, the northern elephant seal was reduced to as few as 20 individuals in the 1890s and now seems to have extremely low genetic diversity: no allozyme polymorphism at 24 loci from a sample of 159 seals from five colonies (Bormell and Selander 1974) or at 43 loci from a sample of 67 seals from two colonies (Hoelzel et al. 1993). Despite this lack of genetic diversity, the northern elephant seal is thriving now with a total population approaching 200,000. The Mauritius kestrel also passed through a narrow bottleneck, one breeding pair, that sharply reduced its genetic diversity and has now recovered to over 200 pairs (Groombridge et al. 2000). In this case, examination of genetic material in museum specimens confirms that the original population was very diverse despite being confined to a small island. Under special circumstances, passing through a bottleneck might have a positive effect by eliminating all the individuals carrying deleterious recessive alleles, thus purging this allele from the population. Support for this idea came from a captive-breeding program for Speke's gazelle (Templeton and Read 1983), but, recently, this study and the whole concept of purging through inbreeding have been questioned (Byers and Waller 1999, Kalinowski et al. 2000). Even if such benefits do exist, they might be short-lived if a bottleneck left the species so genetically uniform that it was ill-prepared to adapt to future environmental change. Finally, some evidence suggests that we typically underestimate the rate at which new genetic diversity arises through mutation (Jeffreys et al. 1985) or new species evolve through natural selection and reproductive isolation (Hendry et al. 2000).

Cultural Diversity

The sharing of genes between parents and offspring is not the only mechanism by which information is transmitted from one generation to the next. Among many social animals information also moves among individuals and generations through learning, a process often called *cultural transmission*. Methods for exploiting novel food items provide some of the best-documented examples of cultural transmission; one such example occurred when the knowledge that food could be obtained by pecking open the caps of milk bottles spread among the blue tits of England (Fisher and Hinde 1949). The location of migration routes, water holes, food patches, nesting sites, and hibernacula may be learned by young animals following old animals. For example, it is likely that the matriarchs of elephant herds know the location of water in times of drought and can lead their herds there (Moss 1988). If a herd's matriarch died before the information could be transmitted, the cultural diversity of that elephant herd would be diminished, perhaps with disastrous consequences. Loss of cultural transmission has been a problem for some conservationists trying to reintroduce captive-reared animals to the wild. For example, golden lion tamarins released into their native habitat have had problems identifying food and predators, information that they would have learned from other tamarins under normal circumstances (Kleiman 1989).

Among all species *Homo sapiens* has the most diverse culture, and maintaining human cultural diversity should also be of some concern to conservation biologists. Of course, it is hard to imagine conservation biology encompassing humanity's languages, art, music, science, literature, architecture, and so on. When conservation biologists think about maintaining human cultural diversity, they usually focus on the diverse ways in which rural people, especially people who still use traditional technology, interact with the ecosystems in which they live. For example, ethnobotanists are particularly concerned about maintaining the cultures associated with human use of plants for food, medicine, and other purposes (Balick and Cox 1996, Nazarea 1998).

CASE STUDY   ### The Cheetah

Running at speeds up to 112 km/hr the cheetah is the world's fastest sprinter, but it is having difficulty outpacing some problems that threaten it with extinction. Twenty thousand years ago four species of cheetah roamed grasslands in Africa, Asia, Europe, and North America, and as recently as a hundred years ago the remaining species of cheetah was widespread throughout much of Africa and southwestern Asia as far east as India. Today fewer than 20,000 animals remain, largely in southern and eastern Africa, and they are hard-pressed by a lack of habitat with plentiful prey; by lions, which are both competitors for prey and predators on juvenile cheetahs; by poaching; and perhaps by an inconspicuous but potentially serious problem—a lack of genetic diversity.

Stephen O'Brien and a team of colleagues (1983, 1985) used electrophoresis of allozymes to look for allelic diversity at 52 loci in a sample of 55 cheetahs from southern Africa. They found none: polymorphism = 0; heterozygosity = 0. Thinking that perhaps all members of the cat family have low genetic diversity, they sampled allelic diversity at 48 to 50 loci in seven other feline species and found polymorphism (P) to range from 8% to 20.8% and heterozygosity (H) to range from 0.029 to 0.072, typical values for mammals (see Newman et al. 1985). Some further evidence of the cheetah's lack of genetic diversity came from experiments in which small patches of skin were transferred between pairs of cheetahs. Normally, such skin grafts are quickly rejected if they are between unrelated individuals, but the cheetah grafts were rejected slowly (three cases) or not at all (11 cases). Measurements of cheetah skull characteristics also revealed a high level of asymmetry (e.g., the left jaw longer than the right jaw); developmental abnormalities such as asymmetry are often thought to be related to inbreeding (Wayne et al. 1986). In later work the researchers found that cheetahs from east Africa had some genetic diversity (P = 4%; H = 0.014), and with a larger sample of cheetahs from southern Africa ($N = 98$)

they found some polymorphism for one locus (P = 2%; H = 0.0004) (O'Brien et al. 1987). Why are cheetahs one of the most genetically depauperate species ever examined? No one knows for sure, but some evidence suggests that they went through a major bottleneck about 10,000 years ago at a time when many large mammals went extinct (Menotti-Raymond and O'Brien 1993). (For a critique of this idea, see Pimm et al. [1989].)

The cheetah's lack of genetic diversity is of more than academic interest. It is probably linked to two facts: first, samples of cheetah semen had spermatozoal concentrations 7 to 10 times less than those of domestic cats, and second, 70% to 80% of their spermatozoa were abnormal, compared with 29% for domestic cats (O'Brien et al. 1985, 1987). Lack of genetic diversity may also explain a rate of 29.1% infant mortality among captive-born cheetahs, one of the highest rates recorded among captive mammals. Finally, genetic uniformity may explain what happened to a captive population of cheetahs in Oregon, where beginning in 1982 an outbreak of feline infectious peritonitis (FIP) and related diseases killed 27 of 42 cheetahs and afflicted over 90% of the population (O'Brien et al. 1985, Heeney et al. 1990). This disease is not usually particularly lethal to felines; in fact 10 lions living at the same site showed no symptoms of the disease. Perhaps the virus adapted to the particular genotype that all these cheetahs shared, and thus it had a devastating effect.

For better or worse, O'Brien's work sparked considerable controversy, primarily because field ecologists knew of no problems facing wild cheetah populations that could be attributed to low genetic diversity. In contrast, it was eminently clear to them that lion predation and habitat loss were the serious threats to cheetahs. (See Caro and Laurenson [1994], Merola [1994], O'Brien's [1994], Laurenson et al. [1995] for the core of the debate and May [1995] and Kelly and Durant [2000] for some of the aftermath.) Unfortunately, such debates drift toward polar constructs in which the protagonists seem to be saying that it is all genes or all ecology and demography. The truth is seldom so simple. In this case it is difficult to deny the great and immediate importance of habitat loss and lion predation, but it seems fool-hardy to ignore the possibility that the cheetah's impoverished genome may also be an issue—if not now, then in the future when new threats arise.

Summary

Genetic diversity is essentially a measure of the diversity of information a species has encoded in its genes. It is based on the distribution of different alleles among individuals and can be quantitatively expressed as polymorphism (which is based on the proportion of genes that have more than one common allele) and heterozygosity (which is based on the proportion of genes for which an average individual is heterozygous). Genetic diversity is important for three primary reasons: evolutionary potential, loss of fitness, and utilitarian values. Species with high levels of genetic diversity 1) are better equipped to evolve in response to changing environments; 2) are less likely to suffer a loss of fitness because of the expression of deleterious recessive alleles in homozygous individuals, among other problems; and 3) offer plant and animal breeders greater scope for developing varieties with specific desirable traits such as resistance to certain diseases. Genetic diversity can be eroded by some phenomena associated with small populations. First, when a population is reduced to a small size (i.e., it passes through a bottleneck), some genetic variance and uncommon alleles are likely to be lost. Similarly, in populations that remain small for multiple generations, random genetic drift changes the frequency of alleles; this often reduces genetic diversity, particularly when genes are fixed for a single allele. Finally, inbreeding between closely related individuals can diminish genetic diversity. When estimating the effects of these processes on populations, it is important to estimate the effective population size, which is often substantially less than the actual population size. Conservation biologists are also concerned with cultural diversity, the information that many animal species, including humans, pass from generation to generation through learning.

FURTHER READING

Hartl and Clark (1997) give a comprehensive treatment of population genetics, and Hartl (2000) provides a primer on the same topic. For more reading on the interface between conservation and genetics, see Frankel and Soulé 1981, Schonewald-Cox et al. (1983), Chapters 3–6 of Soulé (1986), Falk and Holsinger (1991), Avise (1994), Loeschcke et al. 1994, and Avise and Hamrick (1996). There also is a journal, *Conservation Genetics.*

TOPICS FOR DISCUSSION

Below are genotypes at three loci for a sample of 10 individuals:

Locus	1	2	3
Individual			
1	aa	BB	CC
2	aa	Bb	CC
3	Aa	BB	CC
4	aa	Bb	CC
5	Aa	BB	CC
6	AA	BB	CC
7	aa	BB	CC
8	AA	BB	CC
9	AA	BB	CC
10	Aa	BB	CC

1. What are the frequencies of alleles for each locus?
2. What are the frequencies of genotypes for each locus?
3. What is the polymorphism for this population using the 95% criterion (the frequency of the most common allele < 95%)?
4. What is the average heterozygosity for this population?
5. What would genotype frequencies be at locus 2 in this population if it were in Hardy-Weinberg equilibrium?

6. If individuals 1–6 were females and individuals 7–10 were males, what would be the effective population size of this population?

7. What portion of the genetic variance of this population would be likely to remain after three generations of random genetic drift? (Use the effective population size calculated in the preceding question.)

Answers: 1. Locus 1: a=.55, A=.45; Locus 2: b=.10, B=.90; Locus 3: C=1.00
2. Locus 1: aa=.4, AA=.3, Aa=.3; Locus 2: Bb=.2, BB=.8; Locus 3: CC=1
3. .67 because loci 1 and 2 are polymorphic
4. .17 (0.3Aa + 0.2Bb)/3 = .17
5. bb = .01, Bb=.18, BB=.81
6. 9.6 (4 × 6 × 4)/(6 + 4)
7. .85$[1 − 1/(2 × 9.6)]^3$

The species depicted above are, in order from left to right: *Hibiscadelphus bombycinus,* heath hen, *Cyanea arborea,* Labrador duck, great auk, Carolina parakeet, *Pultenaea pauciflora,* Steller sea cow, blue antelope, *Melaleuca arenaria,* quagga, Israel painted frog, stumptooth minnow, passenger pigeon, Xerces butterfly, Tasmanian wolf or thylacine, and Clear Lake Splittail.

Threats to Biodiversity

The last word in ignorance is the man who says of an animal or plant: "What good is it?" If the land mechanism as a whole is good, then every part is good, whether we understand it or not. If the biota, in the course of aeons, has built something we like but do not understand, then who but a fool would discard seemingly useless parts? To keep every cog and wheel is the first precaution of intelligent tinkering.

—*Aldo Leopold*

When the last individual of a race of living things breathes no more, another heaven and another earth must pass before such a one can be seen again.

—*William Beebe*

. . . the worst thing that will *probably* happen—in fact is already well under way—is not energy depletion, economic collapse, conventional war, or even the expansion of totalitarian governments. As terrible as these catastrophes would be for us, they can be repaired within a few generations. The one process now ongoing that will take millions of years to correct is the loss of genetic and species diversity by the destruction of natural habitats. This is the folly our descendants are least likely to forgive us.

—*E. O. Wilson*

Extinction is forever.

Chapter 6
Mass Extinctions and Global Change

Among all the things I heard as an undergraduate, one stands out. It came from a professor who was fielding a barrage of queries such as, "Do we *have* to type our papers?" "Do we *have* to use this exact format for our citations?" He finally ended this line of questioning with an answer that cut to the quick, "There is only one thing in life that you *have* to do, and that is die. Everything else is optional." Death has an evolutionary analogue—extinction—and evolutionary biologists are confident that, just as death is the inevitable fate of every individual, extinction is the fate of every species. In fact, the fossil record indicates that of all the species that have ever lived on earth, about 99.9% have gone extinct (Raup 1991). It is also reasonable to assume that those that are extant now will eventually meet the same fate. Although extinction is inevitable, it does take different forms. A species may disappear because it evolves into a similar, but distinct, new species, or it may disappear into an evolutionary dead end. A few creatures have persisted nearly unchanged for such long periods that they are popularly called living fossils; for example, the horseshoe crab, genus *Limulus*, has changed little in 190 million years. A more typical "life span" is a million years or so. (Average life spans in the fossil record are closer to 4 million years, but the fossil record is probably biased toward widespread, successful species with longer life spans, and life spans across all species are probably somewhat shorter [Jenkins 1992].)

Although species have been falling to extinction throughout the 3.5-billion-year history of life on earth, the clock of extinction has not run smoothly. There have been at least five periods when huge numbers of species have vanished, leaving behind a greatly impoverished biota. Concern that we may be in the midst of another spasm of extinctions—one of our own making—is, of course, the catalyst behind conservation biology. Before examining the evidence for a human-induced extinction spasm and its likely mechanisms, we need some understanding of the episodes of mass extinction that preceded our arrival on the scene (Sepkoski 1995).

Extinction Episodes of the Past

During the first couple of billion years of life on earth there probably was not a great deal of biological diversity. Life existed in the sea as a set of microbes—bacteria and bacteria-like species—that were sometimes abundant enough to form slimy mats (Schopf and Klein 1992). It is likely that some species flourished for lengthy periods, while others became extinct quickly; however, the scanty fossil record from this period is too limited to document comings and goings. About 1.9 billion years ago eukaryotic organisms arose with their DNA enclosed in a membrane and with organelles in their cells such as mitochondria. Soon after, complex organisms developed that had differentiated cells organized into tissues and organs, but still the fossil record from this period limits our understanding of speciation and extinction. It was not until the Cambrian period, beginning about 600 million years ago, that a great proliferation of macroscopic species occurred and produced a fossil record that allows us to track the rise and fall of biodiversity (Jablonski 1991).

Since the Cambrian period, biodiversity has generally risen, but there have been some notable exceptions (Fig. 6.1). Biodiversity collapsed dramatically during at least five periods because of mass extinctions around the globe. The five major mass extinctions receive most of the attention, but they are only one end of a spectrum of extinction events. Collectively, more species went extinct during smaller events that were less dramatic but more frequent (Raup 1991). We will briefly examine two of the five major extinction events as case studies, starting with the best-known one, the one that saw the demise of the dinosaurs.

CASE STUDY   **The Cretaceous-Tertiary Extinctions**

About 315 million years ago reptiles developed skins and eggshells that were relatively impervious to water loss, and starting about 280 million years ago these were the dominant large animals in terrestrial environments. In popular language this was the era "when dinosaurs ruled the earth," and the images of tyrannosaurs, triceratops, and others are so familiar that they do not need to be elaborated on here. Suffice it to say that there was a wide variety of reptile species occupying many ecological

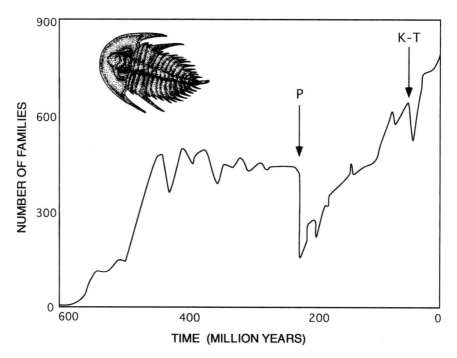

Figure 6.1. The rise and occasional fall of biodiversity as indicated by the fossil record of families of marine organisms. Marine organisms are used as an index of past biodiversity because they have left the most complete fossil record. (They are more likely than terrestrial taxa to leave corpses in places where they might be quickly covered with sediments and thus protected from scavengers and physical disturbance.) The number of families is used as an index of biodiversity rather than species or genera because a single species or genus might be missing from the known fossil record, but the fossil record for families is likely to be nearly complete. (Redrawn from Sepkoski 1982.)

niches. When the age of dinosaurs came to a dramatic end about 65 million years ago, mammals began to flourish, evolving from relatively few types into bats, whales, and the myriad other forms we know today. Paleontologists label this point in the earth's history as the end of the Cretaceous period and the beginning of the Tertiary period, often abbreviated as the K-T boundary. This time was also marked by changes in many other taxa. Overall, about 38% of the genera of marine animals were lost (Raup 1991) with percentages much higher in some groups. Ammonoid molluscs went from being a very diverse and abundant group to being extinct (Marshall and Ward 1996). An extremely abundant set of planktonic marine animals called Foraminifera largely disappeared, although they rebounded later. Among plants, the Cretaceous-Tertiary boundary saw a sharp but brief rise in the abundance of primitive vascular plants such as ferns, club mosses, horsetails, and conifers and other gymnosperms. The number of flowering plants (angiosperms) was reduced at this time, but then began a dramatic increase (Knoll 1984, Stewart and Rothwell 1993).

What caused these changes? For many years scientists assumed that a cooling of the climate was responsible, with dinosaurs being particularly vulnerable because they were ectothermic (i.e., dependent on environmental heat, "cold-blooded" in vernacular language) like modern reptiles. It is now widely believed that at least some species of dinosaurs had a metabolic rate high enough for them to be endothermic. Nevertheless, climatic explanations for the K-T extinctions are not really challenged by the idea that dinosaurs may have been endothermic, because even endotherms can be affected by a significant change in the climate.

Explanations for the K-T extinctions were revolutionized in 1980 when a group of physical scientists led by Luis Alvarez proposed that 65 million years ago the earth was struck by a 10-km-wide meteorite traveling at 90,000 km/hr (Alvarez et al. 1980). They believed that this impact generated a thick cloud of dust that enveloped the earth, shutting out much of the incoming solar radiation and reducing photosynthesis to very low levels. Short-term effects might have included huge tidal waves and extensive fires. In other words, a cavalcade of factors arising from a single cataclysmic event caused the massive extinctions. Initially, the meteorite theory was largely based on a single line of evidence. At locations around the globe, geologists had found an unusually high concentration of iridium in the layer of sedimentary rocks that were formed about 65 million years ago. Iridium is an element that is usually uncommon near the earth's surface, but it is abundant in some meteorites. Therefore Alvarez and his colleagues concluded that it was likely that the iridium in sedimentary rocks deposited at the K-T boundary had originated in a giant meteorite or asteroid. Most scientists came to accept the meteorite theory after evidence came to light that a circular formation, 180 km in diameter and centered on the north coast of the Yucatan Peninsula, was created by a meteorite impact about 65 million years ago (Florentin et al. 1991, Alvarez et al. 1992, Pope et al. 1998).

CASE STUDY   The Permian Extinctions

As dramatic as the Cretaceous-Tertiary extinctions were, they pale in comparison with the massive extinctions that marked the end of the Permian period, 251 million years ago. At this time over half of the 500 families of marine vertebrates and invertebrates capable of forming fossils became extinct, compared with roughly 12% in the four other major extinctions (Sepkoski 1984, Erwin 1994). It is harder to estimate how many species dis-

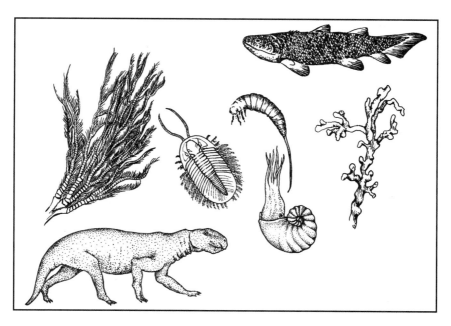

Figure 6.2. Permian organisms included, from left to right, therapsid reptiles, crinoid echinoderms, trilobites, monuran insects, sarcopterygian fish, nautiloid molluscs, and bryozoans.

appeared, but extrapolations suggest that 77% to 96% of all marine animal species were lost (Raup 1979). Particularly hard hit were filter-feeding animals such as corals, brachiopods, crinoids, and bryozoans, especially those living in tropical oceans (Fig. 6.2). The fossil record for terrestrial organisms is more limited, but it suggests that many taxa—most vertebrates and vascular plants, for example—also declined during this period (Knoll 1984, Retallack 1995). Some animals such as the therapsids, a group of reptiles that gave rise to mammals, were hard hit during this event (Stanley 1987). The only known mass extinction event for insects, with eight of 27 orders disappearing, occurred during this time (Labandeira and Sepkoski 1993).

Various causes for the Permian extinctions have been proposed, and no one of them is widely accepted (Erwin 1993, 1994). Meteorite impact is a possibility, but supporting evidence is quite limited (Bowring et al. 1999); massive volcanic activity in Asia might have had some similar effects on climate (Jin et al. 2000). Continental drift may have been indirectly responsible. During the Permian period, almost all of the earth's land mass occurred in a single supercontinent, Pangaea, which had drifted into a position stretching from pole to pole. The climate of this huge land mass was probably quite unstable. Along the coast of this continent a drop in sea level dried out many shallow marine basins, and this could have constituted a critical loss of habitat for marine species. Later, when sea level rose again, many terrestrial species living near the coast may have lost their habitat. Considerable evidence suggests that oceanic oxygen levels may have been low enough to be lethal for some marine organisms (Knoll et al. 1996, Isozaki 1997). Furthermore, an overturn in these anoxic waters could have led to extensive CO_2 poisoning and CO_2-induced climate change. In short, we do not know what caused the greatest extinction event in global history, but it is likely that it involved a complex interplay of many different factors (Erwin 1993, 1994).

Recoveries from Past Extinctions

Close examination of Figure 6.1 reveals a good news–bad news story for anyone seeking a historical basis for viewing contemporary extinctions. The good news is that the overall trend in biodiversity is upward with mass extinctions just temporary setbacks in this pattern. Life has moved from a slimy soup of microbes to magnificent creatures like beavers, honey bees, and saguaro cacti, and the challenges posed by meteorites, volcanoes, and glaciers have not returned us to our primordial roots. Some people have feared that the stupidity of humanity could eradicate life on earth, perhaps through a nuclear holocaust, but this view is rich in human arrogance. Life on earth, in some form, is almost certain to persist despite us (Gould 1990).

The bad news is that these recoveries have required tens of millions of years each (Jablonski 1995). That means that if we are in the midst of a human-caused extinction spasm at the moment, none of us, or our children, or grandchildren, or great × thousands grandchildren will witness the recovery. Indeed, if the human life span on earth is typical of most species, then no *Homo sapiens* will be around to enjoy the return of biodiversity. Some people may find considerable solace in the idea that life on earth will likely go on until our sun ceases to exist. For me, the prospect that life will return to even higher levels of diversity after I am gone is only a modest consolation.

Estimating the Current Rate of Extinction

Ask a group of conservation biologists about current rates of extinction, and they are likely to start rolling off statistics about the thousands of species that are being lost each year. Yet, ask that same group to name ten species that have gone extinct in the last year, and they will probably struggle to name any. Why the discrepancy? There are two major reasons.

First, it is difficult to name species that have become extinct recently because it is hard to say with conviction that a species *has* become extinct.

In fact, strictly speaking, it is impossible. Consider a 1985 phone conversation between a colleague of mine and a man who called to report that he had a flock of passenger pigeons at his bird feeder. My colleague explained that the passenger pigeon became extinct in 1914 and that the man had probably seen some mourning doves. The man was absolutely insistent that he had seen passenger pigeons, not mourning doves, and finally said, "Is it or is it not *possible* that I saw passenger pigeons in my backyard?" My colleague, lapsing from his usual civility replied, "Yes, it is *possible*; it is also *possible* that you saw Martians in your backyard. But neither possibility is very likely." The fact is that a few species have been presumed to be extinct and then later rediscovered after periods longer than 1914–1985. For example, the cahow, a rare seabird that nests in Bermuda, disappeared in 1621 and was not recorded by any scientists until 1906, when a single specimen was discovered; and it was 1951 before a breeding colony was found (Fisher et al. 1969). Many species have never been reported again after they were originally described in the 1800s, yet they are not considered extinct because we do not know if the absence of records is the result of extinction or no one has looked for them. Diamond (1987) has argued that in cases like these the burden of proof should be reversed, and that a species should be considered extinct unless someone has proven that it is extant.

The second issue is that if most of the world's species (perhaps 85% to 99%) have never been described by scientists (see Chapter 3, Species Diversity), then it is very likely that most of the species becoming extinct are also unknown to scientists. E. O. Wilson (1992) has suggested that we call this phenomenon of species becoming extinct before they are described *Centinelan extinctions* after a small ridge in Ecuador called Centinela. Two botanists from the Missouri Botanical Garden, Alwyn Gentry and Calaway Dodson, visited Centinela in 1978 and discovered about 90 plant species that were either endemic to that ridge or found in only a few nearby locales (Dodson and Gentry 1991). By 1986 the ridge had been completely cleared and planted with crops. Most of the 90 species (and who knows how many insects and other species) were gone. If Gentry and Dodson had not visited the area in 1978, we would be completely ignorant of the lost species. This phenomenon also could have taken place in comparatively well studied regions such as Europe and North America, where many rare, inconspicuous species might have disappeared before they were described by scientists (E. O. Wilson personal communication). Both of these phenomena—undocumented extinctions of known species and extinctions of unknown species—are relatively more likely in the marine realm than in freshwater or terrestrial systems (Culotta 1994).

Despite these constraints, it is possible to document some recent extinctions. The most comprehensive list, compiled by staff of the World Conservation Monitoring Centre, lists 90 species of plants and 726 animals that

TABLE 6.1 Numbers of plant and animal species by major taxon listed by the World Conservation Monitoring Centre (www.redlist.org) as having become extinct since 1600.*

Animals		Plants	
Molluscs	303	Mosses	3
Crustaceans	9	Gymnosperms	1
Insects	73	Angiosperms	
Other invertebrates	4	Dicots	83
Fishes	92	Monocots	3
Amphibians	5		
Reptiles	22		
Birds	131		
Mammals	87		
Totals	726		90

*This list includes 40 species that still survive in captivity.

probably have become extinct, at least in the wild, since 1600 (Table 6.1). It is undoubtedly incomplete for all groups, extremely so for invertebrates. It does not even attempt to list extinct fungi, algae, bacteria, and other microbes. The discrepancy between two or three extinctions per year documented in Table 6.1 and estimates of hundreds or thousands of extinctions per year is in large part a result of predictions about what deforestation is doing to the rich biota of tropical forests. Several scientists have predicted the global rate of species extinctions based on the impacts of tropical deforestation (e.g., Reid 1992, Groom 1994), and one group has even estimated losses of populations (Hughes et al.1997). Here we will consider one of the best known examples, that of E. O. Wilson (1992).

Wilson begins with the idea that there is a predictable relationship between the number of species and the area they occupy. This idea was first extensively explored by Wilson and Robert MacArthur in their island biogeography model (MacArthur and Wilson 1967), which is described further in Chapter 8, Habitat Degradation and Loss. Suffice it to say here that the model predicts that the number of species on an island can be estimated from the equation $S = CA^z$, where S is species richness, A is area, and C and z are constants that vary depending on the particular group of species and set of islands. Across many studies, z values often range between 0.15 and 0.35 (see Williamson 1981). A z value of 0.30 corresponds to an easily remembered rule of thumb: if the area in question is decreased by 90%, then the number of species it supports will be halved. Next, Wilson chose an estimate for the rate at which the area of tropical forest is decreasing; he used 1.8% per year, a figure based on 1989 data assembled by Norman Myers (1989b). With a z value of 0.30 this would translate into losing 0.54% of the tropical forest biota per year. At a z value of 0.35 the loss would be 0.63% per year; for z = 0.15 the annual species loss would be 0.27%. To be conservative, Wilson used this lowest reasonable estimate of annual species loss, 0.27%, and multiplied it by a conservative estimate of the number of species in tropical forests, 10,000,000 species, to arrive at an estimate of 27,000 species going extinct per year. This figure is conservative both because of the particular values used and because it is limited to tropical forest species, often estimated to constitute about half the earth's biota. It is also conservative because it assumes that species have fairly broad geographic ranges; if species had very small geographic ranges, cutting 1.8% of the forest would eliminate 1.8% of the species, not 0.27% to 0.63% (see Pimm et al. 1995). It also assumes that some suitable habitat will persist within the range of most species; if we lose 90% of a given type of ecosystem (shifting from 100% to 10%), we will lose half the species tied to that ecosystem, but if we then shift from 10% to 0%, all the remaining species will disappear. (See Kinzig and Harte [2000] for a similar analysis that incorporates the issue of species being endemic to a small area.)

Wilson concludes by noting that one million years is a typical "life span" for a species and that this figure would translate into one species out of every million species becoming extinct each year. This means that we could expect a tropical forest biota of 10 million species to lose 10 species per year under normal circumstances. In other words, the estimate of 27,000 extinctions per year is 2700 times greater than the background rate of extinction. Even if this figure is too high by an order of magnitude (perhaps the current extinction rate is "only" 270 times the background rate), it still provides a handy retort to people who argue, "Why is everyone so worried about species going extinct; extinction is a natural process." Yes, extinction is a natural process, but a human-induced extinction rate hundreds or thousands of times greater than the natural extinction rate is *not* natural.

How long will this rate of loss continue? How many species will be lost in total? These are much harder questions to answer, but Wilson (1992) has ventured a guess. Assuming that the human population will stabilize at somewhere between 10 to 15 billion people in the next 50 to 100 years, and that the loss of ecosystems will stabilize concomitantly, Wilson believes we will lose 10% to 25% of our biota during this period. The 10% figure assumes that we act in a wise and judicious way; even the 25% figure will be optimistic unless population growth and excessive consumption can be curbed.

The Prospect of Global Climate Change

Catastrophic changes in the earth's climate that precipitate mass extinctions may be relatively infrequent, but more modest changes occur quite regularly. These too may be evolutionary bottlenecks that eliminate some species. Scientists believe that the earth is experiencing a significant change in climate now because of human-induced changes in atmospheric concentrations of CO_2, and that these changes may stress the earth's biota (Peters and Lovejoy 1992). In this section we will briefly review the recent history of climate change and its effect on biota, and we will explore some possible consequences of climate change in the next few decades.

Recent History of Global Climate Change

To understand how the earth's climate has been changing during the last 2.5 million years we need to leave earthbound science and delve into astronomy. Everyone is familiar with how climate parameters change from day to night as the earth revolves on its axis, alternately warming one side

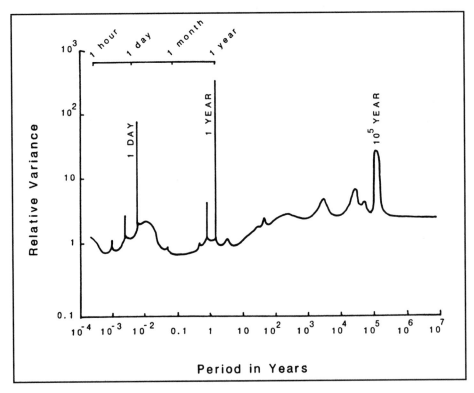

Figure 6.3. A plot of climate variation versus time shows three peaks based on cycles of the earth's movements. The first two peaks at 1 day and 1 year correspond to daily and annual cycles. The third peak at about 100,000 years is generated by three separate cycles described in the text and drawn in Figure 6.4. (Reproduced with permission from Hunter (1992) as redrawn from Huntley and Webb 1989.)

with solar radiation and then cooling it again. Similarly, we all know about how the earth's orbit around the sun and tilted axis generate annual climatic cycles because the Northern Hemisphere is tilted toward the sun during half our orbit (from the March equinox until the September equinox), and the Southern Hemisphere is tilted toward the sun during the other half. These two cycles generate a pattern of climate variance that shapes organisms in many familiar ways; for example, the diurnal-nocturnal behavior of animals and the seasonal growth and death of annual plants (Fig. 6.3). Far less familiar to most people are three other astronomical cycles that also generate climate changes over much longer periods and also engender biotic change (Berger et al. 1984, Imbrie and Imbrie 1986).

The first of these long-term cycles involves the tilt of the earth on its axis. On average the earth is tilted about 23.5°, but this figure varies from about 22° to 25° over the course of 41,000 years (Fig. 6.4A). The second cycle involves the shape of the earth's orbit around the sun. It is not a perfectly circular orbit, and its degree of eccentricity varies on a cycle of about 100,000 years (Fig. 6.4B). The third cycle is called precession of the equinoxes, and it relates to where the earth is in its orbit around the sun when the solstices and equinoxes occur. If the June solstice (the North Pole tilted toward the

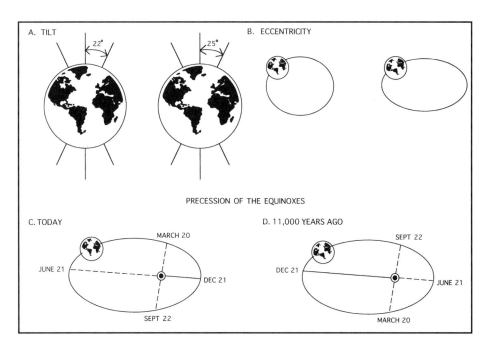

Figure 6.4. Three long-term cyclical changes in the earth's movements collectively generate a 100,000-year cycle of climate. See the text for a description. The shapes in B, C, and D are exaggerated to make the illustrations clearer. (Based on figures in Imbrie and Imbrie 1986.)

sun and the South Pole tilted away from the sun) occurs when the earth is relatively close to the sun and the December solstice (the North Pole tilted away from the sun and the South Pole tilted toward the sun) occurs when the earth is relatively close to the sun (Fig. 6.4C), then the change in climate from summer to winter will be relatively modest in the Northern Hemisphere and relatively pronounced in the Southern Hemisphere. In other words, the distance from the earth to the sun can either accentuate or ameliorate the effects of the axial tilt on the weather, depending on whether you are in the Northern or Southern Hemisphere. Currently, these factors tend to ameliorate seasonal climate change in the Northern Hemisphere, but 11,000 years ago (about half of a double cycle that has periodicities of 19,000 and 23,000 years) the June solstice occurred when the earth was relatively close to the sun, and seasonal changes in the weather were moderated in the Southern Hemisphere (Fig. 6.4D).

Together these three cycles generate a quasi-cycle of about 100,000 years that has produced eight long periods of extensive glaciation followed by brief, warmer interglacial periods during the last 800,000 years. It is easy to think of these changes in terms of temperature zones that move toward the poles during warming periods and back toward the equator during cooling periods. However, in practice the changes are not nearly that simple; three different cycles are involved, and each of these may affect temperature and precipitation patterns somewhat differently. Moreover, when we consider other factors such as variations in solar output, oceanic currents and jet streams that move thermal energy around the globe, or glaciers and CO_2 that influence the balance of solar radiation that strikes the earth versus radiant energy that is returned to space, it is easy to understand why climate changes are so complex.

The complexities of climate changes are especially obscured when we use average parameters for the whole globe, as in Figure 6.5, rather than examining local climate changes. For example, scientists, examining the trace elements and continental dust in deep ice cores from Greenland, have estimated that 11,300 years ago the climate warmed 7°C and that there was a 50% increase in precipitation in just 50 years or less (Dansgaard et al. 1989, Johnsen et al. 1992, Alley 2000). Such extraordinarily rapid changes—sometimes called climate flickers—must have been associated with some local event such as a shift in the location of the Gulf Stream that carries warm waters into the North Atlantic (Lehman and Keigwin 1992). In contrast, Figure 6.5 suggests that average temperatures for the whole globe have increased less than 7°C between the last glacial maximum and the present time.

Response of Organisms to Global Climate Change

Changes in the earth's climate during the last several million years have been much less dramatic than the changes that marked some of the

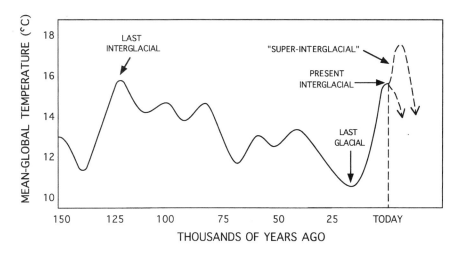

Figure 6.5. Global mean temperature record of the last 150,000 years. (Redrawn by permission from Imbrie and Imbrie 1986.)

earlier mass extinctions, and thus it is safe to say that most species have adapted to them, rather than perished. The easiest way for a species to adapt to climate change is to shift its geographic range to a new area that has the appropriate climate. The simplest response would be moving toward the poles during warming periods and toward the equator during cooling periods. This response is well documented; for example, 18,000 years ago, when glaciers extended south to southern New York, boreal forest and tundra species occurred in Virginia (Webb 1992). In mountainous areas the range shifts could be much shorter (uphill during warming periods and downhill during cooling periods) because moving 100 m in elevation is equivalent to moving about 110 km in latitude in terms of temperatures. Closer to the equator species have been affected more by changes in precipitation than by changes in temperature. In particular, equatorial climates have often been drier during glacial periods, thus shrinking the area of tropical forests and increasing the area of tropical grasslands and deserts (Maslin and Burns 2000, Mayle et al. 2000). Forest species survived such periods as members of relict forests. Even though we are now in an interglacial period and tropical forests have expanded again, some tropical ecologists have argued that they can detect where these climate-change refugia were because they still harbor high numbers of endemic species (Whitmore and Prance 1987). While this refugia hypothesis is not holding up to close scrutiny, at least in tropical regions (Bush 1994, Willis and Whittaker 2000), the effects of climate change on genetic patterns and speciation are of growing interest (Hewitt 2000).

It is not hard to envision whole sets of species—forest communities, grassland communities—shifting across the globe in response to climate

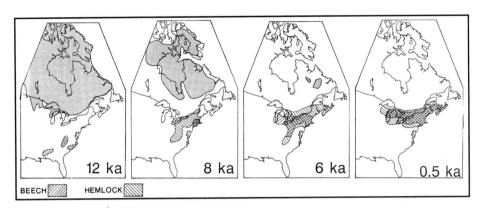

Figure 6.6. Changes in the geographic ranges of American beech and eastern hemlock indicate that these two species are responding to their environment independently of one another. Ka = 1000 years ago. (From Hunter et al. 1988 and Hunter 1990 as redrawn from Jacobson et al. 1987. Reprinted by permission of Prentice-Hall, Englewood Cliffs, New Jersey.)

change, but this image is a bit too simplistic. If we look beyond the distribution of communities and examine the past distributions of individual species, a more complex pattern emerges. In general, most species responded to climate change individualistically, not in lockstep with other species. For example, Figure 6.6 shows that two species that co-occur widely today, American beech and eastern hemlock, did not overlap 12,000 years ago when the climate was different (Jacobson et al. 1987). Conversely, three species with widely divergent ranges today—the black-tailed prairie dog, northern bog lemming, and eastern chipmunk—co-occurred in some areas 23,000 years ago (Graham 1986). If climate change were a simple matter of warming and cooling, species might be more likely to have parallel responses. However, given the complexity of climate change, relatively few range shifts are likely to be closely correlated, especially during very rapid climate flickers (Roy et al. 1996, Bartlein et al. 1997). Likely exceptions to this rule would include parasites and their hosts or herbivorous insects and their preferred plants.

Prospects for Future Climate Change

When we look at the climate record of the last 150,000 years (see Fig. 6.5), some pivotal questions emerge. Will the current interglacial period end soon? Will human alteration of the earth's climate lead us into a "super-interglacial" period? The only thing that is certain is that the climate will continue to change. Predicting when the current interglacial period might end is a daunting task that few people have attempted, but predicting the consequences of an increase in CO_2 has become a major enterprise for scientists. The concern about global warming and CO_2 begins with three observations (Houghton 1997). First, water vapor, carbon dioxide, methane (CH_4), nitrous oxide (N2O), chlorofluorocarbons (CFCs), ozone (O_3), and some other gases are known to allow solar radiation to penetrate the atmosphere and warm the earth's surface, but to inhibit reradiation of energy back into space. This is the so-called "greenhouse effect" and these are *greenhouse gases*. Second, atmospheric concentrations of many major greenhouse gases have been rising. In particular, atmospheric CO_2 concentrations have increased about 30% over preindustrial (ca. 1750–1800) levels, and methane concentrations have more than doubled. Third, mean global temperatures appear to be rising, by about 0.3–0.6°C since 1860.

It is widely believed that our enormous consumption of fossil fuels, generation of CFCs, and devastation of ecosystems that serve as carbon reservoirs are responsible for much of the increase in greenhouse gases and that this increase is causing global temperatures to rise. Some scientists offer other possible explanations; perhaps temperatures are rising because of astronomical factors and the warmer temperatures favor more photosynthesis and more production of CO_2. Nevertheless, almost all climatologists now believe that the first scenario, more greenhouse gases leading to warmer temperatures, is more plausible. Even if they are wrong, the obvious remedies—namely, curtailing fossil fuel consumption and destruction of ecosystems—make sense independent of their effect on greenhouse gases. If more greenhouse gases lead to warmer temperatures, what will this mean? More specifically, if CO_2 concentrations double (a probability in the next 50 years or so) and temperatures rise 1.5–4.5°C (Houghton 1997), what would be the likely impacts of this change on the earth's biota?

Will Organisms Be Able to Adapt to Future Climate Changes?

Given that the earth's current biota has experienced and survived eight glacial/interglacial cycles during the last 800,000 years, most species appear to be quite well adapted to climate change. Indeed, evidence of biotic responses to current climate change is starting to accumulate (Hughes 2000, Inouye et al. 2000). Obviously, this adaptability is good news if we are entering a period of significant climate flux. The bad news is that many species may not be able to adapt to another climate change as readily as they have to past changes for two closely related reasons. First, current populations of many species are already stressed by habitat degradation and loss, by overexploitation, and by other factors that we will discuss in the next four chapters (Myers 1992). Stressed populations are likely to be small, and therefore they have a relatively low chance of producing the dispersing offspring that are a prerequisite for a species shifting its geographic range in response to climate change. Second, because human alteration of landscapes has 1) reduced the total amount of suitable habitat for many species and 2) fragmented landscapes with roads, agricultural fields, and urban areas, the odds of a dispersing individual arriving in a suitable habitat have been reduced. Whether it is a willow seed carried on the wind or a juvenile salamander trudging along, many dispersing individuals will perish because of the long distances and inhospitable environments between patches of suitable habitat.

There are other reasons to be concerned about the ability of organisms to adapt to greenhouse warming. It is possible that global temperatures may increase to levels greater than anything most species have experienced. This would require longer range shifts, and some species might encounter geographic bottlenecks. Imagine terrestrial species living south of the equator in Africa or South America; if they shift their ranges toward the South Pole, they encounter a gradually tapering continent that terminates in the ocean. A different kind of geographic bottleneck is likely to occur along maritime shores because warmer temperatures would melt a portion of the polar ice caps, causing sea levels to rise (Harris and Cropper 1992). In many regions, shoreline species that needed to move inland would find themselves squeezed between the ocean and intensive shorefront development. Similarly, although species living in mountainous areas can move their ranges up the mountains, the mountains get smaller as you go up, and eventually they stop (Murphy and Weiss 1992). This phenomenon may be indirectly responsible for the extinction of the golden toad and the disappearance of 20 other frog species from Monteverde, a tropical mountain site in Costa Rica (Pounds et al. 1999). Warmer ocean temperatures appear to have led to a sharp decrease in the number of misty days during the dry season, and this may have forced some frog species to congregate along streams where they were more vulnerable to lethal parasites (Fig. 6.7) (Pounds and Crump 1987). Some bird species at Monteverde have apparently responded to the climate change by shifting their range upward, but the site only reaches about 1800 m in elevation, thus limiting this opportunity.

It is also conceivable that the *rate* of temperature change resulting from greenhouse warming could be far greater than the rates of change during the other climate shifts of the last 2.5 million years. Rapidly changing temperatures would further tax the abilities of organisms to move their geographic ranges. This issue highlights the need to think of different species individually because their relative mobility differs greatly and, sometimes, in ways that are hard to predict (Clark et al. 1998). Some species are quite mobile between generations but sedentary as individuals; for example, wind-dispersed plants and spiders that travel long distances as seeds or ju-

Figure 6.7. Inhabitants of tropical mountains such as this species of harlequin frog may be particularly sensitive to global climate change.

TABLE 6.2 The ability of organisms to shift their geographic ranges depends on their mobility, both as individuals, and, more importantly, between generations.*

Mobile Between Generations	Sedentary Between Generations
Mobile as individuals	
Migratory, early-successional birds	Migratory, philopatric birds
Insects of ephemeral ponds	Insects of deep lakes
Pelagic fishes	Anadromous fishes
Sedentary as individuals	
Territorial fishes with planktonic larvae	Desert-spring fishes
Early-successional plants; self-incompatible annuals	Late-successional plants; self-compatible perennials
Intertidal molluscs	Terrestrial molluscs

*Compiled with George Jacobson.

veniles, but then stay put for life. Some are relatively mobile as individuals but sedentary between generations; for example, animals that migrate annually, but always return to their natal area to breed such as many salmon species. Some are sedentary both as individuals and generations such as many plants that reproduce vegetatively (Table 6.2).

Summary

The history of life on earth provides both good news and bad for conservationists. The good news is that the diversity of life has been generally increasing despite the fact that extinction is a natural, ongoing process and, occasionally, huge numbers of species have become extinct in a short time. The bad news is that recovery from these mass extinctions takes millions of years; thus if we are in the midst of a human-induced mass extinction now, it is unlikely that any humans will survive to see the recovery. The earth is probably experiencing a mass extinction event currently, but evidence for this does not come from tallying species extinctions because it is nearly impossible to document an extinction event with certainty. This is especially true for species that have never been described. Most estimates of current extinction rates are based on 1) estimates of the numbers of species found in tropical forests, 2) estimated rates at which tropical forests are being lost, and 3) a predicted relationship, based on island biogeography, between the number of species and the area of their habitat.

Global changes leading to mass extinctions may be relatively infrequent, but changes in global climates are occurring constantly. These changes are mainly caused by cyclical changes in the earth's movements around the sun, but recent human activities (e.g., adding greenhouse gases to the atmosphere and degrading the ability of ecosystems to store carbon) are now playing a major role. Species have generally been able to adapt to moderate climate change by shifting their geographic ranges, but under current conditions (small, stressed populations and fragmented landscapes) some species may be unable to shift their ranges and thus may become extinct.

FURTHER READING

Stanley (1987) and Raup (1991) provide very readable accounts of past extinctions; Stanley is better for descriptions, and Raup is better for explanations. See Hallam and Wignall (1997) for more details. Wilson (1992) gives a concise account of past extinctions and a fuller description of his estimate for the current rate of extinction. For a book on rates of extinction, past and current, see Lawton and May (1995), and for a short synthesis, see Pimm (1998). See Imbrie and Imbrie (1986) for the history of ice ages and climate change and Peters and Lovejoy (1992) and Gates (1993) for papers on the effects of climate change on biodiversity. For a comprehensive treatment on global warming, see Houghton (1997); for a more popularized one, see Philander (1998). The lead organization on climate change issues is the Intergovernmental Panel on Climate Change (IPCC) (their website is www.

ipcc.ch). Also see www.nacc.usgcrp.gov for the United States government's approach to climate change and www.climatehotmap.org for a site maintained by several environmental groups. For more on recent extinctions visit the website of the Committee on Recently Extinct Organisms: creo.amnh.org/index.html.

TOPICS FOR DISCUSSION

1. How do you feel about the idea that the earth's biota will probably eventually recover from a human-induced spasm of extinctions after humans are extinct? Is this idea more depressing or consoling to you?

2. The current rate of extinction is probably much higher now than normal, background rates, but this is only half the equation. Is the rate at which new species are evolving likely to be greater or less than normal?

3. How would you allocate conservation funds between 1) efforts to ameliorate human effects on concentrations of greenhouse gases and 2) programs to help biota adapt to climate change?

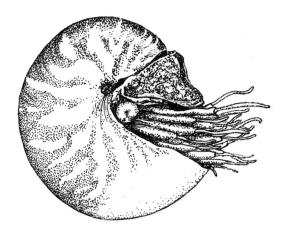

Chapter 7
Extinction Processes

Some species are survivors. Even doomsday images of a world laid barren by a nuclear apocalypse usually portray a few inhabitants—perhaps some cockroaches and dandelions. In contrast, some species seem quite fragile. Consider the dodo, a bird that has become a symbol for extinction, as in "dead as a dodo." The dodo evolved on Mauritius, an Indian Ocean island so remote that no humans inhabited it until 1644. Yet, by about 1681 the dodos were gone, victims of hungry colonists and passing sailors who found the birds tasty (they were like giant pigeons) and easy to catch (evolving in the absence of predators they had become flightless and tame) (Halliday 1978). Of course, the phrase, "dumb as a dodo," is unfair to a species that was probably well adapted to its environment, but it does convey the idea that the dodo was ill-prepared for a significant environmental change.

In this chapter we will seek to understand why some species like the dodo are so vulnerable to extinction, especially human-induced extinction. We will begin with some general observations about characteristics shared by vulnerable species; next, we will review some basic ideas about population structure; and, finally, we will describe a technique, population viability analysis, used by conservation biologists to analyze the vulnerability of small populations.

Why Are Some Species More Vulnerable to Extinction Than Others?

The simple answer to this question is that some species are rarer than others. This answer is intuitively obvious, and there are some convincing data to support it (Fig. 7.1). However, this answer does not increase our

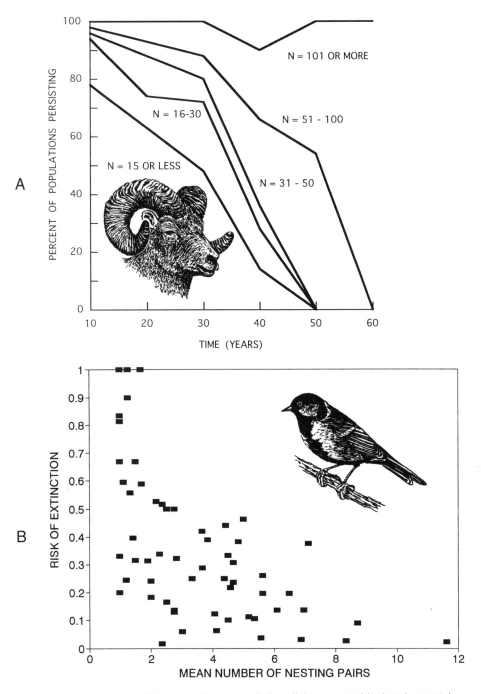

Figure 7.1. (A). A study of bighorn sheep populations living on semi-isolated mountain ranges (primarily in deserts of the southwestern United States) demonstrated that larger populations were likely to persist longer than smaller populations. All populations of less than 50 were extinct within 50 years. (Also see Wehausen 1999 and Berger 1999.) (B). Similarly, a study of the risk of extinction for populations of 62 bird species on 16 small islands off the coast of Britain and Ireland indicated that, in general, species with smaller populations were at greater risk. In this figure, risk of extinction is the reciprocal of the average time to extinction in years. (Also see Tracy and George 1992, Duncan and Young 2000, Eisto et al. 2000, and Vucetich et al. 2000.) (Part A redrawn from Berger 1990; Part B redrawn by permission from Pimm et al. 1988.)

understanding of extinction processes very much. To dig deeper we will address this issue as three subquestions.

Why Are Some Species Rarer Than Others?

This question returns us to Chapter 3, Species Diversity, and our discussion of the three ways in which a species can be rare (Rabinowitz et al. 1981, 1986). Briefly, these were 1) restriction to an uncommon type of habitat, 2) limitation to a small geographic range, and 3) occurrence only at low population densities. Let us examine each of these further. First, some species are restricted to a rare type of habitat because they have evolved special characteristics that allow them to live there and nowhere else; blind, unpigmented cave-dwelling invertebrates, fishes, and amphibians are good examples of this (Culver et al. 2000). Alternatively, some species are probably found in rare habitats primarily because they cannot compete successfully elsewhere. Consider, for example, the various plant species (e.g., McDonald's rock-cress) that are restricted to serpentine rock outcrops where they must tolerate high levels of magnesium. It is likely that most of these species could live in normal soils, but do not because they cannot compete well with other plants in normal soils.

Second, many rare species are confined to small ranges by geographic barriers such as islands surrounded by ocean, or lakes surrounded by land. For example, over 500 species of cichlid fishes (some researchers have estimated 1000) are endemic to Lake Malawi in Africa (Keenleyside 1991). In some cases barriers may be subtle (e.g., a change in temperature or the presence of a competitor), but these can still restrict the range of a species with narrow tolerances (Huston 1994). For example, Gentry (1986) suggested that many Amazonian plant species with small geographic ranges may have evolved in areas with special soil conditions and have not been able to expand their ranges across other soil types.

Third, species may occur at low population densities for a variety of reasons. Body size is a key reason because, all other things being equal, a large organism requires more space than a small one. This is most conspicuous when you consider the extensive home ranges of large animals, but it also applies to plants: you can fit far more lilies than oaks on a single hectare. Organisms may also live at low population densities if the resources they require are scarce and dispersed. The classic examples are carnivores that commonly live at low densities because they are at the apex of their food web and thus must travel over relatively large areas to obtain food. Pelagic marine carnivores that move across vast stretches of ocean such as bluefin tuna are perhaps the best example. Some plant species may occur at low densities because they have a higher fitness when they are not competing with nearby members of the same species (Rabinowitz et al. 1984). Note that numerical abundance and biomass are not the only ways to measure evolutionary success. If we measure success in terms of evolutionary longevity, some rare species may be quite successful in the long run, even though they are not very successful in terms of their current abundance.

Most rare species will fall into only one of these three categories, but some will fit two, and a few may survive despite being rare in all three ways (Fig. 7.2).

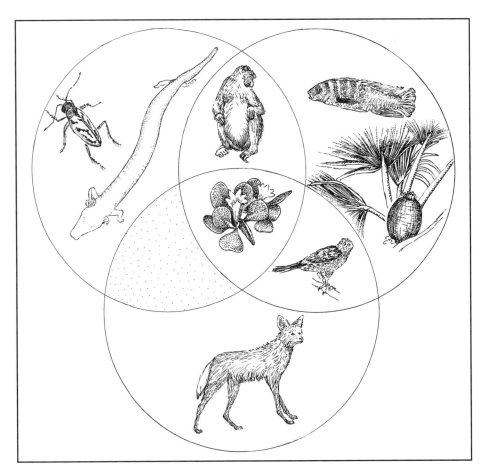

Figure 7.2. There are three basic ways that a species can be rare. Some such as the olm (a cave salamander) and the northeastern beach tiger beetle are confined to rare habitats (*left circle*). Others have small geographic ranges (*right circle*) such as *Pseudo-tropheus heteropictus*, a cichlid fish from Lake Malawi, and the coco-de-mer from two small islands in the Seychelles archipelago. Still others such as the maned wolf occur at low population densities, often because they are large or require resources that are widely dispersed (*bottom circle*). Some are rare in more than one respect. For example, the proboscis monkey lives in mangrove swamps on the island of Borneo, and the Hawaiian hawk lives at low densities on the island of Hawaii. The dwarf naupaka numbers about 350 individuals in four populations growing on beach dunes on the Hawaiian island of Maui. (See Pitman et al. 1999 and Ricklefs 2000 for a recent analysis of these patterns among tropical forest trees in Peru.)

Why Are Rare Species Usually More Vulnerable to Extinction Than Common Species?

The first and probably most important answer to this question is that a rare species has a greater chance of being pushed into extinction by an environmental change than a common species. This is particularly true of species with small geographic ranges, because an environmental event may encompass the species' entire range whether it is a specific catastrophe (e.g., a volcano eliminating an island) or a gradual change (e.g., immigration of a competitive species). About three-quarters of all the animal species known to have become extinct since 1600 were island species (Jenkins 1992). Similarly, species that are confined to a very specific type of habitat may be vulnerable to environmental change. A study of plant species' persistence in Swiss grasslands showed that the habitat specialists were much more vulnerable to extinction than the habitat generalists (Fischer and Stöcklin 1997). Demographic problems can also lead to extinction of small populations; for example, an unbalanced sex ratio can limit the birthrate severely. Finally, small populations, especially ones that have recently become small, are likely to suffer from the genetic problems discussed in Chapter 5: genetic drift, inbreeding, and bottlenecks. Similarly, over long periods the lack of genetic diversity in a rare species may restrict its ability to adapt to a changing environment. We will return to a more detailed review of these factors in our discussion of population viability analysis below, and in the next three chapters we will discuss environmental change in depth.

Why Are Some Species Particularly Sensitive to Human-Induced Threats?

This question is necessary because population size and distribution are not perfect predictors of a species' vulnerability to extinction, especially when human impacts are involved. In 1813 when John James Audubon camped on the shore of the Ohio River, watching a flock of passenger pigeons that stretched from horizon to horizon and took 3 days to pass, he could not have guessed that just 70 years later the species would be decimated and by 1914 extinct (Schorger 1973). Conversely, consider another species from the same region, the Virginia round-leaf birch, which was so rare that it remained undiscovered until 1914 when four individuals were found (Preston 1976). Despite this precarious state—it was lost by scientists from 1914 to 1976—it persists today. Here are four primary characteristics that tend to predispose a species, even one that is not necessarily rare, toward problems with people.

1. Limited adaptability and resilience. Some species have a limited ability to adapt to change or to recover from a disturbance because of their low reproductive capacity (e.g., small number of progeny, long generation

Figure 7.3. The ability of species to survive in the face of environmental change is often correlated with their reproductive capacity.

time, etc.), limited dispersal capabilities, inflexible habitat requirements, and so on. Contrast a humpback whale that can produce only one young every 2 years with various insects (Fig. 7.3). For example, a female fruit fly can lay 100 eggs and have 25 generations per year, theoretically leading to 10^{41} progeny in 1 year. (Figure out how many times a line of 10^{41} fruit flies would reach to the sun and back with one fly for every 2 mm of the 150,000,000 km.) Adaptability is not just an issue of having a large reproductive capacity; some species such as house sparrows and dandelions are able to flourish in our cities, suburbs, and farms simply because their particular physiology, morphology, behavior, etcetera allow them to do so. Unfortunately, they are greatly outnumbered by species whose inflexible habitat requirements, sensitivity to predation or competition, and so on leave them with a limited ability to cope with major human-induced changes.

2. Human attention. Some species suffer because they are singled out for attention from people. In the case of dodos, passenger pigeons, and many other species, being deliciously edible was their Achilles' heel. On the other hand, some species are persecuted because they are very unpopular. Witness what happens to most bats, snakes, spiders, and wild canines (especially, wolves, African wild dogs, and dholes) when they are unfortunate enough to have a close encounter with a human.

3. Ecological overlap. Many species are threatened with extinction because they are tied to the types of ecosystems preferred by people. Humans have thrived in places with fertile soils and benign climates, and organisms that are restricted to these sites have usually lost out to agriculture and cities (Dobson et al. 1997, Wilcove et al. 1998, Duncan and Young 2000). For example, environments that support tallgrass prairies make wonderful farmland, and now the native biota of these ecosystems is often restricted to a handful of overlooked sites like railroad rights-of-way and unmanaged cemeteries (Breymeyer 1990). Similarly, rivers are focal points of human activity because they provide water, transportation corridors, waste disposal, and hydroelectric facilities, and as a consequence, many riverine species (especially, fishes and mussels) are in great jeopardy (Wilcove and Bean 1994, Richter et al. 1997, Wilcove et al. 1998, Dudgeon 2000).

4. Large home-range requirements. Conflicts caused by overlapping habitat will be exacerbated if the organism requires large areas of land for a home range. It is one thing for some asters to find a few square meters of suitable habitat in a human-dominated landscape; it is something else for a wolf pack to find hundreds of square kilometers. Of course, this factor cannot be readily separated from the fact that animals with extensive home-range requirements tend to be rare (i.e., have low population densities) and large enough to attract human attention.

Some species may not fare well in their contact with people in part because they have had little time to adapt to humans and all the challenges they bring: notably, overexploitation, habitat degradation, and introductions of exotic species. This is particularly likely to be true of species inhabiting remote islands that have only very recently (in an evolutionary time frame) been colonized by any large mammal, human or otherwise.

Populations

Up until now we have been discussing species as though they were composed of a single population. It is not that simple. Different species have different population structures, and these have considerable bearing on their vulnerability to extinction. Let us begin with a very broad definition of *population:* a group of individuals of the same species occupying a defined area at the same time. Often the area, and thus the population, are arbitrarily defined by the boundaries of a researcher's study area or by a political unit (e.g., Arizona's population of saguaro cacti). Defining the area in terms of ecological boundaries important to the species in question will make the definition somewhat less arbitrary (e.g., a pond and its population of perch).

Ideally, biologists would usually prefer to define populations with respect to their demography or genetics. In terms of demographics, a sound definition would be as follows: a group of interacting individuals of the same species whose population structure (i.e., age and gender) and dynamics (e.g., mortality and natality) are relatively independent of other groups. From a population genetics perspective, we could distinguish two groups as separate populations if one group has an allele not shared with the other group. Alternatively, we could use the overall distribution of genetic diversity (recall $H_t = H_s + D_{st}$ from Chapter 5, section on Heterozygosity). If D_{st}, the variability among populations, is above some threshold we could call two groups separate populations. There is no widespread rule of thumb for such a value of D_{st} (or its analogues F_{st} or G_{st}), but the work of Sewall Wright (1978) suggests that it would be at least 0.05 or 5% of the overall diversity.

Whether you define populations demographically or genetically, a key issue is how much movement or interchange (typically, by the dispersal of juvenile animals or plant propagules) there is between two groups of organisms; with less interchange it is more likely that two groups will be separate populations (Fig. 7.4). It is generally accepted that exchange has to be very low, less than one breeding individual per generation, to allow two groups to develop unique alleles (Kimura and Ohta 1971). At a somewhat higher rate of exchange, you would find no unique alleles, but would be likely to see differences in the frequency of alleles that would be reflected in the value of D_{st}. Finally, even higher rates of exchange would be necessary to allow two groups to have the same demographic features. Unfortunately, we do not have a good understanding of what these thresholds of exchange

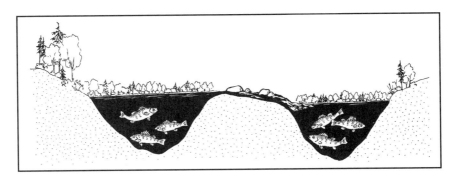

Figure 7.4. If we use an area-based definition of population, the perch in the two ponds are readily recognized as separate populations. From a population dynamics perspective, the perch will be separate populations if interchange is so limited that the populations have different levels of mortality, natality, etcetera. Using a reproductive isolation definition, we can define all the perch in both ponds as a single population as long as at least one breeding individual per generation is exchanged between the two ponds.

are, and, indeed, they are likely to vary considerably among different species (Mills and Allendorf 1996).

This focus on exchanging individuals brings us to an important topic, the spatial structure of populations. Simply put, the chances of two groups being a single population through the exchange of individuals is lower if two groups are far apart and separated by an inhospitable environment. In the next section we will examine what this means from a conservation perspective after reviewing some basic concepts.

Patchy Distributions and Metapopulations

Consider the broad-leaved cattail. This is a species that is found throughout much of the Northern Hemisphere, but it occurs only in discrete patches of habitat—certain types of freshwater wetlands—that are usually only a small portion of the overall landscape. Within their patches of habitat, cattails are often exceedingly abundant, but between these patches there are large stretches of land without any cattails. Cattails are a good example of an attribute that is common to many species: patchy distributions. Patchy distributions are the basis for a model of population structure that has attracted considerable attention in conservation biology circles: *metapopulations.*

In metapopulation terms, each patch of habitat contains a different population of the species in question, and a group of different patch populations is collectively called a metapopulation. To put it another way, a metapopulation is a "population of populations" (Hanski and Gilpin 1991, 1997; Hanksi 1998). Metapopulations exist at a spatial scale where individuals can occasionally disperse among different patches, but do not make frequent movements because the patches are separated by substantial expanses of unsuitable habitat. This rate of movement is usually sufficient to avoid long-term genetic differentiation among patches, but low enough to allow each patch to be quite independent demographically. To avoid some of the ambiguity surrounding demographic versus genetic definitions of populations, groups within patches are often called subpopulations or local populations.

It is important to recognize that not all species that are distributed in habitat patches are composed of metapopulations (Harrison 1994, Hanski and Simberloff 1997). In many cases, perhaps most, the frequency of movement among patches is so great that there really is only one large population with no meaningful subdivision of the population. This is almost certainly the case for highly mobile species, for example, most birds. This point is often obscured by the fact that some of the language of metapopulations—to be presented in the rest of this section—is often used for patchily distributed populations that are not metapopulations (see Dono-

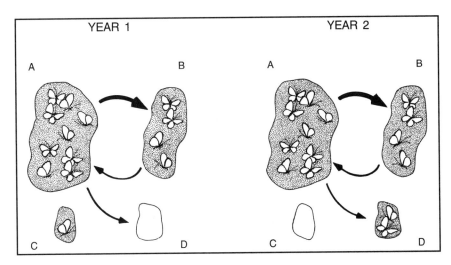

Figure 7.5. A schematic depiction of metapopulations. Occupied patches are shaded; empty ones are unshaded. Arrows represent movement among patches with the width of arrow corresponding to the number of dispersers. Patch A is a source of butterflies because it is a net producer of emigrants, while patch B is a sink because it is a net recipient of immigrants. The butterfly subpopulation in patch C has become extinct, while in patch D a new subpopulation has begun to develop from dispersers that have colonized the patch. Patch A is probably a core subpopulation because of its size and persistence, whereas C and D are satellites. We would need data from more years to say if B is a core or a satellite.

van et al. 1995 and Vierling 2000 for examples of birds in fragmented forests and marshes).

We can summarize the last two paragraphs by thinking in terms of three levels of movements among patches. At high rates of interchange among patches there is effectively only one population occupying all the patches. If there is very little or no movement among patches, then each patch is occupied by a distinct population. At intermediate rates of movement, the patches are occupied by a metapopulation composed of many subpopulations.

Our brief examination of metapopulation dynamics will focus on two types of subpopulations—sources and sinks—and two processes—extinctions and colonizations (Figs. 7.5 and 7.6). Some subpopulations are *sources* because they produce a substantial number of emigrants that disperse to other patches. Some subpopulations are *sinks* because they cannot maintain themselves without a net immigration of individuals from other subpopulations. In other words, some subpopulations are saved from extinction by immigration from other subpopulations; this process has been called the *rescue effect* (Brown and Kodric-Brown 1977). Sinks and sources are useful

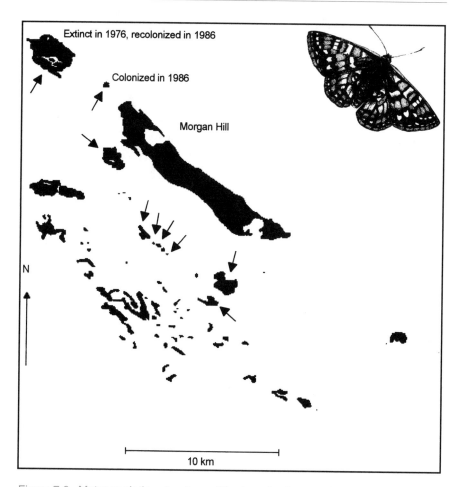

Figure 7.6. Metapopulation structure of the bay checkerspot butterfly. The large shaded area, Morgan Hill, supports a persistent core subpopulation of roughly 100,000 individuals. All the smaller shaded areas denote potential habitats of serpentine soil, but only those marked with an arrow supported subpopulations, numbering tens to hundreds, in 1987. The proximity of these subpopulations to Morgan Hill suggests that they are satellites, and probably sinks, supported by dispersal from the Morgan Hill source subpopulation. A metapopulation perspective has also influenced conservation biology research for species such as Furbush's lousewort (Menges 1990) and the spotted owl (Lamberson et al. 1994). (Redrawn by permission from Harrison et al. 1988.)

concepts, but in practice it is difficult to distinguish them with confidence because it is difficult to monitor the movements of individuals among subpopulations (Hoopes and Harrison 1998). Moreover, a population that is a source one year may be a sink the next year, or vice versa, especially if environmental quality (e.g., food availability) changes.

Despite the balancing effect of immigration and emigration, subpopulations sometimes appear and disappear in a manner often compared with small lights winking on and off in a dark expanse. More formally, these appearances and disappearances are called *turnover* (Hanski and Simberloff 1997). Each appearance is a *colonization* event; for example, when a species of grass colonizes a forest opening after a tornado creates the opening. Each disappearance is a *local extinction* event; for example, when the bullfrogs in a pond are killed by a disease. These processes occur at an ecological time scale and may be quite rapid (e.g., a windstorm drops a swarm of spiders and seeds to colonize a recently burnt grassland) or quite slow (e.g., after the burn, an annual plant species restricted to recently burnt grasslands gradually disappears). These events may be interwoven with the whole pattern of disturbance and succession that operates in a given ecosystem (e.g., the spiders and plants), or they may affect only one or a few species (e.g., the bullfrogs and their pathogens). Subpopulations that persist for relatively long periods are often called *core* subpopulations, whereas those that are more likely to wink on and off are often called *satellite* subpopulations (Boorman and Levitt 1973). Core subpopulations are likely to be large and a net source of individuals, and satellite subpopulations are likely to be small and a net sink, but, undoubtedly, there are exceptions to these generalizations.

To the extent that these small-scale extinction and colonization events are reasonably in balance with one another, they need not worry conservation biologists. However, to state the all-too-obvious problem, the rate of subpopulation extinctions may often exceed the rate of colonizations in the lands and waters dominated by people. In fact, the metapopulation model is particularly applicable to one common form of ecosystem degradation—habitat fragmentation—that we will discuss at length in the next chapter (Hanski 1998, 1999). Habitat fragmentation can reduce vast ecosystems to small, isolated patches and thereby subdivide species that once had large, regionwide populations into much smaller groups. If the species has reasonably good dispersal abilities, these groups may persist as a metapopulation; if it does not, it may disappear from all the habitat patches one by one (Templeton et al. 1990). Even when habitat patches are naturally small and isolated, fragmentation can further reduce the sizes of habitat patches and increase the distances between them, thus making subpopulations smaller, more isolated, and more vulnerable to extinction (Wahlberg et al. 1996). It has also compelled biologists to focus more attention on dispersal, a difficult-to-study process that has traditionally been overshadowed by studies of natality and mortality (Hanski 1999).

In sum, the metapopulation concept offers a useful framework for understanding the dynamics of populations in patchy landscapes, and patchy landscapes are becoming more and more common because of human activity. This said, it would be a mistake to assume that all populations of threatened species conform to the metapopulation concept, or to assume that universal conservation rules can be derived from metapopulation theory (Harrison 1994).

Population Viability Analysis

Conservation biologists often ask, "What is a *minimum viable population?*" Or, more fully: "For this particular population, what is the smallest that it can be and still have a reasonable probability of surviving for sometime into the future?" If you believe in a literal interpretation of the Bible and the story of Noah, the answer is two, one male and one female. However, population biologists have shown that Noah would have to have been extremely lucky or to have enjoyed long-term divine intervention, because most populations this small are doomed to extinction. The technique used to estimate minimum viable populations (MVP) is called population viability analysis, often abbreviated PVA (Soulé 1987). In a general sense, any systematic attempt to understand the processes that make a population vulnerable to extinction could be called a PVA. In practice, the term usually refers to using models to predict the likely fate of a population (Beissinger and Westphal 1998). At their simplest, these models are deterministic predictions of what will happen to a population that has certain rates of natality and mortality. (Box 7.1 begins with an example of a deterministic model; see Doak et al. [1994] and Silvertown et al. [1996] for examples with tortoises and plants, respectively.) The most complex PVA models incorporate metapopulation dynamics, are tied to the particular spatial distribution of the population being modeled, or both (e.g., Lindenmayer and Lacy 1995, Boone and Hunter 1996, Lindenmayer et al. 2000, respectively). Here we will focus on the most common form of PVA models, those that focus on a single population and incorporate an element of stochasticity, or randomness.

To understand the stochastic approach to PVA you have to appreciate the role of probability in the extinction of populations; in many respects PVA evolved out of risk assessment, which is based on probabilities (Burgman et al. 1993). Recall from Chapter 6 that sooner or later all populations become extinct; only *when* and *why* are left to chance. Let us start with "when" by considering two generic predictions. First, the smaller a population is, the greater the probability that it will become extinct in a given span of time. Conversely, the longer the time period being considered is, the greater the probability that a population of a given size will become extinct. Conservation biologists translate these ideas into real-world predictions that usually take one of two forms: 1) a dodo population needs to have at least x individuals if it is to have a 95% probability of surviving for 500 years;

or 2) a population of 25 dodos has an x% probability of surviving 500 years if current conditions persist. (The figures 95% and 500 years are somewhat arbitrary; 90% and 99%, 100 and 1000 years are also used commonly. In reality these models are not nearly accurate enough to distinguish, with confidence, between a 90% and 99% probability of extinction over such long periods.) One key objective of PVA is to replace these x's with good predictions. Another objective is to understand "why" the population will become extinct—what factors will be responsible—because this will give conservationists some guidance on how to direct their management.

Mark Shaffer (1981) identified four interacting factors or processes that might contribute to a population's extinction. He referred to these as stochasticities—uncertainties—to emphasize that they were based on probabilities:

1. Demographic stochasticity is uncertainty resulting from random variation in reproductive success and survivorship at the individual level. The importance of demographic issues is best illustrated with a simple hypothetical example. If 95% of a population of 10,000 frogs were killed by a disease, the 500 remaining could probably rebuild the population over several years. In contrast, if 95% of a population of 100 frogs died, there is a fair chance that the remaining five frogs might be all males, perhaps because females were more susceptible to the disease. Alternatively, they might be two large females and three males too small to mate with them effectively. You can imagine many scenarios; the point is that with very small populations there is a fair chance that extinction will occur simply because of vagaries in the age and sex structure of the population.

In some species it is not sufficient to have a balanced age and sex structure; apparently, there must be a fairly large number of individuals to provide enough social stimulation for reproduction. This is called the *Allee effect* after Warder Allee (1931), who, having noted that species such as red deer and starlings thrive only in social groups, suggested that they require the stimulation of a group to breed. It has been suggested that the extinction of the passenger pigeon, which often nested in huge colonies, may have been hastened because natality dropped after populations became too low to provide enough social stimulation (Schorger 1973). For some species, issues such as group defense against predators or efficiency in finding food may also explain a need for group living. An extreme case of the need for sociality could occur if a species became so rare and widely dispersed that individuals had difficulty finding one another; this may be an issue for whales that travel over immense spans of ocean. It is easy to envision a rare plant (of a species incapable of self-fertilization or vegetative reproduction) failing to reproduce because no pollen ever arrived from another plant (Bawa and Ashton 1991).

BOX 7.1   **Population viability analyses**
James P. Gibbs[1]

Conservation biologists managing threatened populations are frequently confronted with questions such as: "How many tortoises should be maintained in a particular reserve to ensure that a population will be thriving 100 years from now?" "How many caribou should be released on this mountaintop to successfully reestablish a population?" "What recovery objective should be set for this endangered orchid?" Answers to these fundamental questions are rarely intuitive; many variables interact to determine the size of a population and how long it might persist. Often, the only way to gain insight is to develop a mathematical model of the population and to use it to perform a *population viability analysis* or *PVA*.

A population viability analysis is based on a model that relates a *dependent variable* (such as population size) to the *independent variables* that influence it (such as weather, harvest levels, mortality, etc.). The relationship between independent and dependent variables is mediated through the model's *parameters* (such as survival rates and reproductive rates of individuals). In this way, the model permits us to ask whether, for example, a population will rebound if poaching is limited, or if it will be more secure in the future if 200 rather than 100 individuals are reintroduced to an area.

What sets PVA apart from other types of population models? PVA integrates both the magnitude of the model parameters *and* the amount that they vary over time and space. In other words, PVA embraces rather than ignores the variability that we observe in nature, something of great importance to the fates of small populations. To do this, PVA generally involves three steps. First, a single population projection is made over a specified period. The population size at any given time step is a function of both the population size at the previous time step and of values drawn at random from distributions of numbers that describe a model's parameters. For example, mortality rates may vary about a certain average value, but will actually be higher or lower than the average in any given year. The PVA therefore selects a value for mortality at random from the full range of possible values at each time step. Similarly, it does this for all the other parameters in a model. Accommodating natural variability in this fashion provides the realism that makes PVA so useful for studying small populations. The second step of a PVA involves making many such projections (typically 500 or more). Each projection is, of course, unique, usually terminating at different points. The last step in PVA is then to calculate the proportion of all the projections made for which the population reached a certain threshold. Thus, a prediction from a PVA generally has three elements: a population threshold (often zero), a probability (from 0 to 1, or 0% to 100%) that the population will reach that threshold, and an interval of time to which the

[1] State University of New York, College of Environmental Science and Forestry, Syracuse, New York

BOX 7.1   *Continued*

prediction pertains. In aggregate, all the projections provide a good sense of the range of possible fates of the population. This is what PVA is used for primarily—estimating the chance that a population will rise above or below some level under different conditions given the natural variability in the system. It is therefore a specialized form of risk analysis.

An example. Consider the following model of a population in which the total number of individuals (N) can change over a discrete interval of time (from t to $t + 1$) only because of births and deaths in the population

$$N_{t+1} = (N_t \times S) + (N_t \times B \times S)$$

where S = the probability of an individual surviving from t to $t + 1$ and B = the number of offspring produced on average per individual at each time interval. The ($N_t \times S$) component of the equation represents the survival of adults from one time step to the next, and ($N_t \times B \times S$) represents the production of offspring and their subsequent survival.

If $N_t = 10$, $S = 0.5$, and $B = 1$, then the population will remain stable at 10 individuals no matter how far we project it into the future (i.e., $10 = [10 \times 0.5] + [10 \times 1 \times 0.5]$). Also, its probability of extinction is zero (Fig. 7.7). Because this model's parameters do not vary, it is termed a *deterministic model* and it always provides a single, discrete prediction. We know that population parameters are not fixed, however, so this prediction is not very useful. We need to add some elements of variation to the model to make it a *stochastic model* and thus a more realistic assessment of the population's future.

We can first add an element of demographic stochasticity. Rather than simply multiplying the whole population by the survival value (the "average" expectation), we can examine the fate of each individual in the population. At each time step and for each

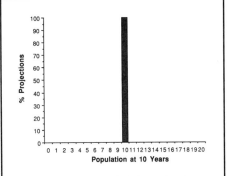

Figure 7.7. Deterministic projections.

individual, we can generate a random number (a "uniform" random number, scaled from 0 to 1, in which each value between 0 and 1 has the same likelihood of being sampled), and we can compare it with the survival value of 0.5. If the random number is ≥ the survival value, the individual lives. If it is < 0.5, the individual dies. (The same process could be simulated simply by flipping a coin.) Under these conditions 5 individuals from an initial 10 will survive in most cases, but frequently 6 or 4 will survive, sometimes 7 or 3, occasionally 8 or 2, rarely 9 or 1, and very rarely 10 or 0. These are the chance or stochastic events

BOX 7.1   *Continued*

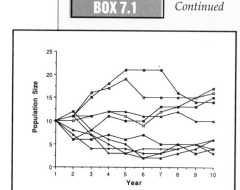

Figure 7.8. Ten projections with survival-related stochasticity.

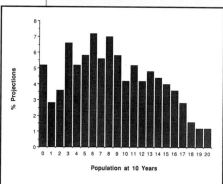

Figure 7.9. Projections with survival-related stochasticity.

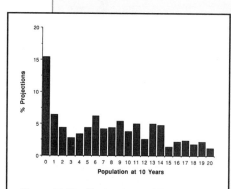

Figure 7.10. Projections with survival- and gender-related stochasticity.

that actually happen in small populations. Next we make 500 separate projections of the population over a 10-year period, the first 10 of which are shown in Figure 7.8. When we add this element of stochasticity, each population projection is different. After 500 such projections, we can look at the frequency of ending population sizes at 10 years (Fig. 7.9) and determine that the probability of population extinction is about 5%. In other words, some 25 projections, of the 500 made, fell to zero by year 10.

Let us add a second source of demographic stochasticity in this small population: gender. We can again use uniform random numbers to determine which individuals are females and which are males. We will assume a sex ratio of 1:1. If a uniform random number (from 0 to 1) is ≥ 0.5, then the individual born is a female; otherwise, it is a male. Similarly, we will assume that only females produce offspring and that they do so only if there are some males alive. This addition of a gender-related stochastic process further increases the population's probability of extinction at year 10 to about 15% (Fig. 7.10).

Finally, we can consider a catastrophic event that occurs on average every 10 years (annual probability of 0.1) and that decimates reproduction when it occurs (i.e., $B = 0$). Uniform random numbers can be used here again: if a uniform random number sampled in a particular year is < 0.1, then the catastrophe occurs; otherwise, it is a normal year, and reproduction remains unchanged. The combination of stochastic demographic forces and occasional catastrophic reductions in reproduction increases the probability of extinction at 10 years to 35% (Fig. 7.11).

To pursue this example, we could add other model parameters and develop a fairly realistic model of a particular population. Manipulation of the values of those parameters would permit an examination of how different

BOX 7.1   *Continued*

management activities might affect the population's persistence. Through such manipulations, we could also identify the key parameters that influence population growth. This is termed a *sensitivity analysis* and is useful for targeting management and research efforts (Lindenmayer et al. 1993).

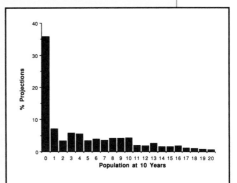

Further Considerations

Without PVAs, many insights about threatened populations are unavailable to us. Population processes are generally too complicated for biologists to understand without the use of such models to synthesize information about all the potential influences acting simul-

Figure 7.11. Projections with survival- and gender-related stochasticities and reproductive catastrophes.

taneously on a small population. Predictions from a PVA are, however, only as reliable as the logic used to construct the population model and the validity of the estimates of the model's parameters. Meager information is available on the natural history of many species, especially those threatened with extinction, and deriving estimates of the key parameters of population models is often quite problematic. Also, we often do not know how the different demographic parameters interact. For example, how does population size affect inbreeding and thereby influence reproductive rates, which might, in turn, affect population growth rates and population size in a form of feedback loop? Biologists are only now learning about such interactions. Also, we must be cautious about the period over which populations are projected. Because we cannot know the state of the environment even 100 years from now, making population projections over a modest interval (10 to 50 years) is most prudent. Finally, PVA should be regarded primarily as a tool for guiding research, management, and policy and for synthesizing knowledge about a species. PVA is an extraordinary tool for understanding how a population works and what influences its ups and downs. There are, however, generally too many uncertainties about the details of the models to permit using them to make definitive statements about the precise fate of particular populations.

Coda

A fine overview of PVA is provided by Beissinger and Westphal (1998). PVA software for microcomputers is available (e.g., VORTEX and RAMAS). For details on these programs (and others) and a review that compares their strengths and weaknesses for particular applications, see Lindenmayer et al. (1993, 1995). Generic risk-assessment software also can be adapted to perform PVAs. Computer programming languages offer the most versatility in performing PVAs, but legitimate PVAs also can be performed using conventional spreadsheet programs.

To predict the effects of demographic factors we need estimates of demographic parameters such as the population's size, sex, and age structure and its natality and mortality rates. Ideally, we would have specific natality and mortality rates for each age group and gender; for example, what is the probability of a three-year-old female surviving until she is 4 years old? How many young will she probably produce? It is often necessary to use rough estimates of these parameters because it is too difficult and expensive to gather the required data.

2. Environmental stochasticity refers to random variation in parameters that measure habitat quality such as climate, nutrients, water, cover, pollutants, and relationships with other species that might be prey, predators, competitors, parasites, or pathogens. At a conceptual level it is easy to understand how these factors are related to a population's probability of surviving. However, translating these relationships into quantitative predictions becomes very complex, and thus most PVAs either do not include any environmental stochasticities, or they include only a few simple factors that are thought to be limiting (Boyce 1992): when environmental stochasticities are incorporated into a PVA, this is usually done by making a link between certain key environmental factors and one or more demographic parameters. For example, a PVA for an amphibian population might include a predicted relationship between annual variation in precipitation and the number of offspring produced that year and the number of individuals surviving (Pounds and Crump 1994).

3. Catastrophes are events such as droughts or hurricanes that occur at random intervals. In a sense they are a form of environmental stochasticity, but they differ in that they are discrete, specific events rather than continuous variation in a parameter such as temperature that is routinely affecting population dynamics. In the context of PVAs, their predicted effect on a population is usually modeled differently. They are predicted to kill a portion of the population outright at some irregular interval rather than having a continuous effect on a parameter such as natality.

4. Genetic stochasticity is random variation in the gene frequencies of a population resulting from genetic drift, bottlenecks, inbreeding, and similar factors (see Chapter 5). These processes are understood well enough in some experimental situations, notably with *Drosophila* fruit flies, to allow population biologists to make quantitative estimates of their effect. Unfortunately, it is probably a significant extrapolation to use numbers based on fruit fly research in PVAs for all the species in which these processes have not been studied. For example, our understanding of genetics would suggest that northern elephant seals should be suffering severely from inbreeding problems because in the 1890s they had apparently been reduced by overhunting to less than 20 individuals (Bonnell and Selander 1974). However, they do not seem to have any genetic problems. Perhaps they are

very lucky; perhaps the potential for inbreeding depression is intrinsically low in elephant seals. Among the four factors listed here, genetic stochasticities probably have the least effect on MVP estimates, especially for short-term predictions (Lande 1988).

These four factors cannot be incorporated into a model in a simple, additive fashion. They all interact with one another in a complex manner that is likely to involve positive-feedback loops (snowballing effects) that collectively constitute what has been described as an extinction vortex (Gilpin and Soulé 1986). For example, one form of environmental stochasticity, habitat fragmentation, can easily curtail dispersal among subpopulations and thus profoundly affect the demographic and genetic structure of a population (Fahrig and Merriam 1994). Consider the extinction of the heath hen, a subspecies of prairie chicken that used to range along the United States' Atlantic coast from Maine to Virginia (Fig. 7.12). After environmental factors (overhunting and habitat degradation) reduced the heath hen to one population on a small island, it succumbed to a catastrophic fire, more

Figure 7.12. A combination of factors drove the heath hen into extinction; they are environmental stochasticity, demographic stochasticity, genetic stochasticity, and catastrophes.

environmental problems (predation and disease), a demographic imbalance (too few females), and perhaps a genetic problem manifested as sterility (Shaffer 1981).

Combining all the various parameters that might affect population viability for a given species into a truly comprehensive model would be nearly impossible, although some very complex and sophisticated models have been created. (See Box 7.1 for a simplified model designed to illustrate some basic elements of a PVA.) Furthermore, if a realistic, comprehensive model could be created, another huge hurdle would remain: obtaining reasonable numbers to plug into the model. Even basic parameters such as age-specific natality and mortality have not been measured for most species and are not easily obtained (Beissinger and Westphal 1998, Reed et al. 1998). Despite these reservations, PVAs do not have to be comprehensive to be useful, as we will see in the case history reported below.

PVA models are best thought of as a method for organizing and enhancing our understanding of the factors that shape a population's likelihood of persistence, as well as for comparing the effects of different management alternatives on relative probabilities of extinction (Possingham et al. 1993, Beissinger and Westphal 1998). As Michael Soulé (1987) wrote, when summarizing one of the first assessments of population viability, ". . . models are tools for thinkers, not crutches for the thoughtless."

One early PVA based solely on genetic factors generated an idea, widely known as the 50/500 rule, which most people would argue has become more of a crutch than a tool. Ian Robert Franklin (1980) estimated that an effective population size (recall N_e from Chapter 5) of 50 was the minimum viable population size required to avoid problems associated with inbreeding and should give a population a reasonable chance of persisting for 100 years or so. For long-term survival, N_e should be at least 500 so that a population could retain enough genetic variability to evolve in step with changing environments. This rule has been abandoned by most conservation biologists for being far too simplistic. At the very least, MVPs will vary greatly among species, and within a species they will vary depending on the particular circumstances facing each population (Lindenmayer et al. 1993). Furthermore, most people lose sight of the fact that N_e is likely to be only 10% to 20% of N, meaning that a 50/500 rule would require actual populations of at least 250/2500 (Vucetich et al. 1997). Nevertheless, the 50/500 rule persists in some circles and is mentioned here because you are likely to encounter it.

Some people fear that estimating MVPs is an invitation for naive or optimistic managers to maintain populations only at this level and no higher, clearly a risky strategy given the uncertainty that surrounds these estimates. They might argue that it is better not to make any estimate than to make one that may not be accurate. Nevertheless, Soulé (1987) appreciates the necessity of providing some guidance to wild life managers, and he has suggested that, at least for vertebrate species, there is sufficient evidence to propose a broad rule of thumb: populations should be in the low thousands if they are to have a 95% probability of surviving for several centuries. This is bad news for larger vertebrates because few current reserves are large enough to sustain thousands of individuals of large species. If these species are to survive, it will require larger reserves and better management of the seminatural ecosystems between reserves. Failing this, the intensive management techniques described in Chapter 12, Managing Populations, may allow some small populations to persist. If not, captive propagation (see Chapter 13, Zoos and Gardens) may be a last resort.

Finally, the focus on populations in this chapter is a good reminder that conservation biologists may at times become too fixated on the global extinction of entire species and thus overlook the slow, incremental loss of populations that is likely to lead to species extinctions (Hobbs and Mooney 1998).

CASE STUDY   The Eastern Barred Bandicoot

After Australia drifted away from the other continents about 45 million years ago, its marsupial mammals were able to evolve in isolation from other mammals, and they came to occupy a broader span of ecological niches than any other order of mammals. Sometimes, the match between an Australian marsupial and its placental counterparts in other parts of the world is quite obvious; for example, the extinct thylacine or Tasmanian wolf had a striking resemblance to canines elsewhere. On the other hand, the various species of bandicoots look like an odd cross between a rabbit (large ears and medium body size) and a shrew (long, pointed snout). In their habits they are more like shrews and other insectivores, although some bandicoot species are quite omnivorous. Many species of bandicoot have become extinct or declined precipitously since the European settlement of Australia, principally because overgrazing has eliminated cover for them, and thus they

are vulnerable to introduced predators such as cats and red foxes. One species, the eastern barred bandicoot, is in grave danger of extinction on mainland Australia (it is still reasonably secure on Tasmania) and has been the subject of a population viability analysis by Robert Lacy and Tim Clark (1990).

In 1989, at the time of the Lacy and Clark PVA, 150 to 300 eastern barred bandicoots (henceforth, we will just call them bandicoots) remained in the state of Victoria near the city of Hamilton. Bandicoot populations should be able to withstand considerable mortality because their reproductive rate is among the highest of any mammal their size. They have a gestation period of 12 days; young are weaned at 60 days, and the interval between births is 70 to 90 days; young breed for the first time at 4.5 months; and litter size averages 2.2 young. Despite this fecundity, the Victoria bandicoot population declined about 25% per year during the 1980s. High mortality rates were almost certainly responsible for the decline, but Lacy and Clark had no independent measures of mortality; therefore they used estimates back-calculated from the observed fecundity rate and overall population decline. These estimates were 50% mortality between 0 and 3 months of age, and again 50% from 3 to 4.5 months, 37% between 4.5 and 6 months, and 25% every 3 months for adults. Environmental stochasticity was included in the model by assuming that the carrying capacity of the bandicoots' environment and environmental effects on mortality rates varied randomly among seasons (every 3 months) over a modest range. Lacy and Clark incorporated the possibility of catastrophes by including in their model a 3.4% chance per year of a drought that eliminated all reproduction, and a 5.6% chance per year of a flood or fire causing 25% mortality.

After running their model for 1000 simulations, Lacy and Clark concluded that the Victoria bandicoots would be extinct in 10 years under current conditions. This estimate increased only to 20 years under more optimistic scenarios such as no catastrophes occurring during this period. The PVA by Lacy and Clark was pivotal in demonstrating to Victoria's conservation agencies that the eastern barred bandicoot was in dire straits. This was particularly true because agency personnel, like most people, found it easy to ignore the threat of a random event like a drought until it was explicitly included in a model (T. Clark, personal communication).

A parallel PVA was used to evaluate the effectiveness of various management options, including 1) reducing predation risk by providing more cover (shrubs planted between a double line of fencing), 2) controlling predators (primarily feral and pet cats), and 3) modifying road designs to slow vehicles and thus reduce roadkills (Maguire et al. 1990). The PVA indicated that only a management plan that incorporated all three elements was likely to avert extinction. The PVA also indicated that the probability of extinction of the population would increase if some individuals were removed to establish a captive-breeding program, but this effect could be reduced by removing juvenile animals from places where their chances of being killed by predators was great anyway.

To date the recovery efforts for the eastern barred bandicoot of Victoria (see Seebeck 1990) are a mix of successes and failures. The bad news is that the original Victoria population has nearly disappeared. The hopeful news is that captive propagation has proven quite effective, and these animals have been used to establish seven new populations some of which seem to be persisting reasonably well (T. Clark, personal communication). In any case, they are faring better than they would have in the absence of a PVA that catalyzed management action.

Summary

Some species are more vulnerable to extinction than others. Rare species are particularly vulnerable to extinction, especially those that are rare because they are confined to a small geographic range such as a single island or lake. The processes that can drive a rare species into extinction include changes in the environment (broadly defined to include physical features such as climate, as well as interacting species such as predators, competitors, and pathogens), demographic effects, and genetic problems. Some species are threatened with extinction, even though they are not intrinsically rare, because of conflicts with people. These include species that inhabit the types of ecosystems used by people, require large areas of habitat, are likely to be exploited or persecuted by people, or are ill-prepared to adapt to human-induced changes.

To understand extinction processes we need to understand population structure, especially metapopulation structure in which populations are subdivided into semi-isolated subpopulations occupying patches of habitat in a matrix of nonhabitat. In this context, a key question becomes whether the rate at which new subpopulations are created by colonization exceeds the rate at which existing subpopulations are lost to extinction. This has become a major problem because natural ecosystems have been extensively destroyed and fragmented by human activities.

Understanding extinction processes has been facilitated by a process called population viability analysis that uses simulation models to assess the long-term viability of a population. PVAs estimate minimum viable populations (MVP), the smallest population that has a high chance of persisting for an extended period. PVAs are based on estimating the probabilities surrounding environmental, demographic, and genetic factors that can influence a population's likelihood of persistence. They are valuable tools for facilitating our understanding of extinction, but should be only one of the tools used for making decisions about managing populations.

FURTHER READING

For more ideas on why some species are vulnerable to extinction, see Terborgh and Winter (1980), Pimm et al. (1988), and Jablonski (1991). Three volumes on metapopulations are worth examining: multiauthored compilations edited by Hanski and Gilpin (1997) and McCullough (1996) and a book by Hanski (1999). Useful books on population viability include Soulé (1987), Burgman et al. (1993), and Beissinger and McCullough (in press). For shorter treatments see Boyce (1992), Lindenmayer et al. (1993), Caughley (1994), Beissinger and Westphal (1998), and Noon et al. (1999). Most PVAs have involved animals; see Menges (2000) for papers on plant population viability. To try your hand at a PVA for the eastern barred bandicoot see the case study above, and check out Chapter 4 of Gibbs et al. (1998).

TOPICS FOR DISCUSSION

1. Consider each of the three major ways to be rare (limited geographic range, restriction to rare habitats, and low population densities) and discuss how organisms that exhibit each kind of rarity are likely to be affected by the four major risks facing populations (environmental, demographic, and genetic stochasticities, and catastrophes). It may be helpful to construct a 3 × 4 matrix and fill in the cells.
2. Large carnivores have many features that make them particularly sensitive to human disturbance. What are they? Although greatly reduced, most large carnivore species are still extant. What features have saved them from extinction?
3. Under what circumstances would a species that existed as a single population be less vulnerable to extinction than a species that existed as a metapopulation? Under what circumstances would a metapopulation be less vulnerable?
4. What do you think are the primary strengths and weaknesses of population viability analyses?

Chapter 8
Ecosystem Degradation and Loss

With light streaming out of our cities at night, with roads and power lines etched across most landscapes, any visitor from another planet would be well aware of human activities long before arriving on earth. A conservation biologist might argue that *Homo sapiens* is only one of many millions of species that constitute life on earth, but there is no denying that people are a dominant life-form. As we have captured more and more of the earth's resources, allowing our population and biomass to grow larger and larger, many other species have declined or even disappeared. Indeed, if you measured some overarching parameters such as global biodiversity, global biomass, or global productivity, you could probably build a good argument that we have degraded the *overall* ability of the earth to support life. This seems particularly likely given that it has been estimated that about 40% of the photosynthesis on earth is currently being appropriated by people (Vitousek et al. 1986, 1997).

In this chapter we will examine the various ways in which people diminish the earth's ability to support a diverse biota. To begin, we need to make some distinctions that may seem to be splitting hairs, starting with *habitat* versus *ecosystem.* A habitat is the physical and biological environment used by an individual, a population, a species, or perhaps a group of species (Hall et al. 1997). In other words, at the species level we can speak of blue whale habitat and sequoia habitat, and perhaps waterfowl habitat. However, if the group of species is too broad, the term becomes so general as to be almost meaningless. What does "wild life habitat" mean if virtually every environment supports wild organisms? Even a parking lot will have microbes and small invertebrates living in the cracks in the pavement. An ecosystem is a group of organisms and their physical environment (see Chapter 4) such as a lake or a forest, and it may or may not correspond to the habitat of a species. A forest ecosystem may constitute the sole habitat of a squirrel, but a frog's habitat might include both the forest and a lake, and a bark beetle's habitat might only be certain species of trees in the forest.

We can also make a distinction between degradation and loss of habitats or ecosystems. *Habitat degradation* is the process by which habitat quality for a given species is diminished; for example, when contaminants reduce a species' ability to reproduce in an area. Ideally, habitat quality would be estimated using parameters that are closely tied to population viability and evolutionary fitness, parameters such as reproductive rate and survivorship. In practice, these parameters are difficult to measure, and therefore ecologists often only measure population density on the assumption that there are more individuals in good habitat. However, this assumption may not always be valid because, sometimes, large, crowded populations occur in marginal habitat where they cannot reproduce (Van Home 1983).

When habitat quality is so low that the environment is no longer usable by a given species, then *habitat* loss has occurred. The line between habitat degradation and loss will often be unclear. For example, if degradation proceeds to a point that the local population of a species stops reproducing, but some individuals can still be found (e.g., juvenile animals dispersing in search of suitable habitat in which to breed, or a few old trees that survive, but whose seeds never survive), is this habitat loss or severe degradation? Sometimes, these differences can be clarified if we describe the types of habitat use more explicitly, for example, by referring to breeding habitat, foraging habitat, winter habitat, and so on.

Habitat loss or degradation for one species will probably constitute habitat gain or enhancement for some other species. For example, cutting a forest is likely to degrade or destroy habitat for a squirrel, but the resulting early successional ecosystem is likely to be new habitat for at least one butterfly species. All other things being equal, conservationists would usually choose not to cut a forest to create butterfly habitat, but often the choices are not straightforward. Restoration of wetlands in the Everglades may eliminate some habitat for the endangered Florida panther (which now use drained wetlands) while creating more habitat for wood storks, snail kites, and other species that are also of great concern for conservationists (Jewell 1998). Furthermore, the particular hydrological regime used in these wetlands could tend to favor either wood storks (which need periods of very limited water to concentrate their food in residual pools) or snail kites (which need long, wet periods) (Bancroft et al. 1992, Beissinger 1995, Curnutt et al. 2000). Conservation management often involves balancing multiple objectives.

Ecosystem degradation occurs when alterations to an ecosystem degrade or destroy habitat for many of the species that constitute the ecosystem. For example, when warm water from a power plant increases the temperature of a river causing many temperature-sensitive species to disappear, this is ecosystem degradation by a conservation biologist's definition. An ecosystem ecologist might focus on changes in ecosystem function such as a reduction in productivity rather than on structural attributes such as the abundance and diversity of biota. Some people might argue that not all human-induced changes to an ecosystem are necessarily degradation. If the temperature-sensitive fish species lost because of a power plant were replaced by a popular species of game fish, then anglers might argue that the ecosystem had been enhanced. Even some conservation biologists, who usually see any significant change from natural conditions as degradation, might not be too critical of the power plant if its outflow became habitat for manatees, an endangered species that requires warm refugia in the winter.

Ecosystem loss occurs when the changes to an ecosystem are so profound and when so many species, particularly those that dominate the ecosystem, are lost that the ecosystem is converted to another type. Deforestation and draining wetlands are just two of many processes that destroy ecosystems.

Let us consider a hypothetical example to illustrate these distinctions. Imagine a small forest park on the edge of city in which there are many dead and dying trees (i.e., snags). The park manager might decide to make this forest safer for walkers by removing all snags near paths. This would degrade the park's value as a habitat for the many species that require snags such as woodpeckers and termites. If the manager were very thorough and cut down every snag in the park, regardless of its location, this would constitute habitat loss for snag-dependent species. Assuming snag-associated species were more than a trivial portion of the forest's biota, then loss of snags would also lead to ecosystem degradation. Removing the forest to create a golf course would constitute ecosystem loss.

There are many ways to degrade or destroy habitats or ecosystems, and in this chapter we can only provide a broad overview. We will begin with two sections on things we add to natural environments: 1) substances that contaminate air, water, soil, and biota; and 2) physical structures such as roads, dams, and buildings. The third section covers some of the ways we modify physical environments by eroding soil, consuming water, and changing fire regimes. In the fourth, fifth, and sixth sections we will review three major processes by which ecosystems are destroyed or severely degraded—deforestation, desertification, and the various processes afflicting wetlands and aquatic ecosystems (e.g., draining and filling). We will not focus on two of the major causes of ecosystem loss and species endangerment: conversion of ecosystems to urban areas and agriculture (Flather et al. 1998, Thompson and Jones 1999, Czech et al. 2000). Their direct effects are so unsubtle that they do not require much elaboration; we will discuss how to mitigate their impacts in Chapter 11, Managing Ecosystems. Finally, we will discuss fragmentation, a process by which ecosystem destruction can isolate the biota of those ecosystems that remain intact. For the sake of simplicity we will cover each issue independently, but realize that in the real world many problems occur simultaneously and interact with one another.

Two special forms of ecosystem degradation—overexploitation of biota and introduction of exotic species—will be covered in Chapters 9 and 10, Overexploitation and Exotic Species. Note that all of these sundry threats are direct, proximate causes of loss of biodiversity. As with so many problems, the ultimate cause is human overpopulation and overconsumption, but we will reserve discussion of this topic until Part IV, The Human Factors. One deadly enterprise merits special mention here: war. The human dimensions of war's tragedies are all too familiar, and it takes but a moment's reflection to extend its images—ravaged lands, shattered bodies—to all biota. As you read the following chapter, realize that virtually all of the activities described here can become part of a war machine with dire and far-reaching consequences (Westing 1980, 1984). Indeed, long after a war is over, elephants, rhinos, and any large, marketable animal will continue to suffer from the widespread distribution of weaponry.

Contamination

One might define a pollutant or contaminant as a substance that is where we do not want it to be. This suggests that substances often do not stay where we put them; they move. There are three main media that can move pollutants—air, water, and living organisms—and we will structure our overview of the topic by focusing on air pollution, water pollution, and pesticides. Note that there is overlap among these media; for example, acid rain begins as air pollutants and ends up contaminating a lake or causing increased concentrations of heavy metals in biota. Pesticides can be distributed by air or water, but we will focus on those that move from organism to organism in a food web.

Air Pollution

Every day huge quantities of materials are lofted into the atmosphere from our vehicles, factories, and homes. Nitrogen oxides and sulfur oxides combine with water to form nitric and sulfuric acids, the basis of acid rain. Chlorofluorocarbons (CFCs) and halons rise to the upper atmosphere where they reduce the concentration of ozone, allowing more harmful ultraviolet radiation to reach the earth's surface. Closer to earth, ozone and a suite of other chemicals form toxic clouds called smog.

Through extensive research we know that these and other forms of air pollution have impaired the health of people and domestic plants and animals (Rubes et al. 1992, Holgate et al. 1999). We know less about the effects of air pollution on wild species, but given the basic similarity in the physiology of domestic and wild species, it is likely that they are also affected (Barker and Tingey 1992, Lovett 1994). For example, air pollution near Mexico City is apparently responsible for the decline of some tree species living in the surrounding mountains (Tovar 1989), potentially including some rare pines. In some parts of the Netherlands, songbirds are laying thin eggs and losing many clutches because acid rain has reduced the abundance of snails that are the birds' main source of calcium (Graveland et al. 1994). Chronic effects that diminish an individuals health and vigor, and thereby reduce reproductive success or longevity, are probably more common than acute effects that kill organisms directly. Consequently, the effects of air pollution may be rather subtle. They may contribute to the decline of populations without being the primary cause.

Although chronic effects may be the major consequence of air pollution, acute effects do occur and in some cases lead to localized extinctions. In Europe air pollution is suspected to be the cause for the disappearance of many mushroom-forming fungi (Jaenike 1991). Severe air pollution has even killed the majority of plant species downwind from some factories (Fig. 8.1). No doubt many animal species also become extinct in these zones, but it would be hard to know if they were directly eliminated by air pollution or simply disappeared because of the loss of plant species.

Even species living far from the source of air pollution may be affected. Notably, recent declines in some remote amphibian populations (Alford and Richards 1999, Houlahan et al. 2000) might be linked to air pollution because of its effects on acidity of aquatic ecosystems, global climate, and ultraviolet radiation. For example, several studies have shown certain amphibian species, especially those living at high altitudes, to be vulnerable to ultraviolet-B radiation (e.g., Blaustein et al. 1997, Broomhall et al. 2000), and at least one paper has directly implicated climate change in the loss of many frog species at a Costa Rican site (Pounds et al. 1999; see Chapter 6). Similarly, one of the major threats to coral reefs can probably be traced to global climate change induced by air pollution; unusually warm water temperatures are thought to be a primary cause of "bleaching," the massive death of coral polyps (Goreau et al. 2000).

Water Pollution

The list of substances with which we pollute aquatic ecosystems is very diverse. It includes innocuous materials such as mud and plant matter

Figure 8.1. Fumes from a copper smelter killed most of the vegetation in the Copper Basin, Tennessee. This photo was taken in 1945, about 25 years after the fumes were controlled. (USDA Forest Service photo.)

that may become contaminants only when they reach such high concentrations that they smother the bottom of aquatic ecosystems or use up all the oxygen as they decompose. The list also includes chemicals such as nitrates and phosphates that are important nutrients for aquatic plants, but can lead to an excessive growth of plants, upsetting the balance of an aquatic ecosystem. On the other hand, there are chemicals such as dioxin that endanger life at concentrations so low that they are measured in parts per billion. Some pollutants are routinely discharged into aquatic ecosystems from factories and sewage treatment plants. Others enter in a catastrophic deluge after an accident such as the rupture of an oil tanker. Still others such as sediments, pesticides, and fertilizers often seep in gradually carried by the runoff from our agricultural fields, lawns, and streets. When pollutants originate from broad areas, these places are called *nonpoint sources,* in contrast to specific sites (e.g., factories), which are called *point sources.* It may surprise you to know that nonpoint-source pollution, usually involving sediments and nutrients and not highly toxic chemicals, is considered the leading threat to endangered freshwater species in the United States (Richter et al. 1997).

Not surprisingly, aquatic species and ecosystems are more threatened

TABLE 8.1 Changes in species richness of some invertebrate taxa in the Rhine.*

	Upper Rhine		Middle Rhine		Lower Rhine	
	1916	1980	1916	1980	~1900	1981–1987
Gastropoda (snails)	8	4	8	5	11	10
Lamellibranchiata (mussels)	11	4	10	4	14	7
Crustacea (crustaceans)	3	2	3	2	3	13
Heteroptera (true bugs)	2	1	1	0	1	1
Odonata (dragonflies)	2	1	1	0	3	2
Ephemeroptera (mayflies)	11	4	3	0	21	2
Plecoptera (stoneflies)	13	0	12	0	13	0
Trichoptera (caddisflies)	11	5	11	2	17	5
Total	61	21	49	13	83	40

*From Borseliske et al. (1991).

by water pollution than are terrestrial biota. On a local scale, there are many lakes, streams, rivers, and bays where water pollution has eliminated so many species that it would be fair to say that the aquatic ecosystem has been destroyed, even though a body of water and a handful of species remain. One of Europe's largest rivers, the Rhine, exemplifies this problem; along substantial stretches the natural biota has been severely altered by pollution (Table 8.1) (Broseliske et al. 1991).

Elimination of a species from a single water body may mean global extinction because many aquatic species are found in a single lake or river system, having evolved in isolation from their relatives in nearby water bodies. One of the most interesting examples of this comes from Lake Victoria in East Africa, home to hundreds of endemic cichlid fish species (Seehausen et al. 1997). Separation among these closely related species is highly dependent on females choosing mates of the correct species; however, with growing eutrophication the lake's turbidity is increasing, and the females cannot distinguish the colors they need to see to choose the correct mates. Consequently, cichlid diversity is declining in eutrophic areas of the lake. Even if the population occupying a water body is not an endemic species, it is likely to be genetically different from populations in other places.

Water pollution is less likely to cause global extinction of species in marine ecosystems than in freshwater ecosystems for two related reasons. First, marine ecosystems are often vast and thus difficult to pollute in their entirety. (There is some validity to the saying, "the solution to pollution is dilution," even though it should not be a universal prescription for pollution abatement.) Second, many marine species have relatively large geographic ranges, making it unlikely that their entire range would be so polluted as to be uninhabitable (Carlton et al. 1991, Culotta 1994).

Even though water pollution may not be responsible for the global extinction of marine species, it still can have a profound impact on marine biodiversity (Norse 1993). Many marine species have been locally extirpated by pollution; for example, when coral reefs are smothered in silt or overrun with macroalgae because of excessive nutrients and eutrophication (Meesters et al. 1998). Water pollution can also upset the equilibrium of marine food webs, for example, when an excess of nutrients causes the explosive growth of plankton known as the red tide.

Pesticides

To capture a large portion of the earth's resources people must compete against other organisms, and pesticides are one of our preferred tools for doing this. We use enormous quantities of insecticides and rodenticides to kill animals that would eat our crops, herbicides to kill plants that would compete with our crop plants, and fungicides to kill fungi that would decompose our food and fiber. Worldwide, over 50,000 different pesticide products with active ingredients weighing over 2.6 million metric tons are used each year (World Resources Institute et al. 1998). Some of these pesticides are relatively benign. They kill only a small group of target organisms, they are used in limited areas (e.g., food storage facilities), and after use they quickly break down into harmless chemicals. Unfortunately, very few pesticides meet all these criteria, and some such as the notorious DDT wreak havoc on a broad set of nontarget organisms for a long period over large areas.

Croplands strewn with corpses can mark the aftermath of pesticide use, but more often the effects are not seen until much later and in more subtle ways. One example of this has garnered considerable attention in recent years: pesticides and related chemicals that mimic the action of the female sex hormone, estradiol (Colborn et al. 1996, National Research Council 1999). Sterility, delayed sexual maturity, abnormal sex organs, and an array of other problems have been attributed to these contaminants that are characterized as "endocrine disruptors" or "hormonally active agents." Long-term, insidious effects of pesticides are well documented because some of them can persist in the tissue of living organisms, accumulating in one individual, and passing on to other individuals through a food web. The most infamous example involves a set of chemicals known as chlorinated hydrocarbons (which includes DDT, many other pesticides, and some chemicals that are not pesticides such as PCBs, polychlorinated biphenyls). They are soluble in fat and can take years, even decades, to break down. This means that they pass from prey to predators up a food chain and can concentrate in top predators, a process known as *biomagnification* (Fig. 8.2). Populations of several predatory birds (ospreys, brown pelicans, bald eagles, peregrines, and others) were dramatically reduced by chlorinated hydrocarbons during the 1950s and 1960s. Use of these chemi-

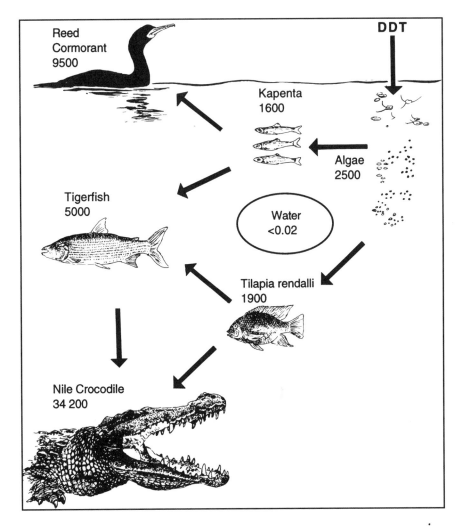

Figure 8.2. Persistent pesticides and similar compounds accumulate in the tissues of one species and then are passed up the food web to other species where they become more concentrated. This process is called biomagnification or bioamplification. In this figure DDT has entered the food web of Lake Kariba in Zimbabwe and reached its highest levels in top predators such as crocodiles, tiger-fish, and cormorants. Numbers are parts per billion of DDT and its derivatives in the fat of the species illustrated. (Redrawn by permission from Berg et al. 1992.)

cals has been sharply curtailed in many wealthier countries, and this has allowed populations of these birds to recover somewhat (Sheail 1985). However, use of chlorinated hydrocarbon pesticides continues in many less-developed countries, and, because of their persistence, a wide variety of chlorinated hydrocarbons continue to contaminate the environment of places where they have been banned (Berg et al. 1992, Douthwaite 1995).

Accounts of the negative effects of pesticides typically focus on species that are most similar to us—birds and mammals—because we tend to be more concerned about their welfare, and because toxic effects on these species may portend toxic effects on us. When DDT was discovered in human milk, environmentalists produced a poster showing a mother nursing her child. The label on her breasts read Unfit for Human Consumption. A poster showing all the nonpest insects killed by insecticides would not have been very effective. Nevertheless, it is likely that the most serious effects of pesticides are on those organisms that are most closely related to the target species. Consider the insect order Lepidoptera (butterflies and moths), which includes many pest species, as well as many endangered species. It seems reasonable to assume that attempts to control pest lepidoptera with insecticides would jeopardize some rare lepidoptera, although in practice this has not been well documented to date (New 1997, Pimentel and Raven 2000). Loss of nontarget insect populations may have far-reaching consequences. In particular, there is growing concern about the loss of pollinating insects and the consequences this may have for a wide range of plants that require animal pollinators, for the other animals dependent on those plants, and for human food production (Allen-Wardwell et al. 1998). Quite a few studies have investigated how killing insects may affect insectivorous birds; some effects have been documented, but catastrophic impacts have not been noted (Witham and Hunter 1985, O'Connor and Shrubb 1986, Holmes 1998).

Roads, Dams, and Other Structures

Flying in a plane, you can easily see the hand of humanity; most landscapes are crisscrossed with roads, railroads, fences, and utility corridors and dotted with buildings, dams, mines, parking lots, and many other structures. The total area covered by such structures is significant (about 2 million km^2 worldwide; about 1.4% of the land area [Vitousek et al. 1986]) and represents a loss of habitat for virtually all wild species. Looking beyond the immediate footprint of these structures, one can see that a much larger area is affected. For example, Forman (2000) estimated that roads and their adjacent impact zones cover 20% of the area of the United States. Thus we can list "construction of human infrastructure" along with deforestation, desertification, and other processes that destroy entire ecosystems, all of which we will discuss later in this chapter. In this section we will focus on the consequences of adding these and other structures to the biota of entire landscapes, especially on animals that move across landscapes.

Roads

The most ubiquitous structures created by people are roads, and while roads facilitate the movement of people, they can also serve as impediments to the movements of many animals (Forman and Alexander 1998). Some roads have curbs or lane dividers that are an absolute barrier to small, flightless animals such as amphibians, small reptiles, and various invertebrates. More commonly animals are capable of crossing a road, but may be run down in the process (Fig. 8.3). In a two-year study of a 3.6-km stretch of highway in Ontario, Canada over 32,000 vertebrate carcasses were found (Ashley and Robinson 1996). Most of the mortality fell on amphibians and reptiles; mass migrations of these species to and from breeding sites make them especially vulnerable (Fahrig et al. 1995). Nevertheless, just the mor-

Figure 8.3. Roads act as filters to the movements of many animals, especially because of collisions such as the one that killed this tayra in Belize. (M.L. Hunter photo.)

tality of birds (62 species; 1302 individuals) and mammals (21 species; 282 individuals) in this study would extrapolate to billions of carcasses on the world's road system without even attempting to measure the mortality of amphibians, reptiles, and invertebrates. Most of the individual animals killed on roads may be of common species that are in no danger of extinction, but even a few road deaths can be of great consequence for an endangered species. For example, in Spain, road collisions accounted for 17% of mortality, second only to poaching, in a study of the endangered Iberian lynx (Ferreras et al. 1992). For Florida scrub jays, another imperilled species, territories adjacent to roads are population sinks because of traffic-induced mortality (Mumme et al. 2000). For some species roads are a psychological filter; individuals are apparently reluctant to cross them even though physically capable of doing so. A study in Germany found that even a narrow forest road closed to traffic represented a major barrier to small mammals and beetles (Mader 1984). Grizzly bears are apparently reluctant to approach even the vicinity of roads (McLellan and Shackleton 1988). If organisms are unable or unwilling to cross a road, then the populations on either side of the road may become isolated from one another (Gibbs 1998), and this may have repercussions as discussed in Chapter 7.

A second major problem associated with roads is the access they provide to people who may overexploit organisms or destroy whole ecosystems. The roads penetrating formerly remote areas of the Amazon Basin, allowing access by settlers who raze the tropical forests, are a particularly lamentable example of this phenomenon. A more subtle example comes from research on the population viability of large carnivores such as wolves. Some evidence suggests that when the density of roads exceeds roughly 0.4 to 0.7 km of road per square kilometer of habitat, encounters between hunters and wolves—many of which end in a dead wolf—become too common for wolves to persist (Thiel 1985, Fuller et al. 1992, Mladenoff et al. 1995).

Roads may also provide access to exotic organisms that can disrupt native populations (Parendes and Jones 2000). Usually, these will be species carried, intentionally or not, by people traveling along the highway. Sometimes, exotic species will move along the road by themselves. In particular, weedy exotic plants seem to use the disturbed ground of roadsides to invade a landscape (Frenkel 1970, Greenberg et al. 1997). Finally, roads have a variety of physical and chemical attributes that are likely to affect adjacent aquatic and terrestrial ecosystems. These include various substances such as dust, sediment, salt, heavy metals, hydrocarbons; a sunny, windy, warm microclimate; blocking surface water runoff; and more (Haskell 2000, Jones et al. 2000, Trombulak and Frissell 2000). One of the most annoying physical aspects of roads for human observers—traffic noise—can reduce bird population densities in a band hundreds of meters wide (Reijnen et al. 1995).

Dams

The damming of streams and rivers has destroyed many aquatic ecosystems, flooding ecosystems upstream of the dam and changing water flows to downstream ecosystems. We will return to these issues in a later section; here the focus will be on the barrier effects of dams. Many animals move up and down rivers during the course of a year, or during their life cycle, searching for the best places to forage or breed. Some of them can fly or walk around dams (otters, mergansers, mayflies, etc.), but for totally aquatic species dams can be very significant barriers. Moving downstream these animals are likely to be churned to death or at least highly stressed in turbines (Benstead et al. 1999, Karieva et al. 2000). Moving upstream they encounter an insurmountable wall that may or may not have a fish ladder around it. Even if there is a fish ladder, only a portion of the population will be able to find their way up it. The reservoir behind a dam may also impede movement, especially if it has been stocked with exotic, predatory fish. Of course, fish are the best-known victims of dams, especially anadromous fish such as salmon that move long distances between riverine spawning areas and marine foraging areas (Fig. 8.4). Some salmon populations have been completely eliminated, largely by

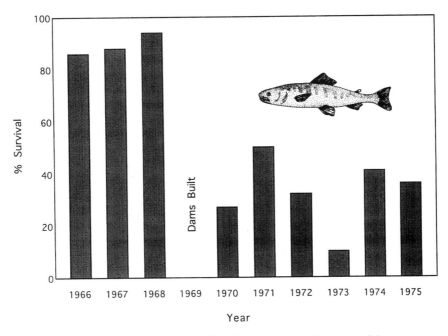

Figure 8.4. Survival of wild, juvenile, chinook salmon migrating toward the sea before (1966–1968) and after (1970–1975) completion of two dams on the Snake River in Washington. (Redrawn by permission from Raymond 1979.)

dams, despite millions of dollars spent building fishways, trucking fish around dams, supplementing populations with hatchery-reared stocks, and so on (Meffe 1992). A study of eight rivers in Sweden suggested that the effects of dams and reservoirs on shoreline plants are also shaped by dispersal issues: water-dispersed species with a limited ability to float were strongly affected by damming (Jansson et al. 2000a; also see 2000b).

Other Barriers

Some landscapes are dissected by barriers specifically designed to inhibit the movement of animals. Notably, rangeland fences stretch huge distances, controlling the movement of both livestock and wild large mammals. For example, in Botswana, thousands of kilometers of fences have been erected to comply with regulations that require livestock to be isolated from wild ungulates that might harbor diseases if it is to be exported to European Community nations. These fences have had catastrophic consequences for native ungulates, especially wildebeest, because in Botswana's arid landscapes it is critical that native ungulates be able to move to areas that have water during the dry season (Williamson et al. 1988). Similarly, in the western United States wild ungulates that formerly moved between summer and winter ranges have sometimes found their traditional routes severed by livestock fences (Kie et al. 1994).

Utility corridors also dissect landscapes, potentially isolating the organisms on either side. Rights-of-way may impede the movements of some organisms simply because they represent a break in the natural vegetation. For example, a forest herb that spreads by means of vegetative reproduction would be impeded by the dry, sunny environment of a power line running through a forest. Pipelines and irrigation canals also have the potential to be direct barriers. In the best-known example of this issue—the Trans-Alaskan Pipeline and caribou migrations—the problem was largely avoided by elevating the pipe (Curatolo and Murphy 1986).

Most bird species can readily fly over human-made barriers, although some forest birds are very reluctant to venture into the open, and some of the large, flightless birds (e.g., emus and ostriches) are as easily stopped by fences as are large mammals. Unfortunately, birds are often killed by flying into human structures. Large numbers of migrating birds collide with power lines, antennas, lighthouses, windmills, and similar structures (Janss 2000); even local movements can result in a collision with a large window.

Trash and Other Things

In this final section on human-made structures we will list some of the other things people make and then add to the natural environment that are detrimental to other organisms. Much of this material is trash, things discarded by people, perhaps intentionally or perhaps not. Lost or discarded fishing gear is a major hazard (O'Hara 1988). The worst offenders are probably lost gill nets—often called ghost nets—which can drift for months or years, still catching fish, diving birds, seals, and other creatures. It is difficult to estimate the extent of this mortality, but with about 21,300 km of nets (enough to reach more than halfway around the world) set nightly to catch salmon and squid in the North Pacific alone, the total loss is likely to be enormous (Laist 1987). Even a single strand of monofilament fishing line discarded by an angler can ensnare an animal and kill it. Fishing sinkers made of lead and lead shot discharged by waterfowl hunters accumulate on the bottoms of water bodies where they are likely to be swallowed by bottom-feeding birds and cause lead poisoning (Sanderson and Bellrose 1986). This form of lead poisoning can also pass up the food chain in a manner analogous to biomagnification of pesticides (Vyas et al. 2000). Lead shot in carcasses often poisons scavengers such as California condors and various species of eagles. One of the major causes of death among sea turtles appears to be marine debris (Bjorndal et al. 1994). In particular, turtles mistake plastic bags and balloons for jellyfish and end up clogging their digestive tracts with plastic (Committee on Sea Turtle Conservation 1990). Plastic in any form is rather insidious because it can persist in the environment so long before decomposing. The plastic rings that hold six-packs of beverages have been particularly criticized because so many of them end up in natural places where people picnic and camp.

Some of the problems we cause by putting human-made objects into natural environments would be hard to predict. Consider a seemingly innocuous item, red plastic insulators for electric fences. It turns out that large numbers of hummingbirds mistook the insulators for flowers and electrocuted themselves until the manufacturer withdrew the product. Street lights on beaches may make people feel safer, but they can disorient hatchling sea turtles when they emerge from their nests and make an already perilous trip to the sea even more dangerous (Witherington and Bjorndal 1991). Given that six of the seven species of marine turtle are endangered to varying degrees, any added source of mortality may be of some consequence.

Lastly, we can list the forms of motorized transport that travel across our lands and waters without using roads, often crushing plants, colliding with animals, and compacting and eroding soil. Collisions with motor boats are a major source of mortality for manatees in Florida (Marmontel et al. 1997). In deserts and on beaches off-road vehicles are a threat to sedentary or slow-moving species such as plants, hatchling birds, and desert tortoises. Ironically, fences are a common way to control off-road vehicles (Brooks 1995), and these have their own ecological problems unless carefully designed.

Earth, Fire, Water

In this section we will consider some of the ways people modify physical environments that may have negative consequences for biota. We will focus on three issues—soil erosion, changing fire regimes, and water consumption—that usually degrade ecosystems without destroying them.

Soil Erosion

Soil erosion is a natural process, an inevitable consequence of wind, rain, and gravity. The problem is that the rate of soil erosion is often greatly accelerated by human use of ecosystems (Fig. 8.5). Agriculture, overgrazing by livestock, timber harvesting, and road and building construction are the main ways in which we remove the soil's protective layer of vegetation and tear it up with machines and hooves.

Soil erosion is a double-edged sword. Not only does it produce sediments that can blanket other ecosystems, leading to some of the water pol-

Figure 8.5. Soil erosion has profoundly degraded ecosystem productivity in many regions, although it is most noticeable in mountainous areas, as in this photo from the Himalayas. (M. L. Hunter photo.)

lution problems discussed above (Cohen et al. 1993), but it also degrades the productivity of the land from which the soil is eroded. This latter problem can take a very long time to rectify; in tropical and temperate agriculture systems, it takes several hundred years for a centimeter of soil to form (Pimentel 1992). A centimeter of soil can be lost in a few hours during a torrential rain; more often, it will disappear over the course of years, making soil erosion a subtle problem that is easy to overlook and hard to measure extensively (Trimble and Crosson 2000). As Lester Brown has written. "Society can survive the exhaustion of oil reserves, but not the continuing wholesale loss of topsoil."

When a terrestrial ecosystem loses soil and its productivity is diminished, what are the consequences for the ecosystem's biota? This is a difficult question, in part because there is no simple relationship between productivity and biodiversity. Some highly productive ecosystems support a very diverse biota (e.g., tropical rain forests), and some support relatively few species (e.g., salt marshes). In the short term, most species are likely to be more affected by the agent of soil disturbance—the plow, the chainsaw, etcetera—than by the subsequent soil erosion. In the long term, diminishing the productivity of an ecosystem for several centuries could be one more stressor that pushes a species that is dependent on that type of ecosystem a bit closer to extinction.

In some severe cases, ecosystems can be highly degraded and species extirpated by soil erosion. For example, on Round Island in the Indian Ocean, rabbits and goats introduced to provide a food source for passing mariners removed most of the vegetation and this led to severe soil erosion. Two species of reptiles became extinct, and ten species of plants, three reptiles, and a seabird were at risk until the exotic herbivores were removed and erosion was brought under control (North et al. 1994). We will return to soil erosion below in our discussion on desertification.

Fire Regimes

Few phenomena can match the ability of a large, hot forest fire to totally transform a natural ecosystem in a short time. Volcanoes, nuclear bombs, and large meteorites could readily match a fire, but, thankfully, these are rare events. The apparent devastation wrought by severe fires has led to concerted efforts to control all fires. Smokey the Bear's "Only you can prevent forest fires!" is one of the best-known phrases in the United States' advertising media.

Unfortunately, the campaign has been too successful in many respects, especially when humans reduce the frequency of fires in ecosystems where they are a natural phenomenon (Leach and Givnish 1996). For example, most natural grasslands and shrublands, and some types of forests (e.g., certain eucalypt forests in Australia and pine forests of the southeastern and southwestern United States) are adapted to experiencing low-intensity fires at frequent intervals (Whelan 1995, Bond and van Wilgen 1996). Consequently, their vegetation changes dramatically without fires to inhibit the influx of fire-intolerant species. Furthermore, when low-intensity fires are suppressed, fuel can accumulate and any fire that does get started is likely to be very intensive. One of the best-known examples of the consequences of removing fire from a fire-dependent ecosystem comes from Michigan where fire suppression in jack pine forests led to a shortage of young jack pine stands, the sole habitat of the rare Kirtland's warbler (Probst and Weinrich 1993). The Kirtland's warbler almost became extinct before its dependency on fire-regenerated stands was recognized and the U.S Forest Service began a fire management program to provide its habitat needs.

On the other side of the coin, humans have often burnt ecosystems quite deliberately. This can be an ecological problem if the frequency and intensity of the fires are too great. For example, after Polynesians occupied New Zealand about a thousand years ago, they used fire to reduce the area of forest from about 80% to about 50% with significant consequence for the islands' biota (King 1984). Undoubtedly, the very earliest humans realized that fire often promotes grassy vegetation, and therefore they set fires to produce food for their preferred prey animals and later for their livestock. These practices continue in many places to this day and, when overdone, can be a problem. This is most evident when fire is used as a tool to clear forest for agriculture, as is happening in many countries with burgeoning human populations, an issue we will discuss below. It might also be a problem in semiarid environments where frequent burning allows little opportunity for the soil's organic matter to develop and thus can contribute to desertification (Savory and Butterfield 1999).

Water Use

Every year people directly use 4430 cubic kilometers of water (Postel et al. 1996); that's over 700,000 liters per person. Some of it we drink. Far more of it we use to irrigate our crops and lawns, to bathe, to flush our toilets, to manufacture sundry products such as paper, and to cool our power plants. To be specific, an estimated 65% is used for agriculture, 22% for industry, 7% for domestic purposes, and 6% is lost to evaporation from reservoirs (Postel et al. 1996). Some of this water is returned to an aquatic ecosystem; most of it is returned to the atmosphere through evaporation and transpiration. When large volumes of water are removed from aquatic ecosystems, their biota is likely to be affected. Not surprisingly, the effects are most dramatic in arid regions. Desert springs, streams, and wetlands are usually rare and fragile ecosystems, often containing unique species that have evolved in isolation (Minckley and Deacon 1991). Obviously, if most of their water is removed, these ecosystems will be degraded (Moyle and Williams 1990, Contreras-B. and Lozano-V. 1994). Consider Ash Meadows, a desert wetland on the border of Nevada and California, which is maintained by several dozen springs and is home to over twenty endemic species of fishes, plants, and invertebrates (Beatley 1977). Maintaining Ash Meadows and its unique biota has required a protracted battle to minimize water extraction for agricultural and residential developments.

Water scarcity can be an issue even in places where there is a great deal of water. Most of the southern tip of Florida is essentially one huge wetland—the Everglades—which covers many thousands of square kilometers in a sheet of water. Yet, the Everglades is so shallow, and the demands on its water from farmers and coastal communities are so great, that the whole ecosystem is being profoundly changed by a scarcity of water (Davis and Ogden 1994). Notably, the number of herons, egrets, and other wading birds has declined sharply in part because a reduction in freshwater input has reduced the productivity of estuarine parts of the Everglades.

Deforestation

* Forests cover less than 6% of the earth's total surface area.
* Forests are habitat for a majority of the earth's known species.
* Forests are being lost faster than they are growing.

These three facts highlight why many conservation biologists believe that deforestation is probably the most important direct threat to biodiversity. In this section we will first review some of the causes, and then some of the consequences, of deforestation.

Causes of Deforestation

Forests tend to grow in places with reasonably fertile soils and benign climates, not too dry and not too cold. These also tend to be good places for people to live and grow crops. Consequently, many forests have been removed to make way for our agriculture, homes, businesses, mines, reservoirs, and so on, roughly 10 million km² since the beginning of agriculture (Miller and Tangley 1991). This process has slowed or even stopped in some areas that were extensively deforested many years ago such as Europe, China, and eastern North America. In some of these places, the demand for forest land is less because the human population has stabilized, or because the local economy has shifted from agriculture (the single biggest cause of deforestation) to industry. In other places there are simply few forests left to remove. Unfortunately, deforestation continues at an alarming pace in many tropical regions. The statistics vary—an area the size of Switzerland every year, nearly 50,000 ha every day, and so on—but the basic fact remains: forests are disappearing, especially tropical forests (Miller and Tangley 1991, FAO 1993, Skole and Tucker 1993, Nepstad et al. 1999, United Nations Development Programme et al. 2000). The fundamental reasons for the current spate of tropical deforestation are twofold. First, human populations are increasing rapidly in most tropical areas. Second, most of these people are poor, and clearing forest to open a small plot where crops can be grown is often their only choice for survival.

Unfortunately, these farmers are often trapped in poverty because the lands they clear are not really suitable for agriculture in the first place. After only a few years the soil's fertility is drained, and they must move on to another site and clear more forest. The process of clearing a small patch of tropical forest, growing crops for a few years, and then moving on to another site is called shifting cultivation and is sustainable when human populations are low and the abandoned site is allowed to return to forest. It has taken place for thousands of years, probably since the beginning of agriculture. However, when populations are too high, then people stay at a site too long or return to a previously used site too soon. Alternatively, they may sell the land to a cattle rancher. Particularly in Latin America, much of the tropical forest initially cleared for subsistence agriculture ends up as low-quality rangeland for cattle. These excessive uses of a site are likely to degrade the soil so badly that, even when it is abandoned, it will probably take several centuries, or even millennia, for a reasonably natural forest to return. Tropical forest soils are notorious for being easily degraded and difficult to reforest (Lal 1987).

In many people's eyes timber harvesting is another major cause of deforestation. For example, Pimm (1991, p.136) wrote "... consider the ultimate form of external environmental disturbance—total destruction of the habitat, such as might result from logging of a forest, or an asteroid collision, or a nuclear holocaust." This viewpoint needs to be scrutinized, however. A forest can be profoundly disturbed by severe fires or windstorms, but in time the forest will be restored by ecological succession. Similarly, when a forest is clearcut, it will eventually return to a forest again if it is given enough time and freedom from additional disturbances such as plows and cattle and real estate developers. It may or may not resemble a forest that was disturbed by natural phenomena, but it will be a forest.

Time is the critical issue here. Calling a clearcut forest deforested is probably appropriate only if its recovery will take significantly longer than recovery from a natural disturbance. Or to put it another way, a site can be called deforested if it has been converted to a nonforested ecosystem that will persist for a significant period. Consider Siberia, where huge tracts of virgin boreal forest remain, but ambitious harvesting plans are being implemented. If forests do not grow back on these sites reasonably soon, this will constitute large-scale deforestation, but if forests return after an interval that is not much longer than recovery after a fire, then this cutting should not be called deforestation. Note that logging can significantly affect a forest even if only a small portion of the trees are removed. This issue will be covered in Chapter 9, Overexploitation.

Consequences of Deforestation

The extraordinary species diversity of forests is based on a number of factors (Hunter 1990); here are four key ones. First and most basically, the environmental conditions that forests require—some soil and a reasonably benign climate—are favorable to life in general. Contrast the places where forests grow to a tundra or desert. Second, the durability of wood means that forests contain an enormous reservoir of organic matter, and this material represents food and shelter to a large set of invertebrates, fungi, and microorganisms. Just two families of wood-boring beetles—long-horned beetles and metallic wood borers—contain twice as many species as all the bird, mammal, reptile, and amphibian species combined (Hunter 1990). Third, the strength of wood makes forests taller, more three-dimensional, than other terrestrial ecosystems. The height of a forest means that it contains many different microenvironments from the sunny, windy foliage at the top of the canopy to the cool, damp recesses of a crack in the bark of a tree trunk or in the soil itself. Each of these different microenvironments may support a different set of small creatures. Fourth, forests are dynamic ecosystems, frequently changing through the processes of disturbance and succession, and many of these changes are marked by differences in species composition.

Among all forests, the most diverse are the tropical rain forests. Indeed, many biologists believe that half of all the species on earth may occur in tropical rain forests (Wilson 1992). Our knowledge is too limited to corroborate this statement (as was explained in Chapter 3), but we can consider many fragmentary bits of supporting evidence. For example, 43 species of ants have been found on one tree in a Peruvian tropical forest, about equal to the number that occur in all of Great Britain, and 1000 tree species were found collectively in ten 1-ha plots in Borneo, far more than occur in all of the United States and Canada (Wilson 1992). The reasons for the extraordinary diversity of tropical forests are complex and not well understood. Suffice it to say here that the four factors mentioned above probably play a role (for example, tropical rain forests are taller and have larger reservoirs of organic matter than many other types of forest), as well as other factors such as long-term climate change. See Terborgh (1992), Huston (1994), and Kricher (1997) for further discussion of this subject.

Some scientists have argued that the threats to tropical woodlands in Latin America (also called tropical dry forests) are more critical than the threats to tropical rain forests because the biota of tropical woodlands approaches the richness of tropical rain forest biota, and far fewer tropical woodlands remain (Janzen 1988a, Mares 1992). Latin American tropical woodlands have been largely overlooked by conservationists because most of them were already converted to rangeland and cropland before the advent of the environmental movement.

Needless to say, when people convert a forest to another type of ecosystem, most of the forest-dependent species are lost from that site for some period. If the forest regrows after a few years, the consequences may be relatively minor, analogous to the loss of habitat that occurs after a natural disturbance. Too often, the loss of habitat is so long-lasting and extensive that species are driven to extinction. It is easy to name forest-dwelling species that are threatened with extinction largely because of deforestation—giant pandas, tigers, gorillas, and many many more—but, of course, these are just the tip of the iceberg (Fig. 8.6). With most of the earth's biodiversity residing in insects and other small organisms, and with many, perhaps most, of these small species living in tropical forests where they remain unknown to science, we can only make gross estimates of the likely impact of deforestation (Lawton et al. 1998). As we saw in Chapter 6, estimates of extinctions based on the impact of tropical deforestation were the foundation for Wilson's (1992) prediction that we are currently in the midst of an extinction spasm. Fully acknowledging the extent of our uncertainty, it is still clear that a large portion of the earth's biodiversity is found in tropical forests and that these forests are being lost to deforestation at a very high rate (Pimm and Raven 2000). Consequently, all conservation biologists believe that protection of tropical forests must be a high priority.

Thus far we have focused on the biological consequences of deforestation, but through changes in the physical environment, deforestation can have effects far beyond the edge of the forest. We have already discussed the importance of soil erosion as a source of sediment that can contaminate aquatic ecosystems. On a global scale, forests affect the earth's climate by acting as reservoirs of carbon, and when they are cut, much of the carbon moves into the atmosphere as carbon dioxide, the major greenhouse gas (Houghton et al. 2000). More locally, because much of the water vapor in the atmosphere above a forest is maintained by evaporation and transpiration, when a forest is cut, rainfall may decrease. This makes the hot, dry conditions of a deforested site even hotter and drier.

Figure 8.6. Many well-known species are in jeopardy because of deforestation, but they are outnumbered by many smaller, less well known species. Pictured here are the giant panda, mountain gorilla, ivory-billed woodpecker, Homerus swallowtail butterfly, medusa tree, and Puerto Rican boa.

Desertification

When you envision a barren, nearly lifeless landscape, do you think of deserts? This image ignores the myriad species that flourish in desert ecosystems, but the fact remains that, generally speaking, fewer species can survive in arid environments than in more humid ones. Therefore it is of great concern to conservationists that the extent of arid land—currently about 35% of the earth's land surface—is apparently increasing because of human activities (Mainguet 1999). In particular, grasslands and woodlands (relatively dry forests in which tree crowns do not meet to form a continuous canopy) are being degraded until they are dominated by sparse, relatively unproductive vegetation (Fig. 8.7). This process is called *desertification*. Many definitions of desertification emphasize a loss of agricultural productivity, whereas conservation biologists are more likely to focus on

Figure 8.7. Desertification can profoundly degrade or destroy grassland and woodland ecosystems.

desertification causing a loss of habitat for many species. Of course, these two factors are linked in various ways. For example, as agricultural productivity declines, people exploit more and more land to meet their needs, thus eliminating more and more habitat for wild species.

Causes of Desertification

In most parts of the world desertification is closely associated with overgrazing (Schlesinger et al. 1990, Noss and Cooperrider 1994, Kerley and Whitford 2000). Too many cattle, sheep, goats, and other livestock consume and trample too many plants, and this alters the species composition and structure of the vegetation and reduces the overall biomass. With few plants to protect the soil and with many animal hooves breaking and compacting the soil, erosion is likely to increase. The excessive burning of grasslands, usually to provide fodder for livestock, may further exacerbate the effects of overgrazing (Savory and Butterfield 1999).

Cultivation is also a major cause of desertification (Khogali 1991). Soil erosion and replacement of natural vegetation by crops are obvious and widespread consequences of cultivation. In irrigated croplands two other problems can arise: salinization and waterlogging (Thomas and Middleton 1993, Contreras-B. and Lozano-V. 1994, Mainguet 1994). Salinization is common when irrigation is used in arid environments; large volumes of water evaporate, leaving behind sodium chloride and other salts that can reach toxic concentrations. Sometimes, farmers try to solve this problem by using enough water to leach the salts lower into the soil, but this practice can eventually raise the water table to the surface, producing waterlogged soils. Cutting trees in woodlands, usually for fuelwood, can also contribute to desertification.

According to some authors, climate change is a major cause of desertification (Dodd 1994). It is indisputable that, over the long term, whenever cyclical changes in the earth's orbit have led to warmer and drier conditions, some grasslands and woodlands have become deserts (see Chapter 6). Thus it is not surprising that the relative importance of long-term climate change and short-term droughts, natural erosion, and human-induced causes of desertification is a complex and controversial topic (Thomas and Middleton 1994). Some argue that anthropogenic factors such as overgrazing are paramount; others argue for climate change; and the truth may lie between these poles and involve their interaction (Ward and Ngairorue 2000). Some authors would simply like to define desertification to include only human-induced degradation. After considering these issues at length, the United Nations Environmental Programme defined desertification as "land degradation in arid, semiarid, and dry subhumid areas resulting mainly from adverse human impact" (Helldén 1991). This definition does not exclude a role for climate change, but it clearly implies that desertification is a problem open to human solutions.

Consequences of Desertification

Desertification and its consequences are often overlooked until they become extreme, in part because it is harder to recognize the work of hungry livestock (the cumulative impact of thousands of small bites) than the work of a hungry chainsaw (Fleischner 1994). A deforested site often looks like a disaster, but an overgrazed ecosystem where grasses have been replaced by unpalatable brush may not look degraded to the untrained eye. Ecosystems that are vulnerable to desertification may not match the wealth of biodiversity of forests, but they do have a large set of unique species that merit the attention of conservation biologists, including such well-known species as African elephants, cheetahs, black-footed ferrets, both black and white rhinos, great bustards, and African wild dogs.

In decrying the loss of grasslands and woodlands to desertification, it is important not to imply that deserts lack biodiversity value. Thousands of species are found in deserts, and many of them are highly endangered: desert tortoises; Asian and African wild asses; sundry species of cactus; and a variety of antelopes such as the addax, scimitar-horned oryx, and Arabian oryx, to name some of the better-known taxa.

It is instructive to think of a continuum of decreasing ecosystem biomass and productivity from forests to woodlands to grasslands to deserts. Ecosystems that already fall in the desert part of this continuum are still vulnerable to desertification and are being pushed further down the continuum of decreasing productivity and biomass. This perspective raises the possibility that some species adapted to the lower end of this continuum might benefit from desertification by having larger areas of habitat (Whitford 1997). This may be true of some common, highly adaptable species, but the rare species of greatest concern to conservation biologists are likely to be habitat specialists that need natural, undisturbed deserts. They are unlikely to survive in degraded ecosystems that are a human-created facsimile of natural desert.

One reason why the consequences of desertification for biodiversity have been significant is that relatively few ecosystems that are vulnerable to desertification have been protected as parks. This is partly because people are slow to recognize degradation of these ecosystems. It is also partly because these lands usually lack the amenities—lakes, mountains, forests—that people seek for outdoor recreation. A notable exception to this generalization comes from eastern and southern Africa, where tourists visit grassland parks to see the spectacular suite of large mammals. The status of United States grasslands is more typical; only about 0.6% of them have been set aside in national parks and wild life refuges (Hunter and Calhoun 1995). Conservationists have tried to rectify this situation, but attempts to establish a large prairie reserve in the United States—large enough that bison could truly roam—have been thwarted by opposition from ranchers. Of course, establishing some grassland parks would not be a complete solution; wiser management of all grassland ecosystems must be the goal.

Draining, Dredging, Damming, Etcetera

Swamps and marshes, bogs and fens, lakes and ponds, rivers and streams, estuaries and the ocean, and more: there is a wide variety of ecosystems—freshwater ecosystems, marine ecosystems, and wetlands—in which water is a medium for life, not just an essential nutriment. Similarly, there is a wide variety of ways in which people destroy these ecosystems by changing their hydrology (Fig. 8.8). We will begin by reviewing some of these methods.

Figure 8.8. A complex of aquatic ecosystems before and after human alterations. In the lower right a housing development that was previously surrounded by dikes is being extended by filling the wetland. Nearby, the channel is being dredged. Upstream the river has been channelized and the adjacent wetlands ditched. A tributary on the right side of the main river has been dammed to create a reservoir. In the real world it would be highly unusual to have all these activities in a small area.

Filling a wet depression with material until the surface of the water table is well below ground is an obvious way to turn a wet ecosystem into a dry one. This method is usually too expensive to use for creating agricultural land, but it is routinely used to create house lots, airports, parking lots, and other high-priced land. Small, shallow wetlands are particularly vulnerable to being filled.

Draining a wet ecosystem (i.e., lowering the water table by moving the water somewhere else) is a common practice. In its simplest form it involves digging ditches that allow the water to drain away. Under the right circumstances, this method can be used to drain large areas relatively easily. In some environments—such as seasonally wet meadows—underground pipes, called tiles, can be buried to provide drainage. Occasionally, water is actually pumped out of a wet ecosystem at great expense.

The primary impetus for both draining and filling is to acquire more land that is useful for human enterprises. The single biggest use for land created in this fashion is agriculture, except in urban and suburban areas where housing developments, shopping malls, and other projects are often the key issue. Occasionally, sites are drained to improve their ability to produce timber, and in some countries peatlands are drained so that the peat can be mined for fuel. Clearly, it is easier to drain or fill a shallow basin than a deep one, and thus wetlands are far more vulnerable to these losses than are lakes, rivers, and estuaries.

Dredging involves digging up the bottom of a water body—the mud and a host of mud-dwelling creatures—and depositing the material elsewhere, often in a wetland that someone wants filled. The goal is usually to maintain a shipping channel in a river or harbor; the ecological result is a scarred bottom and sediment pollution. Sometimes, the sediments contain high concentrations of toxins that are returned to the food web after dredging.

Channelizing rivers and streams means making them straighter, wider, and deeper and replacing riparian (shoreline) vegetation with banks of stone or concrete. This conversion from a complex of natural riverine communities to a barren canal may meet engineering objectives, usually flood control, but is obviously an environmental calamity.

Damming rivers and streams can profoundly change ecosystems both upstream and downstream (Nilsson and Berggren 2000). First, upstream of a dam, a flowing-water ecosystem (the technical term is lotic) is converted to a standing-water (lentic) ecosystem, and wetland and upland ecosystems will also be flooded and thus become part of a reservoir. Wetlands are especially likely to be extensively flooded because their elevation is often close to that of a nearby river. Additionally, many reservoirs are subject to dramatic fluctuations in water level depending on changing demands for electricity and water. This means that the shores of reservoirs are often quite barren because relatively few species can cope with being inundated and then exposed in this manner (Nilsson et al. 1997, Jansson et al. 2000a). Second, downstream of a dam, floodplain ecosystems are likely to be replaced by upland ecosystems if the dam minimizes or eliminates the seasonal floods that are critical to the maintenance of these ecosystems. In the river itself, species are likely to be challenged by flow rates that are very unnatural: too much short-term fluctuation in response to demands for water or electricity, or not enough annual fluctuation in response to rainy and dry seasons. Also, the water temperature may be too warm (if drained from the top of the reservoir), or too cold (if drained from the bottom of the reservoir) (Vaughan and Taylor 1999). A third issue, dams as barriers to the movements of aquatic species, was discussed above in the section on roads and dams as barriers.

Diking consists of constructing earthen banks, usually called dikes or levees, constructed along the edges of water bodies to prevent flooding. Given that floods are natural phenomena vital to the maintenance of many types of ecosystems, diking can easily destroy ecosystems, especially because it is often linked with developing land for other purposes.

Obviously, the world's oceans and seas are too large to be converted to other types of ecosystems by filling, draining, dredging, channelizing, damming, or diking, but they are not completely immune to these processes. The bays, coves, and inlets that line the edges of oceans and seas (often these are estuaries where salt and fresh waters meet) are small enough to be affected by these processes, especially filling and dredging. Furthermore, sometimes our attempts to control currents in these areas by building breakwaters, jetties, and other structures can end up changing marine ecosystems profoundly. For example, shortsighted attempts to maintain sandy beaches by building jetties often end up accelerating beach erosion and sand deposition somewhere else.

To discuss the consequences for biodiversity of filling, draining, dredging, channelizing, damming, and diking, we will focus on the two groups of species that are most vulnerable to these processes: those associated with wetlands and rivers.

Consequences for Wetland Biota

Stemming the loss of wetlands has become a major goal of conservationists for two basic reasons: the rarity of wetlands and their ecological value. Wetlands cover a relatively small portion of the earth's total surface, roughly 1% to 2%, and this portion is decreasing (Harcourt 1992). In the conterminous United States, an estimated 53% of wetlands were lost between the 1780s and 1980s (Dahl 1991), and worldwide figures are probably roughly comparable (Dugan 1993, Mitsch and Gosselink 2000). These facts alone make it imperative to protect remaining wetlands, given a goal of protecting biodiversity at the ecosystem level (see Chapter 4). Furthermore, wetlands are often keystone ecosystems, playing critical roles in a landscape through hydrological processes, biomass production and export, removal of contaminants from polluted water, and so forth. (See Mitsch and Gosselink 2000 for a review of this topic.)

At the species level of biodiversity, wetlands are important because they are habitat for a diverse biota comprising three groups of species. First, there are species that are primarily aquatic (such as many species of fish and insects) that can use the pools of water often found in wetlands. Some may be permanent residents; some may be visitors, coming only at high tide, or during spring or monsoonal high waters.

Second, many terrestrial species are facultative users of wetlands with a portion of their population found in wetlands. Wetlands can be particularly important refugia for terrestrial species that are sensitive to human interference; this is because wetlands tend to be too wet for humans to hunt, plow, or extract trees and too dry for them to access by boat. For example, the mangrove swamps in the mouth of the Ganges River, the Sunderbans, harbor one of the world's largest remaining tiger populations.

Finally, there are many thousands of species that are uniquely adapted to the interface of wet and dry environments found in wetlands. These include whole families of plants (cattails, water lilies, bur-reeds, and many more) and insects (e.g., predaceous diving beetles, water boatmen, and several families of damselflies and dragonflies) that are almost exclusively found in wetlands. Among vertebrates, most amphibians and turtles are wetland species.

Throughout the world the loss of wetlands has pushed many species toward extinction. Nine of the world's 15 species of cranes—birds that require wetlands for breeding and often foraging—are in jeopardy. In recent years herpetologists have been alarmed by precipitous drops in many frog populations, and wetland loss is the primary cause (Houlahan et al. 2000).

Consequences for River Biota

Rivers and streams are often likened to the arteries of a landscape, and this metaphor is apt from both an ecological and economic perspective. It is hard to imagine human history without rivers—bringing water to our croplands and homes, driving waterwheels and turbines, providing a transportation network, and carrying away our wastes. Think about how many of the world's cities are located on a river, and you will appreciate their pivotal role.

Unfortunately, being the focus of so much attention has left many rivers badly degraded by water pollution, channelization, and dredging or converted to reservoirs by dams (Wilcove and Bean 1994). The victims of this scarcity of clean, free-flowing rivers do not draw much public attention because they are chiefly fish, molluscs, and insects, not the birds and mammals that galvanize public support (Allan and Flecker 1993). Scores of riverine fish species are threatened with extinction, but most of them are minnows and other small species that are seldom seen. Only a few economically important fish species such as various salmon species are likely to garner much attention. Even lower on the list of public popularity are mussels, crayfish, and other invertebrates, even though hundreds, perhaps thousands, of species are endangered by river degradation and conversion. One analysis of North American crayfishes and unionid mussels estimated that 63% of the crayfish species (198 of 313) and 67% of the unionid mussels (201 of 300) were either extinct or at some level of risk (Master 1990). Another analysis of the freshwater fauna of North America demonstrated that the recent (since 1900) extinction rate of these animals was about five times greater than that of terrestrial vertebrates and that this difference was likely to persist in the future (Ricciardi and Rasmussen 1999). A similarly dramatic story could be told for Asian rivers—home to over half of the world's large dams (over 15 m tall) and a large portion of the world's freshwater crabs, snails, turtles, crocodilians, river dolphins, and fishes (Dudgeon 2000). For example, there are 105 families of freshwater fishes in Asia compared with 74 in Africa and 60 in South America. For many taxa we do not even have enough information to evaluate rarity or endangerment. For example, 388 algal species were recorded in one stream in southern Ontario (Moore 1972), but few streams have been inventoried this thoroughly, and thus virtually all of their algal species could be eradicated without documentation of their disappearance.

Fragmentation

When early explorers of wild regions found a high vantage point from which to scan the terrain, they often wrote of a "sea of green" to convey the unbroken vastness of the forests and grasslands they traversed. A modern traveler, looking down from a plane, is likely to describe a typical landscape as a "patchwork quilt"—a mosaic of pastures and croplands, woodlots and house lots and parking lots. The process by which a natural landscape is broken up into small parcels of natural ecosystems, isolated from one another in a matrix of lands dominated by human activities, is called *fragmentation*. It is a major focal point for conservation biologists, both because it has degraded many landscapes and because many nature reserves have become isolated fragments or are in danger of becoming so (Saunders et al. 1991).

BOX 8.1   Island biogeography theory

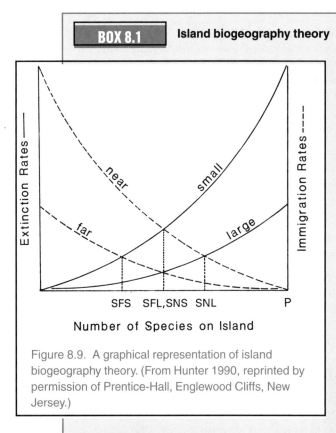

Figure 8.9. A graphical representation of island biogeography theory. (From Hunter 1990, reprinted by permission of Prentice-Hall, Englewood Cliffs, New Jersey.)

The fundamental idea of MacArthur and Wilson's (1967) equilibrium theory of island biogeography is that the number of species on an island represents a balance between immigration and extinction. The rate of immigration is determined largely by how isolated an island is; the more isolated, the lower its immigration rate. This is represented in Figure 8.9 with the curve for remote islands (far) being lower than the curve for islands that are near the mainland (near). Extinction rates are a function of island size; populations on large islands tend to be larger and thus less vulnerable to extinction. In Figure 8.9 the extinction curve for large islands is lower than the curve for small islands.

For any given island there is an extinction rate and an immigration rate that will balance one another and keep the number of species relatively constant. In this example, the numbers of species for four equilibria are represented as follows: *SFS*, number of species on a far, small island; *SFL*, far, large island; *SNS*, near, small island; *SNL*, near, large island. *P* is the total number of species that could potentially immigrate to the island from a nearby landmass.

Fragmentation captured the interest of conservation biologists for two other reasons. First, it was recognized as a major issue at about the same time that conservation biology was emerging as a new discipline; in other words, it was new ground for conservation biology to plow. Second, it appeared to have a theoretical foundation in an intriguing body of ideas and observations known as island biogeography (Box 8.1). It seemed reasonable to assume that the effects of isolation on the biota of oceanic islands might provide a model for understanding the effects of isolation on populations inhabiting patches of natural ecosystems that were isolated in a sea of human-altered land.

Most conservation biologists have come to recognize that the applicability of island biogeography theory to fragmentation issues is somewhat limited, primarily because fragmentation "islands" are not nearly as isolated for most species as true oceanic islands (Zimmerman and Bierregaard 1986, Doak and Mills 1994; also see Harrison and Bruna 1999 and Debinski and Holt 2000). Nevertheless, island biogeography does provide a conceptual foundation for understanding fragmentation and is the origin for two important ideas. Small fragments (or islands) have fewer species than large fragments, and more isolated fragments have fewer species than less isolated fragments. We will begin by considering these two ideas further.

Fragment Size and Isolation

There are three main reasons why large fragments have more species than small fragments (Fig. 8.10). First, a large fragment will almost always have a greater variety of environments than a small fragment (e.g., different types of soil, a stream, a rock outcrop, an area recently disturbed by fire), and each of these will provide niches for some species that would be absent otherwise.

Second, a large fragment is likely to have both common species and uncommon species (i.e., species that occur at low densities), but a small fragment is likely to have only common species. This idea is easy to grasp when we consider species that have large home ranges; for example, it means that we are unlikely to find a bear in a tiny fragment. However, it also applies to species that have rather limited home ranges, but still avoid small fragments. Certain small birds such as ovenbirds and scarlet tanagers have home ranges of only a hectare or two, but are usually not found in forest fragments less than 10 ha in size (Robbins et al. 1989). Species that do not occur in small patches of habitat are called *area-sensitive species* and are often of concern to conservationists. Furthermore, uncommon species that are not area-sensitive (i.e., that can find habitat in a small fragment) are also unlikely to occur in a small patch by chance alone. This last point is a subtle one that is often overlooked (Haila 1999), but it is easily explained with an example. Imagine there was an uncommon tree species that had an average density of one individual per 1000 ha; all other things being equal, a 100-ha sample plot would have a 1:10 chance of containing this species, but a 10-ha plot would have only a 1:100 chance. This sampling effect, added across many species, would mean that a small fragment would have fewer species than a large fragment simply because it is a smaller sample. To adjust for this phenomenon, fragmentation studies should focus on number of

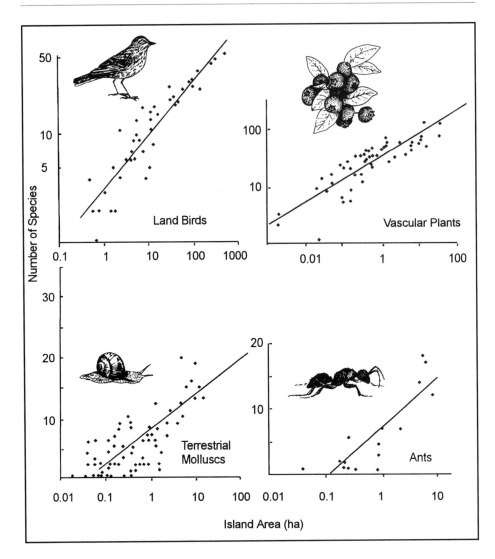

Figure 8.10. The number of species in a sample plot or on an island increases as area increases, but the steepness or slope of the curve varies considerably among taxa. Note that in these graphs for taxa on islands in the Baltic Sea some of the y axes are linear and some are logarithmic. All of the x axes are logarithmic. Recall from Chapter 6 that these lines are described by the formula $S = CA^z$, where S = number of species, A = area, and C and z are constants. (Redrawn from Järvinen and Ranta 1987.)

species per unit area (e.g., Rudnicky and Hunter 1993), but most only report the number of species in each fragment.

The third major reason that small fragments have fewer species than large fragments is a key component of island biogeography theory. Simply put, small fragments will, on average, have smaller populations of any given species than large islands, and a small population is more susceptible to becoming extinct than a large population. This idea was a key point in the preceding chapter.

Fragments that are isolated from other, similar patches by great distances or by terrain that is especially inhospitable are likely to have fewer species than less isolated fragments for two reasons. First, relatively few individuals of a given species will immigrate into an isolated fragment. Immigrating individuals are important both because they can "rescue" a small population from extinction and because they can replace a population that has already disappeared (Brown and Kodric-Brown 1977). Second, species that are mobile enough to use an "archipelago" of small habitat patches to collectively comprise a home range are less likely to use an isolated fragment simply because it is inefficient to visit it. Consider a great blue heron that travels from its colony to forage in several different wetlands; an isolated wetland, far from the colony and other wetlands, is less likely to be visited (Gibbs et al. 1991).

Causes of Fragmentation

The fundamental cause of fragmentation is expanding human populations converting natural ecosystems into human-dominated ecosystems. Fragmentation typically begins when people dissect a natural landscape with roads and then perforate it by converting some natural ecosystems into human-dominated ones (Fig. 8.11). It culminates with natural ecosystems reduced to tiny, isolated parcels. Thus fragmentation involves both reducing the area of natural ecosystems and increasing their isolation. As the single largest user of land, agriculture is the proximate cause of most fragmentation. Fly over a landscape where agriculture prevails, and you will find the natural ecosystems reduced to isolated bits of nature. For most terrestrial species, a large expanse of cropland is a barrier nearly as effective as a stretch of water. Of course, other human land uses, notably urban and suburban sprawl, can also leave natural ecosystems in isolated patches, but their total area is much more limited than that of agriculture. Some writers use "fragmentation" to describe any process that breaks up extensive ecosystems, including natural events such as fires, whereas other writers restrict the term to human-induced changes. In any case, human activities are the major cause of fragmentation in most landscapes.

Sometimes, it is unclear whether human land uses cause fragmentation. Consider clearcutting forests; if this leads to the forest's being converted to farmland, then clearcutting obviously contributes to fragmentation. However, if the clearcut site is allowed to undergo succession and return to forest, this may or may not constitute fragmentation, depending

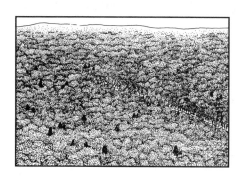

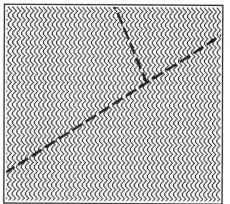

Dissection

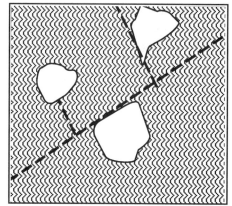

Perforation

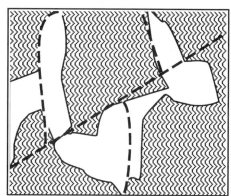

Fragmentation

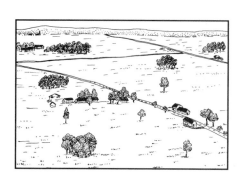

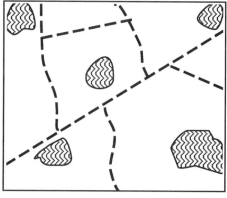

Attrition

on whether the clearcut is extensive enough to constitute a significant barrier to the movement of plants and animals (Haila 1999). Of course, this will vary from species to species. A slow-moving, moisture-loving slug is far more likely to be deterred by a clearcut than most birds that can fly across a clearcut in a few seconds. Similarly, at what point on the continuum of desertification does fragmentation occur? Most animal species may be able to cross a landscape degraded by moderate overgrazing, but could be blocked if the vegetation is so sparse that it does not even provide a modicum of cover. A plant whose seeds are dispersed long distances by wind or animals may cross a desertified barrier easily, whereas a short-dispersal plant may be incapable of crossing the barrier in one trip and unable to establish a population halfway across in the degraded habitat.

Our understanding of these subtler, potential forms of fragmentation is somehat limited because the majority of fragmentation studies has taken place in landscapes dominated by agriculture with forests reduced to islands. Studies in landscapes where forests or grasslands form the matrix are relatively uncommon (Rochelle et al. 1999). Intuitively, it makes sense that the greater the difference between the natural ecosystem and the human-dominated ecosystem, the more likely it is that the human-dominated ecosystem will isolate the biota of the natural fragment. This suggests that fragmentation effects can be minimized by limiting our manipulations of managed ecosystems (e.g., forests managed for timber production or grasslands managed for livestock production) so that they resemble natural ecosystems as closely as possible. We will return to this idea in Chapter 11, Managing Ecosystems.

Consequences of Fragmentation

Ecosystem destruction is the driving force behind fragmentation, and thus it is inevitable that fragmentation is associated with negative effects on biodiversity. The reason that fragmentation elicits so much special concern from conservationists is that its consequences are much greater than we would anticipate based solely on the area of ecosystems destroyed. Notably, remnant ecosystems that seem to have escaped destruction may no longer be available for area-sensitive species that cannot use small patches of habitat. Most prominent among these are large predators that need extensive home ranges to find enough prey (Crooks and Soulé 1999). Some small

◄ Figure 8.11. People usually initiate fragmentation by building a road into a natural landscape, thereby *dissecting* it. Next, they *perforate* the landscape by converting some natural ecosystems into agricultural lands. As more and more lands are converted to agriculture, these patches coalesce and the natural ecosystems are isolated from one another; at this stage *fragmentation* has occurred. Finally, as more of the natural patches are converted, becoming smaller and farther apart, *attrition* is occurring. (Terminology from R. Forman, personal communication, and 1995; also see Collinge and Forman 1998.)

species that use small home ranges also avoid small habitat patches, for example, birds (Robbins et al. 1989) and beetles (Gibbs and Stanton 2001). This may occur because they require the microclimate characteristic of the interior of large habitat patches, or because they select habitat patches large enough to support other members of their species (a type of loose coloniality) (Stamps 1991), or because of their interactions with other biota as predators, prey, or competitors (Gibbs and Stanton 2001).

In highly fragmented landscapes, it is difficult for individuals (usually juvenile animals, seeds, or spores) to disperse to another suitable patch of habitat. If immigration and emigration are very limited, then the individuals occupying a fragment may effectively constitute a small independent population. As we saw in Chapter 7, small populations are more likely to disappear. Furthermore, if a population does disappear, a low immigration rate will mean it takes much longer to establish a new population. Even if fragmentation only leads to partial isolation, this may change one large population into a metapopulation, which may also affect population viability and persistence. The dispersal of fire is also an issue; fragmentation has greatly disrupted natural fire regimes in regions where fires once swept across the landscape (Leach and Givnish 1996).

The migration of animal species that travel between habitats seasonally could be impeded by fragmentation (Hunter 1997). In practice, this is likely to be a problem mainly for species that walk rather than fly such as large mammals that travel up mountains in the spring and down in the fall, or amphibians that migrate to and from spring breeding pools. Similarly, over long periods the climate changes described in Chapter 6 require species to shift their entire geographic ranges. In a fragmented landscape this may be difficult for sedentary species with limited dispersal abilities such as many plant species (Peters and Lovejoy 1992).

Finally, one consequence of fragmentation is based on a simple rule of geometry: the perimeter length of a patch changes as a linear function, whereas its area changes as a square function. To take a simple example, a 4×4 km patch has a perimeter of 16 km and an area of 16 km^2, and if we decrease it to 2×2 km, its perimeter halves to 8 km, but its area decreases fourfold to 4 km^2. This means that as fragmentation makes patches smaller and smaller, their ratio of edge to interior increases disproportionately (Fig. 8.12). Similarly, if we define a zone in the patch that is within a certain distance of the patch's edge, the relative area of this edge zone will also increase disproportionately as the patch gets smaller. Finally, although fragmentation does not necessarily affect the shape of a patch, it should be noted that another rule of geometry (a circle is the shape with the shortest perimeter) means that the farther a patch's shape departs from circular, the longer its edge will be.

Why is it important that small patches have relatively more edge habi-

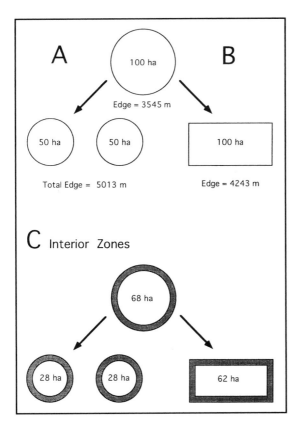

Figure 8.12. Three principles of geometry that affect the edge-to-area ratios of patches: (A) small patches have relatively longer edges than large patches. (B) Patches that are less circular in shape have longer edges than circular patches. (C) The interior zone of a small or non-circular patch is relatively small compared with that of a large, circular patch. (In these patches the shaded edge zone is 100 m wide.)

tat and less interior habitat? First, the physical environment near an edge is different. For example, in a woodlot surrounded by fields, the forest edge zone will usually be windier, drier, and warmer in the summer and colder in the winter than the forest interior, and this may prevent some species, especially plants, from using this zone (Chen et al. 1992, Matlack 1993, Matlack and Litvaitis 1999). A study of forest fragmentation in Amazonia has found much higher levels of tree mortality near edges (Bierregaard et al. 1992). Second, species associated with the surrounding disturbed habitat may penetrate the edge zone. These are likely to include exotic species (e.g., competitors such as weeds and predators such as cats, rats, and people) that we will discuss in Chapters 9 and 10, Overexploitation and Exotic Species. The best-documented problem associated with edges is that birds nesting near forest-farmland edges often experience unusually high levels of nest predation (Paton 1994). Furthermore, in many parts of North America, nesting near edges brings an added challenge from brown-headed cowbirds, a brood parasite that lays its eggs in other birds' nests causing the loss of its hosts' offspring (Brittingham and Temple 1983). Sometimes, it is hard to tease out the relative importance of patch size, distance to edge, and the overall loss of natural ecosystems in a region (Hartley and Hunter 1998, Fauth 2000, Kurki et al. 2000). Nevertheless, it seems clear that whenever we have a natural ecosystem surrounded by a disturbed ecosystem, the natural ecosystem is going to experience some disturbing effects, what Dan Janzen (1986) has called "the eternal external threat." The width of these "impact zones" will vary greatly, from tens of meters in the case of microclimate issues to kilometers in the case of poachers invading a protected reserve.

Case Study   Madagascar[1]

The island of Madagascar lies only 400 km from the cradle of human evolution in Africa, yet for thousands of years after *Homo sapiens* had spread throughout most of the world, Madagascar remained undiscovered by people. Madagascar was not, however, isolated from all primates. At least 50 million years ago some primitive primates colonized the island, perhaps floating to the island on a tree swept to sea in a flood. Eventually, they evolved into dozens of species represented in modern times by four families: lemurs, dwarf lemurs, indris, and the aye-aye (Fig. 8.13). Having split away from Africa about 80 to 160 million years ago, Madagascar was an isolated haven for evolution in many life-forms besides primates. Seven families of plants, five of birds, and five of mammals are restricted to Madagascar and nearby isles; overall roughly 80% to 90% of all the native plant and animal species are endemic to the island. The biota is rich, as well as unique. For example, Madagascar has about 12,000 plant species compared with Europe's 12,500, even though Madagascar is only about the size of France. Frogs provide a more impressive comparison; Madagascar has about 250 species (all but two of them endemic), while only 25 frog species inhabit all of Europe. Madagascar's great climatic and geologic diversity is probably the main reason for its biotic diversity. The island is subdivided into many regions with profoundly different topography, geology, soils, and weather patterns. Collectively, these provide a diverse array of environments from rain

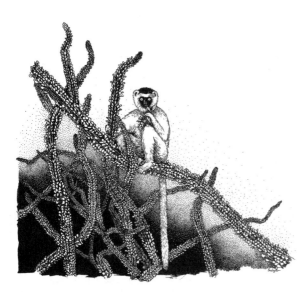

Figure 8.13. Madagascar is home to many unique species such as the Verreaux's sifaka and the didierea trees it often climbs.

forests to semiarid lands dominated by didiereas, an endemic group of spine-covered plants vaguely reminiscent of cacti.

Madagascar's isolation came to a rather abrupt end less than 2000 years ago when human colonists arrived, probably from both Africa and Southeast Asia, and began shaping the land to their needs. The Malagasy people set fires to produce fodder for large herds of cattle and cleared the forest for "slash and burn" agriculture. It is unclear how extensively forested Madagascar was when people arrived. Some ecologists have assumed that some type of forest or woodland covered the whole island; others believe that some parts of central Madagascar were grasslands. In either case, virtually all types of ecosystems are highly degraded today, and only about 10.4% of the island is forested: 5.8% rain forest, 2.9% spiny woodlands, and 1.7% other types (Nelson and Horning 1993). The rain forests along the island's eastern escarpment have suffered less loss than many other areas, presumably because they are not very amenable to human occupation. However, they are disappearing now. Originally, they covered about 11.2 million ha, but by 1950 only 7.6 million ha remained, and in 1985 only 3.8 million ha remained (Green and Sussman 1990). Presumably, the loss of forests and resulting siltation has also affected freshwater and coastal ecosystems (which include many mangrove swamps and coral reefs), but this has been little studied. We do know that Madagascar's unique freshwater fish fauna is severely threatened by deforestation, overfishing, and exotic species (Benstead et al. 2000). One port, Mahajanga, was lost after 100 million m^3 of sediment were deposited in 25 years. Many of Madagascar's most striking species—elephant birds, giant lemurs, and giant tortoises—disappeared quite soon after human colonization. The fact that many of the species were relatively large suggests that overhunting played a role in their demise too, and we will return to this issue in the next chapter. The bottom line in all of this is that the growing numbers of Malagasy people and cattle have made ecosystem degradation and loss almost inevitable. (There were about 2.5 million people in 1900, 4 million in 1950, and 15 million in 2000; it is generally estimated that the Malagasy's keep about one head of cattle per person.)

Conservationists throughout the world have set their sights on Madagascar because the stakes are so high (we have so many unique taxa to lose) and the threats so enormous. Here lies some ground for optimism. Ambitious projects to protect some key examples of various ecosystems, to foster ecotourism and other forms of sustainable development, and to improve land-use practices throughout the island are under way with sponsorship from a diverse array of national and international organizations.

[1] This section was primarily based on Jolly (1980), Jolly et al. (1984), Dewar (1984), Jenkins (1987), Reinthal and Stiassny (1991), Groombridge (1992), Quammen (1996), Goodman and Patterson 1997), and personal communication with Eleanor Sterling.

Summary

Ecosystems and habitats (the physical and biological environment used by a particular species) are routinely degraded, and sometimes destroyed, by human activities. These activities are the most critical threat to biodiversity. Contamination of air, water, soil, and organisms by pollutants is a major form of degradation. Pollution can range from relatively innocuous materials such as sediment that smothers the bottom of a stream to extraordinarily toxic chemicals that are lethal at small doses. Sometimes, populations are eliminated outright by pollution, especially by pesticides; more often, pollution represents a stress that reduces population fitness. People also construct many physical structures that may degrade habitat quality for certain species. Roads are the best-known examples; they impede the movement of some organisms, and, worse still, some organisms are run down by vehicles. Moreover, dams and fences are likely to be absolute barriers to the movements of some species. Ecosystems can also be degraded by altering physical processes. For example, people commonly 1) accelerate soil erosion, which causes silt pollution and decreases site productivity; 2) decrease the frequency of fire in ecosystems where it is a natural event, or increase the frequency of fire where it is uncommon; and 3) remove too much water from ecosystems where it is needed.

Deforestation is a major form of ecosystem destruction that has profound consequences for biodiversity because forests cover less than 6% of the earth's total surface area yet are habitat for a majority of the earth's known species. Deforestation has slowed in many temperate regions, but tropical deforestation continues at an alarming pace, especially given the incredible biodiversity tropical forests harbor. Many arid and semiarid ecosystems are being degraded and even destroyed by a process called desertification, primarily the product of overgrazing by livestock and unsound cultivation. Myriad species occur in these environments and are at risk because of desertification. Many aquatic ecosystems have been destroyed by profound changes in their hydrologic regime imposed by filling, draining, dredging, damming, channelizing, and diking. Rivers and wetlands have been especially vulnerable to these alterations. For this reason, and because they represent a small portion of the earth's area, the species tied to these ecosystems are in considerable jeopardy.

Fragmentation is the process by which a natural landscape is broken up into small parcels of natural ecosystems isolated from one another in a matrix of other ecosystems, usually dominated by human activities. Fragmentation can diminish biodiversity because small, isolated patches of habitat have fewer species than larger, less-isolated patches. This is true because 1) small patches have less environmental heterogeneity than large patches, 2) some area-sensitive species and uncommon species are unlikely to be found in small patches, 3) small patches have small populations that are more vulnerable to local extinction, 4) immigration into populations occupying isolated patches is limited, and 5) isolated patches are less likely to be used by species that routinely travel among patches. Besides affecting biodiversity by reducing patch size and increasing isolation, fragmentation also creates more edges between different types of ecosystems. These edge zones represent degraded habitat for many species.

FURTHER READING

Many books review the various ways the earth has been degraded by pollution; one of the better textbooks is Miller (1999); for statistics and an overview, see the periodic reviews published by the World Resources Institute and its collaborators; WRI is accessible on the web at www.wri.org. For books on the degradation and destruction of various types of ecosystems, see Mainguet (1994) on desertification, United Nations Development Programme et al. (2000) and Schelhas and Greenberg (1996) on deforestation, Mitsch and Gosselink (2000) on wetlands, Boon et al. (2000) on rivers, and Norse (1993) on marine ecosystems. No comprehensive recent review on fragmentation is available, but Harris (1984) and Burgess and Sharpe (1981) are classics, and Rochelle et al. (1999) is a useful compendium. Quammen (1996) is a very readable account of island biogeography and fragmentation. For insights into prehistorical degradation see Redman (1999).

TOPICS FOR DISCUSSION

1. Do you think that, whenever people significantly change an ecosystem from its natural state, this constitutes ecosystem degradation? Recall the example of ecosystem degradation used at the beginning of the chapter—a power plant warming a river, causing temperature-sensitive species to disappear. Would you consider this ecosystem degradation if all the species that disappeared were common and they were replaced by a larger number of species, all of them native, including one that is an endangered species?

2. Describe some reasonable thresholds at which habitat degradation can be considered habitat destruction, or ecosystem degradation can be considered ecosystem destruction.

3. List some characteristics of various species that might be predisposed to be sensitive to contamination.

4. Describe some ways in which different types of barriers (e.g., roads,

dams, fences, agricultural lands) might affect different types of movement (e.g., migration, dispersal, and movements around a home range).

5. Outline some of the fundamental similarities and dissimilarities between deforestation and desertification.

6. Why are lakes less vulnerable to ecosystem destruction than rivers?

7. Fragmentation is usually considered to be a terrestrial process. Can you think of circumstances under which it may affect aquatic ecosystems? (It may be helpful to refer to Fig. 8.10.)

Chapter 9
Overexploitation

Very few things are as poignant and gripping as the remains of a dead creature. The carcass of a slaughtered elephant can move people to action far more readily than the eroded land on which it died. Even a truckload of logs is more likely to catch people's attention than the fumes generated by the truck. Of course, people kill other organisms all the time. It's just that the most provocative examples—killing other sentient beings, especially mammals and birds—are well hidden from most of us. The closest we come to killing is swatting flies, weeding a garden, or giving a dog a flea bath. Even at the grocery store, with its huge arrays of dead plants and animals and their products, we are unlikely to think about the organisms that die to feed us. Intellectually, most people can accept the killing of other creatures for human well-being until it gets out of hand, until people start overexploiting other species. Then, our emotions join with our intellect to decry this threat to biological diversity.

There is a tendency to think that overexploitation (which we can define as human overuse of a population of organisms to an extent that threatens its viability or radically alters the natural community in which it lives) is a relatively new phenomenon. It is a romantic notion that throughout most of our span on earth we have lived in harmony with nature. This view is a bit naive, as we will see in some examples of past overexploitation.

The Long History of Overexploitation

After the most recent glaciation the grasslands of central North America harbored an extraordinary array of large mammals. The diversity of antelopes, horses, cheetahs, giant ground sloths, mammoths, mastodonts, and others easily rivaled the large mammal fauna of Africa today (Fig. 9.1). However, about 11,000 years ago, at the end of the Pleistocene epoch, they disappeared; 34 genera of large mammals became extinct in less than 1000 years, while 40 more became extinct in South America (Martin 1984, Martin and Steadman 1999). This is a massive die-off when you consider that only 20 large mammal genera had become extinct in North America over the previous 3 *million* years. Is it a coincidence that so many large mammals went extinct shortly after the time that humans, crossing from Siberia to Alaska, probably first arrived in the Western Hemisphere? Paul Martin, an anthropologist, thinks not and has argued in many articles and books that overhunting was primarily responsible for these extinctions (e.g., Martin 1984, 1986). Martin's critics have argued that the extinctions were primarily the result of significant climate change (e.g., Graham and Lundelius 1984, Webb 1984). If climate change were responsible, we might expect many small mammals, primarily rodents, to have become extinct in the region too; however, only four North American genera became extinct at the end of the Pleistocene, compared with 46 genera over the preceding 3 million years (Martin 1984). On the other hand, perhaps small mammals are less susceptible to climate change than large mammals. We cannot definitively say

Figure 9.1. Some scientists believe that human overexploitation was responsible for the extinction of many large North American mammals about 11,000 years ago. Shown here, from bottom left, clockwise, are a stegomastodont, giant ground sloth, bush-antlered deer, cheetah, mammoth, and an American horse.

which answer is correct, and the truth may lie in the middle (Marshall 1984), but many scientists believe that the end of the Pleistocene saw a massive overkill by the first human inhabitants of the Americas.

The best evidence that overhunting by early people has eliminated some species comes from islands. On many remote islands, birds evolved in the absence of mammalian predators, sometimes losing their ability to fly in the process. When people arrived on these islands, they found easy prey. For example, when Polynesians, now known as Maoris, arrived in New Zealand about A.D. 1200, the islands had 11 species of moas, a group of flightless birds ranging in size from a turkey to far larger than an ostrich (Fig. 9.2) (Anderson 1989). By the time Europeans colonized the islands in

Figure 9.2. In this 1903 photo two Maori medical students pose beside a reconstruction of a moa. (Photo by A Hamilton. Reproduced courtesy of the National Museum, New Zealand.)

the 1700s, the moas were all gone along with five species of rail and six waterfowl species. Indeed, some evidence suggests that all the moas were extinct less than 100 years after Polynesian colonization (Holdaway and Jacomb 2000). The demise of the moas and other birds undoubtedly was hastened by forest clearing and other changes wrought by the Maoris, but the abundance of moa remains at Maori village sites makes it clear that hunting was a major factor.

On small islands throughout the Pacific, scores of birds are known to have become extinct after the arrival of Polynesians (Steadman 1995). In the Hawaiian islands 44 species of endemic land birds out of 82 became extinct between the arrival of Polynesians and the arrival of Europeans (Olson and James 1984). Again, habitat changes were undoubtedly important, but it is likely that overhunting was a major problem, especially for various species of flightless geese, ibises, and rails. As we saw in the preceding chapter, on Madagascar the loss was not limited to birds. The arrival of people 1500 to 2000 years ago caused the extinction of two giant tortoises, a bear-size giant lemur, a small species of hippopotamus, many other mammals, and several elephant birds, some of which rivaled the largest moas in size (Dewar 1984, Burney and MacPhee 1988).

More recently, the history of the United States provides many striking examples of overexploitation (Trefethen 1964, Mowat 1984, Matthiessen 1987, Wilcove 1999). During the colonial period beaver, turkey, and white-tailed deer were nearly eradicated from the coastal plain, and as the frontier moved farther west, the wave of exploitation followed. The arrival of railroads in the 19th century provided easy access to large urban markets for game animals harvested on the frontier (Fig. 9.3). Market hunting led to the demise of the passenger pigeon, arguably one of the most abundant birds ever to have lived, and took the American bison from extreme abundance to extreme rarity. The heath hen, Carolina parakeet, Labrador duck, and great auk were hunted into extinction. The great whales pursued around the world by Yankee whalers may not recover. Overcutting of trees throughout the country has not driven any tree species to extinction, but it has widely degraded the ability of forests to support a diverse biota. Currently, the worst forms of overexploitation may be happening through global overfishing. This is partially disguised by the fact that we are still able to harvest huge quantities of marine species; only on closer inspection does one notice that the predatory fish that used to dominate catches are being replaced by species further down the food chain (Casey and Myers 1998, Pauly et al. 1998).

Note that our long history of overexploitation can never be used to justify current overexploitation. It would be akin to justifying humans killing one another by pointing to our long history of war and smaller conflicts.

Figure 9.3. Commercial exploitation for urban markets has devastated populations of many species. (Photo from the William Hornaday Collection of the U S. Library of Congress.)

Types of Exploitation

Commercial Exploitation

Money "makes the world go round" and is the driving force behind most exploitation of wild life. Significant sums of money are involved because of the importance and diversity of products obtained: food, fiber, fuel, medicine, building materials, and more (recall Chapter 3). When we think of people who make a living selling wild life, we often think of small, independent entrepreneurs: fur trappers, loggers, clam diggers, and others. In practice, the scale of commercial exploitation of wild creatures ranges from children selling berries by the roadside on a Saturday afternoon to some of the world's largest multinational corporations logging trees and government-owned fleets combing the seas for fish.

Unfortunately, commercial exploitation of wild life can easily become overexploitation for at least eight reasons.

1. The potential market for wild products is enormous. Indeed, with a global economy, once a wild product enters commerce, there are over six billion potential consumers (Fitzgerald 1989, Hemley 1994). The major markets for rhino horns and elephant ivory obtained in Africa are in the Far

TABLE 9.1 World trade in wild life.*

Primates	20,000–25,000 live
Birds	450,000–600,000 live
Reptiles	800,000–one million live, farmed reptiles
	400,000–600,000 wild-caught, live reptiles
	1–6 million skins and skin pieces
	300,000 manufactured products
Ornamental fish	500–600 million (freshwater and marine species)
Corals	775–1100 metric tons of live and raw coral
	1.5–1.6 million raw and live coral items
	7,500–40,000 carvings
Orchids	65,000 wild-collected orchids
	917 million artificially propagated live plants
	39,000 wild-collected roots
	300,000 artificially propagated roots
Cacti	20,000–40,000 live plants
	30,000–60,000 seeds
	340,000–500,000 parts and products

*The data represent minimum estimates of the annual trade in some wild species. They include both species collected in the wild (e.g., most marine fish caught for the pet trade) and wild species propagated in captivity and then traded internationally (e.g., most freshwater fish). Total declared value of wild products is estimated to be 5 to 8 billion U.S. dollars annually, excluding timber and fisheries products.
Source: TRAFFIC (USA), a program of the World Wildlife Fund.

East. Coral collected in the Philippines is destined for Europe and North America (Table 9.1) (Fig. 9.4).

2. People who exploit wild life for financial gain, like everyone else, have an enormous desire for money. First, they need food, clothing, and shelter; then a car, a second car, and a second home; and then status and power. This is in sharp contrast to subsistence-based exploitation as we will see below.

3. Domestic substitutes for wild products are not identical and often sell for less. People can usually sell wild berries for more than cultivated ones, wild (slowly grown) wood for more than plantation-grown wood, venison for more than beef, and pheasants for more than chickens.

4. The market price of a wild species usually increases as it becomes rarer, and this will precipitate greater exploitation and will make the wild species even rarer. For example, at the end of the 19th century the demand for hat feathers pushed egrets into the most remote regions of the southeastern United States, but hunters pursued them relentlessly as the price of

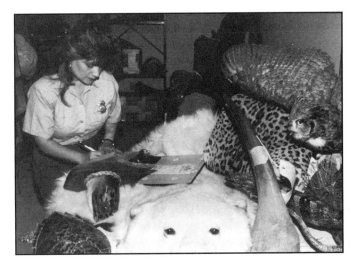

Figure 9.4. Government agents attempt to confiscate illegal wild life trade items. (Photo by John and Karen Hollingsworth, U.S. Fish and Wildlife Service.)

decorative plumes rose to twice their weight in gold (Bent 1926). This vicious cycle is exacerbated by the desire of people to have what their peers do not have: perhaps a shawl woven from shahtoosh, the neck fur of Tibetan antelopes, or a Brazilian rosewood guitar. The royalty of medieval Europe purchased unicorn horns (actually narwhal tusks) for 20 times their weight in gold (Lopez 1986).

5. **Wild resources are often communal resources,** owned by no one and everyone. This means that the costs of overexploitation are shared by many people, not just the person who is abusing the resource, while the benefits are obtained by the exploiter. This is what Garrett Hardin (1968) has called the "Tragedy of the Commons." This dilemma commonly applies to aquatic species because individuals do not usually own the wild life of lakes and seas, whereas in terrestrial systems landowners usually own the plants and sometimes the animals. In many countries the major landowner is the government (national, regional, or local), and the private individual is relatively free to overexploit. (We will return to the tragedy of the commons in Chapter 16, Economics.)

6. **Wild life is often found in remote places** where laws and social constraints do not operate effectively. It is much easier to use wild life irresponsibly on the high seas or in a remote forest than under public scrutiny.

7. **Commercial users often have the capital to purchase expensive technology** for collecting wild life in large quantities: for example, seagoing vessels for fishing and whaling, logging machinery, and even helicopters with which to poach elephants and rhinos.

8. The disparity among national currencies makes it profitable to exploit rare species around the world. Expansion of the global marketplace through increased transportation and lowering of trade barriers means that overexploitation is likely to occur whenever there is a large difference in the buying power of currencies. For example, the strength of the Japanese yen against the U.S. dollar has driven the dockside value of a single bluefin tuna to over $20,000 for a United States fishing boat. That is over $20 per pound even before the costs of shipping, handling, auctions, wholesalers, and retailers are added to what the consumer must pay. At such high prices, U.S. consumers would not pay for tuna, and the market would dry up. However, in Japan, where a cup of coffee can cost $15, bluefin tuna still seems reasonably priced, and Japanese consumers eat it regularly. They thus provide an incentive for U.S. fishers to continue to pursue bluefin tuna even when they have become quite rare. Likewise, the live-fish market for coral reef groupers (Serranidae) in China has stimulated Indonesian fishermen to dynamite reefs, hoping to stun one of these rare, large fish (a highly destructive practice that destroys entire reefs). It is now possible for fishers to attain half of their annual salary catching a fish that is valued by a distant country with a strong currency. This effectively limits the traditional check to overexploitation, which dictates that effort will decline as the exploited species becomes rare.

Subsistence Exploitation

Most rural people exploit wild life to directly meet some portion of their personal needs for food, clothing, fuel, and shelter (Fig. 9.5) (Prescott-Allen and Prescott-Allen 1982, Robinson and Redford 1991, Robinson and Bennett 2000). Among some rural people—especially those who are more affluent—these activities, like a Saturday spent fishing or gathering mushrooms, are just supplemental to the household economy. They are motivated primarily by recreational needs and secondarily by subsistence needs. On the other end of the continuum, some rural people obtain virtually all of their life requisites by gathering and hunting wild species. Worldwide, most rural people fall in the middle of this range, obtaining a moderate portion of their needs from the wild, especially fuel and building materials.

In contrast to commercial exploitation, the scale of subsistence exploitation is limited by the number of people living in places where they have access to wild life and by their levels of consumption (items 1 and 2 in the preceding section). (We will pass over the question of whether bartering with wild products within a local community constitutes subsistence or commercial exploitation.) This is not to say that subsistence use cannot lead

Figure 9.5. Subsistence use of wild plants and animals is very important for many rural people.

to overexploitation (witness the moas), only that it is less likely to lead to overexploitation than commercial use.

Recreational Exploitation

Many people routinely use wild life just for the fun of it. For example, among adults in the United States 39% use wild animals recreationally; i.e., there are an estimated 14 million hunters, 35 million anglers, and 63 million "wildlife watchers" (people who participate in outdoor activities that focus on viewing wild animals) (U.S. Fish and Wildlife Service 1997). When we think about recreational exploitation of wild creatures, hunting and fishing come to mind first, perhaps because killing animals is considered the ultimate form of exploitation. Much has been written about the pros and cons of these sports from a conservation perspective (Mitchell 1982, Mighetto 1991, Liddle 1997). On the one hand, sport hunters and anglers have badly overexploited some populations, for example, extirpating game fish from some small lakes. On the other hand, in many countries sport hunters and anglers have contributed huge sums of money to conservation through license fees and taxes on their equipment, and these monies have been used for purchasing habitat, hiring wardens and biologists, and so on (Kallman et al. 1987). Often these funds are used for self-serving purposes such as stocking streams with hatchery-reared trout, but most have benefitted a broad spectrum of game and nongame species to some degree. Also, funds spent by hunters and anglers for lodging, food, and guide services go a long way toward developing local support for conservation in rural areas, especially in developing nations (Lewis and Alpert 1997), and they often far exceed those spent by "wildlife watchers" (U.S. Fish and Wildlife Service 1997). Incidentally, some of the worst cases of overexploitation come from hunters who pursue smaller prey such as butterflies and molluscs (New 1997). Naturalist collectors are notorious for going to great lengths to add rare species to their collections.

Turning to the naturalists who simply seek contact with wild life for viewing or photography, they too, like hunters and anglers, exploit wild creatures, although their activities are usually called "nonconsumptive" (Edington and Edington 1986, Liddle 1997). Shy animals will be frightened; small plants and animals will be trampled. A well-known anecdote among bird-watchers recounts how a large group of birders gathered at a marsh to search for black rails, an extremely shy bird that is usually seen only when flushed at close quarters. The birders lined up and swept across the marsh, but no rails were flushed. After everyone else had left, one birder recrossed the marsh and spotted a black rail under a tuft of grass, crushed to death. In Africa, tourists have made life difficult for cheetahs by gathering whenever a cheetah is found on its kill (Fig. 9.6). These clusters of vehicles attract the attention of other predators that may drive the cheetah from its food (Edington and Edington 1986). One factor in the global decline of coral reefs is curious divers simply touching corals (Hawkins et al. 1999). Some effects may be quite subtle; for example, just having people nearby can influence how an animal spends its time, shifting from resting and foraging to monitoring humans (Galicia and Baldassarre 1997, Liddle 1997, Steidl and Anthony 2000). We will return to some of the pros and cons of ecotourism in Chapter 16, Economics.

Incidental Exploitation

Not all exploitation is deliberate; often in the process of exploiting one species, other species are incidentally exploited as well. This phenomenon is so common in fishing that there is a specific term for this unintentional mortality: *bycatch*. The best-known example of this involves setting nets around schools of tuna and drowning dolphins in the process, a practice that has been sharply curtailed in recent years because of the popularity of dolphins. Unfortunately, other forms of fishing continue to kill seabirds and marine mammals; indeed, incidental mortality in gill nets is the major

Figure 9.6. Even nonconsumptive use of wild life can be harmful, as when tourists attract hyenas to the kill site of a cheetah.

threat to the world's most endangered marine cetacean, Mexico's vaquita (D'Agrosa et al. 2000). In gross terms, trawling for shrimp may be the most destructive form of fishing; commonly 90% of the catch, by weight, is by-catch of unwanted species that are dumped overboard dead (Bricklemyer et al. 1989) (Fig. 9.7). Most of these species lack the charisma of dolphins, but because shrimp trawling has killed many Kemp's ridleys, a highly endangered sea turtle, United States trawlers must now have a TED (turtle exclusion device) to allow turtles to escape (Crowder et al. 1994). Trawling is particularly destructive when it scours the sea bed, obliterating the structural diversity created by kelp, sponges, and other species (Watling and Norse 1998). Traps on land can also be nondiscriminating: for example, gorillas are occasionally caught in snares set to catch duikers (small forest antelopes), and giant pandas are caught in musk deer snares (Schaller 1993, Noss 1998).

Indirect Exploitation

The term "indirect exploitation" could be used to cover a wide set of human activities that indirectly kill other organisms: the roads, fences, antennas, and so forth described in the preceding chapter; the introductions of exotic species that we will cover in the next chapter. Perhaps the clearest

Figure 9.7. Most of the animals killed by shrimp trawlers are thrown overboard, and they include endangered species such as this loggerhead turtle. (Photo by Michael Weber, Center for Marine Conservation.)

case of indirect exploitation involves our domestic animals and their exploitation of other species. We have already discussed the effects of livestock overgrazing. Predation by domestic animals, especially house cats, is another example. One study of domestic cats indicated that the average cat kills at least four birds per year (Churcher and Lawton 1987); that number multiplied by at least 100 million cats (there are 58 million in the United States alone) suggests that cat predation may approach one billion birds per year.

Consequences of Overexploitation

The most basic consequence of overexploitation is rather obvious; if we remove too many individuals from a population, we may subject it to all the problems of small populations discussed in Chapter 7 (Fig. 9.8). In this section we will consider some of the more subtle effects of overexploitation.

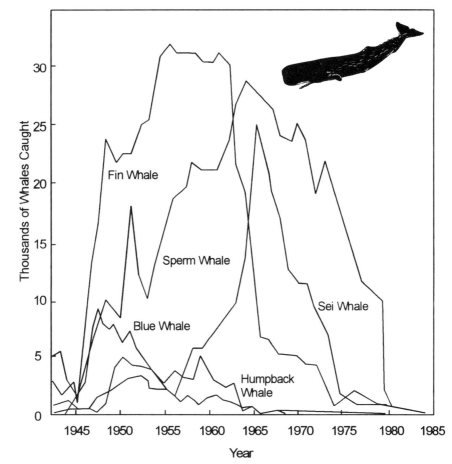

Figure 9.8. This graph shows how whalers have overexploited a series of great whales, starting with fin and blue whales and then switching to sperm and sei whales. (Redrawn by permission from Miller 1992.)

Population Effects

Not all the individuals in a population are equally susceptible to exploitation; their vulnerability may be influenced by their size, age, sex, phenotype, where they are, and when they are there. Consequently, the structure of a population, particularly, its age, sex, and genetic composition, can be changed by exploitation. Let us examine some brief examples.

Age: In many fisheries, the most profitable fish to catch are the largest, oldest individuals, but these individuals also have the highest reproductive capacity. Consequently, the effects of overfishing are exacerbated because decisions on when and where to fish and what kind of equipment to use

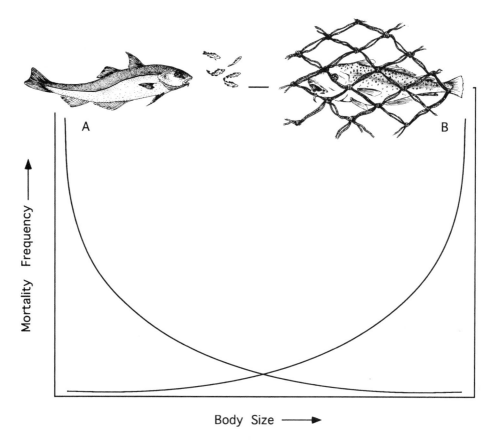

Figure 9.9. Mortality resulting from human fishing tends to increase as fish become larger (*line B*), whereas natural mortality is greatest when fish are small (*line A*). This mismatch may exacerbate the effects of overfishing. (Graph based on personal communication with Robert Steneck.)

(e.g., net mesh size) are often directed toward the most fecund members of the population (Fig. 9.9). The fact that this pattern of mortality is very different from natural mortality is especially worrying. A mismatch in age-specific mortality between natural predators and humans can also occur in animal populations that are subject to hunting because hunters often select animals in their prime rather than the young or old that are easier to kill (Solberg et al. 2000). Finally, loggers tend to harvest trees when their growth rates are starting to decline rather than at an age when natural mortality is common (Hunter 1990).

Sex: Among many mammal species, males are more exploited than females because they are bigger and thus more desirable and because they often travel over larger areas, making contact with people more likely. Consequently, exploited mammal populations often have a sex ratio that is more skewed toward females than is the case in unexploited populations. The effect on population viability may be minimal because most mammals are polygynous (i.e., one male will mate with multiple females), but there could be important exceptions. Off the west coast of South America, preferential hunting for male sperm whales led to a shortage of males that still persisted nearly 20 years after whaling ended. More importantly, this shortage of males was blamed for the low pregnancy rate among females (Whitehead et al. 1997). At least one population model has also shown that male-biased harvesting can jeopardize a population (Ginsberg and Milner-Gulland 1994).

Genetic Structure: Preferential harvest can also act as a form of artificial selection and change the genetic makeup of a population (Laikre and Ryman 1996). Some forests are subjected to a form of overexploitation called high-grading in which the best trees (e.g., those having the best form) are cut and the worst (e.g., diseased individuals) are left behind. It is widely assumed that high-grading is likely to alter a population's genetic structure to some degree, but, surprisingly, this issue has received relatively little attention from forest geneticists. One study from Ontario found a roughly 25% overall loss of alleles after harvesting with over 80% loss among rare alleles (Buchert et al. 1997). Overfishing has altered the genetic structure of many salmon populations by allowing some small males, which spend little or no time foraging at sea and thus are less likely to be caught by commercial fishing vessels, to become a large portion of the population (Gross 1991). These small males are able to pass on their genes by "sneaking" access to females rather than fighting for access with the large males that have returned from the sea.

It is not likely that a change in the age, sex, or genetic structure of a species caused by differential exploitation could by itself cause the extinction of a species. However, it could certainly exacerbate other factors, like a small population size, and could make extinction more likely. Recall from Chapter 7 that demographic stochasticity was a significant threat to small populations and recall from Chapter 5 the issue of effective population size.

Ecosystem Effects

The effects of overexploitation can ripple throughout an entire ecosystem if the exploited species has a key ecological role as a dominant species or a keystone species. To take an extreme example, if you cut all the pines in a pine forest, you will no longer have a forest ecosystem, at least until succession restores the forest.

For a more moderate example, consider some of the potential problems that may ensue from partially logging a forest such as alterations to the physical structure of the vegetation. Notably, large trees are likely to be less common because, in a managed forest, trees are cut when their growth rate begins to decline, and this is often long before they reach maximum size. Similarly, trees of commercially valuable species may become scarce in a partially logged forest. Both tree size and species are important habitat attributes for many animals, ranging from an eagle seeking a suitable nest site to a bark beetle looking for a spot to carve its tunnel (Palik and Engstrom 1999). Another problem can arise because dead or dying trees are often uncommon in managed forests where trees are usually cut before they are too susceptible to disease. This may create a shortage of habitat for a huge number of invertebrates, fungi, and microorganisms that use the dead wood of snags and logs; woodpeckers and other cavity-nesting vertebrates that we commonly associate with snags are just the tip of the iceberg (McComb and Lindenmayer 1999) (Fig. 9.10).

This is a very incomplete list of potential consequences of timber harvesting. (For a fuller treatment of these issues, see Hunter 1990, 1999.) The bottom line is that we cannot remove a substantial portion of the population of a dominant species without affecting the rest of the ecosystem to some degree. Although we have focused on forests here, this principle will apply to any ecosystem, such as overexploiting the grass in a grassland ecosystem through excessive livestock grazing.

Overexploiting a species that is relatively uncommon, but has a key-

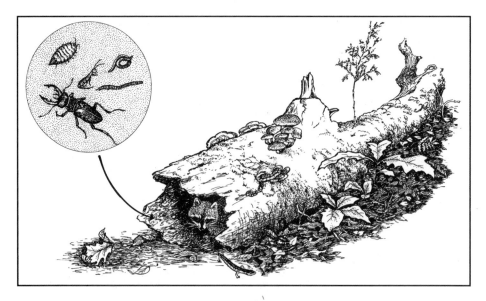

Figure 9.10. Intensively harvested forests may have few dead trees and logs that are a critical microhabitat for many species.

stone role, also will have profound effects for the rest of the ecosystem. For example, sea otter populations have been overtrapped along several stretches of the Pacific coast, and this has allowed populations of their prey, notably, sea urchins, to flourish (Duggins 1980). The abundance of sea urchins has limited recruitment of kelp, and as a result entire kelp bed ecosystems, with a large set of dependent species, have disappeared. Another layer of complexity has been added to this story in parts of Alaska where killer whale predation on sea otter populations has also allowed kelp forests to develop (Estes et al. 1998). It is likely that killer whales switched their attention to sea otters because their traditional prey, seals and sea lions, were less available, perhaps as a result of overexploitation of fish by humans. A similar example comes from many coral reefs where overfishing of herbivorous fish has led to a bloom of macroalgae that are smothering the coral reef (Culotta 1994). Owen-Smith (1989) has suggested that humans eliminated very large keystone herbivores (species weighing over 1000 kg such as mammoths and mastodonts) from northern Eurasia and the Americas at the end of the Pleistocene and thereby profoundly changed the distribution and forage quality of the vegetation. These changes may have led to the extinction of many smaller herbivores, and he warns that if today's elephants and rhinos are lost, a similar cascade of extinctions could occur. In Australia, Flannery (1995) has advanced a controversial idea that human extirpation of large herbivores roughly 40,000 years ago increased the vegetation biomass and thus provided more fuel for the fires that have shaped so much of Australian ecology ever since.

We must be particularly vigilant to recognize the loss of keystone species in ecosystems that superficially appear to be intact. In a provocative paper, "The Empty Forest," Kent Redford (1992) writes about the vast stretches of Amazonian forest that seem to be undisturbed, but that are almost devoid of large mammals and birds because of overhunting. He speculates about what this may mean in the long term because of the ecological roles of these species as seed dispersers, herbivores, and so on.

Even when the indirect, ecological effects of overexploitation do not reverberate through an entire ecosystem, they still may deserve the attention of conservationists. For example, overexploitation of a palm, *Phoenix reclinata*, in Kenya may have a deleterious effect on an endangered monkey, the Tana River crested mangabey, because seeds and fruits of this palm constitute up to 62% of the mangabey's monthly diet (Kinnaird 1992).

Some Final Perspectives on Exploitation

It is easy to condemn the overexploitation of wild life, and conservationists should do so with vigor and conviction, but we must be careful to focus on *over*exploitation and not exploitation per se, for as consumers of wild life we all exploit wild life. To take a particularly relevant example, the vast bulk of trees harvested in the world come from seminatural forest ecosystems, not plantations, and that is generally good; it means more habitat for wild life. However, it also means that by reading this book you are probably exploiting wild trees. In other words, we have to be responsible consumers, not just critics of the people who make their living from the use of wild life. When I tell students in my international conservation class that in some countries elephant poachers are shot on sight without the due process of law, they nod as if they think that this is an unfortunate but justifiable policy. They are shocked when I then suggest that it might be equally justifiable for U.S. customs officials to shoot American tourists returning from abroad with ivory souvenirs. The role of consumers in overexploitation is captured nicely in a quote from the actress Gina Lollobrigida, shortly after she purchased seven new fur coats: "What can I do? The tigers in my coat were already dead. . . . If I don't buy the coats, somebody else will."

In particular, biologists who condemn overexploitation should not forget that their profession has many skeletons in the closet (literally and figuratively) from past activities. For example, beginning in 1884 several museum-organized expeditions sought to find the last northern elephant seals without success; finally, in 1892 they found seven and collected six of them (Busch 1985). Fortunately, some seals were apparently overlooked, and the species has recovered.

It is also important to remember that killing plants and animals is not the only way to exploit them. The market for pets is enormous, and millions of live animals, especially fish, birds, and reptiles, are caught in the wild and sold every year (Andrews 1990, Beissinger and Snyder 1992). Live plants, particularly orchids and cacti, are also in great demand. Obviously, from the perspective of a wild population it does not matter whether an individual is dead or alive when it is removed. Indeed, the trade in live organisms can be more deleterious because many individuals die between the time of capture and the time they arrive at their ultimate destination, and thus a larger number needs to be acquired initially. Some of the worst examples of this involve young primates, in great demand for medical research, that are often captured by shooting their mothers from the treetops. Of course, many of us are alive today because of medical research, and so again, we should condemn such practices, but must tread carefully to avoid hypocrisy. The key solution to these dilemmas revolves around careful management of exploited populations, an issue that we will cover in Chapter 12, Managing Populations.

Finally, we should not lose sight of the fact that the loss of ecosystems is typically a much more important threat to biodiversity than overexploitation. This chapter has focused on many exceptions to this generalization, but the truth of the following remains:

> The law doth punish man or woman,
> That steals the goose from off the common,
> But lets the greater felon loose,
> That steals the common from the goose.
>   Anonymous 1764

CASE STUDY

The Gulf of Maine
Robert S. Steneck[1]

Sailing across the Gulf of Maine today you can see a vast ecosystem that appears little changed after thousands of years of human use. However, this illusion would soon disappear if you could slip beneath the surface and see the gulf through the eyes of a marine creature. Both coastal and offshore marine communities of the Gulf of Maine have been changed profoundly over the past several hundred years because of the virtual elimination of large predatory fish.

As long as 8000 years ago, the "Red Paint People" lived year-round on the coast of Maine catching marine fish no more than a short canoe trip from shore. The middens left by indigenous people over the next several thousand years indicate that large fish such as Atlantic cod (some estimated to exceed 100 kilograms) remained sufficiently abundant to comprise over 80% of the bone volume of the middens (Carlson 1986). When the first Europeans explored the Gulf of Maine, it was the abundance of large fish that so impressed them (Caldwell 1981). The northern half of Juan Vespucci's 1526 map of the New World was identified as Bacallaos—which is Portuguese for "land of the codfish." In 1602, Bartholomew Gosnold named Cape Cod for the myriad fish that "vexed" his ship. Captain John Smith reported three important facts in 1616: 1) that cod were abundant along the coast, 2) that native Americans already knew this, and 3) that the cod in Maine were two to three times larger than those found elsewhere in the New World. In the early 1600s, seafood from the Gulf of Maine had a larger share of the market in Europe than it does today. At that time, 10,000 men were employed fishing for cod in New England (Caldwell 1981); by the 1880s, three times that number were employed in Nova Scotia alone (Barnard 1986). Late 19th century advances in ships and fishing technology greatly increased fishing effectiveness. This may have been the zenith of the codfish industry.

Since the 1800s, cod and other large-bodied predatory fish have declined in abundance and size until they have become virtually absent from coastal habitats (Witman and Sebens 1992, Steneck 1997, Steneck and Carlton 2000). This decline is evident in published charts of coastal fishing grounds. The continuous near-shore fishing grounds charted in the 1800s were reduced to small discrete patches by the 1920s (Rich 1930) and today are gone. In addition to declining abundances, average fish body size also has steadily dropped over the past several decades. For example, codfish sizes decreased from an average of about 80 cm in 1950 (Bigelow and Schroeder 1953) to 30 cm in the late 1980s (Ojeda and Dearborn 1989).

There is growing evidence that coastal marine landscapes have changed as a result of the loss of the large predatory finfish (Fig. 9.11). Today, mobile benthic invertebrates (e.g., Menge and Sutherland 1987) and small, commercially unimportant finfish (Wahle and Steneck 1992, Steneck 1997) are highly conspicuous and appear to be the most important predators in coastal zones of the Gulf of Maine. Experiments indicate that adult crab, lobsters, and sea urchins live today in coastal habitats without significant threats from predators (Wahle and Steneck 1992, Steneck 1997, 1998).

Figure 9.11. The decline of large, predatory fish in the Gulf of Maine (e.g., codfish large enough to prey on adult lobsters) has dramatically affected the entire marine community.

Furthermore, the absence of predators allows more lobsters to live in areas with little shelter than was possible when predators were abundant. This expansion of habitable areas for lobsters may have contributed to the currently thriving lobster industry, which in recent years has repeatedly exceeded its record harvest set in the 1880s.

The hyperabundance of sea urchins was also probably the result of populations growing unchecked by predators (Steneck 1998). At high densities their grazing denudes coastal zones of most erect, fleshy seaweeds such as kelp and thus reduces coastal productivity and habitat structure for other organisms (e.g., Bologna and Steneck 1993). However, in the late 1980s sea urchins themselves became targeted for their roe, which is highly valued in Japan. In a shockingly short period—about a decade—the carpets of sea urchins disappeared. Since sea urchins are the dominant herbivore in the system, the reduced grazing resulted in the establishment of kelp forests and shag-carpet tangles of red algae (Steneck and Carlton 2000).

What we have observed in the Gulf of Maine is called "fishing down the food web" (Pauly et al. 1998). That is, top predators such as cod are often the first fish targeted because they are highly valued. As that trophic level declines, its prey become more abundant and become the new target. The loss of higher trophic levels in food webs causing increases in lower levels is known as a "trophic cascade" (e.g., Steneck 1998). Fishing down the Gulf of Maine's food web has resulted in such a trophic cascade (Steneck 1997), which continues to this day with new, emerging fisheries for distant markets.

Recently, a variety of intertidal and subtidal seaweeds have been harvested for food and fertilizer, and a market for small, herbivorous, periwinkle snails has developed. There is growing concern that sea urchins, snails, and seaweeds may be incapable of sustaining the escalating pressures on them. When species richness is high, overfishing one species may be compensated for by another in that trophic level, but in the Gulf of Maine recent glaciation has left some trophic levels with only a few taxa occupying similar niches. Consequently, entire trophic levels are being functionally reduced, causing whole marine communities of hundreds of species (most of them noncommercial) to be profoundly altered. Some of these changes are predictable such as increases in prey when predators are reduced. Other changes are entirely unpredictable. For example, as the seaweed community changed, the habitat architecture also changed. What was once a featureless, encrusting, calcareous, algal-dominated bottom became a shag carpet of red algae. This is an ideal habitat for settling crabs that would have died from fish predation without the algae in which to hide. Swarms of baby crabs live in the algal shag carpet, and they consume virtually all of the settling urchin larvae. So, despite there no longer being any harvesting of sea urchins (because there is nothing to harvest), the population has not returned—it is locked in an alternate, algae-dominated, stable state.

Fishing down food webs in the Gulf of Maine has resulted in hundreds of kilometers of coast now having virtually one trophic level (seaweed). This and other examples worldwide of fishing down marine food webs (e.g., Pauly et al. 1998) indicate large-scale overexploitation. The escalating rate of overexploitation is also alarming. Whereas prehistoric indigenous Americans may have had thousands of years of sustainable harvests, we currently seem unable to have sustainable harvests and relatively stable marine communities for more than a few decades or even a few years.

[1]School of Marine Sciences, University of Maine, Darling Marine Center, Walpole, Maine.

Summary

Exploitation of wild plants and animals is a fundamental human activity, although when it involves killing sentient species, especially birds and mammals, some people are uncomfortable with the idea. When exploitation becomes overexploitation (i.e., when our use of a population seriously threatens its viability or radically alters the natural community in which it lives), everyone should be uncomfortable with the idea, even those who readily accept the idea of killing other organisms. Human overexploitation has a long history, especially on islands, but that is no excuse for the abuses that persist today. The worst of these involve commercial exploitation, particularly because the market demand for wild organisms is enormous and the rarer a species becomes the more it is worth. Subsistence use of wild organisms is limited by the number of people living in rural areas and their needs, but still has the potential to threaten populations. Overexploitation also can result from incidental exploitation (catching species accidentally while harvesting other target species) and recreational exploitation (e.g., hunting, fishing, and, under some circumstances, nonconsumptive activities such as bird-watching). Besides reducing population size, overexploitation also can have deleterious effects on the age, sex, and genetic structure of populations, and, when directed against keystone or dominant species, it can negatively affect whole ecosystems. Finally, when condemning overexploitation, it is important to think about the consumers of wild species—all of us—as well as those who earn their living harvesting wild life.

FURTHER READING

See Burney (1993) for a brief overview of prehistoric overexploitation and Martin and Klein (1984), Flannery (1995), and MacPhee (1999) for more detailed ones. For the historic period, Verney (1979), Mowat (1984), Matthiessen (1987), and Wilcove (1999) are interesting reading. Liddle (1997) gives a comprehensive account of the impacts of recreational exploitation. Safina (1998) provides a particularly compelling account of overexploitation in the sea. Checkout the websites of Traffic, a group that monitors wild life trade (www.traffic.org) and the Bushmeat Crisis Taskforce, a group focused on commercial exploitation of wild animals for meat (www.bushmeat.org).

TOPICS FOR DISCUSSION

1. Assuming that exploitation can be carefully controlled, should commercial exploiters of wild life (people doing it to make a living) have precedence over recreational exploiters of wild life (people doing it for fun)? Which group usually has precedence in practice? Why?
2. Many laws have been passed to regulate overexploitation, and we will review these in Chapter 12, Managing Populations. Try to think of some practices that might minimize the effects of overexploitation on the age, sex, and genetic structure of populations, as well as the effects of overexploitation on entire ecosystems.
3. What steps can consumers of wild life take to make sure their consumption is not contributing to overexploitation?

Chapter 10
Exotic Species

Koalas and Australia, sequoias and California, tsetse flies and sub-Saharan Africa—the native flora and fauna of different parts of the world can be as distinctive as the various languages, cuisines, and religions that mark the diversity of human cultures. To some extent these relationships are based on the special habitat requirements of koalas, sequoias, and tsetse flies, but this is not the whole story. Although most species are continually shifting their ranges, particularly in response to climate change, these movements are often impeded by barriers. The barriers may be as subtle as a change in temperature or salinity or as sublime as an ocean. They may be relatively short-lived like the sea-level changes that have separated North America from Asia and South America, or they may be relatively enduring like the isolation that has defined the Hawaiian islands since they began to rise from the ocean floor 27 million years ago.

Some barriers are very effective, making it a rare and chance event for an individual to cross the barrier—a seed carried on the wind for thousands of kilometers, a lizard or a crab clinging to a floating log—and far rarer still for some of these potential colonists to establish a population. If the isolation persists, new genes will arise and new species will probably evolve. Consequently, isolation is a critical factor that shapes the biodiversity of a place. It filters the biota of other places, allowing only a subset to become established locally, and it fosters the creation of new elements of biodiversity.

In the last couple of thousand years, isolation has been diminished for many species. The worldwide movement of people, especially with the rise in maritime shipping in the last few hundred years, has created a new agent for moving biota around the globe and especially to formerly remote islands. To put it another way, the rate at which biological communities are reshuffled as species move in and out of them through geographic range shifts has been greatly accelerated by human activities. Some species have been carried as passengers, others as stowaways. For example, on long voyages Vikings carried caged birds to release on the assumption that if a bird flew off and did not return, land was near (Long 1981). No doubt, the Viking ships also had a large retinue of rats, lice, and fleas on board, plus barnacles and algae clinging to the hull.

Exotic is the adjective most commonly used by conservation biologists to describe a species living outside of its native range. However, many botanists refer to *alien* or *adventive* plants, and you will also encounter the terms *introduced species, nonindigenous species,* and *nonnative species* for both plants and animals. *Invaders* or *invasive species* are terms used for exotic populations that are expanding dramatically. Biologists use these terms if a species is outside of its natural geographic range (i.e., the geographic range it would occupy without human interference). Most nonbiologists are likely to call a species "exotic" if it is from a different nation or state, but, of course, political boundaries are irrelevant from a biological perspective.

How Do Species Move?

Stowaways

A large portion of the species that have been transported around the globe were stowaways, species that we would have gladly left behind. The Norway rat, house mouse, and black rat (often called the ship rat) come to mind first. In human terms, these three species cause billions of dollars in losses each year; they have also been major culprits in the extinction of many species, particularly on islands (Fig. 10.1). Conservation biologists often overlook microorganisms as exotic species, but the stowaways we carry in our bodies have had extraordinarily profound effects; for example, disease organisms carried by European explorers and colonists have decimated native peoples around the world (Crosby 1986, Diamond 1997). Similarly, disease organisms afflicting wild life and domestic plants and animals have been spread far and wide by our activities (the rabies virus and chestnut blight to name just two of many).

Stowaways often go unnoticed because they are small and inconspicuous. Many insects have been spread widely, traveling as eggs and pupae on food, logs, and other objects. European earthworms probably arrived in New England in soil clinging to the roots of apple trees and other plants. (Most people do not realize that virtually all earthworms are exotic in Canada and the northern United States [Hendrix 1995]; presumably recent glaciers eradicated any native earthworms.) The roadsides of North America and New Zealand are dominated by plants from Europe such as dandelions, plantains, and certain thistles, most of which probably arrived as

Figure 10.1. Ships have spread Norway rats and black rats to virtually every corner of the earth, even remote islands, where they have caused hundreds, perhaps thousands, of extinctions and billions of dollars of losses for humans.

seeds in packing material or hay carried to feed livestock during voyages. A German researcher scraped the mud off a single car on four occasions during one growing season and 3926 seedlings of 124 species germinated in these samples (Schmidt 1989). Probably the greatest flood of exotic organisms involves small marine organisms—plankton and the planktonic offspring of larger species—that arrive by the millions in the ballast water of ships (Carlton 1985, Ruiz et al. 1997, 2000, Ricciardi and MacIsaac 2000). Vessels departing for an oceanic crossing with little or no cargo take on huge volumes of seawater for stability. On arriving at their destination they discharge the water and along with it millions of small creatures. This is probably how the infamous zebra mussel arrived in the United States and Canadian Great Lakes from across the Atlantic (Johnson and Padilla 1996,

Ricciardi et al. 1998). Some marine organisms (algae and barnacles, for example) may be transported clinging to ship's hulls; however, with modern vessels, ballast-water stowaways are far more abundant and diverse. Carlton and Geller (1993) found 367 taxa of marine organisms in samples of ballast water collected from Japanese cargo ships arriving at an Oregon port.

It is one thing for governments to regulate deliberate introductions; it is far more difficult to control accidental introductions of stowaways. Thus the stowaway problem will probably increase as global transportation systems develop. Among the 205 exotic species discovered in the United States since 1980, most of them probably arrived as stowaways (Office of Technology Assessment 1993).

Subsistence and Commerce

Most deliberate attempts to mingle the world's biota have been motivated by our need for food, especially familiar food. Some of the earliest examples of this practice come from the Pacific Ocean where Polynesians explored far and wide, bringing with them pigs, dogs, chickens, yams, sweet potatoes, bananas, and more (Diamond 1997). Colonists everywhere have brought their own domestic plants and animals with them, and often sent new plants and animals back to their homelands. Early during the European colonization of the New World, potatoes, tomatoes, corn, and turkeys went back to the Old World on the same ships that were introducing horses, pigs, wheat, bananas, and many other species.

Species used for food dominate the list of planned introductions, but other needs have also prompted introductions. Exotic tree species have been planted widely as sources of lumber, fiber, and fuel, sometimes growing better than they did in their native environment (Richardson 1998). For example, Monterey pine, an uncommon species that is little used for lumber in its native California, is a prized plantation species in Australia and many other countries. Conversely, Australian eucalypts are common in California and elsewhere.

The consequences of these introductions are usually quite localized as long as these species remain domestic. However, some of these species escape into the wild; they become *feral*. Horses, donkeys, and pigs are now feral in many places in the New World, causing problems we will describe below. Sometimes, domestic animals have been released with the specific intent of establishing feral populations. Sailors released goats, pigs, and rabbits on many remote islands that had no native mammals so that they could shoot fresh meat on future visits.

Occasionally, wild species are imported for commercial or subsistence purposes. Red squirrels were introduced to Newfoundland to provide food for pine marten, a valuable species for fur trappers. Nile perch were introduced to Lake Victoria to bolster commercial fisheries (Witte et al. 1992a, Schindler et al. 1998). Escapees from fur farms have established populations in areas outside their native range: nutrias in the southeastern United States and Great Britain, American mink and raccoon dogs in Europe (Putman 1989).

Recreation

Sport hunters and anglers have been very active in the planned introductions of exotic wild species. Anglers have been particularly ambitious in this regard, carrying fish by the bucket and truckload to water bodies all over the world (Moyle et al. 1986, Rahel 2000). In California 50 of the 133 freshwater fish species are not native to the state, and sportfishing was the leading impetus for most of these introductions (Moyle 1976b). It is probably fair to say that people have dumped new species of fish into almost every water body that they visit regularly, including both game fish and smaller species introduced to provide food for game fish. One of the most disturbing examples of fish being introduced for sport involves the Green River in the southwestern United States (Holden 1991). When fish biologists decided to introduce rainbow trout and kokanee salmon in 1962, they first poisoned the river with rotenone to rid it of native "trash" fish (i.e., fish of little sport or commercial value). Some of these so-called trash fish such as the Colorado squawfish, roundtail chub, and bonytail chub are now listed as endangered species.

Among terrestrial creatures, game birds have been favorites for introductions. In Hawaii alone 75 different species of game birds (chiefly, Galliformes, i.e., pheasants, quail, partridges, etc.) have been introduced, although only 17 species were successfully established (Long 1981). One of the world's most popular game birds, the ring-necked pheasant, is now more common in Europe and North America than in its native range in Asia. Wild and domestic pigs are the same species, and between domestic individuals going feral and wild individuals being introduced by hunters, pig hunting is possible throughout much of the world. Over 50 species of large mammals have been imported to ranches in Texas to allow hunters to bag exotic animals (sometimes called Texotics) without traveling to exotic lands (Bolen and Robinson 1999) (Fig. 10.2). Ironically, in the 1960s there were probably more blackbuck, a beautiful little antelope with impressive horns, in Texas than in their native range on the Indian subcontinent. Fortunately, native blackbuck populations have rebounded since then. Many exotic big-game mammals are so valuable that they are kept within well-maintained fences, and it is debatable whether they are living in the wild. However, some species such as the aoudad, a wild sheep from north Africa, have many self-sustaining populations on open range.

Figure 10.2. Game fishes, birds, and mammals like the blackbuck have been introduced widely by anglers and hunters.

Whimsy or Aesthetics

> I'll find him where he lies asleep
> And in his ear I'll holler "Mortimer."
> Nay I'll have a starling shall
> Be taught to speak nothing but "Mortimer."

These lines, spoken by Hotspur in Shakespeare's *King Henry IV,* Part I, are the reason why there are millions of starlings in North America. In the 1890s Eugene Scheifflin, a New York man who loved both birds and Shakespeare, decided that it would be fun to introduce to the United States all of the bird species named by Shakespeare (Laycock 1966). Most of his attempts failed, but with the starling, mentioned only once by Shakespeare, he was overwhelmingly successful.

Acclimatization societies—social groups whose sole purpose was to introduce new species—were quite popular among European colonists during the late 19th and early 20th centuries. Indeed, in New Zealand, the Auckland Acclimatization Society is still active, although it changed its name to the Auckland Fish and Game Council to recognize its broader interests and in deference to the negative side of introducing exotics. To a large degree these groups were motivated by a love of nature and nostalgia for the species they left behind in Europe, and European songbirds were their favorite subjects. On the whole they were not very successful, with notable exceptions like the starling, but in New Zealand they had, from their perspective, good luck. A naturalist traveling through New Zealand today will see far more songbirds native to Europe than New Zealand songbirds.

Importing plants because of their ornamental beauty and importing animals as pets could be classified as motivated by aesthetics or commerce or recreation. Most ornamental plants have not succeeded in escaping from gardens, but quite a few have such as the multiflora rose, an Asian plant that is now abundant in North America. Notwithstanding myths about alligators in the New York City sewer system, most exotic pets soon die when they escape or are released from captivity. However, there are many exceptions; for example, several species of parrots and tropical fishes are breeding in the wild in southern Florida.

Science

To study species closely scientists often establish breeding colonies in their laboratories. Sometimes these species are from outside their native range, and sometimes they escape. The gypsy moth is probably the most notorious example of this. It is now widespread in forests of the United States after escaping in 1869 from the lab of a scientist who imported it from Europe, hoping to develop a silk industry in New England (Forbush and Fernald 1896). Not far away a visiting scientist at the Marine Biological Laboratory at Wood's Hole, Massachusetts, released a species of sea squirt, *Botrylloides diegensis,* in 1973, and the species has now usurped space on hard marine substrates throughout southern New England (Carlton 1989).

The interplay between scientists and exotic species also raises the prospect of creating and distributing whole new "species" through genetic engineering. Could a supertomato ever turn into a superweed? This technology raises some significant concerns that have many parallels to the exotic species issue (Regal 1993, Parker and Karieva 1996).

Biological Control

Many exotic species have been introduced to control other exotic species that were introduced earlier. Sometimes, this practice works quite well, even though it is making the best of a bad situation. Notably, entomologists have been able to completely control scores of exotic insect pest species, and partially control many more, by visiting the native range of the pest species; finding a predator, parasite, or pathogen that attacks the pest; and then introducing this species (DeBach and Rosen 1990).

Unfortunately, poorly planned introductions often make a bad situation worse (Howarth 1991, Miller and Aplet 1993, Simberloff and Stiling 1996, Myers et al. 2000, Strong and Pemberton 2000). Rats and rabbits introduced to islands can reach plague proportions, but introducing their predators (e.g., stoats, ferrets, and weasels in New Zealand and mongooses in Hawaii and the West Indies) was worse than useless (King 1984). The rats and rabbits proved largely immune to the predators, but the predators wrought havoc on other species, notably ground-nesting birds. In Australia,

Figure 10.3. Not all exotic predators have sharp claws and teeth. This is a predatory snail, *Euglandina rosea*, that has decimated viviparous tree snails on Moorea, a Pacific island.

red foxes were introduced to control introduced rabbits; the rabbits thrived, but the foxes contributed to the decline of the native marsupials. On Moorea, a South Pacific island, a predatory snail, *Euglandina rosea*, was introduced in 1977 to control the giant African snail, which had been introduced earlier to provide escargot for French colonists, but then began wreaking havoc on crop plants (Fig. 10.3). Today, the African snails persists, but *Euglandina* has eradicated seven species of viviparous tree snails from the island, leaving six species surviving only in captivity and one globally extinct (Murray et al. 1988, Cowie 1992).

The difference between successful and unsuccessful biological controls may depend on introducing exotic predators, parasites, or pathogens that are completely dependent on the host species that you are trying to control.

Habitat Change

When we think of exotics we usually think of species actually transported by people, deliberately or accidentally, but we could also include species that were able to expand their ranges themselves because of human

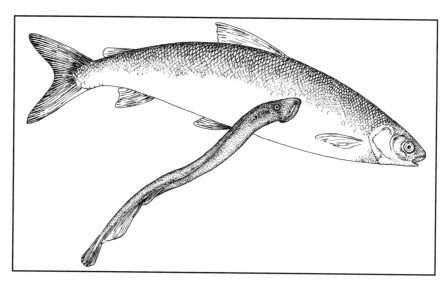

Figure 10.4. After sea lampreys used the Welland canal to bypass Niagara Falls and enter the upper Great Lakes, two of their host species, the deepwater cisco and blackfin cisco, became extinct. (Smith 1968.)

changes to the environment. For example, construction of the Welland Ship Canal allowed sea lampreys, a parasitic fish, to bypass Niagara Falls and invade the upper Great Lakes (Smith 1968) (Fig. 10.4). When the Suez Canal was opened in 1869, it permitted many species from the Red Sea to invade the eastern Mediterranean. Similarly, construction of a sea-level canal across the isthmus of Panama could allow a large-scale exchange of Pacific and Atlantic species.

Under this definition, the coyote, mallard, brown-headed cowbird, and a host of prairie plants (especially members of the aster and grass families) are exotic species in the eastern United States because opening the eastern forests for agriculture allowed them to expand their ranges from the west (Brothers 1992). In the case of the coyote, this process was facilitated by our extirpation of wolves, which can compete with coyotes. Given that species are continually shifting their geographic ranges in response to climate changes, it may require careful study to say whether a range shift is primarily tied to human alterations of the environment or is primarily a response to natural changes.

Impacts of Exotic Species

Look out the window, and there is a good chance you will see more exotic species than native ones: exotic grasses, shrubs, trees, perhaps an exotic bird on the sidewalk, or an exotic fly on the window. Many exotic species are living in environments so completely manipulated by people that their direct impacts on native biota are not very severe. Unfortunately, there are thousands of exceptions to this generalization. Indeed, at least one assessment of the problems facing endangered species in the United States identified exotic species as the single most pervasive issue, affecting 305 out of 877 listed species (Czech et al. 2000). Equally dramatic is an estimate of the total economic and environmental cost of exotic species in the United States: 137 billion U.S. dollars (Pimentel et al. 2000). In this section we will review some of the ways in which exotic species jeopardize other species and whole ecosystems.

Predators and Grazers

It is easy to understand the impacts of exotic species when an introduced species kills and eats native species. A particularly infamous anecdote about this comes from Stephen's Island, an islet between the North and South Islands of New Zealand. A lighthouse keeper stationed there in 1894 kept a cat, and as a hobby he prepared study skins from the birds his cat killed and mailed them to the British Museum of Natural History. Some time later a letter arrived from London telling him that his cat had collected a species new to science, the Stephen's Island wren, but by then the wren was extinct, apparently wiped out by a single cat (Fig. 10.5). A more contemporary example comes from Guam, where the brown snake, acciden-

Figure 10.5 . Perhaps the most ironic victim of an exotic species was the Stephen's Island wren, apparently wiped out by a single cat brought to the island by a lighthouse keeper.

tally introduced from Australia or New Guinea, apparently extirpated eight or nine of the island's thirteen native species of forest birds plus three to five lizards (Fritts and Rodda 1998). Probably the most dramatic loss of vertebrate species in historic times involves an exotic predator. In East Africa's Lake Victoria roughly 200 endemic fish species have been lost to extinction following a population explosion of the exotic Nile perch (Witte et al. 1992a, Seehausen et al. 1997). These losses have had profound effects on the entire trophic structure of the lake ecosystem (Fig. 10.6) (Witte et al. 1992b, Goldschmidt et al. 1993). Finally, predation by exotic fishes is one important ingredient in the stew that is causing a global decline in many amphibian species (Knapp and Mathews 2000).

From an economic perspective, introduced insects that consume crop plants are among the most destructive exotic pests. Witness the Mediter-

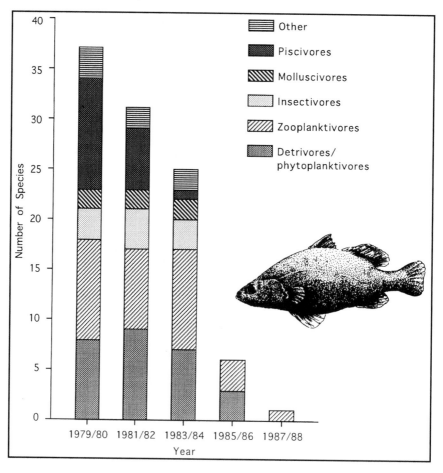

Figure 10.6. The introduction of Nile perch to Lake Victoria has led to the extinction of approximately 200 species of fish endemic to the lake and significant changes in the lake's food web. (Redrawn by permission from Witte et al. 1992b.)

ranean fruit fly, the boll weevil, the corn borer, and other insects that cause billions of dollars in damage despite massive campaigns to control them (Pimentel et al. 2000). From a biodiversity perspective, the most destructive exotic herbivores have probably been generalist species such as goats, pigs, and rabbits introduced to islands (Coblentz 1978, Coblentz and Baber 1987). Two biological treasures—the Galapagos and Hawaiian archipelagoes— are particularly poignant examples of what exotic herbivores can do to islands (Schofield 1989). Because many islands have evolved a unique flora of species that are not adapted to being preyed on by large herbivores (e.g., no thorns), island plants have been hard hit by mammalian herbivores. Furthermore, because plants are dominant species in most ecosystems, the consequences of overgrazing by exotic herbivores can easily extend well beyond the plants that are being eaten. Recall Round Island from Chapter 8, where introduced rabbits and goats degraded the vegetation so badly that the whole island was eroding into the sea. Two species of reptiles became extinct, and three others, as well as ten species of plants, were at risk before the rabbits and goats were removed (North et al. 1994).

Parasites and Pathogens

Exotic parasites and pathogens have a tremendous potential to afflict native biota; try thinking of them as incredibly abundant tiny predators feeding on the protoplasm of other species, sometimes with lethal consequences. The history of human diseases, especially smallpox and measles, provides plenty of examples of what the introduction of an exotic pathogen can do (Crosby 1986, Diamond 1997). Suffice it to say that European colonists killed far more people in Australia and the Americas with their diseases than with their guns. Throughout Europe and North America the chestnut blight has invaded from Asia, reducing the American and European chestnuts (which were once major components of temperate deciduous forests) to a few sickly specimens incapable of reproducing except by sprouts (Griffin 2000). Two introduced diseases afflicting birds, avian malaria and avian pox, are suspected to have played a major role in the extinction of several Hawaiian birds (Van Riper et al. 1986).

In most cases exotic parasites and pathogens arrive in or on exotic hosts, not all by themselves. For example, avian malaria probably arrived in the Hawaiian islands in the early 1900s carried by exotic birds imported from Asia, although its primary vector, the mosquito *Culex quinquefasciatus* was introduced in 1826 (Van Riper et al. 1986). Consequently, keeping out exotic plants and animals is probably the most effective way of keeping out their parasites and pathogens. There is at least one exception to this generalization: ballast water discharges can introduce a cocktail of exotic marine microbes (Ruiz et al. 2000).

This issue also argues for careful scrutiny of species introduced as biological control agents (Myers et al. 2000, Strong and Pemberton 2000). For example, over 100 species of parasites, pathogens, and predators have been imported to the United States in an attempt to control gypsy moths. Many of these are likely to afflict a wide spectrum of butterflies and moths, and it has been shown that some species such as the cecropia moth do suffer high mortality because of species introduced for biological control (Boettner et al. 2000).

Competitors

The effects of exotics as competitors are most conspicuous with plants and other sedentary species. Some exotic species (e.g., kudzu, zebra mussels, purple loosestrife, water hyacinth) can become so extremely abundant that competition is evident in terms of a basic resource, space. Of course, competition for space is closely tied to competition for a variety of other resources such as water, nutrients, light, and so forth.

When exotic species closely resemble native species, it is likely that they will compete directly with one another, and sometimes the exotic species will be dominant. For example, gray squirrels and American mink from North America have displaced red squirrels and European mink from large areas of Europe (Usher et al. 1992), and Argentine ants are replacing native ants in many parts of the world (Human and Gordon 1997). Sometimes, competition for a single resource can be the key issue, as when European starlings displace parrots from nest cavities in Australia (Pell and Tidemann 1997) or eastern bluebirds in the eastern United States. Even species that are very different taxonomically may be brought into competition for a single resource. In New Zealand exotic wasps consume large quantities of honeydew secreted by a scale insect *Ultracoelostoma assimile* and have contributed to the decline of kakas, an endangered species of parrot that are highly dependent on the honeydew during summers (Beggs and Wilson 1991). Honey bees, native to Europe, have been introduced widely and are known to compete with native insects and birds for nectar and pollen. Moreover, they are not as likely to pollinate many native plants because their morphology and behavior are different from the pollinators with which the plant evolved (Paton 1993, Aizen and Feinsinger 1994a, 1994b) (Fig. 10.7).

Hybridization

Some introduced species are so closely related to a native species that they can interbreed and produce hybrids. Consider the mallard, a duck that has been introduced widely by sport hunters and whose range has been ex-

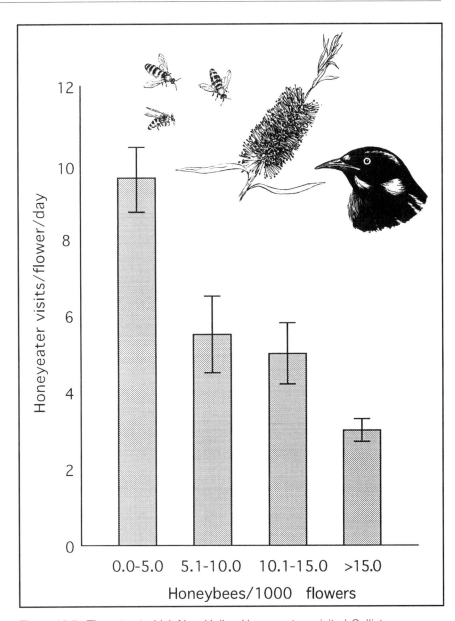

Figure 10.7. The rate at which New Holland honeyeaters visited *Callistemon rugulosus* flowers decreased as the abundance of exotic honey bees increased. (Redrawn by permission from Paton 1993.)

panded by conversion of natural ecosystems into agricultural lands. In captivity mallards have interbred with at least 40 other species. In the wild, mallards have interbred with both ducks that are usually recognized to be distinct species (e.g., Pacific black ducks and American black ducks), and with ducks that are sometimes considered separate species and sometimes subspecies (e.g., Mexican ducks and Hawaiian ducks) (Browne et al. 1993). Some of these ducks are declining, and it is feared that they could eventually disappear, replaced with mallards and mallard hybrids. Similar stories could be told for many rare fishes (Moyle 1976b, Echelle and Echelle 1997), mammals (Greig 1979), and especially plants (Ellstrand 1992, Levin et al. 1996) and for the movement of genes from domestic species to their wild relatives (Ellstrand et al. 1999). See Rhymer and Simberloff (1996) for a review.

This process is often called genetic swamping because the genes of one species come to dominate a common gene pool, largely excluding the genes of the second species. You could argue that genetic swamping is simply the result of natural selection and that two species that interbreed readily when brought into contact were not true species in the first place. On the other hand, if taxonomists had identified the two groups as distinct species, it probably means that they were relatively isolated and morphologically distinguishable. Thus, given a longer time to evolve in isolation, it is likely that they would have become incapable of interbreeding. In other words, by introducing near relatives, or even genetically different populations of the same species, we may be curtailing the process that produces new species. In a sense, we may have caused a species to become extinct shortly before it came into existence.

Ecosystem Effects

The consequences of a biological invasion can reach far beyond the individual species that must cope with a new predator, competitor, pathogen, or parasite. Invading species can alter a variety of ecosystem properties such as productivity, nutrient cycling, natural disturbance regimes, and soil and vegetation structure (Vitousek 1986, Macdonald et al. 1989, Ramakrishnan and Vitousek 1989, Norton 1992, Rabenold et al. 1998). Recall that the Nile perch invasion of Lake Victoria disrupted the trophic structure of the lake (see Fig. 10.6) and that the rabbits and goats on Round Island precipitated soil erosion that profoundly degraded the island ecosystem. New Zealand stream ecosystems with populations of exotic brown trout have a lower density and biomass of insects and a higher biomass of algae compared with streams with native fishes (Townsend 1996). Exotic plants can also change entire ecosystems in many ways (Vitousek 1986, 1990, Mack et al. 2000). For example, nitrogen-fixing exotic plants can significantly alter the soil chemistry of the environments they invade; fire-prone exotic plants can allow fires to burn more extensively; floating aquatic weeds can blanket aquatic ecosystems, profoundly changing water chemistry; exotic plants with deep root systems and high rates of transpiration can lower water tables; and changes in vegetation structure can profoundly alter the habitat of animals (Steenkamp and Chown 1996).

Success Rates

Why are some species more successful invaders than others? Why are some ecosystems more susceptible to invasion than others? These questions have long fascinated ecologists because in answering them we may gain insights into the basic structure of ecosystems, especially the interactions among species (Elton 1958).

The vast majority of individual plants and animals that are transported to a new site soon perish unless they are carefully nurtured by people (Williamson and Fitter 1996a, Mack et al. 2000). Consequently, exotic species that are domesticated tend to do reasonably well, but wild exotics usually perish. For example, an acclimatization society in Cincinnati released 3000 birds of 20 species, and none survived (Long 1981). A somewhat different story emerges when we look at comparable data from islands (Coblentz 1990). For example, among 162 species of birds known to have been introduced to Hawaii, about 70 species are persisting, while in New Zealand 36 birds species have persisted out of 140 introduced species (Long 1981).

Why would it be easier to establish a new species on an island? One likely answer is that most islands have relatively few species (for reasons explained in Chapter 8), and that means an introduced species has fewer competitors, predators, parasites, and pathogens with which to cope (Elton 1958). A corollary to this idea is the possibility that, because many island species have evolved in an impoverished biota, they are less efficient at being competitors or at avoiding being prey (Huston 1994). The idea that low species richness might predispose an ecosystem to being invaded has also been suggested in the case of marine ecosystems (Cohen and Carlton 1998) and has been supported in some controlled experiments (Stachowicz et al. 1999, Symstad 2000). On the other hand, some studies have not found diverse communities to be more resistant to invasion, especially when confounding factors were considered such as nutrient availability and spatial scale (Moyle and Light 1996, Lövel 1997, Levine and D'Antonio 1999, Stohlgren et al. 1999).

Turning to some other possible explanations for variation in the success rates of different exotics, one simple factor is based on the number of individual invaders. An analysis of exotic birds in New Zealand found that introductions using over 100 individual birds were more likely to be successful than small ones, probably for all the reasons we discussed in Chapter 7 (Green 1997, also see Duncan 1997). Similarly, Cohen and Carlton

(1998) suggested that estuaries that receive a lot of shipping traffic are more likely to be invaded by stowaways. There is experimental support for this idea too; Levine (2000) found that the number of propagules was a key factor in the likelihood of exotic plants colonizing sedge tussocks.

Exotic species seem to be particularly common in disturbed ecosystems (Elton 1958). For example, the flora of roadsides is often dominated by introduced plants, and degraded aquatic ecosystems are often dominated by introduced fishes (Moyle 1976b, Baltz and Moyle 1993, Moyle and Light 1996, Parendes and Jones 2000). Intuitively, it makes sense that native species would be best able to withstand the pressures of exotic species if their habitats are undisturbed. In fact, some have argued that exotic species can only invade disturbed environments (Fox and Fox 1986), but there are notable exceptions to this generalization (King 1984, Hobbs 1989, Rejmanek 1989). An interesting variation of this disturbance idea comes from an analysis (Case 1996) in which the single best predictor of successful invasion by exotic birds was how many native species had been driven into extinction by human activities. In fact, the number of species gained through invasion was often close to the number lost to extinction.

If disturbed ecosystems are particularly vulnerable to invasion, it seems logical to predict that species that are adapted to disturbed ecosystems (what ecologists would call early-successional colonizers) will thrive as exotic species (Bazzaz 1986). Similar predictions can also be made; for example, that invaders will tend to be abundant species, tolerant of a wide range of conditions with a high reproductive potential (Rejmánek 1996, Williamson and Fitter 1996b). As generalizations they hold up quite well, but exceptions are common (Ehrlich 1986).

One observation about invading species is not intuitively obvious: species originating in Europe have been especially successful as exotics. Charles Darwin remarked that European plants were very common in the American countryside in a letter to the botanist Asa Gray: "Does it hurt your Yankee pride that we thrash you so confoundedly?" Jane Gray, Asa's wife, responded in kind, observing that American plants were, "modest, woodland, retiring things; and no match for the intrusive, pretentious, and self-asserting foreigners" (Crosby 1986). Alfred Crosby (1986) has argued that the success of European species is the result of coevolution and synergism. European humans, their domestic plants and animals, their weeds and pests, and their pathogens coevolved in Europe. Consequently, when European people began exploring the globe and profoundly disrupting native ecosystems with their guns, plows, steel axes, livestock, and diseases, these ecosystems were opened up to a whole suite of invading species. For example, European livestock arrived on distant shores with seeds in their guts and clinging to their hair, seeds of species that would thrive in the agricultural landscapes created by European farmers. (See Niemelä and Mattson

1996 for an entomological perspective.) A new twist on this story of the invasiveness of European plants has emerged with the discovery that at least one European plant, the diffuse knapweed, exhibits *allelopathy* in North America (i.e., it releases chemicals into the soil that suppress neighboring plants), but not in its native range (Callaway and Aschehoug 2000). Apparently, at least some of the species that share its native range have evolved a tolerance to its chemicals.

In summary, there are some broad patterns to the relative success of different introduced species and to the relative ease with which different ecosystems are invaded; however, exceptions are common and there is much room for improving our knowledge (Vermeij 1996, Mack et al. 2000). Understanding these patterns is important from a conservation perspective because it would allow us to predict: 1) which species are most likely to be invasive and thus should be the focus of efforts to prevent their importation and release (Reichard and Hamilton 1997), and 2) which ecosystems are most vulnerable to invasion and thus require careful monitoring to detect and eliminate exotics as soon as they appear (Higgins et al. 1999).

Irony

Living in the northeastern United States, it is easy for me to lament the bad luck that brought Norway rats and the AIDS virus to these shores and the capricious stupidity that brought starlings. However, am I a hypocrite for condemning these species and planting corn and peppers from Mexico in my garden? Why do some government agriculture and health agencies spend millions of dollars on quarantines to keep out exotic pests, while some fish and game agencies spend millions raising and releasing game fish and birds in places outside their native range? Of course, the answers to these questions come down to values and to the truth of the well-known aphorism from George Orwell's *Animal Farm*, "All animals are created equal, but some animals are created more equal than others."

It is usually easy for conservation biologists to write off agricultural lands and backyard gardens and to strongly condemn any exotic species living in a natural or seminatural ecosystem. Nevertheless, difficulties do arise. Think about the following examples.

In California's Angel Island State Park there are groves of Australian eucalypts that park managers wanted to cut so that the sites could be restored to native grasslands, shrublands, and oak woodlands (Westman 1990). Environmentalists protested the decision because the eucalypts provide important habitat for native animals, notably migrating monarch butterflies. What do you think?

A subspecies of the greater prairie chicken, the heath hen, lived along the

Atlantic seaboard from New England to Virginia until the early 1900s when it succumbed to habitat loss, overhunting, and other problems. With some parts of the heath hen's habitat now protected, should we replace it by introducing a subspecies of the greater prairie chicken that still survives in the Midwest?

In the high-elevation shrublands of Hawaii's Haleakala National Park, exotic ring-necked pheasants and chukar partridges are the dominant birds (Cole et al. 1995). Should they be removed from the park even though their diets suggest that they are filling the ecological niche of extinct birds?

Given that wild horses and burros are exotic species in the United States, should the federal government continue spending millions of dollars caring for them, instead of letting them join the many domestic equines that are used for pet food? Do the facts that wild equines lived in North America until quite recently (a little more than 10,000 years ago) and may have been pushed into extinction by Pleistocene hunters make it more acceptable to let feral horses and burros run free (Martin and Burney 1999)?

If you could wave a magic wand and eliminate any exotic population, are there any nondomesticated species mentioned in this chapter that you would spare? What about Canada's exotic earthworms?

Because people have been moving species for millennia, it is often difficult to tell whether a species is exotic or native without careful study of archeological, historical, geographic, ecological, and genetic evidence. Species are generally assumed to be native unless proven otherwise, and thus there are probably far more exotic species than we realize. Should we undertake research to identify and control these hidden exotics?

Finally, and most importantly, do you accept the general idea—exemplified by the case history from Clear Lake told in Chapter 2—that an ecosystem that has more species following biological invasions is less desirable than a natural, uninvaded ecosystem with fewer species?

CASE STUDY   Exotics in New Zealand[1]

About 65 to 90 million years ago a relatively small piece of the earth's crust broke away from the rest of Gondwana, the ancient southern continent, and went drifting off to the eastward by itself. New Zealand's departure left it with only a subsample of the species then inhabiting Gondwana, and later, as new life-forms evolved elsewhere, very few of these ever made it across the Tasman Sea that separates Australia and New Zealand. For example, while mammals were becoming a dominant group in the rest of the world, New Zealand was colonized only by a few mammals that could fly or swim, namely, some bats, seals, and cetaceans. This long period of isolation allowed New Zealand's biota to evolve into many new species, uniquely adapted to a biological environment that was profoundly different from the rest of the world. It was generally a benign environment with abundant rainfall, mild temperatures, and rich soils, a land free of many of the competitors and predators found elsewhere. Moas and eagles seemed to fill the niche of large mammalian herbivores and carnivores, respectively, while large, flightless insects seemed to be the ecological equivalents of small mammalian herbivores. Many species were uniquely New Zealand's; for example, over 80% of the native plants are endemic.

New Zealand's biota remained sheltered by isolation for a very long time, until about 700 to 800 years ago—yesterday on the time scale of evolution—when Polynesian colonists arrived. Colonizing a land that had been devoid of virtually all mammals for 60 million years, you certainly could argue that humans were an exotic species in New Zealand. However, let us focus on the other species introduced by people. The Polynesians, whose descendants are called Maoris, brought some species with them, and one, the kiore or Polynesian rat, had a profound effect, causing local or total extinction of many insects, land snails, lizards, frogs, bats, and birds (Fig. 10.8). The kiore's impact is difficult to appreciate today because, with the beginning of European colonization in the late 1700s, it was reduced to being just the vanguard of a mammalian invasion that ultimately involved 54 species. These range from the small and inevitable (e.g., house mice, Norway rats, black rats) to the large and improbable (e.g., various species of deer and wild goats). Several species of marsupials came from nearby Australia; many more mammal species arrived from Europe. The invaders include herbivores, notably the brush-tailed possum, that have devastated

Figure 10.8. The kiore or Polynesian rat probably had a profound effect on many native New Zealand animals, including a group of large flightless insects called wetas.

forests; they include carnivores, notably ferrets, stoats, and feral cats, that have devastated the native fauna. Most of the introduced species never became well established, but 14 of them did, and now the overall abundance of mammals is relatively high compared with other parts of the world. Some exotic mammals have penetrated the most remote, uninhabited corners of New Zealand and thus are an exception to the generalization that exotic species usually become established only in disturbed ecosystems.

A naturalist traveling in New Zealand today will see few wild or feral mammals because most of them are shy and nocturnal. The predominance of European birds and plants across most of the countryside is what strikes visiting naturalists. The smaller New Zealand birds that survived hunting by the Maori (recall the preceding chapter) have, for the most part, been pushed into residual patches of habitat by deforestation and exotic predators. Indeed, quite a few species such as the saddleback, stitchbird, and black robin survive only on a few small islands where conservation biologists have been able to eradicate exotic mammals, especially rats. In their place one sees blackbirds, chaffinches, goldfinches, and many other exotics, chiefly from Europe.

Most of New Zealand's forest ecosystems have been converted to open lands by Maori farmers and European sheepherders and now support roughly 45 million sheep. Thus it is hardly surprising that exotic, early-successional plants are a dominant part of the vegetation. Consider this quote from Julien Crozet, an early explorer: "I planted . . . wherever I went—in the plains, in the glens, on the slopes, and even on the mountains; . . . and most of the officers did the same." Exotic trees grown in plantations and exotic grains, fruits, and vegetables occupy significant parts of the landscape too. Even the plant most people associate with New Zealand, the kiwi fruit, is an exotic species. Natural forests are relatively free of exotic plants, but many ecosystems that appear natural—floodplains, lakes, and sand dunes, for example—have large numbers of exotics. Currently, New Zealand has about 2300 native species of vascular plants and 2071 wild exotics. Perhaps more importantly, it also has over 18,000 species in cultivation, and the process of invasion is still continuing. Numerous exotic insects have also arrived with the exotic plants and caused their share of problems; recall the kaka-versus-wasps story told above. From an invertebrate conservation perspective, the best-known losses center on spectacular, giant, flightless insects such as various wetas that have been eliminated by rats.

New Zealand conservationists are engaged in a valiant effort to make the best of a bad situation. They have set aside the vast majority of their remaining natural ecosystems, and they have undertaken many ambitious campaigns to eradicate exotics from some smaller islands and to restore them as microcosms of the unique ecosystems that used to cover the main islands. Lately, they have even carried this restoration campaign to the two main islands with some notable success. Some of the most impressive stories of conservation biology in action have come from New Zealand; we will review one of them in Chapter 12, Managing Populations.

---

[1] This account was distilled primarily from Crosby (1986), King (1984, 1990), Towns et al. (1990, 1997), Wardle (1991), and David Norton (personal communication.)

Summary

Isolation has been a critical factor in shaping the evolution and distribution of species, but human activities have often broken down the barrier of isolation, allowing exotic species to occupy areas outside of their natural geographic ranges. Many species have been moved by accident, for example, as stowaways in ships and as parasites or pathogens on other organisms deliberately moved by people. Motivations for deliberately moving species to new areas include commerce, subsistence, recreation, science, attempts to control exotics established earlier, and simple whimsy. Some species have been able to extend their natural range because of human-induced habitat changes, and these may also be considered exotic species. The effects of exotic species have been diverse and profound, especially on islands. Many populations have been decimated, in large part because of predation, competition, disease, parasitism, and hybridization associated with exotic species. Some entire ecosystems have been altered. Exotic species seem to be particularly successful at invading islands and disturbed ecosystems.

FURTHER READINGS

Many books have been written about exotic species, including recent treatments such as Bright (1998) and Cox (1999). Five that I particularly recommend are Elton (1958), a classic; King (1984), an excellent case study focusing primarily on the introduction of mammals to New Zealand; Crosby (1986), which examines the European invasions of Australia, New Zealand, and the Americas; Office of Technology Assessment (1993), a thorough review of the impacts of exotics in the United States; and Sandlund et al. (1999), which covers biological invasions globally. For a short overview, see Mack et al. (2000), and for a popularized treatment, see Devine (1998). We will briefly discuss control of exotics in Chapter 12; for a more thorough review, see Dahlsten (1986) and Office of Technology Assessment (1993).

TOPICS FOR DISCUSSION

See the section entitled "Irony" for some thought-provoking questions to discuss.

Part III

Sulzer

Maintaining Biodiversity

Unless another large meteorite slams into the earth between the time these words are written and when you read them, it is reasonable to trace most threats to the earth's biodiversity back to human causes. Because of this, some people feel that the best way to diminish our effect on biodiversity is to leave it alone. In other words, we could simply arrest our population growth, reduce our use of resources, and withdraw from large stretches of the planet, leaving the other biota to operate without us. This plan would substantially diminish the overall threat to biodiversity, but it is not realistic. In practice, we need to work with existing social, political, and economic systems, trying to change them from within, trying to make them more compatible with existence of all life on earth (the subject of the book's last section, Part IV). Societies can be changed over decades or centuries; unfortunately, this is not fast enough. We must also attack the problem of maintaining biodiversity directly and quickly because species are being lost now. In Part III we will examine the things that can be done on the ground, in the field, out in the wild places, to maintain biodiversity by managing ecosystems (Chapter 11) and populations (Chapter 12). In Chapter 13 we will discuss the role zoos, aquaria, and botanical gardens can play in maintaining biodiversity, especially their role as insurance against the possibility that our efforts in the field may not succeed. In Chapter 14 we will examine ways to set priorities for these activities.

Chapter 11
Managing Ecosystems

Conservation biologists are fairly skilled at looking at the big picture, at seeing forests, not just trees. They understand that we cannot maintain genetic diversity without maintaining species diversity and that we cannot maintain species diversity without maintaining ecosystem diversity. They know that we cannot think about a species in isolation; we have to be concerned about the whole suite of interacting species and environmental features that constitute its habitat. As Shakespeare's Shylock, the merchant of Venice, said "You take my life when you take the means whereby I live."

When biodiversity advocates think about managing ecosystems, they usually think first about reserves. In particular, they are likely to focus on protecting a cluster of ecosystems that are representative of the region's ecological diversity and thus are likely to contain a large portion of a region's species. This is the coarse-filter strategy of maintaining biodiversity (recall Fig. 4.6). In the first section of this chapter, we will consider the strategies conservationists employ to protect natural ecosystems (i.e., ecosystems that are little changed by people) by establishing and managing reserves.

Most conservationists also recognize that protecting some exemplary natural ecosystems is not enough. We must look beyond the boundaries of reserves to the ecosystems that form the larger matrix in which reserves are imbedded. These are often seminatural ecosystems—ecosystems that have been modified by human activities such as logging, fishing, and grazing livestock, but that are still dominated by native species. Methods for integrating biodiversity maintenance with natural resource management in these modified ecosystems constitute the second section of this chapter.

The third and fourth sections of this chapter deal with cultivated ecosystems (largely agricultural land) and built ecosystems (urban areas and other places intensively used by people). In these areas the main biodiversity issue is usually how to keep these ecosystems from exporting problems such as exotic species and contaminants to natural and seminatural ecosystems.

In the final section of this chapter we will delve into restoration ecology, a relatively new discipline that focuses on methods for restoring the structure and function of ecosystems degraded by human activities.

Protected Ecosystems

The idea that some places should be set aside from the usual gamut of human uses goes back at least 3000 years to Ikhnaton, king of Egypt, and probably earlier to sacred mountains and groves unrecorded by history (Alison 1981). It is hard to know why such places were selected for protection and exactly what types of protection were enacted. Yellowstone National Park, whose establishment in 1872 marks the beginning of the modern era of establishing reserves, was selected for its natural beauty, but protection existed only on paper until 1886 when the U.S. Army was charged with keeping poachers out. In this section we will consider three issues regarding protecting ecosystems: selecting ecosystems to be protected; designing a reserve for those ecosystems; and managing a reserve after it is established. Natural places protected from most human activities may have many names—parks, refuges, sanctuaries, wilderness area, preserves, and more (Table 11.1). Sometimes, these different names reflect different management goals and strategies, and, sometimes, they simply reflect the ambiguity of language. We will use "reserve" as a generic term for areas in which natural ecosystems are protected from most forms of human use.

Reserve Selection

Traditionally, the selection of reserves has been driven by aesthetics and recreation because people love to visit spectacular places—lakes ringed by forested slopes, snow-covered crags, wind-swept beaches—and enjoy the beauty of nature. Of course, conservationists have also promoted the importance of reserves as habitat for wild life, and many places are protected because they harbor an unusual diversity and abundance of wild life (e.g., the Serengeti plains of Tanzania and Kenya) or a species that is uncommon and spectacular (e.g., the redwoods and sequoias of California). A few reserves have been established for species that are uncommon but not very spectacular. For example, in Texas, several hundred hectares are managed as a reserve for the Houston toad, an endangered species that looks more like a lump of mud than something that would appear on the cover of

TABLE 11.1 The World Conservation Monitoring Centre recognizes six basic categories of protected areas.*

Category Ia	Strict nature reserve: protected area managed mainly for science. (1423 units covering 978,000 km²) *Definition* Area of land and/or sea possessing some outstanding or representative ecosystems, geological or physiological features and/or species, available primarily for scientific research and/or environmental monitoring.
Category Ib	Wilderness area: protected area managed mainly for wilderness protection. (654 units covering 939,005 km²) *Definition* Large area of unmodified or slightly modified land, and/or sea, retaining its natural character and influence, without permanent or significant habitation, which is protected and managed so as to preserve its natural condition.
Category II	National Park: protected area managed mainly for ecosystem protection and recreation. (2233 units covering 3,994,440 km²) *Definition* Natural area of land and/or sea, designated to (a) protect the ecological integrity of one or more ecosystems for present and future generations, (b) exclude exploitation or occupation inimical to the purposes of designation of the area, and (c) provide a foundation for spiritual, scientific, education, recreational, and visitor opportunities, all of which must be environmentally and culturally compatible.
Category III	Natural monument: protected area managed mainly for conservation of specific natural features. (409 units covering 191,189 km²) *Definition* Area containing one, or more, specific natural or natural/cultural feature that is of outstanding or unique value because of its inherent rarity, representative, or aesthetic qualities or cultural significance.
Category IV	Habitat/species management area: protected area managed mainly for conservation through management intervention. (3622 units covering 2,450,973 km²) *Definition* Area of land and/or sea subject to active intervention for management purposes so as to ensure the maintenance of habitats and/or to meet the requirements of specific species.

TABLE 11.1 *Continued*

Category V   Protected landscape/seascape: protected area managed mainly for landscape/seascape conservation and recreation. (2418 units covering 1,051,465 km²)

Definition   Area of land, with coast and sea as appropriate, where the interaction of people and nature over time has produced an area of distinct character with significant aesthetic, ecological and/or cultural value, and often with high biological diversity. Safeguarding the integrity of this traditional interaction is vital to the protection, maintenance, and evolution of such an area.

Category VI   Managed resource protected area: protected area managed mainly for the sustainable use of natural ecosystems. (1995 units covering 3,598,619 km²)

Definition   Area containing predominantly unmodified natural systems, managed to ensure long-term protection and maintenance of biological diversity, while providing at the same time a sustainable flow of natural products and services to meet community needs.

*Categories I to III are clearly reserves as we are using the term here. The 1997 United Nations estimates of the number of each different type of protected area and their total area appear in parentheses. The data generally apply only to areas protected by national governments, not areas protected by states, provinces, counties, private organizations, and so on.

National Geographic (Lowe et al. 1990). We will discuss managing the habitat of single species in the next chapter, Managing Populations. Here the primary focus will be on protecting ecosystems as a strategy for maintaining multiple species, while acknowledging that it is also important to think about maintaining ecological or evolutionary processes, especially in the long term (Cowling and Pressey, in press).

All reserves—even those selected for their scenic qualities—encompass ecosystems or portions of ecosystems and thus maintain habitat for a variety of species. However, natural resource managers cannot be content with a haphazard approach because it will lead to an array of protected ecosystems that is incomplete, an array that provides little or no habitat for many species. Yet, how can we systematically protect the habitat of most species if relatively few species have been described by scientists to date (Chapter 3), and if, even in relatively well studied regions such as Europe, we know little about the distribution of most known species? Obviously, one strategy is to do the best we can with whatever species distribution data are available while striving to obtain more information (Balmford and Gaston 1999, Margules and Pressey 2000). An important complement to this strategy, or even alternative, lies with the coarse-filter approach to maintaining biological diversity (Chapter 4) and its assumption that most species, known and unknown, will be protected if a reserve system contains a representative array of the region's ecosystems.

Classifying Ecosystems: To maintain the biodiversity of a region using a coarse-filter approach we need a classification system that defines different types of ecosystems quite finely. In other words, it is not sufficient simply to define a Forest Ecosystem, a Lake Ecosystem, a Marsh Ecosystem, and so on, because there are many different kinds of forests, lakes, and marshes with different biota. For example, the biota of a warm-water, acidic lake would show little overlap with that of a nearby cold-water, alkaline lake.

An ecosystem classification system designed to be used with the coarse-filter strategy should be based on both the physical environment (e.g., soil and climate factors) and on the species that dominate the ecosystems. In practice, classifications, particularly of terrestrial ecosystems, are usually weighted toward dominant species (e.g., oak-pine forests, spruce-fir forests) because it is often easier to recognize the distribution of conspicuous species than the distribution of physical features. There are two problems with relying primarily on dominant species; first, dominant species are often successful species that are able to thrive in a variety of environments, and thus their distribution may mask factors that shape the distribution of other species. For example, in the northeastern United States spruce-fir forests grow on the tops of mountains (because high altitudes are colder) and on the bottoms of valleys (because cold air settles there), and thus there are differences in their biota (Hunter 1991). If we established a reserve that contained only mountaintop, spruce-fir forests, then species found only in valley-bottom, spruce-fir forests would be left out of the reserve. The second problem with defining ecosystems on the basis of their dominant species is that species are continuously changing their range in response to climate change (Chapter 6). Consequently, it is better to focus the coarse-filter strategy on the physical environment as the arena that holds biological diversity rather than on the dominant species that happen to occupy the arena at this time (Hunter et al. 1988, Nichols et al. 1998).

In some landscapes it is difficult to delineate ecosystems because a strong, uniform gradient is the dominant factor. Consider the changes in altitude, and hence climate, going up the side of a mountain; these are profoundly important, but often create a biological continuum with no obvious breaks. From a conservation perspective the solution is simple: try to protect reserves that encompass the entire gradient (Noss 1987b). Similarly, there may be some danger in delineating ecosystems sharply if the ecotones between adjacent ecosystems end up getting overlooked, because natural ecotones may be areas of high species richness and even focal points for the evolution of new species (Smith et al. 1997).

Inevitably, we have to decide on the boundaries of the geographic region within which we will try to maintain a representative array of ecosystems. Ideally, this would be based on ecological factors such as climate, soils, and topography as portrayed in Figure 4.3. (Sometimes, such bioregions or ecoregions are also called ecosystems, for example, when people refer to the "Greater Yellowstone ecosystem.") In practice, regions are often defined politically. Consequently, you are likely to hear conservationists say "We've protected a good example of a warm-water, acidic lake in Norway" (or Minnesota or Smith County), but relatively seldom will you hear someone say "We've protected a warm-water, acidic lake in the Southern Coastal Lowlands Bioregion" (or some other ecologically defined area). Conservationists recognize the problems with using political boundaries, but they usually work for government agencies and other organizations whose purview is limited by political boundaries. Nevertheless, this is slowly changing as more and more conservationists try to take a bioregional approach (Noss 1993, Olson and Dinerstein 1998).

Filling the Gaps: If we use the coarse-filter perspective to examine the array of reserves that already exist, significant gaps that need to be filled often become apparent (Pressey 1994, Pressey et al. 1996, Armesto et al. 1998). In particular, high-altitude ecosystems are common in reserve systems because they are appreciated for their scenery and are of marginal value for most economic endeavors. On the other hand, areas with fertile soils and benign climates are uncommon in reserve systems because they are in demand for agriculture; indeed most such areas were already converted to agriculture before people began creating substantial reserve systems. Nepal's distribution of protected areas shows an interesting pattern (Fig. 11.1). There is a large area of protected land above the altitude of human habitation and at low altitudes but little at intermediate altitudes (Hunter and Yonzon 1993). The reserves at low altitudes are the anomaly; they were originally established as hunting reserves for the royal family at a time when malaria kept the human population density low. Malaria is under control now, and human populations have increased enormously, putting considerable strain on the reserves; at least, however, the reserves are still protecting tigers, rhinos, and thousands of less well known species. Ironically, marine ecosystems have long been poorly represented in reserve systems despite their aesthetic, recreational, and ecological values, although this is finally starting to change (Tisdell and Broadus 1989, Allison et al. 1998); this deficiency can be traced to our lack of sensitivity to things that happen underwater.

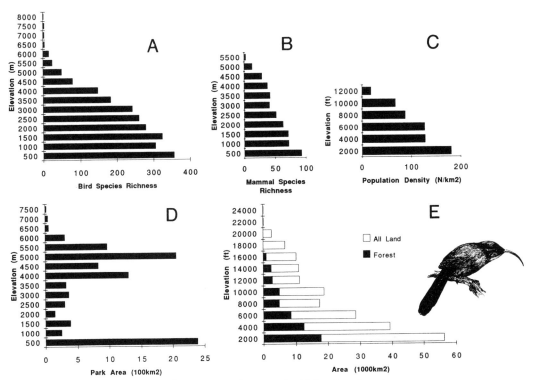

Figure 11.1. In Nepal there are few protected areas at middle elevations because, historically, most of the people lived in these areas. Many species are found exclusively in the ecosystems characteristic of these altitudes, and thus this is an important gap in the network of existing reserves.

Analyses of the gaps in reserve networks can work at different scales. At the global scale, the question arises regarding which continents, biomes, or biogeographic areas are underrepresented (Olson and Dinerstein 1998). In other words, where should international agencies focus their efforts to fill gaps? At the regional scale, which types of ecosystems are missing from the network of reserves? Addressing these issues will be a high priority for national, state-provincial, and local agencies. Gap analyses tend to focus on ecosystems and their large-scale equivalents (e.g., biomes or bioregions), but many of them focus on the distribution of individual species as we will see in the next section.

Centers of Diversity: The world's species are not distributed uniformly. There are some obvious *"hot spots"* such as tropical forests and coral reefs that have unusually large numbers of species. Other places can be called hot spots because they have a wealth of endemic species; Madagascar, the Cape region of South Africa, and southwestern Australia are good examples. Not surprisingly, many conservationists believe that these places should be a high priority for establishing reserves (Myers 1988, 1990, Myers et al. 2000).

Biogeographers and taxonomists can provide a general sense of where centers of diversity and endemism might exist, but to explore the issue systematically requires a Geographic Information System (GIS) (Fig. 11.2). GIS, remote sensing, and related technologies open the door to various quantitative techniques for selecting reserves (see Margules 1989, Bedward et al. 1992, Scott et al. 1993, Jennings 2000 and references cited therein for more on these techniques). Some of these techniques use computer models to facilitate selecting sets of reserves that complement one another (i.e., limit overlap in the species covered) or even offer the most cost-effective solution for

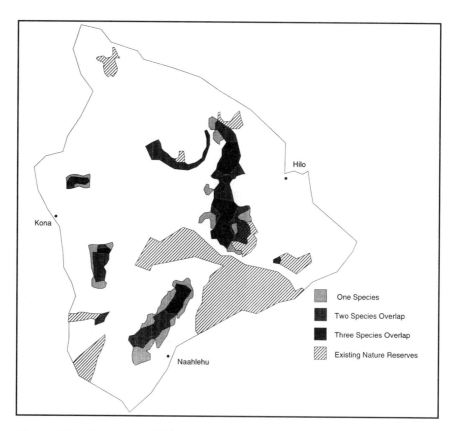

Figure 11.2. Conservation biologists have used Geographic Information Systems (GIS) to combine maps representing distributions of many different species and existing reserves (layers of information) into composite maps. In this figure (redrawn by permission from Scott et al. 1993), a composite map based on the ranges of just three species of Hawaiian finch shows that the existing reserves did not coincide well with the areas of finch diversity. See Scott et al. (1993) for a description of these techniques.

meeting a specific conservation goal (Williams et al. 1996, Ando et al. 1998, Margules and Pressey 2000). This kind of work has revealed some weakness in the hot spot concept. For examples, studies of bird distributions in Great Britain (Prendergast et al. 1993) and Australia (Curnutt et al. 1994) found relatively little overlap between hot spots of species richness and endemism (defined here by birds with relatively small geographic ranges). Similarly, hot spots of species richness for different taxonomic groups (e.g., butterflies versus birds) often do not coincide (Prendergast et al. 1993, Gaston 2000). Perhaps in response to some of these problems, Norman Myers, who is widely credited with the hot spot idea, now describes hot spots in terms of two features: areas with exceptional concentrations of endemic species and experiencing exceptional loss of habitat (Myers et al. 2000).

How Many to Select: Nature reserves are very popular among many people but not necessarily among ranchers, loggers, and other people who use large areas of land for their livelihood. Consequently, the issue of how many areas need to be protected is frequently debated. Clearly, one small representative of each type of ecosystem in each region is not sufficient because it would be too small to protect viable populations of many species, especially animals with large home ranges, and it would be vulnerable to a catastrophic disturbance. Unfortunately, there may be a considerable gulf between what is ecologically desirable and politically feasible (Musters et al. 2000). The World Conservation Union recommends that at least 10% to 15% of the total area of each ecosystem type be protected. This is significantly more than the area currently protected for most types of ecosystems; overall about 6.4% of the earth's land surface is well protected (Categories I to V of Table 11.1), and the distribution is very imbalanced among ecosystem types (United Nations Development Programme et al. 2000). Furthermore the 10% to 15% figure was based on a rather generic recommendation that the extent of the world's protected areas (about 4% to 5% at that time) "needs to be at least tripled" (World Commission on Environment and Development 1987). Recommendations from other sources have ranged from 5% to 99.7% with a rough convergence on 50% depending on the goals and the ecosystems or taxa being considered (Noss and Cooperrider 1994, Soulé and Sanjayan 1998). Obviously, there is no one correct answer. For example, the minimum area for a network of reserves would depend on whether they were surrounded by seminatural ecosystems or built and cultivated ecosystems (Gascon et al. 1999). This leads us into a new issue, reserve design.

Reserve Design

Reserve selection is inevitably followed by reserve design: deciding how large the reserve should be, where its boundaries should lie, and other issues. Many ideas about reserve design can be traced back to a 1975 paper in which Jared Diamond made an analogy between reserves and islands and proposed six design features for reserves based, in part, on island biogeography theory (Fig. 11.3):

1. A large reserve will hold more species than a small reserve because of the species-area relationships described in Chapter 8.

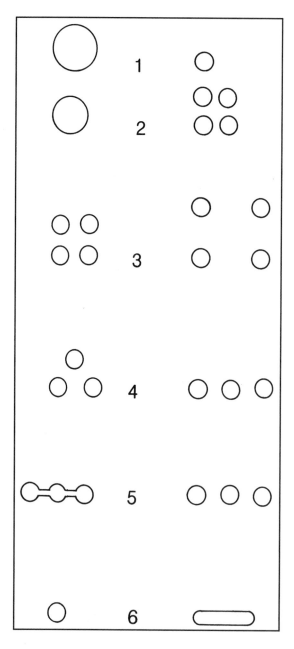

Figure 11.3. Schematic representations of design principles for nature reserves. In each pair the design on the left will probably have a lower extinction rate and thus may have higher species diversity. (Redrawn by permission from Diamond 1975.)

2. A single large reserve is preferable to several small reserves of equal total area, assuming they all represent the same ecosystem type.
3. If it is necessary to have multiple small reserves, they should be close to one another to minimize isolation.
4. Arranging small reserves in a cluster, as opposed to a linear fashion, will also facilitate movement among the reserves.
5. Connecting the reserves with corridors will make dispersal easier for many species.
6. By making reserves as circular as possible, dispersal within the reserve will be enhanced, and the negative effects of edges (see Chapter 8) will be minimized.

These ideas were soon widely accepted even though a number of the points have been challenged (Ambuel and Temple 1983, Blouin and Connor 1985) and one—that a single large reserve is better than several small ones of equal total area—generated a heated controversy. We will address these points and others in three sections on reserve size, landscape context, and connectivity.

Reserve Size: Conservationists prefer large reserves to small reserves for two main reasons. First, large reserves will, on average, contain a wider range of environmental conditions and thus more different types of ecosystems and thus more species than small reserves. Additionally, some species will be absent from small reserves because they require large home ranges (e.g., large carnivores), or simply because they live at low densities and by chance alone are unlikely to be in a small reserve (e.g., many rare plants). In both cases, these are species that are likely to be high priorities for conservation. (See "Fragmentation" in Chapter 8 for further discussion of these ideas.)

Second, large reserves are more secure and easier to manage (at least per unit area) than small reserves for three reasons: 1) large reserves have relatively large populations that are less likely to become extinct (recall Chapter 7); 2) large reserves have a relatively shorter edge than small reserves and thus are less susceptible to external disturbances such as invasions of exotic species and poachers (recall Fig. 8.12); and 3) large reserves are less vulnerable to a catastrophic event such as a volcanic eruption, hurricane, or oil spill because most catastrophes cannot disturb an entire reserve if it is large enough. All three of these factors, especially the second one, make large reserves easier and cheaper to manage per unit area.

The issue of natural catastrophes needs to be clarified. It is important that natural disturbances such as fires be allowed to shape reserves (we will return to this issue below when we discuss reserve management). This means that reserves need to be large enough not to be profoundly changed by a single disturbance event. This concept led Pickett and Thompson (1978) to suggest that reserves should be larger than the *minimum dynamic area,* the smallest area that would hold an array of patches representing different stages of disturbance and succession. For example, if a landscape was characterized by fires covering 1000 hectares, a reserve for this landscape should be many thousands of hectares to contain a series of patches representing burns of different ages.

Reserve size was central to a well-known debate that erupted shortly after Diamond's paper was published, a debate known by the acronym SLOSS, Single Large or Several Small (Diamond 1976, Simberloff and Abele 1976a, 1976b, 1982, Terborgh 1976, Whitcomb et al. 1976, Boecklen and Gotelli 1984, Simberloff 1986). The controversy began when Daniel Simberloff, Lawrence Abele, and others expressed some doubt about Diamond's second principle. They do not believe that there is a simple, universal answer

---

BOX 11.1   **Single large reserve or several small[1]**

The answer to the "single large or several small" debate probably lies somewhere between these two alternatives, but to illustrate the fundamental difference between them, Table 11.2 depicts the two extreme cases. Diamond's approach would be supported if Scenario 1 described the real world. Each successively larger reserve contains all the species of the smaller reserves plus additional species that have more stringent minimum area requirements. There is a predictable gradient among the species. At one end of the gradient, daisies are found in all the reserves; at the other end hawks need so much land that they can survive in only the 240-ha reserve. In this situation if you were given $1,200,000 to save forests from being turned into parking lots and if land cost $5000 per hectare, you should buy the 240-ha reserve and thus maintain 224 species. For the same amount of money you could buy reserves D, E, and F, but you would protect only 199 species.

Scenario 2 describes a situation that would definitely favor the Simberloff approach. Again, large reserves have more species, but each reserve has a unique set of species, a more or less random selection from the species pool. There is no pattern of adding new species with more stringent area requirements because all the species have an equal ability to survive in limited areas, or because the uniqueness of each set masks any tendency for some species to occur only in large areas. Here, the best approach would be to buy reserves A, B, C, D, and E; they would harbor 709 species and cost just $750,000. The G reserve would still cost $1.2 million and only have 224 species.

In summary, the fundamental difference between the two scenarios depends on whether the abilities of different species to colonize and survive in a small, isolated area are equal or whether they vary along a gradient in a

to the question: if you have a finite amount of money, should you buy one large nature reserve or several small ones of equal total area? Defenders of Diamond's model have sometimes reacted as though the first design principle—large reserves are better than small reserves—was under attack and have even accused the opposition of advocating the dismembering of nature reserves (Simberloff and Abele 1984, Willis 1984). In Box 11.1 this question is explored in detail; suffice it to say here that no consensus on the correct answer has been reached beyond an ambiguous compromise position: "Nature reserves should be as large as possible, and there should be many of them" (Soulé and Simberloff 1986). The SLOSS debate has quieted down considerably, partly because people got tired arguing about a question for which there was no clear answer, and partly because they re-

---

BOX 11.1   *Continued*

fixed and predictable fashion. Clearly, neither of these scenarios describes the real world, but which is more accurate? The primary consideration is probably environmental homogeneity. Diamond's assertion that one large reserve is superior to several small ones explicitly assumes that all the reserves represent the same type of environment. However, is it realistic to assume that different reserves can represent the same type of environment? Critics claim that, at least at a microenvironmental scale, there will always be differences among reserves and that these will be reflected in the biota.

Conceptually, it would be easy to determine which scenario is more accurate; we would measure total species richness for a variety of different size islands and determine what portion of their biota is unique or shared with other islands. In practice most studies have been limited to one or two taxonomic groups. The critics of Diamond's view have taken their best supporting evidence from studies of plants and invertebrates, species that usually have small home ranges and are quite sensitive to microhabitats (Simberloff and Abele 1976a, Higgs and Usher 1980, Järvinen 1982, Simberloff and Gotelli 1984, Nilsson et al. 1988, Quinn and Harrison 1988). Diamond's defenders have been largely ornithologists and thus in tune with species that may be less sensitive to microhabitats and more sensitive to area than plants and invertebrates. Overall, the limited field evidence does not seem to clearly favor either side (Simberloff and Abele 1982), especially because the data are confounded with questions of environmental homogeneity and how long a patch must be isolated before the extinction rate stabilizes.

---

[1] From Hunter (1990)

TABLE 11.2 A hypothetical series of seven progressively larger reserves.*

Patch Size (ha)	Number of Species	Number of New Species	Accum. # of Species	Representative Species
Scenario 1				
A (10)	119	—	119	Daisy, etc.
B (10)	119	0	119	Daisy, etc.
C (20)	137	22	137	Daisy, sparrow, etc.
D (40)	159	16	159	Daisy, sparrow, snake, etc.
E (70)	175	24	175	Daisy, sparrow, snake, robin, etc.
F (130)	199	25	199	Daisy, sparrow, snake, robin, squirrel, etc.
G (240)	224	25	224	Daisy, sparrow, snake, robin, squirrel, hawk, etc.
Scenario 2				
A (10)	119	—	119	Daisy, etc.
B (10)	119	119	238	Sparrow, etc.
C (20)	137	137	375	Ivy, grackle, etc.
D (40)	159	159	534	Trillium, blackbird, tortoise, etc.
E (70)	175	175	709	Lily, toad, rabbit, shrew, etc.
F (130)	199	199	908	Holly, snake, warbler, mouse, pine, etc.
G (240)	224	224	1,132	Robin, lizard, frog, squirrel, fox, hawk, etc.

*The series is described with the area of each reserve (column 1), the total number of species in each reserve (column 2), the number of new species added to the series total by each reserve (column 3), and the accumulative number of species in the series (column 4). The last column gives a hypothetical sample of the species found in each reserve. In Scenario 1 each reserve has all the species of the smaller reserves plus some new species. Each reserve has the same area as the total of the three preceding reserves. Species numbers were calculated from $S = CA^z$ with $C = 75$ and $z = 0.2$; this might roughly approximate the number of vascular plant and vertebrate animal species in a temperate forest.

alized that, in practice, reserve size will usually be determined by political and fiscal realities, not ecological models of how species are distributed.

Landscape Context: Although it is common to think of reserves as sacrosanct refuges—islands of nature isolated in a sea of human-altered ecosystems—this is not an accurate view. The boundaries of reserves are permeable walls, and many things move across them (Janzen 1986). Air and water pollution, exotic species, livestock, and poachers are some of the negative factors that can impinge on reserves from outside. On the positive side, reserves often export clean air and water and are a source of individual organisms that can bolster low populations outside the reserve. For example, proponents of marine reserves have shown that fishing outside reserves is improved because breeding stocks in the reserves produce offspring that are caught outside the reserve (Bohnsack 1993, Roberts and Polunin 1993, McClanahan and Mangi 2000, Paddack and Estes 2000). Some of the movements into a reserve are positive too, especially because many reserves are so small that they would probably lose their populations of larger animals if they were truly isolated. These populations can persist because individuals are regularly exchanged with ecosystems outside the reserve thereby avoiding genetic isolation. Gascon et al. (1999) found the fauna of a secondary forest (regrown in abandoned cattle pastures) overlapped substantially with the fauna of primary forest remnants. In short, reserve designers must pay careful attention to what will lie outside a reserve when deciding where to put its boundaries.

One obvious idea is to design reserves so that they will be buffered from the most harmful human activities as much as possible. Ideally, a reserve would probably be a central core buffered by concentric circles of ecosystems with decreasing degrees of naturalness; for example, seminatural forests managed for the large trees required for lumber, then plantation forests managed for small trees used for pulp or fuel, and then agriculture (Mladenoff et al. 1994). The nearest urban area or industrial complex would be many kilometers away. Collectively, the whole tract would constitute what Noss and Harris (1986) call a "multiple-use module," or MUM, or what Grumbine (1990) would call a "greater ecosystem." See Kremen et al. (1999) for an example of designing a national park in Madagascar with a multiple-use buffer zone.

Reserves are easier to buffer if they are fairly circular because a circle has less edge per unit area than any other shape. Keoladeo Ghana National Park in Bharatpur, India, one of the world's premier bird reserves, is surrounded by a high brick wall about 35 km long. However, if the 29 km^2 reserve were circular, the wall would only be 19 km long and far easier to patrol and maintain.

Buffering is also easier if the reserve boundaries correspond with certain natural boundaries such as shorelines and ridge tops. Watershed lines are often excellent reserve boundary lines because a reserve that fully occupies a single watershed will have relatively few problems with water quality and quantity, and it will be a cohesive unit of habitat for many aquatic species (Fig. 11.4). In practice, reserve boundaries are more likely to follow a political or ownership boundary than a natural boundary (Newmark 1985, Wilcove and May 1986). In an interesting twist on buffering, many reserves are located along international frontiers to provide a strategic military buffer in case of war. The most conspicuous example of this is the de

Figure 11.4. The reserve depicted in the center of this drawing illustrates many desirable features, although it is fairly small for ease of illustration. It encompasses a wide range of ecosystems spanning elevations from river level to mountaintop. It fully occupies a watershed by lying within natural boundaries, the watershed line and river shore, and is fairly circular in outline. It is buffered by seminatural forests from plantation forests, and by plantation forests from agriculture. It is connected to other reserves by natural vegetation along both the mountain slope and the river shore.

facto reserve that now exists in the demilitarized zone between North and South Korea providing habitat for two very rare birds, the Japanese and white-naped cranes, as well as for many other species (Kim 1997).

Connectivity: In a Panglossian "best of all possible worlds," reserves would be so large that they would adequately protect even the most demanding species, or they would be completely surrounded by carefully managed seminatural ecosystems through which species could easily move from reserve to reserve. In the real world, very few reserves are large enough to protect their complete biota. An analysis of the mammal fauna of reserves in western North America indicated that only one complex of four adjacent reserves totaling 20,736 km² (Banff, Jasper, Kootenay, and Yoho in Canada) had not lost a mammal species (Newmark 1987, 1995), and a similar pattern was found among Tanzanian parks (Newmark 1996). Moreover, as human populations increase, pressure will mount to convert seminatural ecosystems to agriculture and other intensive uses, further fragmenting landscapes and making it extremely difficult to keep reserves connected through a matrix of seminatural ecosystems. In the face of these realities, conservation biologists often stress the importance of maintaining connectivity among reserves, perhaps with *corridors,* linear strips of protected land (Beier and Noss 1998).

Four basic kinds of movement need to be maintained (Fig. 11.5) (Hunter 1997). First are the daily movements most animals make among the patches of preferred habitat that comprise their home range. These are relatively small-scale movements, and most reserves are large enough to encompass them except for wide-ranging species like large carnivores. The daily movements that many colonial birds and bats make between their roosts and foraging sites can also be quite long, but these species are usually able to fly over areas that are not reserves.

Second are the annual migrations many animals make between winter and summer ranges, or dry season and wet season ranges. The lengths of these movements vary from a few hundred meters for some amphibians and insects to thousands of kilometers for some birds and marine animals. For migration over intermediate distances, for example, herds of large mammals moving between high-altitude summer range and low-altitude winter range, connecting reserves could be of critical importance if the reserves are not large enough to encompass the entire annual range. For this

Figure 11.5. When we think of maintaining landscape connectivity, there are four basic types of movement to consider: home range, migration, dispersal, and geographic range shifts. A carnivore patrolling its territory is often influenced by connectivity, as are herds of large mammals like these chamois that must migrate to lower elevations in winter. From a long-term conservation perspective, probably the most important movements are dispersal (such as among a metapopulation of frogs), and geographic range shifts (which are usually driven by dispersal processes such as seeds being carried by fruit-eating animals).

reason, conservationists have proposed protecting land between two national parks in Tanzania, Lake Manyara and Tarangire, to allow zebras, wildebeest, and other antelopes to move to Lake Manyara in the dry season (Mwalyosi 1991). Similarly, the complex migrations of resplendent quetzals, involving four different life zones in the mountains of Central America, has led to specific recommendations for reserve system design (Powell and Bjork 1995).

Third are the dispersal movements that young animals and plants (the latter usually as seeds, spores, or pollen) make away from their parents. Dispersal movements are a major concern for conservation biologists because they are vital to keeping the organisms of a reserve "connected" with conspecifics living in other reserves, or outside the reserves. Imagine a reserve with 10 tigers. As long as tigers are freely dispersing in and out of the reserve, the reserve's tigers are part of the whole region's tiger population, say 300 tigers, and they are relatively safe from the problems that afflict small populations. In contrast, if tigers are no longer able to disperse into the reserve, the reserve's 10 tigers constitute an isolated, and very vulnerable, population. Of course, dispersal ecology varies greatly among species: some species can easily disperse long distances over any terrain (e.g., fungi spores), but others cannot; some species can persist in small isolated populations with no immigration (e.g., fish species confined to a single spring or cave), but others cannot (Hansson et al. 1992). Unfortunately, we know relatively little about dispersal because it is a difficult phenomenon to study (Chepko-Sade and Halpin 1987, Gaines and Bertness 1993). Nevertheless, population biologists are quite certain that dispersal is a critical issue in the viability of many endangered species (Chapter 7), especially animals, and thus maintaining dispersal routes across otherwise fragmented landscapes is a major goal of conservation biologists.

Fourth are the range shifts that species make in response to climate change, moving back and forth across continents at time scales measured in thousands of years (Chapter 6). No refuges are large enough to accommodate continental-scale movements, but conservationists have considered linking reserves with continental-scale corridors, or at least having reserves arranged as stepping-stones across a continent (Hunter et al. 1988). In mountainous areas, species can respond to climate change by shifting their altitude; therefore linking reserves at different altitudes would deal with this issue in montane environments.

Naturally, the design of a connection should depend on the kinds of organisms and the types of movements it was intended to accommodate. A connection designed to accommodate short-range movements by relatively mobile animals may only need to provide some cover or the right microclimate. To take an extreme example, eastern chipmunks will move among isolated woodlots along a barbed-wire fence with a narrow strip of grass and herbs (Henderson et al. 1985), while, conversely, ringlet butterflies will move among forest clearings using open, grassy tracks (Sutcliffe and Thomas 1996). Connectivity in the context of marine reserves may mean locating reserves strategically with respect to oceanic currents that transport organisms, especially larvae and propagules (Roberts 1997, Allison et al. 1998). A connection designed to allow continental-scale movement by organisms that are relatively sedentary (e.g., terrestrial snails and many plants) would have to provide habitat in which the species could live and reproduce, because it might take multiple generations for a species to move. In this case, the connection needs to provide a range of environments just like a reserve and needs to be wide enough to minimize the negative consequences of edge effects. Unfortunately, designing connections is still based more on simple principles like these than on empirical data (Hobbs 1992, Simberloff et al. 1992), although more and more relevant studies are accumulating (Beier and Noss 1998). Some commonsense approaches suggest themselves, especially, the "piggybacking" of connections onto other efforts to maintain linear strips of natural ecosystems such as hiking trails and strips of riparian vegetation retained to protect water quality. Riparian zones are particularly attractive in this context because they typically form a natural landscape network and have so many other ecological values. In some Australian agricultural landscapes, roadside verges are being maintained in native vegetation to facilitate movement of some animals (Saunders and Hobbs 1991); this is rather ironic, given that roads are a major cause of fragmentation (Chapter 8). For an example of designing a connected network of reserves see Hoctor et al. (2000).

The most common manifestation of the connection idea—protecting corridors between reserves—has attracted critics (Simberloff and Cox 1987, Knopf 1992, Simberloff et al. 1992, Knopf and Samson 1994). They particularly question the cost-effectiveness of corridors. A strip of land 0.5 km wide by 50 km long is likely to be much more expensive to purchase and difficult to protect than a compact area of the same size because it will cross many ownerships. Furthermore, corridors are particularly vulnerable to external disturbances because of their shape, and they may even facilitate the spread of diseases (Hess 1994) and exotic species from one reserve to another. Perhaps the most convincing argument in favor of corridors is that natural landscapes are far more "connected" than those heavily shaped by humans (Beier and Noss 1998). How well this argument stands up in the real world of limited monies for conservation is an open question. This argument also leaves unanswered the question of which will maintain connectivity more effectively: a narrow, well-protected corridor or a broad swath of seminatural ecosystems that stretches among reserves (Hunter 1997).

Reserve Management

Once a reserve has been selected and its boundaries laid out, the hard work begins, for you cannot simply "lock the gate and throw away the key." Here we will review a few of the many problems that make reserve management a challenging career.

Human Visitors: Most reserves are open to visitors; indeed, most reserves would not exist if they did not provide opportunities for outdoor recreation. Unfortunately, the number of human visitors can be overwhelming. Great Smoky Mountains National Park receives over 6 million visitor-days per year, roughly five times as many as the Statue of Liberty (U.S. National Park Service 1999). This means that reserve management encompasses all the problems that accompany entertaining large numbers of people: proliferation of roads, air pollution, sewage disposal, plant trampling, soil erosion, and so on. Simply put, reserve management is, first and foremost, people management.

Because most reserves are not routinely open to hunting, cutting trees, and so on, it is often assumed that controlling direct exploitation of wild life is not an issue. In fact, few reserves are closed to absolutely all forms of exploitation. One widespread exception is sportfishing. Reserve managers usually allow visitors to fish even in reserves where hunting is strictly forbidden, presumably because fish are generally out of sight and do not have the charisma of mammals and birds, and, unlike hunters, anglers pose no danger to other visitors. This acceptance of fishing carries over even into reserves that are specifically established to protect marine ecosystems, very few of which are closed to all fishing. For example, only 10% of the area of the world's largest marine park, the Great Barrier Reef Marine Park in Australia, is closed to fishing, and this is a high figure compared with many marine reserves (Dugan and Davis 1993). A second common exception to the "no resource exploitation rule" is allowing people to gather deadwood for firewood, either while camping or picnicking in the reserve or for use at home. A growing appreciation of the ecological importance of deadwood (McComb and Lindenmayer 1999) has not significantly curtailed this tradition.

To be successful, reserve managers must always foster the good will of local people, but in developing countries the people who live near a reserve are often too poor to spend a weekend enjoying its recreational amenities. To give these people a vested interest in the reserve, managers often allow some limited forms of exploitation. In Chitwan National Park in lowland Nepal, local people are allowed to enter the park once a year for two weeks during the dry season to collect dead grass, some of which stands 5 m tall (Lehmkuhl et al. 1988). Traditionally, they used the grass to thatch roofs, and, like bamboo, for construction, but now most of it is sold to a paper mill for pulp. This grass harvest generates some good will, but it does come at a cost in terms of small logs stolen from the park for firewood (Eric Dinerstein, personal communication). Such activities become much more controversial if the exploited resources are birds, mammals, and live trees as opposed to fish and dead plants (see Schwartzman et al. 2000 and follow-up articles and Bruner et al. 2001). Local people will also be favorably disposed toward a reserve if they can derive an income by providing services for visitors. Unfortunately, in many developing countries, tourist facilities are owned and operated by people who live far from the reserve, in cities or even overseas. For example, when a European or American tourist pays several thousand dollars to visit Africa's spectacular parks, most of that money never goes to Africa at all, and extremely little reaches the people who live near the reserve. This remains a fundamental problem with linking ecotourism and conservation.

Natural Disturbances: Fires, floods, hurricanes, insect outbreaks, and earthquakes are some of the many unpredictable natural events that can shape reserve management. In the past, reserve managers often viewed such events as unmitigated catastrophes that upset the balance of nature they were trying to protect. More recently, most reserve managers have come to understand that disturbances, especially those that initiate ecological succession, are often critical in maintaining the natural structure and function of ecosystems, and that suppressing disturbances can soon degrade a reserve. This revelation has not made the job of reserve managers any easier. Indeed, it has made it more difficult because the public does not understand the ecological role of natural disturbances, particularly fire, and will often question the wisdom of a reserve manager who does not try to suppress disturbances.

Some disturbances cannot be controlled (volcanic eruptions, earthquakes, hurricanes, tornados), but reserve managers still have to decide what to do after the disturbance. Should they replant vegetation, stabilize eroding slopes, and so forth? A purist might say "No, let it be." But what if the eroding slope was filling a stream with silt, a stream that was the last known habitat of a species of mayfly? Could you write off the mayfly ascribing its extinction to a natural event? What if a hundred years ago the mayfly was found in many streams and is confined to the reserve because of water pollution and dams elsewhere?

Wild fires present terrestrial reserve managers with many difficult decisions because, on the one hand, they are viewed as utterly disastrous by most people, but on the other hand, they are essential elements in many ecosystems (Baker 1992, Niklasson and Granström 2000). The dilemma is heightened by the fact that fires are, to some extent, controllable. Reserve managers cannot simply shrug their shoulders and say "It's out of my hands" because small fires can be put out, and the movement of large fires can often be controlled with firebreaks. Conversely, reserve managers can set fires, choosing locations and weather conditions that will allow them to determine how large and hot a fire will become.

Fire frequency is a key issue for reserve managers. Sometimes, fires happen at fairly regular intervals when sufficient fuel accumulates; sometimes, fires occur only at long, unpredictable intervals determined by droughts; if both fuel buildup and droughts need to coincide, then the frequency of fire may be neither totally random nor predictable. Often, reserve managers do not know what the natural fire frequency is for their reserve, and, anyway, it will change over time as the climate changes (Clark 1988, Veblen et al. 2000). If fire frequency is quite short (e.g., in many grasslands and woodlands where only a few years elapse between fires on average), reserve managers will probably have many opportunities to let natural fires burn or to set fires. In ecosystems that tend to burn at longer intervals (every several decades or centuries) it is tempting to suppress fires. This was the policy in Yellowstone National Park from 1872 to 1972, and some ecologists have blamed this policy for the severity of the 1988 fires, which burned over 321,000 ha in the park. It makes sense that a long history of suppressing fires could lead to an artificial buildup of fuels, but in this case the Park's suppression policy may not have contributed to the 1988 burn. By analyzing fire-scarred tree rings and other information ecologists have determined that fires comparable with those in 1988 also burnt the area in the early 1700s (Romme and Despain 1989).

Water Regimes: Reserve managers often find themselves embroiled in an argument over water. Usually, the issue is relatively straightforward: the supply of water is limited, and someone wants to reduce the reserve share and allocate more water to irrigating crops, turning power turbines, or flushing toilets. Sometimes, things are more complicated. For example, managers of the Everglades National Park seeking to restore some semblance of the park's natural water regime—a broad sheet of freshwater that flows slowly south from central Honda through the park—have encountered an ironic problem (Davis and Ogden 1994). Restoring some of the Everglades' flow has reduced water availability in an area outside the park where water was stored for use by people. This area had become prime habitat for the Everglades snail kite, an endangered subspecies, but now its quality as kite habitat has been reduced. This led to significant disagreements between managers of the park and biologists responsible for managing the kite (Bennetts et al. 1994, Curnutt et al. 2000). Manipulating water regimes of wetlands is also a major activity for natural resource managers who wish to maximize waterfowl production by providing optimum mixtures of water and vegetation (Payne 1992). These waterfowl sanctuaries are important habitat for many species, but it could be argued that conceptually they are closer to the modified ecosystems we will discuss in the next section than to nature reserves.

Water management on reserves is also an issue in arid lands, where reserve managers have a long tradition of digging wells to provide water for wild animals. These artificial water holes tend to increase the abundance of animals overall, especially during droughts. They also make it much easier for visitors to watch wild animals. Think about all the African nature films you have seen with elephants and lions coming and going from a water hole. Some arid reserve managers question the wisdom of digging wells (Ayeni 1977, Thrash 1998). If artificial water holes increase wild animal populations, what are the effects on other species—plants that the animals graze or animals that are not dependent on water holes? What are the effects of concentrating animals on disease transmission and social structure? In Kruger National Park in South Africa a sharp decline in the population of the rare roan antelope was linked to the creation of artificial water holes that increased the populations of zebra and wildebeest with which the roan competed (Harrington et al. 1999).

Exotic Species and Abundant Native Species: Many reserves have populations of exotic species that reserve managers would like to eliminate: burros in the Grand Canyon, Brazilian peppers in the Everglades, and rats in the New Zealand Alps to name just three (Houston and Schreiner 1995). Similarly, some reserves have very large populations of certain native species that managers would like to sharply reduce. Notably, many small reserves have large numbers of herbivorous mammals such as deer because the reserve is too small to harbor large carnivores, and these herbivores wreak havoc on the reserve's flora (McShea and Rappole 2000). In some wetland reserves, geese have become a problem by moving huge quantities of nutrients from the surrounding farmland, where they feed, to the wetlands where they roost (Post et al. 1998).

Eliminating exotic species or reducing the population of a native species are challenging tasks because of both logistical and political constraints. Logistically, controlling a successful species can be exceedingly difficult, as we will see in the next chapter, Managing Populations. Suffice it to say here that the scope of the problem is suggested by the billions of dollars farmers spend to control weeds and pests.

Political difficulties are also nearly inevitable, especially if most people are fond of the species in question. Public affection has curtailed many programs to control appealing creatures such as deer, burros, and horses. Public opposition can also be catalyzed by aversion to the proposed methods. Shooting birds and spraying plants with herbicides are sure to provoke a negative reaction, whereas destroying bird eggs and digging up plants may not.

Although these issues present daunting challenges, reserve managers can overcome them. New Zealand biologists have learned how to eliminate rats and other exotic mammals from islands that are the only remaining habitat for many bird, reptile, and insect species eliminated from the main islands. They started poisoning and trapping campaigns on some very small islands (fractions of a hectare) and have been progressing to larger and larger islands, some well over a thousand hectares (Towns et al. 1990, Cowan 1992, Empson and Miskelly 1999).

What is Natural? Fire regimes, water regimes, management of abundant native species, and many other issues facing reserve managers often lead to the question, what is natural? Typically, the question arises after some more specific questions are asked first. How does the current density of deer on this reserve compare with what it was 200 years ago? Is 200 years ago the right benchmark to be using just because that is when people with modern technology began to colonize this region? Or should it be 10,000 years ago, before there were any humans here? This is a complicated issue that quickly moves into philosophical debates about the role of humans in ecosystems (Hunter 1996, Angermeier 2000). Suffice it to say here that many people would take a purist view in which "natural" ecosystems have virtually no human influence. From this perspective, natural reserves should be managed to minimize human influences as much as feasible. On the other hand, many people would argue that humans and ecosystems are so inseparable that it is folly to define "natural" in a way that excludes humans. From this perspective it is reasonable to manage reserves for whatever condition society deems desirable. For example, many European reserves strive to maintain traditional land-use practices (e.g., livestock grazing regimes) that were common before the advent of industrial agriculture and forestry (Sutherland and Hill 1995).

Modified Ecosystems

It is likely that an astute observer could detect human-induced modifications in all the world's ecosystems. Some we have modified beyond recognition; in others, perhaps deep-ocean bottoms, it would be fairly difficult to detect our influence. In this section we will focus on just a narrow set of modifications, those that modify ecosystems through management for three commodities—wood, livestock, and fish—but still leave the ecosystem in a seminatural condition. These activities present important opportunities for conservation biologists to work collaboratively with their fellow natural resource managers, especially foresters, range managers, and fisheries managers. They offer vast expanses of land and water because most of the earth's terrestrial ecosystems and virtually all of its aquatic ecosystems

Figure 11.6. Conservationists cannot afford to adopt a siege mentality, protecting reserves and ignoring the rest of the landscape. (The idea for this figure was shared by Eduardo Santana, but its originator is unknown.)

are seminatural ecosystems open to natural resource utilization. To ignore these areas would be extremely shortsighted (Fig. 11.6). They may never be pristine ecosystems, but they can support a multitude of species, including some species that are often deemed highly sensitive to human activities such as wolves and grizzly bears (Haight et al. 1998, Mace and Waller 1998).

Forestry

Three facts from Chapter 8 bear repeating here: forests cover less than 6% of the earth's total surface area; forests are habitat for a majority of the earth's known species; forests are being lost far faster than they are expanding. Let us add a fourth fact: most forests are not in reserves; they are available for logging. This fact brings both good news and bad. The bad news is that logging can seriously threaten biodiversity in those areas that remain forested. The good news is that logging does not have to be a serious threat, and that forests that are producing a valuable commodity are less likely to be eradicated to make way for other land uses. Here are three ideas for integrating forest management and maintenance of biodiversity extracted from two books on the subject (Hunter 1990, 1999).

Age Structure: It is difficult for people, with a life span measured in decades, to fully appreciate the life and death of trees whose lives span centuries, sometimes millennia. Yet trees do die, of course. In some forests, trees tend to die a few at a time leaving small holes in the forest canopy in which young trees can grow. These forests will have trees of several different ages, and they are called uneven-aged. Other forests are even-aged because most of the trees originated after some disturbance event (e.g., a crown fire or clearcut) killed most of the previous generation.

Age structure is a critical issue because the biota of an old, even-aged forest is not the same as the biota of a young, even-aged forest (Fig. 11.7). Even at the scale of an individual tree, an old tree provides habitat for a different set of species than a young tree. Consequently, maintenance of biodiversity requires having a balanced age-class distribution. This means having either 1) uneven-aged forests (in places where trees usually die a few at a time), 2) landscapes with many different even-aged forests—some young, some middle-aged, some old—(in places where large-scale disturbances typically initiate succession on a large area), or 3) some combination of these two (in some landscapes large-scale disturbances produce even-aged forests at intervals of several hundred or even thousands of years, but most of the time small-scale disturbances are predominant [Lorimer and Frelich 1994, Seymour and Hunter 1999]). Having a balanced age-class distribution is also essential to meet a major goal of timber managers: producing a continuous supply of wood. Unfortunately, this is not the end of the story.

A conflict arises between maintaining biodiversity and timber production because trees usually grow to an optimal size for cutting long before they die of natural causes. This means that old trees and old forests are uncommon, or even absent, in most areas managed for timber production. The most famous example of this conflict comes from the Pacific Northwest where the remaining, old-growth, Douglas fir forests are both critical habitat for the spotted owl and many other species, and a commodity of great value to the timber industry (see the case study in this chapter and Spies and Turner 1999).

There is a second story embedded within this story about the age structure of forests. When a tree eventually dies, it continues to have ecological value because a unique and very diverse set of species are dependent on dead and dying trees (McComb and Lindenmayer 1999). These range from woodpeckers and the broad array of other vertebrates that use tree cavities to the myriad invertebrates, fungi, and bacteria that reduce dead-

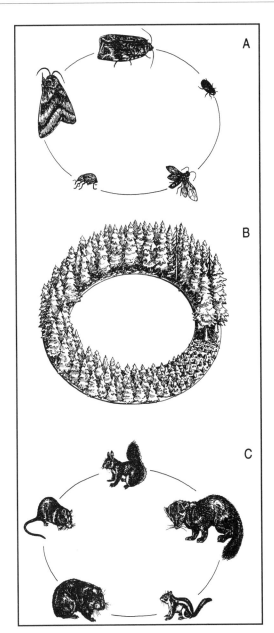

Figure 11.7. The assemblage of species associated with a forest changes as the forest undergoes a cycle of succession and disturbance. Even a single, old tree will support a different biota than a small tree, perhaps because it is taller or its bark more fissured. (From Hunter 1990, reprinted by permission of Prentice-Hall, Englewood Cliffs, New Jersey.)

wood to its organic constituents (recall Fig. 9.10). Furthermore, in many forests fallen trees are also important habitat for another generation of trees. They provide reservoirs of nutrients and moisture for seedlings and are sometimes called nurse logs. Few trees die and are left to rot if a forest is being managed for maximum timber production, and this can be a major problem for all the species dependent on this unique microhabitat.

The conflict between the need for timber production and the need for old and dead trees can be resolved, or at least diminished, in two basic ways. The most obvious is to forego some timber cutting, to allow some trees to age and die. This can take place on many scales. At the smallest scale, it means identifying some individual trees that will be allowed to grow old and die. This is simple in forests where trees are being individually selected for cutting. It is more difficult, but still possible, when large groups of trees are clearcut. At an intermediate scale, forest ecologists sometimes advocate leaving small groups of trees that can age and die; for example, a quarter-hectare patch on every 10 hectares of forest, or strips of trees along streams and lakes. Riparian strips offer two other benefits: protection of the aquatic ecosystem from soil erosion and climatic changes, and travel corridors. Finally, at the largest scale, foregoing logging on entire forests and landscapes returns us to the previous section on protected ecosystems.

The second way to ameliorate this conflict is to defer cutting until the trees are larger and thus allow them to provide habitat for old-forest species for a longer time. For example, cutting an even-aged forest when it is 125 years old rather than 80 years old would, over time, provide 10 times as much habitat for species that require forests older than 75 years (50 years of habitat versus 5). This may mean targeting a different product—wood for lumber and plywood rather than paper pulp and fuel. Products from large, old trees tend to be more valuable than products from small, young trees, so the cost of this approach may not be prohibitive. Silvicultural techniques for stimulating trees to grow bigger (e.g., fertilizing and thinning) can also be useful because organisms are attuned to the size of a tree rather than its actual age. Of course, growing bigger trees does nothing for all the species that need dead trees if the trees are still cut before dying. Forest managers have sometimes remedied a shortage of dead trees by killing live trees, but this tends to be a short-term solution. In the long run it is best to let some trees die of natural causes.

Spatial Patterns: When mature trees die, they leave an opening that can range in size from the canopy gap left by a single windthrown tree, to many thousands of hectares in the case of boreal forest fires (Spies and Turner 1999). Similarly, the scale of logging operations can range from cutting single trees scattered throughout a forest to clearcutting large swathes. Many conservationists favor small-scale cutting because removing single trees distributed over a large area seems much less disruptive than cutting all the trees in one place. However, some would argue that it is more important to match the scale of cutting to the scale of natural disturbances. This would mean cutting individual trees in all-aged forests where trees die one at a time, but it would also mean cutting tracts of even-aged forest in blocks that match the sizes of the natural disturbances that initiate succession (Hunter 1993).

The following hypothetical scenario will make this difference clearer. Imagine an isolated village in which wood is the only source of fuel and the villagers need to cut 1000 trees each year. Near the village is a 1000-ha forest that has 100,000 mature trees and (to keep things simple) the villagers have three choices: 1) cut one tree from each hectare, 2) cut all 1000 trees in a single clearcut of 10 ha, or 3) cut ten 1-ha patches each containing 100 trees. Option 1 would have the least impact in the short term and thus be favored by many conservationists. However, what if this type of forest routinely experiences large-scale natural disturbances, and the trees in this forest are only able to regenerate in openings larger than the size of a single tree crown? (This is true of many tree species that live in even-aged forests; they are called shade intolerant.) In this case many conservationists would propose option 3, ten small patch cuts. However, if you recall Figure 8.12, you will realize that option 3 would fragment the forest more than option 2. It is quite possible that the native biota of this forest will thrive best under option 2. This is especially true if roads are built to extract the wood, because as a generalization, small-scale cutting over a large area requires a relatively long road network.

This scenario was constructed to show that the obvious solution is not necessarily the right one; small-scale cutting is not always preferable to large-scale cutting. This said, conservationists' concerns about clearcutting are usually well founded. There are many forests that are being clearcut because it is the most expedient way to remove trees even though it bears no resemblance to the natural disturbance regime of that particular forest. It is far harder to find forests that should be subject to large-scale disturbances, but that are being logged with small cuts. Furthermore, unless sensitively undertaken, clearcuts may have little resemblance to fires and windthrows, in particular because these natural disturbances usually leave significant numbers of live and dead trees in their wake (Franklin et al. 1997).

Species Composition: Some tree species are more profitable to grow and cut than others. Some tree species have wood so valuable that a single tree is worth tens of thousands of dollars; some trees can grow over 10 m in five years. These differences encourage foresters to try to control the species composition of a site by planting seeds or seedlings of desirable species or controlling undesirable species (e.g., through thinning or herbicides). Not surprisingly, these manipulations can have negative consequences for the forest's other biota. To take a simple example, all the species dependent on acorns will suffer if a forest's oaks are replaced by pines. The effect is likely to be considerably greater if the planted trees are exotics: plantations of Australian eucalyptus trees are found on every continent but Antarctica, and many of these plantations have impoverished floras and faunas.

From a biodiversity standpoint the solution is simple. Foresters should favor the tree species that are native to a particular forest. If a nonnative species must be used, then a close relative is preferable, for example, another species of oak rather than a pine or eucalypt.

Techniques for controlling species composition present an important opportunity for people who wear two hats: conservation biologist and forester (Palik and Engstrom 1999). They can use these methods to shift the species compositions of forests that have been altered by previous management toward their natural composition. We will return to this issue later in the section on restoration ecology.

Livestock Grazing

We are all familiar with the image of cattle trudging across a dusty plain with a cowboy riding herd, and this image does contain two main elements of the world's livestock enterprise: cattle and grassland ecosystems. Yet, many other species are used as livestock, and they graze in a diverse array of uncultivated terrestrial ecosystems, collectively called rangeland. This section is relevant in some degree to sheep, yaks, and llamas on alpine meadows; reindeer on the tundra of Lapland; dromedaries and goats in the deserts of the Middle East; and the various species that are grazed in woodlands (i.e., forests open enough to have a well-developed stratum of ground vegetation). This said, however, we will focus primarily on grasslands and cows.

Compared with forests, grasslands have been given scant attention by ecologists and conservationists, and, consequently, we have a comparatively poor understanding of these ecosystems and an even poorer understanding of what livestock grazing does to them (Noss and Cooperrider 1994), although our knowledge is improving steadily (Tainton 1999). This means that ideas for how to manage grasslands for both livestock and biodiversity are rather speculative and may be confirmed or refuted by future research. Nevertheless, some ideas seem intrinsically obvious because they are based on the logical premise that rangeland management will be more compatible with biodiversity if it maintains ecosystems that are similar to natural ecosystems. Here are some ideas, largely distilled from Noss and Cooperrider (1994).

Native Grazers: One obvious tactic is to use species of livestock that are as close as possible to the species that are native to a particular ecosystem. For example, consider the evolutionary-ecological relationships of the cow, which is thought to have been domesticated from aurochs, a largely forest-dwelling bovine from Eurasia that became extinct in the 1600s (Clutton-Brock 1981). Cattle are clearly more at home in Eurasia than in Australia, where kangaroos and other marsupials were the only large mammalian grazers for at least 20 million years. In North America some people have argued that cattle are a reasonable substitute for American bison (buffalo) because they are fairly close relatives. No doubt they are a better substitute for bison than are goats or sheep, and grazing by cattle may be preferable to no grazing by large mammals at all (Milchunas et al. 1998), but there are some important differences between cattle and bison (Noss and Cooperrider 1994). Notably, cattle need access to water more than bison do, and thus in semiarid landscapes they concentrate in riparian zones, where they often overgraze the vegetation (Fig. 11.8).

To a limited extent this pattern of favoring natives exists already: Asian elephants, reindeer, Bactrian camels, dromedaries, llamas, alpacas, yaks, and water buffalo are all used primarily within their native ranges. Moreover, there is a growing interest in game ranching or farming, i.e., raising undomesticated large mammals such as bison in North America, or eland in Africa within fenced areas (Teer et al. 1993).

Finally, human desire for meat could be met by game cropping: the systematic and, it is hoped, sustainable harvest of wild (neither domesticated nor captive) larger mammals, birds, and reptiles (Hudson et al. 1989, Robinson and Redford 1991, Kerley et al. 1995). Cropping hippos is a widespread example in Africa (Lewis et al. 1990). Game cropping is not livestock management, but it can involve managing rangelands (e.g., by providing water holes) and thus fits within this section.

Natural Grazing Patterns: Another tactic is to use the spatial and temporal patterns of native grazers as a model for livestock grazing systems. For ex-

Figure 11.8. Cattle are not necessarily the domestic equivalent of wild ungulates. For example, they need more access to drinking water than bison.

ample, many native grazers visit an area for a short time, graze it intensively, and then do not return for a year or longer (McNaughton 1993). In contrast, livestock is often allowed to graze an area continuously as long as there is some food and water. When livestock managers do rotate herds among different areas, the emphasis is usually on providing the livestock with more forage rather than on maintaining a seminatural ecosystem (Pieper et al. 2000). It is particularly important to control the spatial distribution of livestock because they tend to gravitate toward and overgraze precisely those places that are most important to the native biota, the relatively uncommon spots with ample water and the most fertile soil. Livestock abundance also needs to be tightly controlled because populations of native herbivores are likely to be relatively low compared with livestock (McNaughton 1993). This is true because most native herbivores are more likely to be limited by predators and the seasonal scarcity of food and water than domestic animals are.

The key issue is to avoid overtaxing the plants' ability to grow and reproduce because overgrazing can profoundly change the vegetation (and thus the entire biota) (Noss and Cooperrider 1994). Moreover, once these changes have occurred, simply removing the livestock will not necessarily lead to the restoration of the original vegetation. Overgrazing can be difficult to assess because one of the most critical processes happens underground where perennial grasses and forbs (vascular plants that are neither woody nor grasslike) must replenish their carbohydrate reserves during each growing season. If grazing curtails this process too much, these plants will be replaced by other species that are less vulnerable to overgrazing, either because they are less palatable to grazers, or because they are more tolerant of being grazed. For further information on grazing systems and overgrazing see Fleischner (1994), Brown and McDonald (1995), Bock and Bock (1999), Rambo and Faeth (1999), and articles in *Ecological Applications* 3(1) (1993).

Natural Disturbance Regimes: Like forests, grasslands are shaped by natural disturbance regimes such as fires, floods, droughts, and tornados. Fire is the most important of these on most rangelands, and ecologically sensitive range management must provide for the continuation of a natural fire regime. In many grassland and shrubland ecosystems, if fire does not occur quite frequently, trees will invade and transform the site into a woodland or forest ecosystem. The similarity and differences among grassland fires, grazing, and mowing is a complex topic that can prove quite controversial when managers propose substituting one for another (Collins et al. 1998, Swengel 1998, Panzer and Schwartz 2000). For example, in Europe it has been suggested that the current rarity of natural fires and native large herbivores means that livestock grazing or mowing are needed to maintain habitat for many open-land species (Pykälä 2000).

Predators and Competitors: The interests of range managers and conservation biologists collide directly over one issue in particular, predator control. Livestock owners are understandably reluctant to share their valuable stock with wolves, lions, snow leopards, and other predators, while, on the other hand, these same predators are flagship species around which conservationists rally. Fortunately, livestock managers sympathetic to biodiversity issues can find ways to minimize the loss of livestock without decimating entire predator populations. Use of guard dogs and selective removal of individual predators that have developed a taste for livestock can be quite effective, for example.

An analogous problem can arise whenever livestock managers feel that native herbivores are competing for scarce forage. Programs to control prairie dogs in North America are a particularly egregious example of this because prairie dogs play keystone roles in grassland ecosystems through their extensive burrowing activity (Miller et al. 2000).

Range Management Techniques: Range managers have a sizable repertoire of management tools that are likely to produce results that are contrary to the well-being of wild life (Holechek et al. 2001). Unwanted vegetation is often removed by dragging a chain between two vehicles, or spraying with herbicides. Exotic species, especially grasses believed to be more palatable or less vulnerable to overgrazing, are introduced. Fences are erected to control the movement of livestock and sometimes wild animals. Water holes are dug and can become a focal point of overgrazing. It is important to remember that these are just tools and that they can be used for positive purposes as well. For example, fences may be necessary to keep livestock out of sensitive riparian zones or from spreading diseases to wild animals. Vegetation control may be the first step in restoring a degraded grassland that has been invaded by shrubs.

Fisheries

Like eating grass in a grassland, catching fish in an aquatic ecosystem may seem like a fairly benign activity, but appearances can be deceptive. As we saw in Chapter 9 and the Gulf of Maine case study fishing can profoundly modify aquatic ecosystems, particularly because many exploited species have pivotal ecological roles as dominant or keystone species. In this section we will examine how fisheries management in seminatural aquatic ecosystems may affect aquatic biodiversity. This topic has received relatively little attention (Wilcove and Bean 1994), and thus some of the ideas presented here are somewhat speculative.

The oceans, lakes, rivers, and other aquatic ecosystems that support fishing are usually publicly owned, and thus a large portion of fisheries

Figure 11.9. Regulating fishing is the primary way that fisheries managers control aquatic ecosystems.

management consists of government agencies managing the people who catch fish, both commercially and recreationally (Fig. 11.9). This means regulating when, where, and how fish are caught, and especially how many fish of what species and sizes. (To keep things simple I will refer just to fish in this section, but the basic principles apply to many other aquatic organisms exploited by people such as shrimp, molluscs, lobsters, and various seaweeds.) The overall goal is usually simple: maximize the sustainable production of desirable fish species. This goal usually means maintaining populations of these fishes at levels that are fairly high, at least 50% of what they would be in a natural, unexploited ecosystem (Newman 1993). Therefore, in theory, sustainable fisheries management could be reasonably consistent with biodiversity conservation as long as one does not focus too narrowly on the target species (recall the issues of bycatch and environ-mental degradation by fishing gear discussed in Chapter 9).

Unfortunately, this is not the end of the story. Fisheries managers are often unable to meet this goal because they cannot adequately regulate fish-ing. The global fish catch has increased steadily from about 17 million met-ric tons per year in 1950 to about 91 million metric tons in 1997 (United Nations Development Programme et al. 2000), and this is clearly not the re-sult of a fivefold increase in the abundance of fish. This trend also disguises an important fact. Many individual species or populations have been over-exploited, and the overall increase has been possible in large part because fishing has switched to less desirable species (Casey and Myers 1998, Pauly et al. 1998). Of course, this switching strategy cannot be employed indefi-nitely, and the Food and Agriculture Organization of the United Nations has long estimated that most fish stocks are being fully exploited or are overexploited (Robinson 1984). The difficulty in restricting fishing is due in part to a inherent mismatch between fishing by people and the natural mor-tality patterns of fish (see Fig. 9.8). Furthermore, regulating fishing is not enough; fisheries managers must also be vocal opponents of water pollu-tion, loss of wetlands, dam construction, and other factors that generally de-grade the environment for fish. In short, managing aquatic ecosystems for biodiversity is usually in tune with the major efforts of fisheries managers. Their lack of success at stemming the tide of overexploitation and habitat degradation may be dismaying, but at least they are trying.

Although the objectives of fisheries managers are in concert with the objectives of conservation biologists much of the time, there are important exceptions (Wilcove et al. 1992). Exotic species provide the most obvious example. From the perspective of a fisheries manager trying to produce large catches of desirable fishes, introducing new species to a water body is often an acceptable practice. From a biodiversity perspective these exotics are an anath-ema (as described in the preceding chapter). Similarly, fisheries managers sometimes try to reduce populations of native, undesirable species—"trash fish"—that compete with preferred species. In its most extreme form this can involve poisoning a lake or river to kill the native fish and then replacing them with desirable species; recall what happened on the Green River (Chapter 10; Holden 1991). Conservation biologists also need to evaluate fisheries man-agement techniques that involve modifying the natural physical or chemical environment of aquatic ecosystems, for example, manipulating water levels, building artificial structures to serve as spawning areas or cover, and adding fertilizer to increase primary production (Cowx 1994). The scale and impact of these modifications is usually quite limited, but in some cases they might have a deleterious effect on biodiversity by altering the habitat of a rare species.

The bottom line is that as long as fisheries managers are attempting to maintain or restore populations of native fishes and their ecosystems, their activities can be endorsed by biodiversity advocates. Sometimes, zealous fisheries managers will attempt something that is likely to degrade biodiversity such as introducing an exotic fish, and then conservation biologists need to question the wisdom of such actions. Unfortunately, the actual track record for maintaining healthy seminatural aquatic ecosystems is poor, which highlights the need for many more aquatic reserves closed to fishing.

Extractive Reserves

The term "extractive reserve" may seem like one of those oxymorons: "soft rock" or "bureaucratic efficiency." It is most commonly associated with areas in the Amazon Basin that have been protected from intrusive forms of land use such as agriculture or commercial logging, but that are still open for limited extraction of resources, for example, collecting nuts and fruits and, especially, tapping rubber trees (Fearnside 1989, Allegretti 1990, Kemf 1993, Salafsky et al. 1993) (Fig. 11.10). This basic idea could be applied anywhere. For example, if a large area of the Arctic were declared off limits to oil extraction and commercial fisheries, but were still open to native people for subsistence hunting and fishing, this area could be called an extractive reserve.

The primary difference between an extractive reserve and a traditional reserve that allows some extraction (e.g., Nepal's Chitwan National Park described above) lies in their goals. An extractive reserve would put produc-

Figure 11.10. Extractive reserves are designed to provide resources for local people while maintaining ecological integrity.

tion of natural resources for local people first, and protection of the ecosystem would be a second, although still very important, goal. A traditional reserve would put ecosystem protection first. The key to making sure that extractive reserves meet both their goals is careful management of the exploited species, the subject of the next chapter. It is also important that in the process of managing the targeted species, others are not inadvertently harmed (Robinson and Bennett 2000).

Ecological Management

The take-home message from this section can be summarized easily: to integrate natural resource management and maintenance of biodiversity manage ecosystems in a way that is as consistent with natural ecological processes as possible. For example, cut trees in a manner that imitates natural disturbances; graze livestock so that they are a surrogate for native herbivores. In other words, use natural ecosystems as a model, a point of departure (Angermeier 2000). Too often managers of these ecosystems use agriculture as a model, and that is fraught with difficulties as we will see in the next section.

Cultivated Ecosystems

Across great sweeps of the earth, the land is a vibrant green testament to photosynthesis, and yet the variety and abundance of wild life is only a shadow of what it should be. These are our cultivated lands, the places where we have replaced natural ecosystems with a sparse assemblage of exotic and native species. Row crops of grains and vegetables are the dominant form of cultivated ecosystem, but we have created many other types of ecosystems to produce food, fiber, or fuel. These include orchards, tree plantations, ponds devoted to aquaculture, cranberry bogs, cattail marshes managed for biomass fuel, and more. Admittedly, drawing a line between a cultivated ecosystem and an intensively managed seminatural ecosystem can be a rather arbitrary decision. A pasture sown with seeds of an exotic grass species and then carefully fertilized and grazed is clearly cultivated, but what if the sown grass were a native species? What if the grass were native to the region, but did not naturally occur on that type of soil? How do we separate tree plantations and intensively managed forests?

The process of turning natural and seminatural ecosystems into cultivated ones is probably the most important proximate cause of biodiversity loss, the ultimate causes being the burgeoning human population and our demand for the products of all these cultivated ecosystems. Consequently, conservationists routinely object to the expansion of cultivated ecosystems.

Beyond this, however, they tend to ignore these places as blank spots on the map of biodiversity, and thus they do not interact much with farmers (here broadly defined to include fish farmers, tree farmers, etc.). This is short-sighted for two reasons that we will examine further: 1) with careful management, some important elements of biodiversity can persist in a cultivated ecosystem; and 2) thoughtful stewardship of cultivated ecosystems can ameliorate their negative effects on surrounding landscapes and minimize their rate of expansion.

Biodiversity in Cultivated Ecosystems

If farmers had total control of their ecosystems, many of them would channel virtually all the resources of a site—energy, water, nutrients—into crop species and a handful of key associates such as nitrifying bacteria and pollinating insects. Witness farmers' efforts to control unwanted species— weeds, pests, vermin. Fortunately for biodiversity, most farmers fall far short of this goal, and some do not pursue it assiduously because they enjoy sharing their land with other species.

The single most important factor allowing wild life to persist in a cultivated setting is the tiny relicts of habitat that receive little or no cultivation (Carroll et al. 1990). These would include a strip of shrubs along a ditch, a patch of trees on a rocky outcrop in the middle of a hayfield, a wet spot in the midst of a plantation, a hedgerow separating two fields, and similar places (Fig. 11.11). They are too small to be managed as independent ecosys-

Figure 11.11. In an agricultural landscape the native biota may persist in small remnant patches that have not been cultivated.

tems, but are large enough to provide refuge to a surprising diversity of wild creatures (Nason et al. 1998, Miller and Cale 2000). Therefore, one of the most important things a farmer can do for biodiversity is to retain these places or even restore and expand them. For example, prairie farmers in North America need to resist the temptation to fill or drain the small potholes that support pintails, avocets, and a large array of other wetland species (Mitsch and Gosselink 2000). Farmers in Europe and elsewhere need to retain hedgerows even though with modern machinery it is now easier to cultivate one large field than two smaller ones (Dowdeswell 1987). Some farmers will actively create these environments; farm ponds are the most common example of this (Baker and Halliday 1999). In Europe habitat for rare plants and butterflies is created by leaving 2- to 3-meter-wide strips called headlands at the edges of fields that are not sprayed with pesticides and fertilizers (Feber et al. 1996, Kleijn and van der Voort 1997). Decisions to allow some land to lie fallow for one or more years, resting before another commercial crop is grown, also creates these patches of habitat, albeit on a short-term, always shifting basis (Henderson et al. 2000).

Habitat remnants are not the whole story in agricultural landscapes (Bignal and McCracken 1996, Chamberlain et al. 2000). The biodiversity of a cultivated landscape is also dependent on the variety of commodities being grown. Not surprisingly, dairy farmers who maintain pastures, hayfields, and feed-corn cropland and who supplement their income with a small orchard, are providing habitat for far more species than farmers who grow nothing but soybeans. Unfortunately, the overall trend has been toward greater specialization. This is particularly noticeable among farmers of developing countries as they shift from an emphasis on subsistence agriculture—growing a diversity of crops to meet most of their personal needs—toward an emphasis on growing cash crops (Bray 1994). The difference between coffee grown under the shade of various trees that provide fruit and firewood versus coffee grown in the open is one example of this phenomenon (Roberts et al. 2000, Wunderle and Latta 2000). The shade coffee supports a much larger native biota and can provide a wider variety of products for the farmer, but commercialization favors sun coffee. A growing movement to make agriculture more ecologically sound (associated with terms such as sustainable agriculture, alternative agriculture, or agroecology) emphasizes using a diversity of crops, including trees, but it remains to be seen if the overall trend toward specialization will be reversed. (See Jackson and Piper 1989, Carroll et al. 1990, Soule and Piper 1992, and Collins and Qualset 1999 about this movement.)

The specific practices farmers employ to cultivate their farms can also have a dramatic effect on wild life (Paoletti et al. 1992). Use of insecticides, herbicides, fungicides, and other types of pesticides is probably the most important example because they are so commonly used and because their effects on targeted and nontargeted species, both on and off the sprayed site, can be so severe. (See the section on pesticides in Chapter 8.) Suffice it to say here that farmers who are concerned about biodiversity will minimize their use of these chemicals. One practice that can rivet the attention of conservationists is farmers' protecting their crops by killing popular vertebrates, for example, shooting kingfishers and herons at a fish farm. Sometimes, this pits farmers against species that are in jeopardy globally but common enough locally to be considered pests by the farmers who have to live with them. Think about the dilemma of an African farmer who lives near a herd of elephants, each one of which eats about 150 kg of vegetation per day (Newmark et al. 1994). In the absence of pesticides, shooting, and trapping, a surprising number of species can find some semblance of their natural habitat in cultivated ecosystems (Elphick 2000).

Even relatively subtle changes in farming practices such as timing can affect wild life; consider two examples from the British Isles (O'Connor and Shrubb 1986) and one from Germany (Johst et al. 2001). British farmers mow grass for hay in June and July, but if it is to be used for silage (stored green and allowed to ferment), it is cut in May. When grasslands are cut for silage, many bird nests are lost, nests that would have produced young if cutting took place later. Consequently, as farmers have switched to silage cutting, some grassland birds have decreased, notably the corncrake, which has become quite uncommon in most of Great Britain and Ireland. A similar timing issue involves the sowing of grains, especially barley and wheat. A shift from spring-sown varieties to autumn-sown varieties has apparently reduced the populations of lapwings, song thrushes, and rooks because these species were dependent on the seeds and soil invertebrates brought to the surface by spring tilling. In Germany, a model of white storks' foraging behavior indicated that there would be much higher breeding success if nests were surrounded by fields that were mowed asynchronously, thus creating a steady supply of newly mown sites, their optimal foraging habitat (Johst et al. 2001). A comprehensive review of farming practices and their potential effects on wild life is beyond our scope here. The basic point is that these practices need to be evaluated and perhaps changed if cultivated ecosystems are to host a wide range of species.

Although most conservation biologists focus on wild biota, biological diversity also includes domestic species in all the myriad forms developed by plant and animal breeders. We will cover their conservation in Chapter 13, Zoos and Gardens, but they raise an interesting issue relevant to this chapter on managing ecosystems: Should conservationists be concerned with maintaining cultivated ecosystems as important elements of biological diversity in their own right, irrespective of their role as habitat for species? The answer for many Europeans is "yes," because they view the countryside as an ecological, cultural, and aesthetic amenity. In some cases

European governments pay subsidies to farmers to encourage them to use traditional methods (Green 1989).

Minimizing the Negative Effects of Cultivated Ecosystems

Many cultivated ecosystems share a landscape with sizable tracts of natural and seminatural ecosystems, and therefore conservationists must strive to minimize the extent to which the cultivated ecosystems impinge on the others. Fortunately, in this regard good farming is good for wild life in some important ways. For example, responsible farmers are vigilant against soil erosion, and this will minimize problems with sediment pollution elsewhere in the landscape. Similarly, conservative, careful use of fertilizers and pesticides will save farmers money and ameliorate problems with eutrophication and pesticide contamination (Matson et al. 1997, Skinner et al. 1997). Minimizing the use of pesticides can also have a direct positive return for agriculture because all farmers are dependent on healthy soils (with their myriad organisms) and many need the assistance of beneficial insects (notably, pollinators and natural enemies of pest species) (Collins and Qualset 1999). *Integrated pest management* (often abbreviated IPM) is a good example of this for it uses natural enemies of pests, specific cultivation practices (e.g., mixing crops), and conservative use of pesticides to achieve pest control. Of course, there are irresponsible farmers and well-intentioned farmers who make mistakes, but at least the goals of farmers and conservationists are in concert in this respect. This is particularly true in the context of sustainable agriculture in which one of the primary goals is to maintain profits for farmers by minimizing costs, and this means limiting soil loss and the expensive use of pesticides and fertilizers (Soule and Piper 1992).

Ironically, one undesirable "export" from cultivated lands can be wild life. Many species are quite successful at living along the interface between cultivated and natural or seminatural ecosystems such as various members of the deer, crow, and kangaroo families and quite a number of small mammalian carnivores such as raccoons and red foxes. Farmers have long been familiar with the losses sometimes inflicted by these species, but their depredations on other native wild life have only recently attracted the attention of conservation biologists. For example, conservationists now understand that white-tailed deer, which thrive in mixed landscapes of forest and farmland, can reach densities at which they have a severe effect on forest plants (McShea and Rappole 2000). Similarly, recall the description of bird-nest predators associated with the edges of agricultural land and forests (Chapter 8). As emphasized in Chapter 2 (see Figs. 2.1 and 2.2), maintaining gamma diversity should take precedence over maintaining alpha or beta diversity, and this will almost always mean favoring the wild life of natural ecosystems over the wild life of cultivated ecosystems. The issue of favoring gamma diversity can be important even when the integrity of adjacent natural ecosystems is not at stake. For example, in open landscapes, planting windbreaks may increase the overall bird species richness of a farm, but it can be at the expense of bird species that are becoming uncommon because they need large blocks of unbroken grassland (O'Leary and Nyberg 2000).

The negative effects of cultivated ecosystems can also be ameliorated simply by limiting their extent. The less land we cultivate, the more room there is for natural and seminatural ecosystems (Box 11.2) (Hunter and Calhoun 1995). Consequently, conservationists should often support efforts to increase the productivity of cultivated lands (Ewel 1991, Sedjo and Botkin 1997). Of course, there are some important pitfalls hidden here. First, the emphasis must be on sustainable, long-term increases in production. Huge areas of once-productive cultivated land are now nearly barren because of shortsighted practices, and both farmers and wild life have suffered. Second, increased productivity cannot be linked to significant exports of pesticides, fertilizers, and soil for obvious reasons. Similarly, productivity gains obtained by increasing fossil fuel use (to run tractors, irrigation pumps, etc.) must be scrutinized. Third, indiscriminate use of high-yield varieties of domestic species has led to the loss of genetic diversity as low-yield varieties have fallen out of favor with farmers (see Chapter 13, Zoos and Gardens). With these reservations clearly in view, we still have ample opportunity to increase the productivity of many cultivated lands through intelligence, innovation, and diligence and thus to achieve net benefits for both people and wild life. In the short term this approach may allow us to convert some cultivated ecosystems into seminatural ecosystems, it is already happening in Europe and the United States (Fry 1991, Dunn et al. 1993). In the long term, increasing human populations will demand these efficiencies, and we will lose everything without them.

Some Economic Perspectives

If society expects farmers, ranchers, fishers, and loggers to adopt practices that are amenable to maintaining biodiversity, what should we offer in return? Our respect? Some money? We will cover many aspects of these issues in Chapters 15 and 16 (Social Factors and Economics), but a quick description of two compensation mechanisms is in order here. Many governments offer various financial subsidies to farmers, and these can be tied to certain practices that are deemed environmentally acceptable (we will discuss one example, the Conservation Reserve Program, in Chapter 16). Such annual payments to "do the right thing" go by many names and

| BOX 11.2 | A triad approach to land-use allocation.[1] |

From the perspective of producing commodities such as food, fiber, and fuel, it is possible to conceptualize a "triad" of three types of land use: 1) cultivated ecosystems where high levels of commodity production are achieved, 2) protected ecosystems with virtually no commodity production, and 3) modified ecosystems in which modest resource use occurs while ecological values are carefully protected. Many environmentalists are reluctant to be advocates of cultivated ecosystems because so much biodiversity has been lost from the conversion of natural ecosystems to cultivated ecosystems. However, in some circumstances it might make sense to switch commodity production from extensive extraction in modified ecosystems to intensive production in cultivated ecosystems so that more land can be set aside in reserves.

The forests of Maine provide a good example, where about 2% have been set aside as reserves, roughly 6% are used for intensive forest management (e.g., tree plantations), and over 90% are used for extensive forestry. Intensive forest management in Maine produces about three times as much wood as extensive management, and thus for every hectare of forest switched from extensive management to intensive management 3 ha could be put in reserves with no net loss in commodity production. In other words, it would be possible to increase Maine's forest reserves from 2% to 10% and to compensate for all of the lost production by increasing the area of intensive production from about 6% to 8.5% (Fig. 11.12). In aquatic ecosystems the trade-offs could be even more dramatic because aquaculture can easily produce 10 times as much fish, often much more, compared with catching fish in seminatural ecosystems. This trade-off would be particularly attractive because the world's aquatic ecosystems are overwhelmingly skewed toward extensive management with very little area allocated to aquaculture or aquatic reserves.

Some conservationists think trades like these make sense. Others believe that we should set aside the reserves anyway and make up for the loss of production by reducing human populations and consumption. No doubt the latter approach would solve the problem, but which approach is more feasible?

[1] Based on Seymour and Hunter (1992) and Hunter and Calhoun (1995).

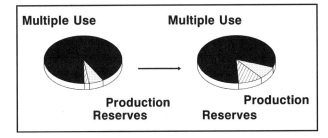

Figure 11.12. The current allocation of Maine's forests from a triad perspective and what the allocation could be if some trade-offs between cultivated ecosystems and reserves were made.

can be offered to anyone who owns or uses the lands and waters, not just farmers (Main et al. 1999). Conservation easements are a one-time agreement to purchase certain property rights from landowners; typically, the landowners can continue their traditional use of the land, but cannot convert it to a more intensive use, especially development as housing, factories, mines, etcetera. Easements and subsidies are widely accepted in conservation circles, in part because modified and cultivated ecosystems are judged to be far preferable to the alternative of having them developed into built ecosystems (Knight et al. 1995).

Built Ecosystems

The final group of ecosystems to consider are easily detected, especially at night. These are the places where people live in great density and where, after dark, our enormous use of energy is manifested by lights readily detected from airplanes and spaceships. In these ecosystems—cities, factories, mines, highways, and the like—human-made structures are dominant, and the hand of nature can be difficult to discern. However, nature is still there, even if it has been reduced to rats and cockroaches hiding in the recesses of a building, a crust of lichens and lichen-inhabiting invertebrates on a bridge abutment, or a line of weeds growing through the cracks in an abandoned parking lot. Some people are reluctant to think of cities as ecosystems, but they do meet our definition: they constitute a physical environment plus interacting populations (Chapter 4). A child feeding pigeons on a city street is participating in an ecological interaction, even though the solar energy in the bread crumbs was fixed by a wheat plant far away (Gilbert 1989, Grimm et al. 2000).

Built ecosystems are not a major focal point for conservation biologists because they are primarily habitat for very adaptable species that are in no danger of extinction. You might guess from their names alone that house finches, house mice, house sparrows, house geckos, and bedbugs are able to survive in close proximity to people. Nevertheless, it would be a mistake to ignore these places completely for at least three reasons that we will consider here.

Habitat for People

In most industrialized countries the vast bulk of people live in urban and suburban environments. This pattern is generally conducive to maintaining biodiversity because, if all of these people were scattered across the countryside, far less land would remain in natural and seminatural ecosystems. Consequently, conservationists have an interest in built ecosystems being pleasant, healthy places so that people will live there. This rationale applies also to recreation. Wild life will fare better if people spend an afternoon at the city park or in their backyard rather than drive to a beach where endangered piping plovers are trying to nest, or worse yet, build a vacation home on the dunes. One way to make urban-suburban life more pleasant is to facilitate positive interactions with wild life: encounters with robins, daisies, and dragonflies rather than rats and ragweed. Such contact may also encourage people to support conservation with their votes and their money and may provide nutriment and inspiration for young conservation biologists (Uhl 1998, Primack et al. 2000).

Biodiversity in Built Ecosystems

Many built ecosystems harbor a surprising variety of wild life, species that cling to any oasis of green in a concrete desert (Adams and Leedy 1991) (Fig. 11.13). Fruit bats roost in a tree that overhangs one of the main streets of Kathmandu. Peregrines wing through the canyonlands of several North American cities searching for pigeons. In Great Britain lands covered with mining and industrial wastes provide habitat for some rare plants that are confined to soils with high concentrations of certain metals (Jefferson and Usher 1986). In the southwestern United States many seminatural ecosystems are dotted with abandoned mine shafts, which represent small, human-built habitats for rare bats and many other species. Of course, most urban species are exceedingly common, and peregrines and eastern barred bandicoots are unusual exceptions to this pattern. Nevertheless, it is important to remember that there are more urban species that merit our esteem than our disdain, more butterflies than cockroaches. (For brevity's sake we will not return to the discussion of intrinsic value in Chapter 2 to address the importance of cockroaches.)

It is interesting to speculate that urban populations of some species may be genetically different from their conspecifics living elsewhere. Perhaps, for example, some urban plant populations are more tolerant of ozone than rural populations of the same species. The famous story of industrial melanism in moths (Chapter 5) suggests that this is not a far-fetched idea and that it may be of practical importance. Any allele that increases the fitness of individuals in human-altered environments has a fair chance of spreading, and it might allow an entire species to persist in our changing world.

Imports and Exports

Built ecosystems interact with other ecosystems in a far-reaching network. Tremendous quantities of energy and matter are imported—notably, fossil fuel, electricity, food, and building materials—often coming from

Figure 11.13. Even in urban landscapes small oases for wild life can be created.

thousands of kilometers away. Tremendous quantities of wastes are exported. Air pollutants travel downwind. Solid wastes travel to open spaces, often nearby or sometimes far away. Most major urban areas are on the shores of rivers or the ocean where currents can carry water pollutants away. Clearly, these imports and exports are of direct concern to natural resource managers trying to maintain biodiversity in the ecosystems where the energy and matter are acquired or where the wastes are disposed.

How to Do It

These three issues can be crystallized into a single goal: making built ecosystems inhabitable for both people and other life forms. Pursuing this goal involves activities that are the cornerstones of environmentalism (pollution abatement, curbing resource use, recycling, etc.) and that need no elaboration here (Smith et al. 1998). (Although, it is worth pointing out that college campuses are ripe for local action [Creighton 1998].) It also requires activities that are a bit closer to mainstream conservation biology—notably, managing the patches of green that dot the urban-suburban landscape (Gilbert 1989). These city parks, backyard gardens, cemeteries, golf courses, and the like conform to our definition of cultivated ecosystems, but they fit here better than in our preceding discussion of farms, because they are so closely linked to built ecosystems. They differ from rural cultivated ecosystems quite significantly because they are managed primarily for their aesthetic qualities rather than commodity production. Sometimes, this means monocultures of exotic species; witness the expanse of lawns that we maintain with liberal inputs of pesticides, fertilizers, and fossil fuels (Bormann et al. 1993). Yet, aesthetic considerations also foster diversity. They encourage people to grow a variety of flowers, shrubs, and trees, and, whether intended or not, a variety of associated animals. Indeed, more and more people are thinking of gardens and city parks as habitat for wild life, not just a pretty place to play croquet. People are replacing lawns with patches of native plants and focusing on plant species that will provide food for birds and butterflies (Kress 1985, Xerces Society 1990, Stein 1993). We do not have space to describe all the techniques for wild life gardening, but there is abundant literature on the subject. Wild life gardening gives everyone an opportunity for hands-on action, even if it is only maintaining a window box. Much of the work outlined here may be in the realms of urban planners and horticulturalists, but conservation biologists have a role too, for example, in pointing out the relationship between ecological connectivity and greenways (vegetated ribbons for walking and biking through urban areas) (Fernández-Juricic 2000).

Restoring Ecosystems

Scan the landscape from any vantage point near the Mediterranean—the Acropolis, Mount Sinai, the seven hills of Rome—and you will witness what thousands of years of human occupation have done.

> . . . in those days the country . . . yielded far more abundant produce. . . . [I]n comparison of what then was, there are remaining only the bones of the wasted body as they may be called . . . all the richer and softer parts of the soil having fallen away and the mere skeleton of the land being left. But in the primitive state of the country its mountains were high hills covered with soil, and the plains . . . of Phellus were full of rich earth, and there was abundance of wood in the mountains . . . [N]ot so very long ago there were still to be seen roof of timber cut from trees growing there, which were of a size sufficient to cover the largest houses; and there were many other high trees, cultivated by man and bearing abundance of food for cattle.

Moreover, the land reaped the benefit of the annual rainfall, not as now losing the water which flows off the bare earth into the sea, but, having an abundant supply in all places, and receiving it into herself and treasuring it up in the close clay soil, it let off into the hollows the streams which it absorbed from the heights, providing everywhere abundant fountains and rivers, of which there may still be observed sacred memorials in places where fountains once existed. Such was the natural state of the country which was cultivated . . . (Critias, 111.b,c,d)

These are not the words of a 20th-century naturalist; they were written by Plato over 2000 years ago (quoted from Forman and Godron 1986). Plato understood what was being lost with a clarity that would be uncommon among most current inhabitants of the Mediterranean basin. It is hard to fully appreciate ecosystem degradation unless you have seen it happening within your lifetime, and much of the Mediterranean Basin suffered its most profound losses long ago. In many other parts of the world, natural ecosystems are being degraded today at a pace so fast that even young conservationists will have some personal experience with these changes.

What can be done about all these degraded ecosystems—the woodlands of the Mediterranean Basin, the deforested lands of Amazonia, the polluted rivers and lakes of the United States? Recall our discussion on global change (Chapter 6), and you will realize that degraded ecosystems will eventually recover. Someday, after the era of *Homo sapiens* has passed, even the hills of the Mediterranean will probably have a flora and fauna as rich as it ever was. Unfortunately, natural recovery processes are likely to be very slow; Figure 6.1 suggests that in the worst cases several million years of evolution might be required. However, we do not have to wait. We can accelerate the recovery process if we wish.

There are many good reasons to restore ecosystems, but biodiversity advocates support restoring degraded ecosystems for one overarching reason (Daily 1995, Dobson et al. 1997). At best, protecting natural ecosystems can only retain what we have, and wisely managing seminatural, cultivated, and built ecosystems can only avoid future degradation. If we want to reverse past degradation, we must think in terms of improving damaged ecosystems. Improvement can mean many different things. For a cultivated ecosystem degraded by erosion it might mean an increase in productivity. For a seminatural forest degraded by excessive logging it might mean restoring its ability to provide habitat for an endangered species. For a protected ecosystem it might mean removing an exotic species so that the ecosystem is closer to its original state. To clarify what improvement means we need to explore the concept further and, in the process, define some terminology.

Some Terminology for Improving Degraded Ecosystems

It is easy to understand ecosystem degradation and improvement if we think in terms of an ecosystem moving through a conceptual space defined by ecosystem structure and function. In Figure 11.14 the filled circle represents a healthy structure (e.g., high diversity) and function (e.g., high productivity), while the empty circle represents the same ecosystem with a structure and function that have been degraded by some human activity. If the degradation process is stopped, the ecosystem will recover over time. Initially, the ecosystem may continue to degrade for a while, especially if severe soil erosion occurs, but, eventually, it will probably move toward its original state because of ecological succession. If the scope of degradation is great, then the recovery may take a long time (centuries or longer)

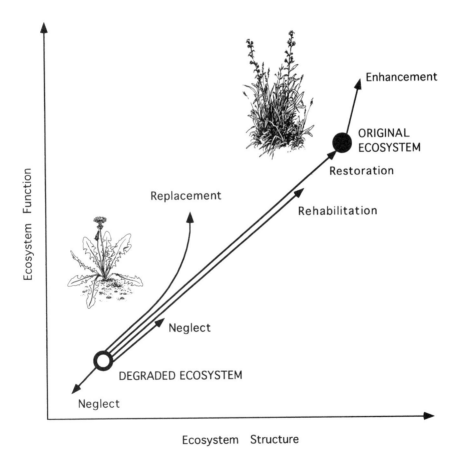

Figure 11.14. A conceptual representation of ecosystem degradation, restoration, and related processes. See the text for an explanation, for each line on this graph there is an italicized term in the text. (Redrawn by permission from Bradshaw 1984.)

and may only be approximate. Restoration ecologists often describe the "let-nature-take-its-own-course" option as *neglect*, a word that clearly shows their preference for active management and improvement. "Recovery" would be a more neutral term.

The type of improvement most in concert with the goals of conservation biology is *restoration*, which means actively trying to return the ecosystem to its original state. Many ecosystem ecologists tend to emphasize function over structure, and thus they would concentrate on restoring an ecosystem's productivity and nutrient cycling. Conservation biologists usually would not be satisfied with this; they would also attempt to restore an approximate replica of the original biota.

Restoration ecology as we have just defined it narrowly is distinct from some closely related activities because ecosystem managers sometimes try to improve an ecosystem without returning it to an approximation of its original state. *Rehabilitation* of a degraded ecosystem means shifting it back toward a greater value or higher use than it is serving currently not necessarily all the way to its original state. "Greater value" or "higher use" can be broadly defined, but usually reflect human instrumental values. Reclaiming a mine site as pasture for livestock rather than restoring it to its former state as a natural grassland would be an example of rehabilitation. ("Reclamation" is another common synonym for rehabilitation.) Sometimes, the goal is *replacement* of a degraded ecosystem by creating a completely new one. Creating a marsh in a mine pit that was formerly a forest would constitute replacement. Replacing terrestrial ecosystems with wetlands is quite common in the United States because laws often compel people who destroy one wetland, to build a road, for example, to create a new wetland somewhere else (Marble 1992).

Finally, *enhancement* is used for any activity that improves the value of an ecosystem, even if the change is rather limited. This term can even include activities that, depending on how you measure value, improve an ecosystem that has not been degraded, as shown in Figure 11.14. We have already discussed one example, installing water holes in desert reserves. Similarly, waterfowl managers often enhance wetlands by putting in water control structures that allow them to maintain the type of vegetation favored by ducks. Conservation biologists will usually be skeptical of enhancing undegraded natural ecosystems because they will wonder what species may be harmed by the manipulation.

The issue of ecosystem restoration, reclamation, and so on often arises when discussing *mitigation* of the impact of a proposed development, especially roads, airports, shopping malls, etc., that will profoundly degrade a site. There are four major forms of mitigation. First and most ideally, the impact should be avoided altogether, for example, by relocating the development to a site that has already been severely degraded. Second, if the impact cannot be avoided, the site should be restored, or at least rehabilitated, after the impact is over, for example, after a mine is exhausted. Third, if the impacts are relatively permanent, another nearby degraded site should be restored to replace the one lost. Fourth, the developer can be required to purchase and permanently protect natural ecosystems, preferably at a ratio of several hectares protected for every one lost.

Six Basic Steps for Restoring an Ecosystem

1. **Set a goal.** Do we wish to restore the preexisting ecosystem, or is it only feasible to rehabilitate the degraded ecosystem? It is important to be realistic, especially because ecosystem restoration can be quite expensive. The total bill for restoring the Everglades will be well into the billions of dollars (Davis and Ogden 1994). Given the dynamic nature of ecosystems (recall Fig. 6.6) and the long history of degradation, we will also need to decide *which* preexisting ecosystem to restore, the one that was present 10 years ago, or 300 years ago. Conservationists often desire to restore ecosystems to a state that existed before the colonization of people, at least technologically advanced people (Angermeier 2000). However, if we choose to restore an ecosystem to an ancient state, we need to recognize that the ecosystem we hope to restore would have changed even in the absence of people. To put it another way, the natural (undegraded by human activities) state of an ecosystem is a moving target because of long-term climate change, species range shifts, and other factors. Finally, once a general goal has been set, that can be translated into a specific set of objectives, usually by comparison with a benchmark ecosystem that exhibits the desired state, or *reference conditions* (Palik et al. 2000). In sum, setting a goal requires answering both ethical (what do we want?) and technical (exactly what does that look like?) questions (Higgs 1997).

2. **Determine a strategy and methods.** Ecosystem restoration is not easy because, to paraphrase Frank Egler (1977), not only are ecosystems more complex than we think they are, they are more complex than we *can* think. This complexity is daunting, but it must not be an excuse for inaction. It does mean that ecosystem restorationists need to do their homework; to understand the ecosystem in question as thoroughly as possible; and to work out a plan of attack with other experts such as civil engineers, landscape architects, horticulturalists, and other specialists, including social scientists who can help ensure community support (Geist and Galatowitsch 1999).

Interestingly, ecosystem restoration can give to the science of ecology as well as take from it because it represents an experimental application of our knowledge of ecosystem function and structure (Zedler 2000). As Bradshaw (1987) put it, "Ecologists working in the field of ecosystem restoration are in the construction business and, like their engineering colleagues, can soon discover if their theory is correct by whether the airplane falls out of the sky, the bridge collapses, or the ecosystem fails to flourish." It is wise to learn from the mistakes of others, and, fortunately, ecosystem restoration has been the subject of many books (e.g., Bradshaw and Chadwick 1980, Cooke et al. 1993, Cairns 1995, Sauer 1998, Whisenant 1999) and is covered in two journals: *Restoration and Management Notes* and *Restoration Ecology*. Note that successful restoration ecology projects also incorporate the social sciences.

Every restoration project is unique. Even steps as fundamental as 3, 4, and 5 are not required in every project, and the specific execution of these steps will always vary.

3. Remove the source of degradation. This step is obvious and critical: you cannot recover from a knife wound until you have removed the knife. We cannot restore a eutrophic lake until we remove the source of excess nutrients. We cannot restore an overgrazed grassland until we have removed much, if not all, of the livestock (Allen and Jackson 1992). In Costa Rica conservationists trying to restore a tropical dry forest must first control fires, a human-generated disturbance in this ecosystem (Janzen 1988b, 1992). These fires severely impede the establishment of woody vegetation, thereby allowing grasses imported from Africa to predominate. In some cases, especially on islands, exotic species themselves are the primary source of degradation, and they must be removed to initiate restoration (Towns et al. 1990, 1997). In other cases, exotic species can be removed later while fine-tuning the ecosystem restoration process. Sometimes, the source of degradation will have disappeared before the restorationist arrives on the scene (e.g., the bulldozers will be gone), but if not this is the first "hands-on" task.

4. Restore the physical environment. In most terrestrial and wetland environments soil is a critical issue. If it is eroding, it must be stabilized; if it has already eroded away it must be replaced. Replacing soil by importing it from elsewhere is expensive and depletes the supply of soil at the other site. Consequently, for large-scale projects it is often necessary to rebuild the soil on site even though this is likely to be a long process. In some cases the existing soil may be so full of contaminants that it needs to be removed and replaced (Smith 1988).

Restoration of an ecosystem's hydrologic regime is often essential, especially in aquatic and wetland ecosystems. We have already mentioned the ongoing attempt to restore millions of hectares of wetlands in southern Florida, chiefly by restoring some semblance of the region's hydrologic regime (Davis and Ogden 1994). On a much smaller scale similar processes are being undertaken through innumerable wetland and aquatic restoration sites. Sometimes, much can be accomplished through changing the management of water control structures (Richter et al. 1997), but, ultimately, the enormous network of dams, dikes, canals, and so on that we have imposed on these ecosystems will need to be redesigned. Building these structures was a huge undertaking, and renovating or removing them will require a similar level of effort. Similarly, restoration of a disturbance regime is often critical; in particular, returning fire to an ecosystem is often an issue as we discussed earlier in this chapter (Lesica and Cooper 1999).

5. Restore the biota. Given time many species would recolonize a suitably restored environment, but this process can be accelerated significantly by translocating populations—collecting appropriate plants and animals and moving them. (The next chapter will cover translocations in more detail.) Plants are usually the priority for restoration projects because they provide habitat for the animals, and many animals are mobile enough to colonize on their own after suitable vegetation is growing (Bradshaw 1983). Simply finding enough suitable specimens for importation can be difficult, especially because we do not want to overexploit the ecosystem where we obtain the colonists. Whenever possible, it is best to work with any organisms that survive on the site rather than undertake the expense and risk of importing new ones. For example, restoring a degraded forest may involve manipulating the age-structure of the current population by selective cutting, a much easier proposition than planting new trees (Frelich and Puettmann 1999, Mast et al. 1999).

Conservation biologists are particularly interested in restoring rare species, but often these will be the most difficult to obtain and establish (Maina and Howe 2000). In the worst cases, some of the ecosystem's original inhabitants will have become extinct. If substitutes of a different subspecies are available, these are generally deemed appropriate for reintroduction as long as they are likely to be adapted to site conditions (Knapp and Rice 1998, Seddon and Soorae 1999), but the issue is more difficult when an entire species is globally extinct. Consider the dilemma of European conservationists who would like to restore a forest complete with a population of aurochs, or a North American longing for a grassland ecosystem with a mammal fauna as rich as that before the Pleistocene extinctions. Under these circumstances most conservationists would argue that we should do the best we can with extant, native species. However, some people have argued that we should invent whole new ecosystems (Baldwin et al. 1994, Martin and Burney 1999), perhaps by introducing species that are ecological equivalents; for example, introducing African antelope to North America to replace those lost during the Pleistocene extinctions.

6. Be patient. It can take many years for reintroduced individuals to grow populations to increase, other species to colonize, and so on. In the

meantime the site should be carefully monitored so that the next restoration project will be based on a larger foundation of knowledge.

A Cautionary Note

Sometimes, promoting ecosystem restoration can have an unintended side effect. The real or perceived opportunity for restoration can make it easier to justify additional ecosystem degradation. If miners promise to replace an abandoned field with a beautiful lake surrounded by a lush forest, they will find it easier to win approval of their proposal. Conservationists need to be conservative on this point because ecosystem restoration has a significant risk of failure even when undertaken with great care and diligence (see Mitsch and Gosselink [2000] for wetland examples). In short, the promised lake ecosystem may turn out to be just a barren body of water. At best, it is not likely to be a perfect replica of a natural lake ecosystem.

CASE STUDY   **Forests of the Pacific Northwest[1]**

Some of the world's most spectacular forests lie in a broad band paralleling the Pacific coast from northern California to southeastern Alaska. Ample rainfall, mild winters, and fertile soils allow trees to grow to prodigious size. These same conditions, plus the wide range of microhabitats created by having exceptionally tall trees and exceptionally large reservoirs of dead wood, support a diverse flora and fauna. From a human perspective, all of this represents a rich lode of natural resources, notably, timber, salmon, and opportunities for outdoor recreation. Unfortunately, it also creates an arena for managing ecosystems in which the stakes are high and the potential for conflicts great.

Humans arrived in this region relatively recently several thousand years ago in the case of people immigrating from Asia across the Bering land bridge; in the 19th century in the case of settlers from the east coast of North America. This relatively short tenure and the overall abundance of natural resources may explain why loggers have not entered a fairly high percentage of the region's forests, relative to other temperate forests in the world. Estimates of this percentage will vary, particularly depending on how we define the region's northern boundary, but may be roughly 5% to 15%. To someone concerned with maintaining biodiversity, these remaining virgin forests are a small legacy that must be carefully protected in reserves. To someone concerned with maintaining the health of the timber industry, these remaining forests represent billions of dollars worth of standing timber, as well as land that can be allocated to growing more timber in the future. There are other perspectives as well—for example, those of people who treasure the region's wild places as a setting for outdoor recreation and those of people who value salmon as a commercial and recreational resource and who recognize the link between healthy forest ecosystems and healthy salmon populations. However, we will focus on the issue of biodiversity versus timber; particularly as it is being addressed in the United States.

Of course, the issue is not usually characterized as biodiversity versus timber; usually, it is seen as spotted owls versus timber. People who are familiar with the situation realize that the spotted owl is essentially a flagship species for environmentalists to rally public attention and a scapegoat for the timber industry to pit against the welfare of people. Legally speaking, the spotted owl is a vehicle for addressing the larger issue of maintaining old forest ecosystems because it is protected under the U.S. Endangered Species Act. This means that its habitat must be protected, and its habitat consists largely of these old remnant forests, often several hundred hectares per pair.

The first response to protecting spotted owl habitat was establishing small reserves (Spotted Owl Habitat Areas, SOHA) around many of the known sites occupied by owls. Soon the inadequacies of this approach became apparent, and the focus switched to identifying areas of many thousands of hectares that would hold several owl territories (Habitat Conservation Areas, HCA) and to maintaining reasonably complete forest

cover between these areas to facilitate owl dispersal. Neither of these approaches specifically considered the needs of species other than spotted owls.

The third approach, devised by a Forest Ecosystem Management Assessment Team and widely known as FEMAT, involved identifying a large set of Late-Successional Reserves and Riparian Reserves designed to protect virtually the entire suite of species associated with old forests, including salmon and other species associated with forest streams. Ostensibly, this was an improvement, but environmentalists were disappointed with the specific plan because it still opened some areas of old-growth forests to commercial logging. Furthermore, it allowed some thinning of stands and salvaging of dead timber in Late-Successional Reserves that would presumably require road access. The FEMAT approach also attempts to improve management of federally owned forests outside of the reserves so that they will provide some habitat for a greater array of species. Specifically, it requires retaining some trees after clearcutting to accelerate the development of vertical structure in logged stands.

The picture painted here applies only to the roughly 50% of forest lands that are publicly owned. The other half of the forest is primarily owned by large timber corporations, and their management is quite different. Virtually all of the old-growth forests have been cut, and the major emphasis is on growing a single species, Douglas fir, on a 40- to 80-year cutting cycle. This usually involves clearcutting a site, planting seedlings, and using various silvicultural techniques to accelerate growth. Management is usually intense enough to consider these forests to be cultivated ecosystems.

It remains to be seen how well the biota of this region will be served by this mixture of natural, modified, and cultivated forests. Certainly, it will fare better than the wild life of places like Europe that have a long history of intensive land use, but it will be compromised to some degree. This compromise seems particularly disturbing to many conservationists because it has largely happened within our lifetimes and is still happening.

[1] This account is distilled from Harris (1984), Hunter (1990), Ruggiero et al. (1991), Forest Ecosystem Management Assessment Team (1993), Franklin et al. (1997), and Spies and Turner (1999).

Summary

Managing ecosystems to maintain biodiversity requires a mixture of different approaches, which include the following: protecting natural ecosystems in reserves; combining biodiversity conservation and commodity production (e.g., forestry and fisheries) in modified, seminatural ecosystems; managing cultivated and built ecosystems to ensure that they efficiently provide for human well-being without having a negative impact on other ecosystems; and restoring degraded ecosystems. For protected ecosystems the key issues are selecting reserves that will protect a representative array of ecosystems (recall Fig. 4.6), designing the reserves (e.g., deciding on their size, shape, and location with respect to other types of ecosystems), and managing the reserves to maintain their natural structure and function (e.g., controlling exotic species and human visitors). Modified ecosystems dominate the earth's surface, and thus it is essential that they provide habitat for most biota in addition to connectivity among reserves. This can be accomplished if these ecosystems are managed in a way that is as consistent as possible with natural processes, for example, managing livestock to imitate the role of native herbivores. Cultivated and built ecosystems do provide habitat for some species, but they are generally not species jeopardized with extinction. Conservationists need to ensure that these ecosystems are safe, enjoyable places for people to live and that they produce most needed commodities so that the pressure on other ecosystems is minimized. Finally, all of the activities described above can only maintain the status quo; if we want to improve an ecosystem that has been degraded by human activities, we must attempt to restore ecosystems.

FURTHER READING

For a grand overview read *People and Ecosystems* (United Nations Development Programme et al. 2000). *Saving Nature's Legacy* by Reed Noss and Allen Cooperrider covers managing North American ecosystems for biodiversity, while Sutherland and Hill (1995) give quite a different European perspective. See Grumbine (1994), Yaffee (1999), Callicott et al. (1999), and Dale et al. (2000) for some conceptual treatments of ecosystem management. For more specific issues, see Shafer (1990) and Wright (1996) on reserves, Hunter (1990, 1999) on forests, Samson and Knopf (1996) on rangelands, Wilcove and Bean (1994) and Boon et al. (2000) on aquatic ecosystems, and O'Connor and Shrubb (1986) and Collins and Qualset (1999) on farms, Spirn (1984) and Gilbert (1989) on urban areas, and Cairns (1995) and Harker et al. (1999) on restoration ecology. "Song of the Dodo" by Quammen (1996) is a very readable account of island biogeography and fragmentation. For some ideas about what you can do on campus see Creighton (1998). See www.nhm.ac.uk/science/projects/worldmap/ for a site focused on the global distribution of biodiversity and reserve selection.

TOPICS FOR DISCUSSION

1. Would you create artificial water holes in arid reserves? Would you remove existing artificial water holes?
2. Would you be willing to convert some portion of a 1-million-ha, semi-natural forest, currently modified by regular logging, into a plantation if an equal portion of the forest were set aside as a reserve?
3. Should we purchase more reserves or manage better the ones we have?
4. Comparing ecosystems modified by fisheries, forestry, or livestock grazing, which do you think pose the most serious problems for conserving biodiversity? In which could the problems be solved most readily?
5. In your region which types of ecosystems have experienced the worst degradation and loss? What steps could be taken to restore them?
6. Under what circumstances, if any, should ecosystems damaged by natural disturbances such as volcanoes be restored?

Chapter 12
Managing Populations

In 1976 there were only seven black robins in the world, and they were slipping into oblivion in their last refuge, a tiny patch of dying forest on top of a sea stack called Little Mangere Island, one of the Chatham Island archipelago about 600 km east of the South Island of New Zealand (Butler and Merton 1992). To save them, the New Zealand Wildlife Service captured all seven and moved them to a more stable patch of forest on adjacent Mangere Island. Yet, on Mangere the robins continued to decline until 1979 when only five survived. A comprehensive, all-out effort to save the species was initiated. This involved supplementing their diets with feeding stations, removing eggs from nests, and transferring these eggs to the nests of others species—foster parents—so that the black robins could lay another clutch. This effort also included erecting artificial nest boxes, controlling parasites and predators, and other techniques that we will discuss later. It was all very complex, laborious, and intrusive, but it worked. At one point the fate of the species depended on a single breeding female—known as Old Blue for her leg band—who proved remarkably long-lived (> 12 years) and tolerant of human manipulation. Now, there are over 250 black robins living on two islands and receiving no regular management. As Don Merton, chief architect of the project, said, "If we can save the black robin from extinction, we can save any species."

This story is an inspiring example of what committed, creative people can do. It is also a dire warning of what may be necessary if we let species descend into such dangerous straits and then have to rescue them. To state the obvious, we cannot take this approach for each of the world's millions of species; ecosystem management must be the backbone of programs for maintaining biodiversity. (Indeed, it played a critical role in the black robin story: habitat restoration on Mangere Island was a critical element that began several years before the robins were transferred from Little Mangere.) Nevertheless, there will be many situations in which it is necessary to manage populations directly because maintaining the ecosystems they inhabit is insufficient. This is particularly true of species that are close to extinction and those threatened by overexploitation.

Realistically, it is not possible to work with every single species that could benefit from direct management. How do we develop management plans for insect species that we have not even classified yet? Despite the difficulties, hundreds of species are being managed now—mainly the vertebrates and vascular plants that are deemed most important ecologically, economically, or aesthetically—and this number will increase.

In this chapter we will review some of the techniques used for managing populations. To manage a population wisely, we must first understand its structure and the factors affecting it, but we will not return to population viability analysis, metapopulation structures, and other topics covered in Chapter 7, Extinction Processes. Here, the focus will be on techniques to use once the problems affecting a population are understood. In broad terms, these are 1) providing resources that may be scarce such as food or water; 2) controlling threats such as predators, especially human predators; and 3) directly manipulating populations, as when individuals are moved to new sites, for example. These techniques can be used to manage all species from very rare black robins to very abundant starlings, but, of course, conservation biologists are usually most concerned about species that are in jeopardy. Therefore we will focus primarily on techniques relevant to recovering small populations.

Providing Resources

The most basic resources that organisms require are energy, carbon, hydrogen, oxygen, nitrogen, and certain other elements: in more familiar terms, food and water. Organisms also need a place to live, a place where the microclimate is not too hot or cold, wet or dry. This place may also need to provide concealment from other organisms, or a substrate to which they can attach themselves and not be swept away by wind, water, or gravity. Conservationists can meet all of these needs for any given species by maintaining the type, or types, of ecosystem it uses as habitat. In other words, all the strategies described in the preceding chapter can be brought to focus on a single species. Sometimes, however, it is sufficient (and more efficient) to provide a key missing resource. This will be the case when an ecosystem is almost suitable habitat, but lacks something that is a limiting factor such as enough hollow trees to serve as nest sites. Here we will review a few examples of these practices. This is not a comprehensive review because the

potential scope of these practices is huge, and they have barely been explored for endangered plants and invertebrates.

Food Energy Plus Nutrients

   People love to feed animals. In the United States alone, roughly
500,000 metric tons of seeds are fed to wild birds each year, and this food
probably improves the survival prospects of many individual birds during
lean winter months (Brittingham 1991). It definitely gives people a great
deal of pleasure. As a tool for helping endangered species, the extent of
food provisioning is much more limited (Archibald 1977a) (Fig. 12.1). One
well-known example involves providing carcasses that are free of contaminants, especially lead shot, to endangered birds of prey such as the California condor, griffon vulture, bald eagle, and white-tailed eagle (Knight and
Anderson 1990, Meretsky et al. 2000). These species benefit from receiving
clean food because they often suffer from contamination and because
young individuals often have difficulty finding food. People often cultivate
specific types of plants as food for animals—for example, seeding abandoned logging roads with clover or planting flower gardens designed to attract butterflies and hummingbirds—and, occasionally, this is undertaken
specifically to assist endangered species. In England, conservationists have

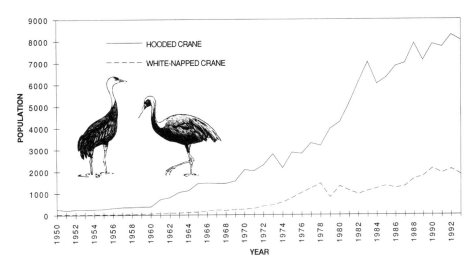

Figure 12.1. The numbers of cranes appearing at Japanese feeding stations increased
markedly over time, suggesting that availability of winter food had limited populations
(Archibald 1977b, personal communication). Of course, other factors could have
contributed to the increase, including the possibility that cranes were simply more
concentrated at feeding stations in later years.

attempted to help a rare moth, the netted carpet, by managing wild patches of touch-me-not balsam, the plant that its larvae consume (Hatcher and Alexander 1994). By thinning forest overstories and removing competing ground vegetation, conservationists hope to maintain the moth by maintaining its host plant. Programs to reestablish a new animal population after the original population has been extirpated (a technique we will review below) occasionally include feeding animals until they are well established. For example, conservationists trying to establish a population of the American burying beetle on an island in Massachusetts provided each transplanted pair with carrion (a small chicken corpse) that the beetles could bury and use as a food cache on which to raise their young (Amaral et al. 1997). In Spain conservation programs for endangered carnivores such as the Iberian lynx include releasing rabbits to augment depleted populations (Calvete et al. 1997).

Besides the major food constituents—carbohydrates, proteins, and fats—animals also need a wide array of macronutrients (e.g., calcium, potassium, and sodium) and micronutrients (e.g., iron, selenium, and iodine), and a scarcity of these can sometimes limit populations (Robbins 1993). In some regions, game managers set out mineral blocks to meet some of these needs, especially sodium requirements, but the efficacy of these practices has not been established.

Feeding wild animals has at least two significant downsides. First, it is likely to foster long-term dependence on people, leaving the animals vulnerable to starvation if feeding is discontinued. Second, feeding tends to concentrate animals and thus may make them more vulnerable to disease and predation (Brittingham 1991). Furthermore, the impact of providing food for endangered species is often not evaluated in the first place (Armstrong and Perrott 2000).

In a broad sense, food includes the energy and nutrient requirements of plants. Plant conservationists have long recognized that managing these resources directly could be an important tool for maintaining endangered plant species if a scarcity of solar radiation or certain nutrients were limiting a population (Stuckey 1967). However, there are few actual cases of conservationists directly managing energy or nutrient resources for endangered plants. Conservationists in Maine have thinned the forest canopy over patches of the small whorled pogonia, a rare orchid, to allow more solar radiation to reach the plants, which seemed to be declining for lack of light (Alison Dibble, personal communication; also see Maschinski et al. 1997a and Eisto et al. 2000). Presumably, some rare plant species would benefit from artificial fertilization, but I have found no examples of this being practiced. In contrast, one botanist, noting that many rare plants are confined to sites of poor fertility, has suggested techniques for *reducing* soil fertility to favor rare species (St. John 1987). For example, cutting sods around *Gentiana pneumonanthe,* a rare heath plant in the Netherlands, removes excessive nutrients and provides a substrate for germination (Oostermeijer et al. 1998).

Water

In arid regions the availability of water is often a limiting factor for many populations. Of course, as we discussed in the preceding chapter, responding to this fact by broadly increasing the availability of water would just replace an arid ecosystem with a less arid ecosystem, and this could decrease overall global diversity. Sometimes, however, it might make sense to provide some drinking water for desert animals. For example, drinking water might be essential to saving an endangered species that is now found only in a small portion of its original geographic range (likely the driest portion) because of hunting and competition with livestock.

Providing water to desert animals is widely undertaken for livestock and occasionally for wild animals (Yoakum et al. 1980, Burkett and Thompson 1994) despite some of the potential shortcomings described in Chapter 11. At its simplest, it involves building a basin to hold water. Usually, some means of obtaining water such as a well or a rainwater catchment is also necessary. More elaborate structures may provide fencing to exclude livestock; escape ramps for small animals that may fall into the basin; and shade, both for the animals' comfort and to inhibit evaporation. Although wild life managers usually build water holes principally for game species such as quail and desert bighorn sheep, they have been constructed specifically for endangered species. In Saudi Arabia water holes are used in programs to restore mountain gazelle and houbara bustard populations (Dunham 1998) (Fig. 12.2).

Figure 12.2. Populations of some desert animals like the houbara bustard can be increased by providing drinking water.

Physical Environments

Each species requires a particular physical environment. It may be as amorphous and common as air or water; thousands of microscopic species drift through life wherever the air or water takes them. It may be as specific as the follicles of your eyelashes, which are probably home to a tiny, elongate species of mite, *Demodex follicularum.* Physical environments provide three basic things. First, many species require a physical environment that provides them with a benign microclimate, perhaps shelter from a chilling wind or desiccating sun. Second, many species require a physical environment that offers concealment from potential predators or grazers or, if they are themselves predators, concealment from their prey. Finally, some species need a particular kind of substrate on which to live to avoid being moved away by gravity, wind, or water. (When biologists, especially zoologists, speak about physical environments that provide shelter or concealment, they often use the term *cover,* but this term would generally not include substrates.)

For many species, a physical environment is provided by other organisms, and thus briar patches offer concealment for rabbits, just as eyelash follicles are a substrate for *Demodex follicularum.* Maintaining Thomber's fishhook cactus and other rare desert plants may depend on maintaining nurse plants, plants of other species that protect smaller species from temperature extremes and herbivores (Nabhan 1987, Suzan et al. 1994) (Fig. 12.3). You might think that these are examples of biological environments rather than physical environments, but the term "biological environment" usually refers to the suite of competitors, symbionts, predators, and so on with which each species interacts.

Providing special physical environments for wild life is a very old management technique. People have been erecting sections of hollow logs in which bees and birds could nest (and from which honey and eggs could be easily extracted) for centuries, probably millennia. Supplying nest environments is especially widespread, probably because it is usually so effective and efficient. Just a small amount of shelter and concealment, enough for a nest, can have a marked effect on an individual's chance of reproducing. The recovery of Puerto Rican parrots, which for many years numbered fewer than 20 wild birds, was helped by creating artificial nest sites and by improving natural nest cavities, particularly by making them deeper, darker, and more waterproof (Snyder et al. 1987). In Turkey and Morocco, new ledges have been built on cliffs to provide nest sites for bald ibises (Hirsch 1977). Rocks were dumped into a Maine lake to create three new spawning beds for a fish, the silvery char—a rare form of arctic char—because water level drops threatened the existing spawning area (Fred Kircheis, personal communication).

Sometimes, conservationists supply additional sites for other activi-

Figure 12.3. Thornber's fishhook cactus requires other species, sometimes called nurse plants, to shelter it from temperature extremes and grazing livestock.

ties such as resting during the day or night, or hibernating for a whole winter. Brush piles are a common way to meet these needs for small mammals and ground-dwelling birds; rock piles (perhaps as hibernacula for snakes) are a less common example. Occasionally, the scale of providing new physical environments is so large that you could argue that a whole new ecosystem has been created; for example, when rocks, junk cars, abandoned oil rigs, and similar items are sunk at sea to make artificial reefs in areas lacking natural reefs. (See the *Bulletin of Marine Science* 44[2] for 51 articles on artificial reefs.) Similarly, managing agricultural landscapes for wild life by creating and maintaining hedgerows can be viewed as ecosystem-scale provision of cover (Dowdeswell 1987).

Interactions

Some populations are limited by the scarcity of another species with which they interact symbiotically. For example, biologists trying to maintain endangered freshwater mussel species have to be concerned with maintaining populations of fish that the mussels can parasitize (Neves et al. 1985). These mussels pass through a life stage (during which they are called glochidia) encysted on the gills or fins of fish, and many mussel species are quite specific about which fish species are acceptable hosts. Similarly, British conservationists discovered that to maintain a rare butterfly, the large blue, they must maintain a population of ants, particularly *Myrmica sabuleti,* because the caterpillars overwintered and pupated in the ants' colonies (New 1991). Unfortunately, they made this discovery only after they had inadvertently eliminated ant colonies in the butterfly's only habitat by prohibiting livestock grazing and thus changing the site's vegetation. The large blue butterfly is now extinct in Britain. (The history of managing populations is rife with tales of mistakes, but these errors must be a call for caution, sound research, and publishing both successes and failures, not inaction.) In some cases people step in to fill a symbiotic role themselves. Concern with the plight of the eastern prairie fringed orchid led to a successful program of hand-pollinating plants, thereby filling the role of sphinx moths (Brown 1994).

*Intra*specific interactions may also demand the attention of conservation biologists. While trying to reestablish colonies of arctic terns, Atlantic puffins, and dark-rumped petrels, Stephen Kress discovered that it was necessary to provide the birds with social stimulation (e.g., Kress 1983, Kress and Nettleship 1988, Podolsky and Kress 1992). Birds were more likely to breed at a site where wooden decoys had been set out and/or where they could hear vocalizations of their species broadcast from tape recorders (Fig. 12.4).

Figure 12.4. Puffin decoys are used to provide a social stimulus for puffins establishing a new colony.

Controlling Threats

In the big picture, human overpopulation, global pollution, deforestation, desertification, and other problems of this magnitude are the principal threats that conservation biologists must meet, but in this section we will focus on smaller-scale problems that can be addressed at the level of population management. For example, here we will not worry about managing energy consumption by the 6 billion people that crowd our planet, but rather how to stop the handful of people who are still poaching giant pandas. We will also consider other species that may threaten a population because of their roles as competitors, predators, grazers, parasites, or pathogens. In many cases these are exotic species, but sometimes they are natives (Garrott et al. 1993).

Overexploitation

The world's best-known poacher is a mythical hero, Robin Hood, and this fact is symbolic of a larger truth. In most people's view unlawful exploitation of wild plants and animals is far closer to illegal parking or speeding than to theft or murder. Indeed, for most species, overexploitation is not even illegal. For plants this is often true even if they are known to be an endangered species (Bean 1983). Most governments consider plants to be the property of landowners because they are immobile (i.e., they do not move from property to property as many animals can), and governments often are reluctant to restrict what people can do to their private property. Diminishing the acceptability of overexploitation requires education and other approaches to social, economic, and political issues that we will discuss in Chapters 15, 16, and 17. Here we will focus on the front lines of what is sometimes called the war on wild life (Reisner 1991).

Biodiversity advocates are usually focused on species that are in such perilous straits that it is best to prohibit human exploitation completely, with the possible exception of nonconsumptive uses such as whale watching. On paper this is simple. We pass a law banning exploitation, and we employ wardens to enforce the law (Sigler 1972). In practice it has all the problems of conventional law enforcement plus some added difficulties. In particular, wardens have to work in remote areas, often with little or no support from the local community. In many places wardens have a long tradition of effectively enforcing laws designed to protect game birds, fishes, and mammals, but they are often reluctant and ill-prepared to take on the added burden of protecting endangered butterflies, plants, reptiles, and so on, even if they have the mandate to do so.

Protecting endangered species can be particularly difficult because the laws of supply and demand dictate that the rarer a species becomes, the more valuable it will probably be, thus offering a greater incentive for poachers to break the law. The classic example of this vicious circle comes with the five species of rhinoceros whose horns are now worth many thousands of dollars per kilogram, in large part because rhinos are now so rare. This crisis has precipitated a dramatic response in some southern Africa nations: conservation officials are capturing rhinos and cutting off their horns (Fig. 12.5). Unfortunately, this solution has many problems: it is expensive, especially because the horns regrow quite quickly; some poachers kill dehorned rhinos out of spite; lack of a horn probably inhibits the ability of mother rhinos to defend their young from predators; and it may affect social dominance (Berger et al. 1993, Berger and Cunningham 1994a, 1994b, 1998, Rachlow and Berger 1997).

Ideally, conservationists would never deal with crisis situations like that of the rhinos. They would work with all the species that are subject to exploitation while these species are still common enough to sustain some appropriate level of harvest. Determining an appropriate level of harvest

Figure 12.5. In some countries conservation officials are dehorning black rhinos to dissuade poachers from killing them.

opens a key issue of *additive mortality* versus *compensatory mortality*. Harvest mortality is said to be compensatory if it does not increase the population's mortality above what it would have been under natural (no harvesting) conditions. If, however, harvest mortality significantly increases total mortality, it is said to be additive. For example, imagine a population of catfish that experiences annual mortality of 20% because of starvation. If we began harvesting 15% of the fish each year and this reduced starvation mortality to 5%, so that overall mortality remained 20%, then our harvesting is inducing compensatory mortality. In contrast, if a 15% harvest increased overall mortality from 20% to 30%, then harvesting is additive. Clearly, in a perfect world, harvesting by humans would be largely compensatory.

Once an appropriate harvest level is determined, there are many ways to achieve that level by limiting exploitation in various ways. If we have enough staff to monitor harvesting closely, we can directly limit *how many* plants or animals of a given species (and perhaps of a given sex or age class) can be harvested from a particular area during a given period. More commonly, indirect methods are used. These could include limiting *who* is allowed to do the harvesting such as only local people, only people who buy a license (perhaps a license expensive enough to be a deterrent), only people selected by lottery, only people who are doing it for sport, or only people who are doing it to make a living. For example, in the United States only Native American subsistence hunters are allowed to harvest bowhead whales, polar bears, and Pacific walruses. We can also limit *when* harvesting is allowed; in many cases harvesting animals is permitted only after the breeding season when there are many young individuals around, many of which are likely to die with or without harvesting. *Where* harvesting occurs is commonly restricted by nature reserves; sometimes no-harvest areas are also established for single species. Limiting *how* harvesting is conducted, the methods employed, can be important. Methods that are likely to kill or injure more organisms than are harvested—incidental harvest—should be eliminated or modified. Fishing with dynamite or poison is an extreme example; simply trawling a net is likely to kill huge numbers of nontarget organisms (see Fig. 9.6) (Kaiser et al. 2000). Sometimes, changes in the equipment are sufficient; for example, modifying nets and traps to allow nontarget species to escape (Roosenburg and Green 2000), or adding warning devices like sonic alarms on fishing nets to keep marine mammals and seabirds away (Melvin et al. 1999, Trippel et al. 1999) . The preceding ideas are just the tip of an iceberg because fish, game, and timber managers have devoted much thought and effort to managing harvests. See Smith et al. (1997) and Strickland et al. (1994) for reviews. Unfortunately, the history of our success in carefully regulating harvests is marked by many dismal failures. Worse yet, some people have suggested that, because biological systems are so complex and unpredictable and our understanding of them is so limited, overexploitation is almost inevitable given human demand for natural resources (Ludwig et al. 1993). For a collection of essays on this topic, some of which are not so pessimistic, see *Ecological Applications* 3(4) (1993).

Indirect Threats by Humans

Most of the harm that wild life suffers at the hand of humanity is indirect and unintentional; it is a by-product of human negligence, ignorance, or apathy. These problems usually operate at the scale of ecosystems (see Chapters 7 and 8), but, occasionally, their effects can be ameliorated through population-scale management. Helping animal populations cope with roads provides a good example. Conservationists and road engineers have collaborated to design many ways to minimize road-crossing mortality for certain species (Fig. 12.6). Tunnels plus fences leading to the tunnels have reduced the mortality of many European populations of frogs and toads traveling to their spring breeding sites (Langton 1989). On a larger scale, underpasses with fences have been installed to reduce road mortality for large mammals in Florida (for Florida panthers in particular) and in Canada (Clevenger and Waltho 2000). In Australia, simply maintaining wide roadside margins has allowed motorists to spot cassowaries (a rare, nearly ostrich-sized bird) and avoid collisions. In Belize a rope bridge over a road is used by howler monkeys.

Other examples of this population-scale approach include building grates over the entrances to bat caves to exclude people (Richter et al. 1993) (Fig. 12.7), designing fences that allow pronghorn antelope through but not livestock (Yoakum et al. 1980), placing silhouettes on large windows to re-

Figure 12.6. Tunnels can allow toads and other amphibians to pass under roads during their spring migrations.

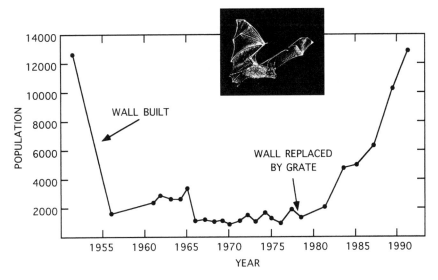

Figure 12.7. Population changes of Indiana bats hibernating in a cave after a stone wall was built to exclude human intruders and after the wall was replaced with a grate. The wall increased temperatures, which increased the bats' rate of fat metabolism; apparently, many did not survive hibernation. (Redrawn by permission from Richter et al. 1993.)

duce bird collisions (Kress 1985), erecting fences to keep people from trampling rare plants (Maschinski et al. 1997b), and restricting boat speeds in areas frequented by manatees (O'Shea et al. 1985). Conservationists trying to save a population of imperial eagles in southwestern Spain were able to alleviate the birds' major source of mortality—electrocution—by putting 9 km of electric line underground and better insulating another 33 km (Ferrer and Hiraldo 1991). Survival of juveniles increased from 17.6% to 80% following their efforts. In another study Spanish biologists discovered that marking thin electric lines (ground wires) with some colorful spirals reduced collision mortality for all bird species by 60% (Alonso et al. 1994).

Consumers

Virtually all organisms are vulnerable to being consumed by other organisms. Normally, this is a fact of life that affects populations but does not threaten them with extinction. Sometimes, however, conservation biologists discover that the impact of consumption on a particular population or species is significant and must be controlled. This is particularly likely to happen if the consumer is an exotic species against which the prey species has evolved few defenses.

In this section we will review a few examples of population management programs that involve controlling consumers. We will define *consumer* broadly as an organism that consumes other organisms. Defined this way, consumers include *predators* (organisms that attack, quickly kill, and consume other organisms, e.g., a crocodile killing a heron or a crow killing a corn seedling); *grazers* (organisms that attack large numbers of prey consuming a part of each one, but rarely killing them, at least in the short term e.g., an antelope grazing on grass or a vampire bat sucking blood from a tapir); and *parasites* (organisms that obtain their nutrients from one or very few host individuals and cause harm but not immediate death, e.g., fleas and viruses). Microbial parasites that cause disease are called *pathogens*. (This classification is modified from Begon et al. 1990.)

Predators: The conservation biology literature is full of examples of predators that have decimated a prey species. Recall Chapter 10 and the toll of species eradicated by exotic rats, cats, and mongooses, especially on islands where many prey species have evolved in isolation from predators. Small islands sometimes present an opportunity for effective population management through the complete eradication of a predator. Starting with some very small islands and learning as they progressed, The New Zealand Wildlife Service is now eradicating rats and cats from islands that are thousands of hectares in size (Towns et al. 1997, Empson and Miskelly 1999).

It is usually easy to decide to remove an exotic predator, but what if a native species is causing the problem? In the 1970s ornithologists working in the Gulf of Maine decided to do something to help tern populations, primarily arctic and common terns, which had declined markedly over the last 40 years (Kress 1983). The terns were declining largely because of predation by gulls (herring and great black-backed) whose populations had increased in response to an increase in food obtained from human sources such as landfills and garbage dumped at sea. Attempts to restore tern colonies were not successful until wild life managers killed gulls nesting on the islands targeted for restoration by distributing pieces of bread containing poison. Surprisingly, this program was not very controversial. Imagine the controversy that would erupt if conservation efforts for African wild dogs and cheetahs led to a campaign to cull the lions that limit their populations (Vucetich and Creel 1999, Kelly and Durant 2000.). Naturally, most conservationists would prefer to employ nonlethal methods for controlling native predators, and often this is feasible (Goodrich and Buskirk 1995). For example, predation on shorebird and turtle nests can be controlled by erecting fences around the nests to exclude foxes, raccoons, and other predators (Fig. 12.8) (Melvin et al. 1992, Yerli et al. 1997), and raven predation on young desert tortoises may be reduced by putting spikes on utility poles to prevent the ravens from using them as hunting perches.

Figure 12.8. Fences around piping plover nests may reduce predation.

On rare occasions, conservation biologists have been forced to choose between species that are both of concern. New Zealand biologists trying to manage Cook's petrels and kakapos (a very rare flightless parrot) on Codfish Island found it necessary to remove the island's wekas, a species of rail that often preys on other birds' nests (Lloyd and Powlesland 1994). Ironically, the weka has disappeared from large sections of its former range, but still it is not nearly as threatened as the kakapo and Cook's petrel.

Grazers: The impacts of grazing animals may be felt rather slowly—one bite at a time—but over a period of a few years they can have dramatic effects on both the species they consume and whole ecosystems. The most striking examples can be found on islands that have been invaded by goats, rabbits, and pigs and in forests that have been defoliated by any number of exotic insects (Chapter 10). Yet excessive grazing often goes unnoticed; for example, most visitors to the forests of eastern North America and Europe would not realize that in many areas the plants are severely overgrazed by native species of deer (Phillips and Maun 1996, Augustine and Frelich 1998).

Grazers are usually more common than predators, and this can substantially increase the difficulty of controlling them. In particular, eradication of most species of exotic grazers, notably insects, is impossible because they are so numerous and resistant to control (Dahlsten 1986). The best we can hope for is to keep their populations low enough to give the threatened species we are worried about a better chance of survival. For limited areas, some large herbivores can be kept at bay with fences, but building and maintaining such fences can be extremely expensive. Erecting 71 km of pig-proof fencing around nine areas within Hawaii Volcanoes National Park cost U.S.$18,600 to $26,700 per kilometer and annual maintenance costs averaged U.S.$1056 per kilometer (Katahira et al. 1993).

As with predators, controlling grazers can become quite controversial. In the western United States millions of dollars are spent every year catching and removing feral horses and burros from ecologically sensitive areas and then caring for them in captivity because the public will not permit the animals to be killed. Although elephants have been massacred in much of Africa, they are so abundant in some southern African reserves that reserve managers must sometimes shoot them because large elephant populations can easily change forest into scrubland and thus may jeopardize many forest-dwelling species (Cumming 1981, Ben-Shahar 1993). Many elephant supporters would prefer to move the elephants rather than kill them, but this is probably not feasible on a meaningful scale. Turning to some much smaller grazers, some plant conservationists have proposed that judicious use of insecticides to maintain rare plants might be in order, but, of course, any use of pesticides stirs up controversy among conservationists (Lesica and Atthowe 2000, Louda and Bevill 2000).

Parasites and Pathogens: Organisms that live on or in another organism, deriving their nutrition from its tissues, will usually fare better (have greater evolutionary fitness) if they do not kill their host. However, it does not always work this way. Many organisms succumb to parasites and pathogens, and many more are stressed to some degree by the demands of supporting a multitude of smaller creatures. Consequently, conservation biologists sometimes need to control the impacts of these usually unseen species (May 1988, Scott 1988, Aguirre and Starkey 1994, Woodroffe 1999).

The simplest way to help a population deal with the threat of parasites and pathogens is to keep it in general good health with adequate food, water, and coven. Vigorous plants and animals are usually able to withstand the effects of parasites and pathogens.

A second approach is to avoid overcrowding, which may both stress the organisms and facilitate the spread of parasites and pathogens. This issue often comes to the fore, or at least it should, while planning population management programs such as feeding, watering, and translocations that

Figure 12.9. Despite taking precautions like wearing surgical masks, researchers spread canine distemper to the only known wild population of black-footed ferret.

may concentrate individuals at unnaturally high densities. "Hands-on" research and management also carries the risk that conservation biologists themselves may spread parasites and pathogens. The world's last-known population of black-footed ferrets was nearly eliminated by canine distemper, probably introduced to the colony by a researcher who had had contact with a sick dog (Thorne and Williams 1988) (Fig. 12.9). Similarly, there is concern that herpetologists might be spreading viruses and fungi among amphibian populations.

Occasionally, wild life managers have tried to vaccinate wild individuals against disease; indeed, this had been done with the black-footed ferrets, obviously without success. Unfortunately, the logistics of catching and vaccinating a large portion of a population are rather daunting. Similarly, treating infected individuals would be difficult and probably unsuccessful. The old adage, "An ounce of prevention is worth a pound of cure," is doubly true for wild organisms (Woodroffe 1999) .

Understandably, programs designed to kill parasites and pathogens do not arouse much concern because there is little public sympathy for these creatures. A purist could argue that parasites and pathogens have just as much intrinsic value as whales and eagles, but this would be a very difficult position to defend, especially if the organism in question affects people (Koshland 1994, Gompper and Williams 1998). Fortunately, this is largely an academic question because it is extremely difficult to totally eradicate a species of parasite or pathogen as long as its host survives.

Competitors

In theory, no two species can occupy exactly the same ecological niche; nevertheless, competition for specific resources—a type of food, a place to nest, or simply space—is often quite intense. Consequently, conservation biologists sometimes find it necessary to tilt the balance toward rare species, lest they lose out to their competitors entirely and become extinct.

Controlling competition is widely practiced by plant conservationists. This can involve a form of competition control known to every gardener— weeding. However, weeding obviously needs to be selective, and hand-removing the various plants that are crowding a rare population is extremely labor intensive (Wester 1994). Sometimes, the labor can be reduced if only a few species, typically exotic species, are targeted for removal. Botanists managing a population of the large-flowered fiddleneck used a grass-specific herbicide to kill competing exotic grasses (Pavlick et al. 1993). More commonly, controlling competition involves regulating the natural patterns of competition that are part of succession (McIntosh 1980). Many imperilled plant species are associated with early-successional communities that are becoming uncommon because of human interference with natural distur-bance patterns. For example, many grassland plant species exist only in en-vironments where frequent fires prevent woody plants from outcompeting herbaceous plants, and some of these species have become uncommon, in part because of human fire control. Consequently, as incongruous as it may seem, managers of grassland plant species often set fire to the populations they are trying to save (Pendergrass et al. 1999) (Fig. 12.10). Of course, they

Figure 12.10. Conservationists often burn grasslands to control the competitors of rare plants such as the northern blazing star.

do this outside the growing season and know that the plant will survive the fire as seeds, roots, or rhizomes. In some cases managers have to employ a disturbance regime that may seem unnatural. Near Cheltenham, England, managers of a tiny reserve (394 m², not much larger than a tennis court) discovered that to perpetuate a very rare buttercup, the adder's-tongue spearwort, they had to allow cattle to graze on a portion of the reserve each year (Frost 1981). Without the disturbance of grazing and trampling, the reserve's star species would be outcompeted by common plants.

Sometimes, conservationists also find it necessary to control competition between endangered animal species and their competitors. The best known example of this comes from efforts to help the Kirtland's warbler compete with brown-headed cowbirds (Mayfield 1977). The cowbirds deposit their eggs in Kirtland's warblers nests where their young usurp food and parental attention, causing the death of the young warblers. (This behavior, which is displayed by several bird species, is called brood parasitism, but it is a form of competition, rather than parasitism, as defined here.) By 1971 the world population of Kirtland's warblers had declined to 201 pairs, but beginning in 1972 a program of trapping about 3000 to 4000 cowbirds per year has helped populations increase roughly fivefold (Solomon 1998). The program to reintroduce American burying beetles mentioned earlier included the removal of several hundred individuals of a similar species (Amaral et al. 1997). Fish biologists are sometimes able to control exotic competitors of endangered stream fishes simply by erecting a barrier that the exotics cannot pass (Rinne and Turner 1991). More dramatically, they sometimes use rotenone or other poisons to kill all the fishes (native and exotics) in a stream and then detoxify the stream and replace the native fishes (McClay 2000).

Many of the interactions between people and wild life we have discussed could be construed as competition, a competition that we usually win, at least in the short term. This competition is both broad (e.g., we compete for space and solar energy whenever we convert a natural ecosystem to a cultivated ecosystem) and narrow (e.g., we compete for food with whales and other marine species by trawling for krill).

Direct Manipulations

Both "manage" and "manipulate" are derived from the Latin *manus*, meaning hand, but manipulate has a stronger link to hands, and it is used here because sometimes it is necessary for conservation biologists literally to put their hands on endangered species. For example, saving the black robin required gathering up the last few birds and moving them to a new island. Such activities are expensive and risky, but when they appear to be the only means to save a species from extinction, they can be justified. We will discuss three topics here: first, translocations—moving organisms from one habitat to another; second, artificial breeding—methods to increase the reproductive output of small populations; and, finally, the interface between population management and maintenance of genetic diversity.

Translocations

People have a long history of moving organisms around the globe; witness all the dubious "successes" described in the chapter on exotic species. Here we will limit ourselves to the intentional movement of organisms for the purpose of maintaining biodiversity in the wild, including both the transportation of wild organisms from one ecosystem to another and the release of captive-bred individuals into the wild. Translocation or relocation can take three basic forms: *introducing* organisms into new sites, places where they did not exist previously; *reintroducing* organisms to environments where they have been extirpated; and *augmenting* or supplementing small existing populations by adding individuals obtained elsewhere.

The first type of translocation may not seem to fit here because, of course, introducing organisms into new sites is how exotic populations are created. Nevertheless, there are some situations in which this can be a useful tool. Notably, some species are native to small areas, usually islands, that are so irrevocably altered that there is no hope of their persisting there. In this case it may make sense to introduce them to a nearby island that is in better condition than their original habitat, but that they are unable to colonize on their own. For example, New Zealanders have moved frogs, flightless insects (the giant weta), and several flightless birds (kakapos; all three species of kiwis; and takahes, a type of rail) from the larger islands of New Zealand, where they were losing out to rats and other exotic species, to small islands where they are probably not native, but where it is feasible to control exotic species (Towns et al. 1997, John Craig, personal communication).

People love to reintroduce species (Fig. 12.11). To return a species to its native range feels like a small, but positive, step in pushing back the oncoming wave of extinction. Some wonderful successes have been achieved. In the early 1900s several North American game species (e.g., wood duck, desert bighorn sheep, white-tailed deer, wild turkey) had been reduced by overhunting to perilously small populations scattered about their native range. Since then, programs to control hunting and to reintroduce these species have allowed them to reoccupy a substantial portion of their original range. In some cases translocations may be the key to restoring an entire ecosystem; for example, moving live colonies of coral onto a dead coral reef may accelerate restoration of the entire ecosystem (Guzman 1991).

Figure 12.11. Translocations can be used to reintroduce a species such as the Pacific barrel cactus shown here to a site from which it has been extirpated.

Unfortunately, the success stories are shadowed by many failures. Summarizing four studies that have spanned many taxa and countries, roughly half of the translocation projects for endangered species were successful (Griffith et al. 1989, Dodd and Seigel 1991, Wolf et al. 1996, Fischer and Lindenmayer 2000). (This estimate is based only on projects with known outcomes; often they are outnumbered by projects for which the outcome is not known.) Turning to some more focused studies, 40 taxa of fishes native to the deserts of North America were transplanted to 407 sites and became established at just 26% of the sites (Hendrickson and Brooks 1991). A review of 15 plant translocations in California, usually transplanting adults, found only four that were judged a complete success (Hall 1987).

Some important lessons have been learned from both the successes and failures (Maunder 1992, Fischer and Lindenmayer 2000), although detailed studies of translocated populations are not common (Armstrong and McLean 1995, Morgan 2000). First, many projects fail simply because the problems that caused the population to become extinct in the first place were still operating. Not surprisingly, reintroduced organisms need undegraded habitat, freedom from overexploitation, etcetera. Second, successful reintroductions often require repeated translocations of substantial numbers of organisms—in other words, a substantial, long-term commitment of money and personnel. A period of careful husbandry (e.g., providing food or water or controlling consumers) may also be required and might lessen the need for large numbers. These are called *soft releases* as distinct from *hard releases* where the organisms are simply transported to their new habitat. Third, individuals obtained in the wild are more likely to survive than offspring from captive populations, especially if the population has been in captivity for several generations. For additional ideas about the biological foundations for successful translocations, see Armstrong and McLean (1995), and for the organizational keys to success, see Reading et al. (1997). Perhaps the most important lesson is that reintroduction projects are risky and expensive, and the best strategy is to avoid having to undertake them in the first place.

Augmenting existing populations is widely practiced by sport hunters and anglers who want to have a large number of prey to pursue and by foresters who want to ensure that there is adequate regeneration of trees after a cut. However, this method has not been widely undertaken to help endangered species. One reason is that a small remnant population would be very vulnerable to any disease carried by the translocated individuals (Viggers et al. 1993, Cunningham 1996). People releasing their pet desert tortoises have probably spread respiratory diseases to wild populations (Jacobson et al. 1991). Genetic issues are also of concern because translocated individuals could introduce "exotic" alleles into the local gene pool, and these might be maladaptive or might displace uncommon local alleles, thereby reducing genetic diversity of the species as a whole (Ellstrand and Elam 1993, Rhymer and Simberloff 1996). On the other hand, for at least one Swedish snake species, inbreeding problems were alleviated by adding new individuals. An island adder population, known to be limited by inbreeding, expanded dramatically after 20 males were added to the breeding population for four breeding seasons and then removed again (Fig. 12.12) (Madsen et al. 1999). A drop in the proportion of stillborn young was the chief factor. A similar story has been reported for a population of greater prairie chickens; translocations led to an increase in egg viability (Westemeier et al. 1998). We will return to these and other genetic issues in a separate section to follow.

Occasionally, translocations are motivated by goals other than establishing a new population or augmenting a declining one. For example, sometimes, when wild life managers are faced with the dilemma of a population that is too large for the available habitat, they will decide to move individuals rather than kill them because of public opposition to lethal control. Similarly, wild life managers sometimes try to return animals to the wild that have been confiscated from people who held them illegally—

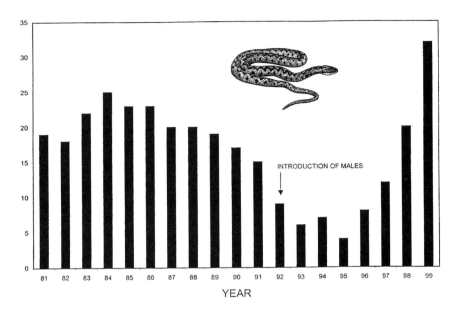

Figure 12.12. Addition of 20 male adders to a declining island population apparently rescued the population from the effects of inbreeding. The graph shows an index of population size (total number of individual males captured; females were not counted because they were much harder to catch) that indicates a slow decline and then a dramatic recovery. The 20 translocated males were not counted in the population estimates. Other data showed an increase in genetic variability after 1996 and a decrease in the number of stillborn young. (Redrawn from Madsen et al. 1999).

notably orangutans and parrots (Yaeger 1997). Again, finding a home for individuals that no one wants to kill or keep in captivity is likely to be a key motive. Most worrisome are those translocation projects that are catalyzed by the need to get a population, usually an endangered plant species, out of the way so that a site can be developed. As with ecosystem restoration projects (Chapter 11), conservationists must be careful to avoid having their work indirectly facilitate ecosystem destruction (Falk and Olwell 1992).

Artificial Breeding

Hands-on manipulations of populations reach their zenith when conservation biologists alter the breeding systems of endangered species to increase their reproductive output. The most elaborate techniques (e.g., artificial insemination, embryo transfer) are generally confined to captive populations and will be covered in the next chapter. Here we will focus on some techniques that have been used with wild populations.

Cross-Fostering and Double-Clutching: Among animals that provide extensive parental care (primarily birds and mammals) reproductive output may be limited more by their ability to provide care than by their physiological capacity to produce young. Under these circumstances, it may be possible to increase reproductive output by using two closely allied techniques, double-clutching and cross-fostering. *Double-clutching* involves removing one set of eggs to induce an animal to produce a second clutch and incubating the initial clutch elsewhere. It has been used fairly often with birds because many bird species are able to produce a second clutch of eggs if their first one is removed—an adaptation to compensate for nest predation. It was a key tool for conservationists who have brought back species like the black robin, California condor, and Mauritius kestrel from the brink of extinction (Jones et al. 1991, Cade and Jones 1993). Double-clutching will not work for most other egg-laying animals (e.g., most reptiles) because in most species the female departs after laying her eggs and will not know whether they have been removed. Double-clutching (more appropriately double-brooding) would not be feasible for most mammals because of the difficulties of caring for newborn mammals and because many mammals cannot soon produce a second brood if their first is lost.

Returning to birds, once the first clutch of eggs is removed, it requires care from a surrogate parent. Sometimes, this will be undertaken by humans, either alone or in concert with a domestic hen. Sometimes, other wild species are enlisted to serve as foster parents; this is called *cross-fostering*. This was done with black robins by transferring some of their eggs to the nests of another species, the Chatham Island tit, thus enabling the black robins to lay a second clutch of eggs (Butler and Merton 1992). Cross-fostering has also been undertaken as a reintroduction technique. For example, by placing whooping crane eggs in the nests of sandhill cranes, conservation biologists tried to establish a group of whooping cranes that would spend the summer in Idaho and then migrate to New Mexico for the winter, a far shorter migration than the northern Manitoba to coastal Texas journey currently undertaken (Drewien and Bizeau 1977). Unfortunately, the Idaho whooping cranes did not breed, a fact that highlights a major problem with cross-fostering: the possibility that a bird raised by a different species will not know which species it is—a major identity crisis. Humans who act as foster parents for endangered species attempt to address this problem by remaining hidden and using appropriate puppets to interact directly with the young animals, but problems can still arise; for example, animals may not develop a healthy fear of humans or other predators (Valutis and Marzluff 1999, Griffin et al. 2000, Meretsky et al. 2000).

Head-Starting: One of the fundamental laws of nature is that little things tend to die quickly. They get eaten or outcompeted by big things. Some species cope with this reality by producing a few, large young and then taking good care of them. Other species try to beat the odds by producing huge numbers of small young that are independent from birth. These latter species offer conservation biologists an opportunity because, if we can reduce mortality during the short period when the young are highly vulnerable to predation, starvation, or desiccation, we can greatly increase the number that survive to adulthood. Techniques designed to increase survivorship of young organisms that do not receive parental care are often called *head-starting*.

Sea turtles provide an important example of this technology, especially since five of the six species are at risk of extinction (Bjorndal 1981, Frazer 1992). During her lifetime a female sea turtle can lay thousands of eggs on beaches, but few are likely to hatch because of nest predation (principally by people and other mammals) and other factors such as storms. Once they have hatched, the young turtles continue to suffer enormous mortality from birds, crabs, and fish. Sea turtle conservationists can reduce this mortality markedly by techniques such as erecting predator-proof fencing around nests; excavating eggs and moving them to a safe place until they have hatched and then returning them to their original nest site; and raising young turtles in captivity until they are large enough to avoid most forms of predation, usually 9 months to a year (Fig. 12.13). (Some people would reserve the term "head-starting" for only the last technique.) While head-starting may be a useful tool under some circumstances, it may have some real problems because of skewed sex ratios of the artificially incu-

Figure 12.13. Gathering turtle eggs and raising them in captivity can reduce predation losses and give young turtles a head start.

bated eggs, maladaptive behavior of hatchlings, and the redirecting of conservation efforts away from fundamental issues such as habitat quality (Frazer 1992, Bowen et al. 1993, 1994, Girondot et al. 1998).

Head-starting has been used for other reptiles (notably crocodilians) and plants (Ferreira and Smith 1987) and could be used for many invertebrates. It reaches its highest level of sophistication with fish, but in this case there is usually another dimension. Typically, the adult fish are captured and manipulated to obtain eggs, or eggs are obtained from a captive-breeding stock. This added dimension moves us one step closer to the type of intensive population management that happens in zoos and gardens, and thus we will cover fish hatcheries in a separate section.

Hatcheries: Raising fish is a big business. Global aquaculture harvests alone total over 33 million tons per year (United Nations Development Programme et al. 2000). (As in Chapter 11 I will refer only to fish for simplicity's sake, but much of this section also applies to various molluscs and crustaceans such as mussels and shrimp.) On top of that we can add many millions of fish that are raised to a certain size then released into the wild, either to be caught by sport anglers, usually soon after release, or to continue growing for commercial harvest when they are much larger. I will not try to review the voluminous literature of fish culture here (see the journals *Progressive Fish-Culturist* and *Aquaculture* for further information); suffice it to say that the technologies developed to raise fish for food can be adapted to raise fish that are at risk of extinction.

This has already happened at some hatcheries, particularly in the southwestern United States where many native fishes are in jeopardy (Johnson and Jensen 1991). There are many more examples at the population level such as rivers that have populations of native salmon species that would probably be absent were it not for salmon hatcheries. Hatcheries are also being established for endangered mussel species (Keller and Zam 1990). Unfortunately, some significant downsides arise when population management becomes so intensive that complex institutions such as fish hatcheries are needed (Meffe 1992). We will see some of these in the next chapter, Zoos and Gardens, and discuss them in Chapter 14, Setting Priorities.

Maintaining Genetic Diversity

Maintaining the genetic diversity of a species is inextricably linked to population management. Someday, we may have gene banking technology that will allow us to maintain genetic diversity independently from the fate of wild populations, but this will be quite a while coming. It is also a rather dismal alternative to having healthy, wild populations.

The most important way to maintain the genetic diversity of a species is very straightforward: have a substantial number of individuals comprising many different populations that occupy the species' entire geographic range and range of habitat types. As we saw in Chapter 5, most populations need to be reasonably large to avoid problems with genetic drift, inbreeding, and bottlenecks (Frankham 1996). Occupation of the entire range is important because populations in different parts of the range may develop unique genetic adaptations to the local environment (Scudder 1989, Lesica and Allendorf 1995). Similarly, there may even be genetic differences among populations occupying different types of environment in the same general area (Blondel et al. 1999). The importance of maintaining genetic diversity by having many different populations depends on how the genetic diversity of a species is distributed (Huenneke 1991, Falk and Olwell 1992). Recall from Chapter 5, Genetic Diversity, that a species' total heterozygosity (H_t) can be partitioned into two components: genetic diversity within the populations that compose the species (H_s), and genetic diversity caused by variability among the populations (D_{st}). Mathematically, this is expressed as: $H_t = H_s + D_{st}$. If a species has a relatively high D_{st}, then it is necessary to maintain many different populations to maintain the species' genetic diversity. Alternatively, if most of the species' genetic diversity exists within each population (i.e., H_s is relatively high), then it is less critical to maintain many different populations.

Because programs for managing populations and their habitats are generally directed toward the goal of having many large, well-distributed populations, they are usually compatible with the goal of maintaining genetic diversity. However, some complexities can arise, especially when direct manipulations are involved. The next four paragraphs describe some examples of potential issues.

It is obvious that reintroduction projects should use individuals that are as genetically similar as possible to the former population to maximize their chances of being adapted to local conditions (Policansky and Magnuson 1998, Montalvo and Ellstrand 2000). Unfortunately, this is not always possible (Seddon and Soorae 1999). When conservation biologists set out to reestablish a population of peregrines in the eastern United States, the native subspecies, the eastern peregrine, was virtually extinct (Barclay and Cade 1983). Lacking any of the native subspecies, they decided to use peregrines from all over the world as breeding stock to create as diverse a gene pool as possible (Fig. 12.14). They assumed that natural selection would favor the assemblage of genes best adapted to conditions as they exist in the region today. Because the reintroduced peregrines seem to be persisting, this approach of maximizing outbreeding appears to have worked. At least one plant reintroduction project, that of the lakeside daisy, was also designed to maximize outbreeding (DeMauro 1993). We do know that maximizing outbreeding is not always the best strategy for creating a new

Figure 12.14. Efforts to replace the eastern peregrine falcon sought to maximize genetic diversity by using individuals from as far away as Australia and Europe.

population from scratch. Recall from Chapter 5 the ibex reintroduction to Slovakia (see Fig. 5.5) (Turcek 1951, Greig 1979). The offspring of mixed-origin ibex (Austria, Turkey, and the Sinai) mated during the fall rather than the winter, as the original ibex had, and therefore their young were born in winter. These young perished, and the population disappeared.

Conservationists involved in direct manipulations are more often concerned about minimizing inbreeding rather than maximizing outbreeding. For example, grizzly bear biologists in the western United States are concerned that the grizzly bear population in and near Yellowstone National Park is so small and isolated that they it may suffer from lack of genetic diversity (Mattson and Reid 1991). Consequently, they have suggested that it may be desirable to translocate some grizzlies into Yellowstone from nearby, but isolated, populations to ameliorate this problem. This is a very artificial solution, but it may be more feasible than providing landscape connectivity that would facilitate dispersal among these populations, especially if only one successful transfer per generation is required (Mills and Allendorf 1996). In the case of the inbred viper population living on a small Swedish island, actively importing new genes seemed to be the only option (Madsen et al. 1999).

At the smallest level of detail, maintaining the genetic diversity of a population can involve regulating the reproductive fitness of specific individuals. Deciding "who gets to mate and with whom" is a routine part of captive-breeding programs, as we will see in the next chapter. It is more difficult to practice with wild populations because we seldom know the genetic makeup of any given individual and have little control over her or his behavior. One form of controlling the reproductive fitness of wild individuals could be practiced: removing (perhaps only temporarily) or killing individuals with undesirable characteristics to limit their contribution to the gene pool. Game managers in Europe often cull animals with undesirable characteristics such as small antlers. For populations of endangered species suffering from severe inbreeding depression, it might be useful to learn how to recognize individuals carrying deleterious recessive genes and then remove them from the population or sterilize them.

Finally, maintaining genetic diversity sometimes requires protecting genetic integrity—specifically taking steps to keep local alleles from being displaced by exotic alleles (Ellstrand and Elam 1993, Rhymer and Simberloff 1996). These steps would include controlling exotic taxa that may hybridize with local organisms (See Chapter 10, Exotic Species). In particular, rare plants that are exposed to large amounts of pollen from closely related common species may lose their genetic integrity and effectively disappear, thereby leaving genetic diversity as a whole diminished (Ellstrand 1992, Levin et al. 1996). Maintaining genetic integrity could also simply mean maintaining the "among-populations" (D_{st}) component of a species' genetic diversity by not breaking down any natural barriers that separate populations (Hogbin and Peakall 1999, Wolf et al. 2000). For example, it could spur conservationists to object to a proposal to connect two isolated lakes with a canal that would allow gene flow between their fish populations (Meffe and Vrijenhoek 1988) (see Fig. 5.1).

CASE STUDY   **The Black Robin**

The opening paragraph of this chapter does not do justice to the extraordinary program of hands-on manipulation that saved the black robin, so in this section we will delve a bit deeper; for the whole story, read Butler and Merton (1992). The history of the black robin begins in 1871 when the species was first described. By this time the robins were confined to Mangere (pronounced MANG'uree) and Little Mangere Islands. They may have once been found throughout the Chatham Islands, but forest clearing by Maori and European colonists and predation by cats and rats left them stranded on these tiny isles. Soon after their discovery they were gone from Mangere too, and for most of the 20th century they were clinging to survival in 9 hectares of forest perched on top of Little Mangere. Their fate seemed sealed when a helicopter landing was cleared on top of Little Mangere to allow people to collect sooty shearwaters for food, and afterward the remaining forest began to die off, apparently because of airborne salt intrusion.

The decision to move the last seven birds to Mangere in 1976 was not an easy one, in part because the program to restore forest on Mangere had not progressed far enough. Furthermore, it obviously was not a sufficient step because by 1979 only five robins remained. This is when the critical decision to undertake cross-fostering was made.

In the first cross-fostering experiments biologists moved robin eggs into nests of the Chatham Island warbler. The warblers proved to be capable egg incubators, but seemed unable to provide enough food for a robin chick. Consequently, robin chicks hatched by warblers had to be transferred back to robin parents for rearing, although not the chick's original parents, who were busy with a new clutch. The limitations of warblers as foster parents were avoided beginning in 1981, when some black robin eggs were

taken 12 km away to South East Island, where there was a population of another potential foster parent species, the Chatham Island tit. The tits proved capable both of incubating eggs and feeding robin chicks adequately, and after these tasks were completed, the chicks were returned to Mangere to join the rest of the robin population. It was later discovered that robin chicks had to have at least some experience being fed by robin parents lest they grow up confused about whether to mate with a robin or a tit. In sum, these manipulations involved translocation coupled with interspecific cross-fostering and then, about 2 weeks after hatching, translocation back with further intraspecific cross-fostering to avoid imprinting on the wrong species. Keeping track of who needed to go where and when was extraordinarily complex, particularly because all the robins and their potential foster parents were not nesting in perfect synchrony. Plastic eggs were often needed to substitute temporarily for the real things to keep parents at the right stage of reproductive activity.

Starting in 1983, adult robins were transferred to South East Island to establish a second population there. This gave the biologists the opportunity to manage the populations' genetic structure, for example, separating close relatives to prevent them from breeding. The black robins seem to have survived extraordinary inbreeding—virtually all of them are descendants of the female Old Blue and a male, Old Yellow—which is likely to have purged any deleterious recessive alleles, but this does not mean that further inbreeding might not be deleterious.

Besides moving eggs, chicks, and adults from nest to nest and island to island, the biologists helped the robins in other ways. They supplemented the diet of robins and surrogate parents by distributing insects at feeding stations. They provided the robins with better shelter by erecting artificial nest boxes and moving the birds' nests into artificial nest boxes when they were not used voluntarily. The nest boxes protected the robins from being crushed by seabirds blundering through the vegetation or evicted by starlings; they made egg transfers easier; and they facilitated control of a major problem, nest parasites. The biologists occasionally killed potential predators such as hawks (Australasian harriers), and they even killed tits to reduce competition for food before the tits' role as surrogate parents came into play.

It all worked. There are about 250 robins living on South East and Mangere Islands, thriving without regular human manipulations, and the future seems bright with plans to reintroduce them to two more islands (Mike Thorsen, personal communication). However, was it worth all the trouble? Some people might say no; and it is certainly true that the black robin is unlikely to contribute to the economic well-being of humanity. On the other hand, the black robins, especially Old Blue, have become an inspiring symbol of what dedicated conservation biologists can accomplish.

Epilogue

I am sure many readers are rather appalled by some of the heavy-handed population management techniques described in this chapter. A part of me is. On the other hand, drastic situations call for drastic solutions. Given a choice between manipulating the lives of a few individual plants or animals or standing aside to watch the evaporation of a long river of evolution, involving millions of years and billions of individuals, I will choose the former. I am not willing to say "Let them die in peace." This said, we do need to question these methods because too often they are merely stanching the flow of blood rather than repairing a severed artery; they are dealing with symptoms rather than their root causes. We will return to this difficult issue in Chapter 14, Setting Priorities.

Summary

If a population is at great risk of extinction, it may not be sufficient to maintain the ecosystem it inhabits. It may be necessary to manage the population more directly by providing resources, controlling direct threats, and undertaking other manipulations. Providing resources can mean supplying food (broadly defined to include energy and nutrients for animals and plants), water, a physical environment (e.g., shelter from climatic extremes and concealment from other organisms), and key interactions with other individuals. Usually, the most critical threat to minimize is human exploitation. This may involve eliminating harvests or at least tightly controlling them so that they represent compensatory mortality not additive mortality. Many indirect human threats (e.g., vehicle collisions, electrocutions) can be avoided or mitigated by changes in the design of human-made structures. Sometimes, it is necessary to control predators, grazers, competitors, parasites, or pathogens that are diminishing an endangered species' chance of survival. Eradicating these species may be desirable, albeit difficult, if they are exotics. Conservationists may decide to translocate endangered species to ecosystems where they have been extirpated (reintroduction), where they are not native (introduction), or where their populations are depleted (augmentation). These techniques are often combined with other direct manipulations such as double-clutching, cross-fostering, head-starting, and hatchery raising that are designed to increase reproductive success. All of these methods are expensive and full of risks and thus best avoided if possible. These methods also raise many questions about how best to maintain genetic diversity. For example, how important is it to maintain the genetic integrity of a population? Also, should outbreeding be used to maximize the genetic diversity of reintroduced populations? The best strategy for maintaining genetic diversity is to have a substantial number of individuals comprising many different populations that occupy the species' entire geographic range.

FURTHER READING

No single book adequately covers managing populations of endangered species, although there are some useful collections of papers for certain taxa such as plants (Elias 1987, Given 1994), desert fishes (Minckley and Deacon 1991), and birds (Temple 1977). Books on certain subtopics of the issue have also been compiled such as plant reintroductions (Falk et al. 1996) and wild life disease (Fowler 1993). Sutherland's (2000) *The Conservation Handbook* is particularly strong for population monitoring and related techniques.

TOPICS FOR DISCUSSION

1. Would you be willing to eliminate a common species of predator from an island where it is native so that a rare species (one of its

prey) could be reintroduced to the island? Would you be willing to eliminate a native plant to permit the reintroduction of a rare competitor? Would you be willing to eliminate a native parasite or pathogen to permit the reintroduction of its host?

2. Imagine that you have been intensively managing a small population for 20 years (providing food, controlling predators, augmenting the population from captive stocks, etc.) and that the population is still highly dependent on your assistance and shows no sign of becoming self-sufficient. Would you consider terminating the program and allowing the population to disappear? Would your decision be different if this was the last wild population (but the species was secure in captivity)? Would it be different if this were absolutely the last population, wild or captive?

3. Would you be willing to permit human exploitation of any common species for which human-induced mortality is compensatory for natural mortality? Why or why not?

4. Imagine that you are managing a genetically distinctive population that, if current problems with inbreeding continue, will be extinct in 5 years. Would you augment the population now with individuals from elsewhere, thereby solving the inbreeding problem but compromising genetic integrity? Or would you wait, hoping that the population might recover by purging itself of deleterious alleles? (Assume that if you take the second alternative, you have an 80% probability of losing the population entirely, but you are 90% confident that you can replace it with a reintroduction from a different population.)

Chapter 13
Zoos and Gardens

In New York, Paris, Delhi, and most other large cities one can find small oases of extraordinary biotic diversity: a patch of tropical forest crowded with orchids and ferns, a coral reef seething with life in a panoply of colors that defies description, or a room reverberating with the calls of birds gathered from far and wide. These are, of course, our zoos, aquariums, and botanical gardens. They are amazing places where skillful husbandry maintains some of nature's most awesome creatures, surrounded by humanity far from their native haunts. These institutions have long served many purposes—recreation, education, research—and in recent years they have become centers for a specialized form of biodiversity conservation, *ex situ* conservation. *Ex situ* conservation is conservation that takes place outside of a species' natural habitat; it contrasts with *in situ* conservation, which takes place within a species' natural habitat. The bulk of this chapter focuses on *ex situ* conservation of wild species, but one section covers a parallel undertaking, conservation of domesticated species. We will begin with an overview of the traditional roles of zoos, aquariums, and botanical gardens.

Changing Roles

Wealthy people such as King Solomon, Montezuma, Louis XIV, and Michael Jackson have collected exotic creatures for millennia, driven by the same basic motivation that drives a stamp collector: it is an entertaining diversion from day-to-day life. Recreational values remain paramount today when most zoos, aquariums, and botanical gardens have become public institutions and, consequently, depend on gate receipts and the good will of taxpayers. This means that they have to be enjoyable places to visit, and, apparently, they are. For example, each year United States' zoos and aquariums alone attract over 100 million visitors, more than the combined attendance at all United States' professional baseball, basketball, hockey, and football games, (Nelson 1990), while global figures are around 600 million, 10% of the earth's population (Sunquist 1995).

There was a time when entertaining people was the only important goal for zoos, aquariums, and botanical gardens, when watching a sea lion balance a ball or chimpanzees attend a tea party were the highlights of a visit. Today, such shows are quite uncommon, and far more emphasis is placed on public education (Fig. 13.1) (Robinson 1988). Organisms are often displayed in settings that simulate their natural habitat, and, occasionally, in multispecies groups. Most exhibits are accompanied by signs that both identify the species on display and describe its natural history. If the species is at risk of extinction, this is usually emphasized, and there is likely to be detailed information about its plight and what is being done to save it. More and more, the central theme of these institutions is conservation, and they are using many channels to communicate this message beyond signs next to exhibits: publications, lecture series, visits to schools, and interaction with the local mass media, to

Figure 13.1. Public zoos, aquariums, and gardens have long emphasized educating visitors, as well as entertaining them.

name but a few. It is difficult to judge how much enthusiasm for biodiversity has been engendered from seeing a segment on the local TV news about the city zoo's newborn gorilla or from viewing a herd of Pere David's deer and reading a sign that the species no longer exists in the wild. This support is probably quite significant, especially given that roughly half of the world's people live within an hour's travel of a zoo, aquarium, or botanical garden. If these institutions did nothing else for biodiversity beyond giving people a tangible link between themselves, the family of pygmy marmosets behind a pane of glass, and the fate of the tropical forests that harbor the remaining pygmy marmosets, their role would be very praiseworthy.

Modern zoos, aquariums, and botanical gardens employ biologists, and if you give biologists daily access to little-known species from all over the globe, they will learn many new things. To put it more directly, conducting scientific research is an important role for these institutions. For many species our understanding of their physiology, diseases, reproductive biology, nutrition, and so on has come primarily from studies on captive populations. Naturally, there are limits to what we can learn about a species outside its natural habitat, especially about ecological interactions. Nevertheless, any information is better than none, and some things learned with captive populations have been vital. For example, capturing wild animals and equipping them with radio transmitters, a key part of studying many wild populations, would be exceedingly dangerous for the animals (and sometimes the researchers) without tranquilizers, many of which were tested and refined in zoos and aquariums. Indeed, the radios themselves and modes of attachment are often first tested in zoos and aquariums. The idea that inbreeding could be a problem for wild animals living in small, isolated reserves was largely generated by detailed analyses of zoo breeding records that revealed that many mammal species manifested inbreeding depression (Ralls and Ballou 1983) (see Fig. 5.4).

In emphasizing the role of zoos, aquariums, and botanical gardens as centers for conservation education and research, I do not mean to ignore all the other institutions that pursue these goals: environmental education centers, natural history museums, universities, and a vast array of other governmental and nongovernmental organizations. The education and research roles of zoos, aquariums, and botanical gardens have been highlighted here because they are often overshadowed by the well-known *ex situ* conservation programs that we will examine next.

Building Arks

Noah's ark has become an overused metaphor for *ex situ* conservation, but it does convey the simplest, most profound justification for the practice:

Figure 13.2. Some species that would probably be extinct today without *ex situ* conservation. Most disappeared from the wild for some period; in some cases (e.g., the nene) a few individuals persisted in the wild, but probably in numbers too small to be viable. Clockwise from the center top they are wisent, red wolf, nene goose, viviparous tree snail, Przewalski's horse, Guam rail, Pere David's deer, *Paphiopedilum delenatii*, black-footed ferret, California condor, Arabian oryx, *Tecophilaea cyanocrocus*, and, in the center, *Franklinia alatamaha*.

many species would not exist today if they had not been taken from the wild and kept in captivity (Fig. 13.2). To be more precise, in 2000 the World Conservation Monitoring Centre's website listed 17 species of plants and 34 species of animals that survived only in captivity. The duration of captivity has varied enormously. Pere David's deer disappeared from the wild nearly 3000 years ago and is not likely to be reintroduced to its natural habitat because the swamps it inhabited in northeastern China were all converted to agriculture long ago (Fisher et al. 1969). In contrast, the black-footed ferret was removed from the wild for less than 5 years, from February 1987 to September 1991, following an outbreak of distemper (Thorne and Williams

1988, Dobson and Lyles 2000). Similarly, we can think of captive populations as insurance against the future loss of wild populations. Certainly, the long-term fate of three species of rhinoceros (black, white, and Indian) is more secure because there are captive populations that are not subject to the poaching that threatens wild populations. A captive population of the Leon Springs pupfish became the only genetically pure form of the species after the wild population was "contaminated" through hybridization with an exotic minnow (Echelle and Echelle 1997).

The other side of the coin is shown in Figure 13.3, which depicts some animals whose last-known individual died in captivity. The most famous of these is Martha, the last passenger pigeon, who died in the Cincinnati Zoo on September 1, 1914, about 10 years after her species had vanished from the wild, less than a century after her kin had numbered in the billions (Schorger 1973). Perhaps some of these species could have been saved by modern *ex situ* husbandry but certainly not all. The last dusky seaside sparrow died in captivity in 1987, despite the advances of modern aviculture (Avise and Nelson 1989). Furthermore, we must remember that there are large numbers of species—blue whales, ivory-billed woodpeckers, and

Figure 13.3. The quagga, thylacine, passenger pigeon, dusky seaside sparrow, and pink-headed duck apparently became extinct when the last known individual died in captivity.

many more—that have never even been maintained in captivity, much less bred there, and likely never will be. The limitations of *ex situ* conservation are particularly evident when we recall that the vast bulk of biodiversity resides with insects and other small life-forms that are seldom kept in captivity. Financial limitations must be considered too; for example, it has been estimated that the capacity of the world's zoos could sustain fewer than 1000 of the over 20,000 species of mammals, birds, reptiles, and amphibians (Conway 1986). Such shortcomings are not an excuse to abandon *ex situ* conservation, but they have to shape a realistic evaluation of its feasibility vis-à-vis *in situ* conservation; we will return to this topic later.

An exhaustive treatment of the techniques used by *ex situ* conservationists lies beyond our purview. Suffice it to say that the technology has become quite sophisticated and that "building spaceships" might be a more appropriate metaphor than "building arks." We will briefly cover just two topics: first, controlling who mates with whom; and second, storing biodiversity in the form of seeds, sperm, embryos, tissue, and similar material.

Studbooks and Pedigrees

A long history of breeding domesticated plants and animals has provided *ex situ* conservationists with a starting point from which they can develop their efforts to maintain captive populations of wild species. Of course, artificial selection techniques—the process by which people have produced roses in myriad colors and dogs in myriad colors, shapes, and sizes—are of little interest to *ex situ* conservationists. Usually, they try to avoid artificial selection on the assumption that someday a portion of their stock will be returned to the wild and subjected to natural selection again. One could argue that two different types of captive breeding should be undertaken: one to maintain populations for exhibition and a second to produce stock for reintroductions. If this distinction is made, then some artificial selection that leads to limited domestication is desirable for populations destined to be kept in captivity but best avoided for those chosen for reintroductions. Unfortunately, it is impossible to avoid totally any selection for domesticity simply because individuals that do not acclimate to confinement to some degree will not produce any offspring (Ashton 1988, Derrickson and Snyder 1992).

Also of great interest to *ex situ* conservationists are practices designed to maximize the retention of genetic diversity. Such practices can be relatively simple (e.g., preventing siblings from mating with one another) or very elaborate, depending on the type of genetic variation to be managed. Most practices require keeping track of each individual's ancestors (their pedigree or lineage). However, when the number of individuals becomes quite large and they are distributed among institutions around the world,

Studbookno	ISIS No.	Sex	Arrival	Birth Date	Birth Place	Sire No	Dam No	Death Date	Zoo
SM/CE 7525	017S	M		01-JUN-75	CHENGDU	UNKN	UNKN	11-JAN-88	NANJING
SF/CE/7526	018S	F		01-JUN-75	CHENGDU	UNKN	UNKN	21-FEB-85	NANJING
SM/00/7603		M	07-JUL-76		WILDCAUGHT			06-APR-79	TORONTO
SM/NN/7621	098S	M		30-JUN-76	NANNING	UNKN	UNKN		NANNING
SF/00/7722		F	29-DEC-77		WILDCAUGHT			22-JUN-84	SYDNEY
SM/00/7723		M	02-NOV-77		WILDCAUGHT			03-APR-78	SYDNEY
SF/CH/7729		F		01-JAN-77	CHINA	UNKN	UNKN	29-JUL-82	MELBOURNE
SF/CH/7819		M		01-JAN-78	CHINA	UNKN	UNKN	O5-DEC-88	AUCKLAND
SF/CQ/7897	019S	F		01-JAN-78	CHONGQING	UNKN	UNKN		NANJING
SM/00/7921		M	25-JUN-87		WILDCAUGHT			17-MAY-88	AUCKLAND

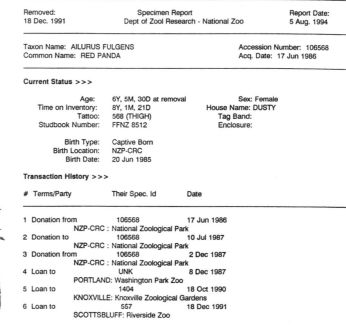

Removed: 18 Dec. 1991	Specimen Report Dept of Zool Research - National Zoo	Report Date: 5 Aug. 1994

Taxon Name: AILURUS FULGENS   Accession Number: 106568
Common Name: RED PANDA   Acq. Date: 17 Jun 1986

Current Status >>>

Age:	6Y, 5M, 30D at removal	Sex:	Female
Time on Inventory:	8Y, 1M, 21D	House Name:	DUSTY
Tattoo:	568 (THIGH)	Tag Band:	
Studbook Number:	FFNZ 8512	Enclosure:	

Birth Type:	Captive Born
Birth Location:	NZP-CRC
Birth Date:	20 Jun 1985

Transaction History >>>

#	Terms/Party	Their Spec. Id	Date
1	Donation from	106568	17 Jun 1986
	NZP-CRC : National Zoological Park		
2	Donation to	106568	10 Jul 1987
	NZP-CRC : National Zoological Park		
3	Donation from	106568	2 Dec 1987
	NZP-CRC : National Zoological Park		
4	Loan to	UNK	8 Dec 1987
	PORTLAND: Washington Park Zoo		
5	Loan to	1404	18 Oct 1990
	KNOXVILLE: Knoxville Zoological Gardens		
6	Loan to	557	18 Dec 1991
	SCOTTSBLUFF: Riverside Zoo		

Figure 13.4. The top part shows a small section of the studbook for red pandas, and below is an ISIS entry for one individual red panda. (Provided by Miles Roberts, National Zoological Park, Washington, DC.)

keeping accurate records is a significant logistical undertaking. See Figure 13.4 for a sample. Despite the barriers, *ex situ* conservationists have organized themselves to create studbooks (pedigree records) for many mammals and birds, and a few reptiles, amphibians, fishes, and invertebrates. There are an estimated 800 to 1000 studbooks in use, but far fewer species are covered because most studbooks cover only the animals kept in a region such as Europe, North America, or Australasia rather than the whole global captive population (Nate Flesness, personal communication). (In other words, many species have multiple studbooks: one for Europe, one for Australasia, etc.) An even larger database exists within ISIS, the International

Species Inventory System, which covers roughly 7500 taxa from 550 zoos (www.isis.org). *Ex situ* plant conservationists are in the process of developing databases for wild plants, but they are not likely to be studbooks per se. This is partly because the breeding systems of plants are such that the "who mates with whom" question is often difficult to answer (consider wind-pollinated plants) and/or less relevant (many plants routinely fertilize themselves or reproduce vegetatively). *Ex situ* wild plant conservationists are often more concerned with keeping closely related taxa from breeding with one another in the artificial proximity of a garden (Ashton 1988).

Even with a studbook, maintaining genetic variation is more easily said than done because most institutions only have a small population of any given species. This forces *ex situ* conservationists to exchange breeding stock regularly despite the risks and expenses of shipping organisms from place to place. In the future it is possible that sperm, rather than whole animals, will be shipped among institutions. This is already the case with gaur, a species of Asian wild cattle (Wiese and Hutchins 1994), but to date artificial insemination techniques have been developed for only a relatively few wild animals (Holt 1994, Holt et al. 1996). Techniques for transferring embryos among individuals of wild species are also being developed. Embryo transfer has even been undertaken between different species, a sort of uterus-to-uterus cross-fostering (e.g., a gaur calf has been born to a domestic cow and a bongo calf to an eland) (Dresser 1988) and, most dramatically, a cow has given birth to a cloned gaur. (For a collection of articles about the brave new world of reproductive technology see the special issue of *Theriogenology* 53[1].) Many zoo biologists remain skeptical concerning whether these high-tech approaches can be developed and effectively implemented for a large suite of species.

In the early days of *ex situ* conservation, breeding programs' efforts focused on avoiding inbreeding. For example, in 1982 zoos in the United States and the former Soviet Union exchanged Przewalski's horses to reduce the extent of inbreeding in both countries (Ryder 1993). However, pedigree information allows *ex situ* conservationists to go well beyond avoiding inbreeding. They can enhance their attempts to maintain genetic diversity by using pedigree information to calculate a measure of relatedness called "mean kinship." Using mean kinship values to decide who should mate with whom helps equalize the distribution of each ancestor's genetic contribution and thereby maintain genetic diversity. That statement is almost as complex as the procedure to which it refers. A simplified explanation will be adequate for our purposes: consider the thousand plus Przewalski's horses now alive, all of which can trace their ancestry back to 13 founders. Imagine that 60% of the members of the current population are direct descendants of mare A and that only 5% are direct descendants of mare B. Using mean kinship values to determine pair formation, more descendants of mare B will be paired for breeding in the future because her genes are underrepresented in the population. There is an unfortunate side effect of these careful breeding programs; large numbers of animals (many of the descendants of mare A in this example) need to be removed from the breeding program, and keeping them alive uses up scarce resources. For example, Sunquist (1995) reported that there were 88 "surplus" orangutans in North American zoos and that keeping them alive through their normal life expectancy would cost $3.8 million. For further information on the procedures used to make these assessments and related techniques for managing *ex situ* populations, see Willis and Wiese (1993), Ballou and Lacy (1995), and Lacy et al. (1995). For an example of how some related procedures shaped a management program for Guam rails, see Haig et al. (1990). To facilitate cooperation among institutions, there are regional programs that analyze studbook data to determine the best pairing; to review one, the Species Survival Plan of the American Zoo and Aquarium Association, see Wiese and Hutchins (1994).

Storing Biodiversity

Zoos, aquariums, and botanic gardens require a great deal of complicated and expensive maintenance, particularly when they are trying to keep large, demanding species like black rhinos, killer whales, and coco-de-mer palms. *Ex situ* conservation would be much easier if species were small, immobile, and did not need to be fed or watered. This kind of thinking has led to *ex situ* conservation techniques directed toward life-history stages that are amenable to storage, particularly microbes, plant propagules (seeds, spores, and vegetative parts), and the sperm and embryos of animals.

Microbes: Most conservation biologists view conservation of microorganisms as an invisible enterprise, riding on the coattails of conservation directed at ecosystems. Nevertheless, microbiologists do strive to maintain the diversity of microorganisms, and their primary technique is *ex situ* storage (Kirsop and Snell 1984). The most common technique is freeze-drying samples, which involves rapid cooling, sealing under vacuum pressure, and then storage at temperatures between $5\,°C$ and $-70\,°C$. For some species cryopreservation is more effective. *Cryopreservation* means storage at extremely low temperatures, commonly in liquid nitrogen ($-196\,°C$) or its vapors ($-150\,°C$). Long-term storage of microbes is usually considered preferable to culturing them continuously for two reasons: 1) it is cheaper, and 2) microbial organisms evolve so rapidly that they would be very different after a period of being cultured.

Animals: Cryopreservation of semen and embryos has become a routine procedure for domestic mammals; for example, every year millions of cows, horses, sheep, etcetera are artificially inseminated using frozen sperm. This technology is more at the experimental stage for wild mammals, and especially birds and other animals, but it does hold some promise as a future method of long-term storage (Grout et al. 1992; Holt 1992, 1994; Holt et al. 1996; Karow and Critser 1997).

Plants: Storage is relatively straightforward for plant species that have seeds (often called *orthodox seeds*) that remain viable when exposed to cold, dry conditions that reduce metabolic activity. For these species, seed longevity can be greatly increased by storing the seeds in chambers that are dry (5% moisture content is often used, but lower might be preferred) and cold (temperatures ranging from merely cool to –196 °C are used) (Frankel et al. 1995, Ellis and Roberts 1998). People have long used cool, dry conditions to store seeds needed for agriculture; for example, O'odham farmers of southwestern North America stored seeds in sealed pots placed in desert caves (Nabhan 1989). On the other hand, some species have seeds (*recalcitrant seeds*) that cannot tolerate desiccation or freezing. Still other species usually reproduce vegetatively and rarely produce seeds. For these species, maintaining pollen, plantlets, and tissue samples may be feasible. Several major institutions are dedicated to storing plant material, but the vast bulk of the effort is directed toward domesticated plants. Only a small portion of wild species are adequately represented in seed banks, but this number is increasing quite rapidly (Mattick et al. 1992, Frankel et al. 1995), despite criticism of this approach (Hamilton 1994).

One drawback to storage techniques has been particularly apparent with seed banks, especially those that do not use cryopreservation. The viability of seeds deteriorates through time, and thus it is necessary to periodically remove them from storage, grow new plants, and then harvest and store the new seeds; this process, an expensive and time-consuming one, is called *growing-out*.

Genetic Material: Storing genetic material is relatively straightforward; one can either freeze tissue at –70 °C and extract DNA at a later date or extract and purify DNA now and store it at room temperature in vials of inert gases (Ryder et al. 2000). This material can provide useful information about the genetic composition of a species in perpetuity. Furthermore, in the wake of the movie "Jurassic Park," and the cloning of a sheep named Dolly and a gaur named "Noah," we must also acknowledge the possibility that scientists may one day be able to reconstruct extinct species from small fragments of DNA (Bawa et al. 1997). The odds of doing this in the foreseeable future are extremely slim, probably very close to zero, with dinosaur genes that have deteriorated for over 60 million years. The odds are somewhat better for a species like the wooly mammoth for which we have tissue, found frozen in Siberian permafrost, that is only thousands of years old, or for an organism such as the quagga that is only recently extinct (Higuchi et al.

1984). Of course, the probability of reintroducing a species "raised from the dead" into a wild ecosystem is even more vanishingly small. When reading the popular press, one sometimes gets the impression that with the advent of cloning and other genetic techniques we can relax a bit in our struggle to save the giant panda and other species, but this prospect shows a stunning ignorance of ecological realities.

The *Ex Situ–In Situ* Interface

Few people are content with the idea that species like black rhinos and California condors could survive in perpetuity in captivity. Ideally *ex situ* conservation is just a stopgap technique until a species can be reintroduced to its native range, after the problems that plagued it have been remedied. Unfortunately, this is easier said than done (Gipps 1991). As you will recall from our discussion of translocations in the preceding chapter, failure rates are high, especially when captive stock is used. Reintroductions have been attempted, or are currently under way, for most of the species depicted in Figure 13.2, but to date none of these has produced an unqualified success, that is, the creation of a secure, free-living, self-sustaining population within its native range and habitat. European bison (or wisent) come close to meeting this definition, but their wild populations are given some special care such as winter feeding (Simberloff 1988). The efforts to reintroduce Arabian oryx to Oman, described below as a case history, were initially successful. The bottom line is that, while it may be possible to reintroduce a species to the wild after it has been confined to captivity for a few generations, it is never easy.

Augmenting wild populations with captive-bred individuals is also a possibility. For example, if the cheetahs of an isolated reserve were known to be suffering from inbreeding, a captive-reared cheetah with a different genotype could be added to the population. Again, ensuring that it survived and became part of the local population is easier said than done, and the specter of introducing a disease always looms. Finally, some *ex situ* conservationists envision a day when they will routinely introduce genetic material from a captive population to a wild population by transferring pollen, sperm, or embryos (Luoma 1987).

The *ex situ–in situ* path is a two-way street. Although many of the inhabitants of zoos, aquariums, and botanical gardens have been reared in captivity, some were removed from wild populations, and occasionally this includes endangered species. Sometimes, endangered species are removed from the wild by conservationists because their chances of survival in the wild are too low; this was the case with black-footed ferrets and California condors. Sometimes, they are removed because they are needed to bolster captive populations and to avoid inbreeding. For example, Nepal exported some Indian rhinos to overseas zoos to increase the genetic diversity of the world's captive population. Ideally, this would be done with individuals who are not breeding members of viable populations (e.g., orphans and individuals whose habitat has been destroyed), but this is typically not the case. In one well-documented case, publicity about establishing a captive-breeding program for the babirusa (a member of the pig family confined to the island of Sulawesi) generated a black market in the species among people hoping to sell animals for the program (Clayton et al. 2000).

All of this requires that *ex situ* conservationists keep their finger on the pulse of what is happening to wild populations and make decisions that complement *in situ* conservation. For example, Foose (1983) noted that in 1980, 539 of the world's 797 captive rhinos (68%) were the southern subspecies of the white rhinoceros, even though this is the only kind of rhino that is moderately secure in the wild. Therefore he proposed that the world's capacity for holding captive rhinos should be directed toward the other species. The good news is that by 2000 the world's capacity for *ex situ* rhino conservation had expanded to support 1100 animals; the bad news is that 730 (66%) of these were still southern white rhinos. Complementing *in situ* conservation has also meant that conservation-sensitive zoos have abandoned the old objective of exhibiting as many different species as possible. They try to focus on a few select taxa whose wild populations can be helped the most through holistic *ex situ* programs that incorporate maintaining healthy populations, education, research, and direct support of *in situ* conservation projects (Hutchins and Wiese 1991, Hutchins et al. 1995). Unfortunately, while the trend is in the right direction, most zoos and aquariums still allocate a disproportionate share of their resources to the species that are deemed most likely to attract the public—notably large mammals—even though these species are usually the most expensive to keep and are difficult to breed. At least one analysis has shown that visitor preferences are not really so narrow, and it is quite feasible to allocate resources in a fashion that would better complement *in situ* conservation (Balmford et al. 1996).

The Controversial Side of Ex Situ Conservation

Ex situ conservation is often highly controversial. Some people do not like it for ethical reasons (Bostock 1993, Norton et al. 1995). They would rather see a species slip into extinction with dignity rather than be subjected to high-tech meddling that will expose some members of the species to the tribulations of captivity. These feelings rise up particularly with animals; it is not obvious how these ethical arguments would play out for a tree like the toromiro, confined to captivity since 1960 with only a remote chance of being reintroduced to the wilds of Rapa Nui (Easter Island) (Maunder et al. 2000).

More common are criticisms by *in situ* conservationists who feel that *ex situ* conservation is too focused on a minority of species, too expensive, and too risky (because of the high incidence of diseases in captivity and the poor success rate of reintroducing captives to the wild) (Snyder et al. 1996). Perhaps most problematic is the danger that captive breeding can become a smokescreen to obscure solutions to the real problems. For example, it has been claimed that the U.S. Fish and Wildlife Service found it easier to support captive-breeding projects for black-footed ferrets and California condors than to tackle the politically difficult problems of prairie dog eradication and lead poisoning, respectively (Snyder et al. 1996). The recent decision by the Chinese government to attempt to clone giant pandas could be seen as an attempt to avoid the thorny issue of habitat conservation.

Arguably the most controversial *ex situ* conservation plan involves the Sumatran rhino, one of the most highly endangered large mammals because of poaching and loss of habitat (Fig. 13.5). In 1984 *ex situ* conservationists initiated a program to establish a captive population, and, ultimately, 40 rhinos were caught and held in captivity at a cost of millions of dollars (T. Foose, personal communication). To date, they still have not bred in captivity, and all but 17 of the 40 have died. Not surprisingly, many critics of *ex situ* conservation have argued that the time and money should have been spent on better management of the remaining Sumatran rhino habitat, particularly because all the other species that share this habitat would have also benefitted from an *in situ* approach (See Rabinowitz 1995 and responses in *Conservation Biology* 9[5].)

Figure 13.5. The Sumatran rhinoceros is highly endangered, and this led to a concerted effort to breed them in captivity, an effort that has proven fruitless to date.

Supporting In Situ *Conservation*

One way to lubricate the friction that can exist between *in situ* and *ex situ* conservationists is for zoos, aquariums, and botanical gardens to become more directly involved in *in situ* conservation and thus escape the constraints of the Noah's ark metaphor. Some of the largest institutions (e.g., the Missouri Botanical Garden, New England Aquarium, Bronx Zoo, and Frankfurt Zoo) have their own field conservation units operating in many parts of the globe. Others have formed a special relationship with a particular reserve. For example, the Minnesota Zoo, which has a large exhibit for Southeast Asian animals, supports conservation programs in Udjong Kulon National Park in Indonesia.

The simplest idea is for zoos, aquariums, and botanical gardens to raise funds for other organizations that undertake *in situ* conservation. Large sums could be raised if each visitor were charged for two tickets, one for regular admission and a second "conservation ticket" costing perhaps 20% of the regular admission. By putting their conservation tickets in collection boxes around the grounds, visitors could direct their contribution to a particular project (e.g., marine mammal conservation next to a seal pool, support for a conservation group in Mexico in a cactus greenhouse, or a "Special Fund for All the Ugly Creatures That Usually Are Ignored" next to a crocodile exhibit).

In the best of all possible worlds it would never be necessary to attempt risky and expensive reintroductions of captive plants and animals. They could stay in captivity, leading safe and sheltered lives and serving as ambassadors for conservation education and research. This scenario is a bit more likely if zoos, aquariums, and botanical gardens become major supporters of *in situ* conservation, not just instruments of last resort.

Conservation of Domesticated Species

With our omnivorous digestive systems, there are many thousands of species we could eat, but only a few hundred are consumed routinely, and only a tiny handful of plants (e.g., wheat, rice, corn, soybeans, potatoes) comprises a large portion of our diet (Prescott-Allen and Prescott-Allen 1990). Our dependence on a few domesticated species has led us to lavish a great deal of attention on them, and this is most apparent in all the varieties we have produced through artificial selection (Fig. 13.6). Some of this genetic diversity can be seen during a trip to the grocery store, amongst the apples and squashes, for example. However, to really learn about this diversity you need to visit a farm and talk to farmers about the varieties that they select for growing. (If they are growing plants, they may use the term *cultivar* for variety; animal farmers are likely to use the term *breed.*)

Figure 13.6. Many domestic plant species come in a startling variety because of the efforts of farmers and plant breeders. Much of this diversity is maintained in seed banks, although some still exists in farmers' fields.

Farmers select varieties that will produce good yields as well as meet consumer preferences. For most farmers living in industrialized countries, this means selecting a variety that will perform well in an environment intensively manipulated with fertilizers, insecticides, herbicides, and perhaps irrigation. Other farmers cannot afford these inputs or simply prefer the low-input style of agriculture known as sustainable agriculture (recall our discussion in Chapter 11). These farmers need to select varieties that will thrive with the local climate, soil, and assemblage of potential pests, producing crops that are reasonably large and have a low risk of failure. Both of these types of farmers need to be concerned with maintaining the genetic diversity of domesticated species (Fuccillo et al. 1997, Maxted et al. 1997, Virchow 1999).

High-tech, high-input farmers need genetic diversity as the basis for developing new varieties that are adapted to ever-changing technologies (e.g., varieties of plants that can withstand more potent herbicides or that lend themselves to mechanical harvesting), ever-changing environments (e.g., insects that have become immune to certain insecticides), and ever-changing consumer preferences. Most plant breeders cater to these farmers, and they have established an international network of repositories for genetic material (often called *germplasm)* (Fuccillo et al. 1997). Typically, these consist of an *ex situ* storage facility or seed bank, as described earlier in this chapter, plus some nearby fields where plant varieties can be cultivated either continuously (for species with recalcitrant seeds) or periodically (for species with orthodox seeds that need to be grown out).

Farmers using traditional practices need genetic diversity in the form of a diverse array of local varieties, usually called *landraces.* This is because local varieties are more likely to be adapted to local conditions and not to need substantial inputs of fertilizers and pesticides (Frankel et al. 1995). Landraces can be maintained in regional germplasm repositories, but there are some disadvantages to this. First, over time a landrace will evolve in response to the conditions at the germplasm center rather than to conditions at its site of origin. This is especially true of recalcitrant seeds that may need continuous cultivation. Second, germplasm material stored at a distant repository is not very accessible to a small farmer.

One way to maintain landraces is to keep them on the farms where they were first developed, essentially *in situ* conservation for domestic species (Oldfield and Alcorn 1987, Vaughan and Chang 1992, Brush 1995, Brush et al. 1995). The problem with this approach is that local farmers are often under considerable economic pressure to replace their local varieties with high-yield varieties. Consequently, it may require a program of financial subsidies and other forms of cooperation to encourage them to continue growing a landrace. Such programs would be a worthwhile investment for high-tech agriculture because the pool of genetic diversity in landraces is an important resource for breeders trying to develop high-yield varieties. Indeed, allowing landraces to disappear undermines the very foundation of genetic diversity of domestic species. Moreover, these programs may be particularly useful because landraces often occur within the native range of wild relatives of the domesticated species, and these are an additional source of genetic material.

Efforts to maintain the genetic diversity of domesticated animals have focused primarily on the studbook and pedigree approach described above and on preserving germplasm from individuals known to have desirable qualities. This work has largely been limited to a few major breeds, but in recent years many people have developed a keen interest in saving rare, local breeds (Jewell 1985, Alderson 1990, Hall and Ruane 1993). To date the maintenance of rare breeds of animals is closer to being a hobby such as collecting living antiques than to a mainstream undertaking supported by the agricultural establishment, but perhaps this will change. At least there are proposals to take a systematic approach to prioritizing breeds for conservation (Ruane 2000).

Finally, we must mention some species that are in danger of falling between the cracks of conservation focused on wild and domestic species: the wild relatives of domestic species. Actually, plant conservationists do a fairly decent job of tracking down these populations and at least collecting their seeds for storage if not undertaking *in situ* conservation (Tewksbury et al. 1999), but the animal conservationists seldom turn their attention to the wild ancestors of pigs, chickens, etcetera (Brisbin 1995).

CASE STUDY   The Arabian Oryx

The political strife of the Middle East tends to color our view of the whole region, and thus many people would be surprised to learn that this is home to a fine model of international cooperation to save an endangered species, the Arabian oryx. Four large antelopes—the Arabian oryx, scimitar-horned oryx, gemsbok, and addax—roam throughout the arid regions of Africa and the Middle East, or, rather, they used to. Human overexploitation, especially since the advent of motorized vehicles capable of taking parties of hunters far into the desert and outrunning herds of antelope, has left all but the gemsbok in grave danger of disappearing from the wild. Indeed, for several years the Arabian oryx did disappear from the wild, and that is the basis of our story summarizing a book by Mark Stanley Price (1989) and later papers by Ostrowski et al. (1998) and Spalton et al. (1999).

Arabian oryx were once found throughout most of the Arabian peninsula, but by the mid 1960s they were confined to a small area of central Oman, and on October 18, 1972, the last known wild herd was eliminated when three animals were killed and three captured alive. Fortunately, 10 years earlier, in 1962, the Fauna and Flora Preservation Society (a British-based conservation group that publishes a journal called Oryx) had launched Operation Oryx to capture some Arabian oryx and start a captive population. Their expedition produced three oryx, which they took to northern Kenya because of the climatic similarity. Over the next 2 years various negotiations netted six more oryx from captives held in London (1), Kuwait (1), and Saudi Arabia (4), and all nine oryx were shipped to the Phoenix, Arizona, zoo. Here the climate was fairly similar to the Arabian peninsula, and the threat of hoof-and-mouth disease, a major risk in Kenya, was minimal. The oryx thrived and reproduced in Phoenix, and beginning in 1972 some individuals were transferred to other United States' zoos to minimize the risk of having all the animals at one site. By 1978 oryx were being shipped back across the Atlantic, and by 1986 there were herds in Morocco, ten Middle Eastern countries, and four European countries totaling over 700 animals. The success of the captive-breeding program may be the result both of good luck (the founder animals probably were not closely related to one another) and of careful breeding management designed to ensure that the genomes of all of the founders were well represented in later generations.

From its inception, the goal of Operation Oryx was to return the Arabian oryx to the wild eventually, and the first steps toward making this a reality began in 1978 when Arabian oryx from the United States were released into large, natural enclosures in Jordan and Israel. A couple of years later a more ambitious and, ultimately, more successful reintroduction project began in Oman. Eighteen oryx were imported into Oman from 1980 to 1984, and 16 of these were integrated into two separate herds with a reasonably natural sex and age composition (two were judged unfit for release). The animals were held at the release site in small pens and then in a 100-ha enclosure for about 1 to 2 years to acclimate them to the area and to one another. After release they were monitored very closely by teams of rangers following in vehicles, a strategy that both protected the oryx from poachers and generated detailed information. The oryx seemed to adapt to their new home quite readily, moving over large areas in search of fresh forage and only rarely returning to the pens to obtain water and food. Their population increased slowly; apparently, inbreeding depression was partly responsible. Therefore some more animals were introduced in 1988 and 1989. By 1996 there were over 400 animals, most of them wild-born, ranging over 16,000 km^2 without any special management. Sadly, just when the project could be deemed a full success, poaching reared its ugly head, driven by a demand for live oryx for private collections. By 1998 the population had crashed to 138 animals with just 28 females, so it was judged necessary to bring them back into captivity. Hopefully, the same fate will not fall on the two populations that were reintroduced in Saudi Arabia beginning in 1990.

If we treat the recent poaching incident as a temporary setback, the project can be judged a success, but it came at great expense. At one point over 40 people were employed by the Oman reintroduction project, and no doubt the captive-breeding program cost far more. An all-out effort to protect the wild Arabian oryx from poachers in the 1960s might have been more cost-effective, but it might have failed. The approach that was taken worked, and it is a testament to the key role that *ex situ* conservation can play.

Summary

Although the original goals of zoos, aquariums, and botanical gardens focused on public recreation, many of these institutions also developed into important centers of public education and biological research. In recent years, much of this research and education activity has acquired a dominant theme: conservation of the world's biological diversity. Major institutions usually go one step further and become directly involved in *ex situ* conservation, specifically, maintaining organisms outside of their natural habitat. Zoos, aquariums, and botanical gardens generally do this through careful husbandry of captive populations, including, among other things, management of breeding systems to minimize loss of genetic diversity and avoid domestication. Some institutions maintain biodiversity *ex situ* by storing seeds, spores, sperm, embryos, and similar material, as well as microorganisms. It is important that *ex situ* and *in situ* conservation programs be carefully integrated with one another so that *ex situ* populations can be 1) insurance against the loss of natural populations; 2) a direct contributor to conservation of wild populations through education, research, and funding; and, if necessary, 3) a source for reintroduction projects.

For one segment of the earth's biodiversity—domestic plants and animals—captivity is their natural state. Nevertheless, it is also necessary to maintain the diversity, especially the genetic diversity, of these species in a proactive fashion. This has traditionally meant storage of germplasm, but, increasingly, it has also involved cooperating with farmers to maintain local varieties or breeds.

FURTHER READING

For popular accounts of the *ex situ* programs of zoos, see Luoma (1987), Tudge (1992), and Wiese and Hutchins (1994). Two periodicals that carry many *ex situ* conservation articles are *Zoo Biology* and the *International Zoo Yearbook*. See Olney et al. (1994) for papers about the interface between *in situ* and *ex situ* conservation and *The Last Panda* (Schaller 1993) to see how this interface has not worked well for giant panda conservation. See Boyd and Houpt (1994) for a profile of a zoo-dependent species, Przewalski's horse. For conserving the genetic diversity of domestic species, key journals are *Ark, Animal Genetic Resources Information, Diversity,* and *Plant Genetic Resource News,* and two books that introduce the issues are Ford-Lloyd and Jackson (1986) and Alderson (1990). The global system for keeping track of captive animal populations, ISIS, can be accessed at www.isis.org. For websites of the European and American associations of zoos and aquariums, see www.eaza.net and www.aza.org, respectively. See www.wcs.org for an organization that works extensively in both the *in situ* and *ex situ* arenas. For a global effort to coordinate seed collections, see www.rbgkew.org.uk/seedbank/msb.html.

TOPICS FOR DISCUSSION

1. If you could design a zoo, aquarium, or botanic garden from scratch, what would it be like? What taxa would it hold? What geographic areas or types of ecosystems would it represent? How would it allocate its resources in terms of captive breeding, research, and education? How would you develop cooperative relationships with *in situ* conservationists? Estimate the annual budget it would take to run your institution. To be realistic, you should plan on millions of dollars per hectare. How would you modify your vision if you had only half of your dream budget?

2. Think about a zoo, aquarium, or botanic garden that you have visited. If you could change one significant thing about it, what would it be?

3. Are there any species that you would be unwilling to maintain in captivity even if it meant their extinction from both the wild and captivity? Why?

4. Do you think that eventually we will know enough about captive propagation and storage techniques such as cryopreservation to maintain virtually all species *ex situ*?

5. Do you think there is a role for private individuals assisting with *ex situ* conservation (e.g., through planting endangered species in their gardens)? What would be some of the pros and cons of this? (See Reinartz 1995 after arriving at your own ideas.)

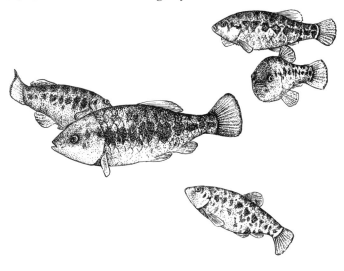

Chapter 14
Setting Priorities

Conservationists understand finite resources, especially the idea that the earth's ability to support a large population of people is limited. Consequently, it is easy for conservationists to appreciate that the resources available for conservation work, chiefly, time and money, are also finite. We can and should decry the myopia of social, political, and economic systems that do not recognize the importance of conservation. Why do we spend far, far more money on medical research and treatment than on controlling environmental pollution when many diseases are merely symptoms of environmental degradation? Yet, in the end we have to work with what society chooses to allocate to us. That means setting priorities, and that is the subject of this chapter. There are many approaches to setting priorities; we will outline six approaches that are quite different but not mutually exclusive.

Levels of Biodiversity

Biosphere, biome, landscape, ecosystem, community guild, species, population, individual gene, allele—it is easy to construct hierarchical organizations for life on earth. In such a hierarchy each level contains more elements of biodiversity than the level below making this one logical and simple way to decide which elements of biodiversity merit primary attention (Noss 1990, Soulé 1991) (Fig. 14.1). If we give priority to protecting a marsh from being destroyed, we protect hundreds, even thousands, of different species that inhabit the marsh. In contrast, if we give priority to protecting a single species, we may be helping only that one species and a few other species with which it is closely associated. This is the essential idea behind the coarse-filter approach to maintaining biodiversity (see Fig. 4.6).

Most conservation biologists recognize the general wisdom of focus-

Figure 14.1. The hierarchical organization of life means that maintaining a single ecosystem will maintain many species, which will in turn maintain an even greater diversity of genes.

ing on ecosystem conservation, but, sometimes, they may seem to favor conservation of a single species. There are two common reasons for this. First, most people find it much easier to identify with species than with ecosystems, with maned wolves rather than with South American grasslands. Consequently, conservation biologists often use a single species as a flagship or umbrella species to garner public support that can be translated into protection for ecosystems (Chapter 3). Second, species are easier to classify than ecosystems. We can usually agree (although exceptions are commonplace) that this toad is a different species from that toad. It is often much harder to agree that this type of stream is a different type of ecosystem from that type of stream (Chapter 4).

Geographic Scales

It is easy to be parochial, to let your perception of the world revolve around your day-to-day life. Conservation biologists need to moderate this tendency by asking, "At what geographic scale is this species or ecosystem at risk?" and then giving priority to those in jeopardy at large scales, especially the global scale. This is the alpha, beta, and gamma diversity perspective of

Chapter 2 and Figure 2.2 again. It merits repetition because it is very important and often ignored. Wealthy countries often spend large sums protecting species that are threatened within their borders but that are globally secure (Hunter and Hutchinson 1994). For example, biologists have labored for over 20 years to restore Atlantic puffins to some islands on the coast of Maine because there are few Atlantic puffin colonies remaining in the United States (Kress and Nettleship 1988). Given that Atlantic puffins number in the millions in Canada, Greenland, Iceland, and Europe, this effort is not a global priority.

On the other hand, it is not a complete waste of time to save species that are only in danger of local extirpation. Protecting species from extinction across their entire geographic range is necessary if a species' complete genetic wealth is to be maintained (Scudder 1989, Lesica and Allendorf 1995). Locally endangered species can also be important 1) in providing an umbrella of protection for elements of biodiversity that are hard to detect and often overlooked such as genetic diversity, invertebrates, and microbes; 2) in maintaining ecological functions (which are largely local-scale processes); 3) in fostering local public support for conservation in general; and 4) in meeting the local needs of people (Hunter and Hutchinson 1994). These arguments for maintaining locally rare species do not outweigh the arguments for giving priority to globally endangered species. Clearly, the earth's biodiversity as a whole would usually be better served if we could take a truly global perspective when setting priorities.

Choosing Areas

When conservation biologists daydream, it is often about winning a huge sum of money at a lottery that they could then use to buy land and establish nature reserves. The conservation literature has many papers on how to spend such money wisely (e.g., Usher 1986, Bedward et al. 1992, Spellerberg 1992, Noss and Cooperrider 1994, Margules and Pressey 2000); here are five criteria distilled from this literature:

1. Size and number. Conservation biologists generally agree that protecting large reserves is more important than protecting small reserves and that protecting many reserves is also important. These goals are potentially in conflict, and this is the basis of the SLOSS (Single Large or Several Small) debate described in Chapter 11. We will not revisit this argument here; suffice it to say that we need both more and larger reserves and that, when forced to choose between these goals, many factors must be weighed.

2. Representativeness. The fundamental assumption of the coarse-filter approach to maintaining biodiversity (see Fig. 4.6) is that a system of reserves should contain a representative sample of a region's ecosystems so that microbes, invertebrates, and other neglected components of biodiversity can be protected (Chapters 4 and 11). Therefore areas selected for a

coarse-filter reserve network would have an array of different ecosystems, and these ecosystems would be typical of the entire region's ecosystems. Deciding what is a typical example of an ecosystem can be rather difficult, especially because of the ambiguities of ecosystem classification. The same patch of forest could be described as an unusual example of a common ecosystem by one person and as a rare, distinct ecosystem by someone else.

3. Rarity. Some areas have ecosystems that are clearly rare or at least uncommon (despite the ambiguities just mentioned), for example, aquatic ecosystems in the midst of an arid region. These should receive priority, as should areas that are habitat for rare or threatened species (Beissinger et al. 1996).

4. Condition. Given a choice between making a relatively pristine area into a reserve versus an area that has been substantially degraded by human activities, conservationists would usually choose the area in the better condition.

5. Feasibility. The practical aspect of protecting an area has many facets that merit consideration when setting priorities. For example, a reserve that is isolated from potential sources of disruption (e.g., one far from the nearest city) will probably be easier to maintain, especially in the long run. On the other hand, conservationists sometimes give higher priority to areas near population centers because these are more likely to be threatened by human activities in the foreseeable future, and it is far more economical to protect land before the onslaught of development begins. Current ownership is also an issue; it is usually easier to create a reserve on publicly owned land than to purchase many, small, privately owned properties.

This is not an exhaustive list of criteria. We could add cost, species richness, fragility, threat, urgency, and more; see Balmford et al. (2000) and Myers et al. (2000) and Sala et al. (2000) for examples and Usher (1986) for a review. Finally, it is important to remember that choosing areas for reserves is only a part of the process of managing ecosystems. The vast majority of ecosystems exist outside reserves, and, sometimes, we must also choose among them for areas to direct conservation action.

Choosing Among Multiple Criteria

Given all the complex issues that need to be considered when setting priorities for conservation action, it is surprising that most conservationists are not immobilized by indecision. What do you do when you are trying to evaluate many criteria simultaneously? One answer is to develop a quantitative evaluation scheme in which different weight is given to various criteria depending on their importance, and a composite score is calculated for each activity you are considering. These techniques are often used for selecting areas for conservation action. Box 14.1 gives a simple example. It is

---

| **BOX 14.1** | A simple example of a quantitative evaluation technique |

If you were responsible for acquiring new protected areas for your state, you would probably develop a systematic way of choosing among potential acquisitions, probably working with a committee representing all the interested parties. First, you would need to make a list of criteria (e.g., representativeness, rarity, size, condition, and feasibility), and then you would need to decide the relative importance of the criteria. You could, for example, consider these five criteria to be of equal importance by giving each one a quantitative weight of 0.2. (It is common to assign weights that add up to 1.) More likely, some criteria will be given greater weight. For example, if you were particularly interested in taking a coarse-filter approach that emphasizes protecting a representative array of ecosystems, you might give representativeness a weight of 0.35. Protecting species and ecosystems that are rare is usually a high priority; thus rarity might be given a weight of 0.25. If you take a long-term view toward maintaining protected areas, the condition of the area at this time will not be overly important, and thus this criterion might have a weight of 0.15. Similarly, if your acquisition funds are replenished every year, feasibility might only receive a score of 0.15 because you can work away at an acquisition project year after year. Finally, although size is important, some important aspects of size are often highly correlated with other criteria (e.g., number of ecosystems or populations of rare species that an area holds); hence, size by itself may receive a comparatively low weight of 0.10.

The next step is to score the areas being considered for all the criteria. As an illustration we will compare two fictional places, Aspen Mountain and Beaver Valley. Aspen Mountain receives a score of 94 for *representativeness* because it contains 28 of the 35 different types of ecosystems that occur in the region; Beaver Valley's score is lower, 84, because it has only 25 types. Because Beaver Valley has populations of three globally endangered species and 11 locally endangered species plus a globally rare type of ecosystem, it has a higher score for *rarity*, 96, than Aspen Mountain, 88, which only supports one globally endangered species and 14 locally endangered species. Roughly 15% of Beaver Valley was heavily cut just last year, but Aspen Mountain only had some light, selective cutting during World War II. Consequently, Beaver Valley's *condition* score is 75, whereas Aspen Mountain receives a score of 98. On the other hand, Aspen Mountain has recently been subdivided and sold to nearly a hundred different landowners, but Beaver Valley is owned by a

BOX 14.1   *Continued*

single company that is willing to sell immediately at a very reasonable price. This difference gives Aspen Mountain a *feasibility* score of 20 and Beaver Valley a score of 100. Finally, Beaver Valley receives a score of 82 for its 5800-ha *size* compared with an 88 for Aspen Mountain, which is 9700 ha.

You are probably wondering how these specific scores were produced. Sometimes, quantitative methods are used (e.g., a score for rarity might be derived from the portion of the region's endangered species that are present and weighted two-thirds on globally endangered species and one-third on locally endangered species). Often, it comes down to a judgment by informed people.

Once scores for the individual criteria are given, deriving an overall score is just a matter of multiplying by the weights and adding as shown in the table below. Do not be fooled by false precision; clearly, 81.4 and 87.8 should be rounded off to 81 and 88, and it is even debatable whether a seven-point spread on a 100-point scale is a significant difference. The bottom line is that these methods should be a tool for organizing and communicating your thinking and not a substitute for good judgment. If your gut feeling is that the method did not work because Aspen Mountain should have come out as a higher priority than Beaver Valley, you can reexamine your thinking with respect to the individual scores and their weighting. You will discover that a key factor in this analysis was the feasibility criterion, even though it has a weight of only 0.15. The beauty of these systems is that they allow you to dissect and analyze what would otherwise just be a gut feeling.

Criteria	Scores			Weighted Scores	
	Aspen M.	Beaver V.	Weights	Aspen M.	Beaver V.
Representativeness	94	84	0.35	32.9	29.4
Rarity	88	96	0.25	22.0	24.0
Condition	98	75	0.15	14.7	11.2
Feasibility	20	100	0.15	3.0	15.0
Size	88	82	0.10	8.8	8.2
Total score				81.4	87.8

important to realize that, like all models, these are just devices for structuring complex ideas and communicating them to others. They should not be used without fully understanding the assumptions behind them, and they should not replace good judgment.

Choosing Species

Blue whales or green sea turtles? Black rhinos or white? One could list dozens of factors to consider when choosing which species should receive priority, but we will address just two overarching questions: "Which species is more valuable?" and "Which species is at greater risk of extinction?"

The first question returns us to Chapter 3 and our discussion of the instrumental values of species. If our primary concern is the welfare of humanity, we should favor species with economic values and, because people are dependent on healthy ecosystems, species with important ecological roles as dominant, controller, or keystone species. If our concern is more equitably distributed among all species, we should still focus on species with important ecological roles because so many other species depend on them. For the same reason, we should give priority to flagship and umbrella species that have strategic value to conservation action. There is no point in denying the charisma possessed by gorillas, whales, and similar species. Whether focused on economic or ecological issues, people usually favor a species with realized value over one whose value is only potential because, as the adage goes, "A bird in the hand is worth two in the bush." Finally, the uniqueness of a species amplifies all other values. If we lose a species like the African elephant, its role will not be easily filled by another species. Therefore we should usually tend to favor species that do not have many closely related species (Vane-Wright et al. 1991, Crozier 1992, Faith 1992, Purvis et al. 2000). (See Balmford et al. [1996], Allendorf et al. [1997], and Ruane [2000] for parallel but rather different exercises in selecting, respectively, zoo collections, salmon stocks, and rare, domestic-animal breeds for conservation.)

The second question, "Which species is at greater risk of extinction?," is also a key issue, especially if you believe that all species have intrinsic value. Intuitively, this seems to be a simple issue: species that are at greater risk of extinction should receive higher priority (see Boxes 3.2 and 3.3). However, some conservationists have advocated a triage approach to dealing with species (McIntyre et al. 1992). *Triage* refers to the idea that there are three classes of war casualties: people who will recover without immediate medical aid; people who will die even if given aid; and people for whom aid is a life-or-death matter. Priority is given to the third group of casualties, and, similarly, priority is given to species that have a reasonable chance of surviving if given attention. Personally, I have difficulty with the triage idea if it means deliberately abandoning a species to extinction; surely, the black robin, described in Chapter 12, would have been lost under a triage system. On the other hand, it seems shortsighted to ignore the feasibility of a project and to allocate vast resources for a fight against nearly impossible odds. For example, one could argue that sending four biologists to southern Brazil to save the Spix's macaw, after it was apparently reduced to a single wild bird, was overreacting to a lost cause (Juniper and Yamashita 1990, ICBP 1991).

Choosing Nations

International organizations have to decide which countries should receive assistance with their efforts to conserve biodiversity. Some of the issues that come to the fore have already been discussed indirectly in other contexts: determining which countries harbor the largest numbers of endangered species or species that are endemic to that country; or determining which countries are in greatest danger of losing their natural ecosystems. Figure 14.2 depicts a system for deciding which nations in the Indo-Pacific region are high priority based on the extent of their current, protected-area system and predictions about the rate of deforestation. Similar systems can be used to assess priorities among biogeographic units such as ecoregions or biomes that reflect species distributions better than political units (Olson and Dinerstein 1998, Sala et al. 2000).

Other issues have little to do with biology. Which countries have sufficient political stability to make ambitious conservation projects feasible? Which nations have the greatest financial need for assistance? Money spent in an unstable country like Rwanda is less likely to be effectively used than in a country like Costa Rica. Conservationists in Swaziland are more likely to need an external subsidy than those in Switzerland. Sometimes, expertise is what is needed. Nations like Saudi Arabia suffer from a shortage of ecologists, not the money to pay their salaries. One analysis recommended nations for conservation effort after explicitly incorporating the estimated cost of creating and maintaining a reserve network covering 15% of each nation's area (Balmford et al. 2000). When cost-effectiveness was added to the formula, countries where conservation is relatively expensive moved down the priority list (e.g., the United States and Australia), while countries such as Peru and Madagascar rose higher.

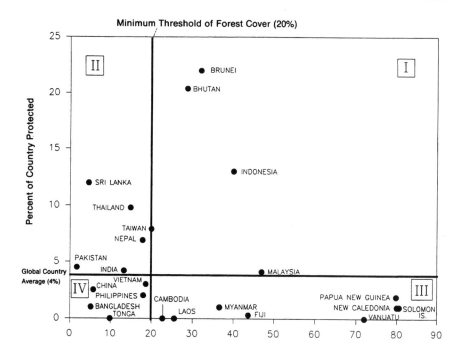

Figure 14.2. Eric Dinerstein and Eric Wikramanayake (1993) used the extent of protected areas and estimates of deforestation to create an index that would guide international conservation organizations in setting priorities among 23 Indo-Pacific countries. They divided the countries into four classes. Category I: Countries with a relatively large percentage (> 4%) of forests under formal protection and that will have a high proportion (> 20%) of unprotected forested areas left in 10 years. Category II: Countries with a relatively large percentage of forest (> 4%) under formal protection, but that will have little (< 20%) unprotected forests left in 10 years. Category III: Countries with a relatively low percentage (< 4%) of forests presently protected. However, under current deforestation rates these countries will still have a relatively large proportion (> 20%) of their unprotected forests remaining in 10 years. Category IV: Countries with a relatively low proportion (< 4%) of forests presently protected. Obviously, Category IV countries require urgent action, while Category II and III countries should be shifted toward Category I status expeditiously.

Choosing Tasks

Projects designed to maintain biodiversity can often be divided into two broad classes of activities: *protecting* ecosystems and species that are threatened versus *restoring* ecosystems and species that have been degraded or locally extirpated. Which is more important? Of course, there is no general answer. How can we weigh restoration of a large ecosystem, say the Aral Sea, against protecting a region's last tiny patch of virgin forest?

Even when the comparison is easier—for example, protecting the virgin forest versus restoring an adjacent degraded forest—many other factors will come into play. Nevertheless, one generalization can be made. It is almost always easier to protect what exists than to restore what has been lost. Consequently, for a given level of effort, the impact of a protection project will usually be greater than the impact of a restoration project.

We can also address the issue of choosing tasks by examining the four basic parts of most complex human undertakings. These are *planning* (figuring out what we want to do and how to do it), *implementation* (doing it), *monitoring* (figuring out what we have done and whether it worked), and *modification* (changing our activity to better achieve our goals). All of these tasks are critical. Some people are great at thinking about things to do, but never get around to doing them. Some people are great at doing things, but may or may not have thought much about what they are doing. Compared with planning and implementation, most people find monitoring boring, but without monitoring there can be no effective modification. In other words, monitoring and modification are often given low priority. This is a mistake. Vast amounts of conservation effort (translocating endangered species, restoring degraded ecosystems, etc.) have been wasted because they were not done correctly the first time and because no one took the time to check the outcome carefully (Goldsmith 1991; Kremen et al. 1994, 1998; Noss and Cooperrider 1994; Elzinga et al. 2001).

The Highest Priority of All

Address the causes of problems, not just the symptoms. Many of the activities described in this book—cross-fostering and double-brooding, maintaining studbooks and seed banks—only address the symptoms of larger, underlying problems. In particular, the peril of endangered species is but a symptom of ecosystem degradation and, ultimately, human overpopulation and excessive consumption (Ehrlich and Wilson 1991, Soulé 1991).

Of course, we cannot devote all of our energy to the ultimate problem of human population and consumption and completely ignore the cascade of symptoms that it produces. However, in setting priorities we must never lose sight of what is a problem and what is merely a symptom of that problem. In *Lives of a Cell*, Lewis Thomas (1974) writes about this issue from a medical perspective, describing much of medical technology as "halfway technology" because it addresses symptoms rather than causes. For example, heart transplant surgery replaces diseased hearts instead of changing the diet and lifestyle problems that produced the diseased heart. There

Figure 14.3. We need to deal with the root causes of the biodiversity crisis. Maintaining biodiversity by limiting human population growth and wisely caring for entire ecosystems is much more efficient than saving critically endangered species. It is analogous to saving lives through public-health programs versus emergency-room surgery.

is an important conservation analogy to this comparison of emergency-room medicine versus the efficiency and cost-effectiveness of public health medicine. Protecting entire ecosystems is good public health compared with the emergency-room tactics of *ex situ* conservation in zoos, aquaria, botanical gardens, hatcheries, or intensive management of single species in the wild (see Chapters 12 and 13) (Fig. 14.3).

CASE STUDY   Vietnam Conservation Areas
Eleanor J. Sterling,[1] Andrew Tordoff,[2] and
Jonathan C. Eames[3]

Have you heard of the saola, giant barking deer, and golden-winged laughing-thrush? If not, you are not alone for these species were unknown to science just 10 years ago, and we still know virtually nothing about them (Fig. 14.4). These and other mammals and birds were all discovered in the mountains separating Vietnam and Laos (Eames et al. 1999, Surridge et al. 1999, Groves and Schaller 2000), making this region a treasure trove for the discovery of new species of mammals unequaled anywhere on earth in recent years. These discoveries were one of several reasons why, in 1998, the Government of Vietnam proposed to increase the protected-area forest network from 1.3 to 2 million ha. To identify where the new forest protected areas should be located, researchers conducted a gap analysis (Wege et al. 1999, Eames and Tordoff [in press]).

Figure 14.4. Several new species of mammals and birds have been discovered and rediscovered recently in the mountains that border Vietnam and Laos and extend farther south into Vietnam. Shown here, from left to right, are the saola, giant barking deer (muntjac), Truongson muntjac, Annamite striped rabbit, black-crowned barwing, and golden-winged laughing-thrush. (Note: Given their recent discovery, the exact status and proper names for these species are a bit uncertain and controversial.)

A gap analysis is a priority-setting technique that provides a preliminary, landscape-scale overview of the distribution and conservation status of species and ecosystems. It identifies "gaps," or vegetation types, ecoregions, species, or other elements of biodiversity that are not represented in a protected-areas network. Gap analyses often use the distribution of vegetation types and selected species (usually, well-known species like mammals and birds) as surrogates for biodiversity in general (Scott and Jennings 1998).

There are several steps in a gap analysis: 1) create maps of an area that show: the distribution of vegetation cover; the distribution of selected species; other selected features of interest in the area such as elevation, slope, aspect, soils, aquatic features, climate, or socioeconomic data; and areas managed primarily for biodiversity; 2) overlay the different maps to identify gaps in the protected-areas system; 3) determine priorities for conservation action by placing the results in the context of other factors—such as ecosystem patch dynamics; habitat quality; population viability; distribution of threatened species; the feasibility of creating a reserve in an area; and the importance of having multiple representations of species or ecosystems throughout their geographic range to protect against catastrophic stochastic events.

In 1998 there were 90 protected areas in Vietnam—10 national parks,

53 nature reserves and 27 cultural and historical sites—covering 1,345,000 ha (equivalent to 4% of the land area of Vietnam). These protected areas were mainly terrestrial forest sites, with a small number of wetland sites; comprehensive, protected-areas networks for wetland and marine sites have yet to be developed.

For the gap analysis, researchers mapped datasets for natural forest types; draft ecoregions; elevation zones; globally threatened species of large mammals, primates, and birds; the existing protected areas; and political provinces. Results showed almost half (575,000 ha) of the existing, protected-areas network encompassed a large area of nonforest land—principally, agricultural land, scrub, and nonnatural grassland.

Researchers identified areas that met representation criteria and then refined their selection by considering the need to include globally threatened species under-represented within the current network; large areas of contiguous natural forest; areas contiguous with other protected areas, including protected areas in other countries; provinces in need of further protection; and existing, well-documented proposals.

As a result of the analysis, researchers recommended 25 expansions to the current, protected-areas forest network, including the creation of 19 new protected areas and the extension of nine existing ones. These expansions would add more than 750,000 ha to the current network. The coverage of evergreen forest in the new network would almost double, to 15%. Protected areas would be established in three political provinces that currently have none, and a large number of globally threatened bird and mammal species would have increased protection in the expanded network.

A more sophisticated gap analysis would go well beyond equitability of representation and would weight ecoregions by variables that affect their importance and priority such as threat level, global uniqueness, regional uniqueness, maintenance of migratory corridors, the potential for effective conservation strategies, etc. (Timmins and Trinh 2000, W. Duckworth, personal communication). Such an analysis would also include datasets on distribution of other animal species—particularly threatened invertebrates—and flora.

[1] Center for Biodiversity and Conservation, AMNH.
[2] BirdLife International Vietnam Programme.
[3] BirdLife International Vietnam Programme.

Summary

There are many ways to set conservation priorities. Efficiency often dictates that we focus on large-scale entities (ecosystems rather than species or genes), especially those that are at risk at a global scale. Choosing specific sites for conservation management involves weighing multiple criteria such as size, representativeness, rarity condition, and feasibility. Selecting high-priority species usually involves some measure of the species' value (broadly defined) and the degree of threat. Prioritizing nations for conservation effort may involve a host of factors such as a country's biological and economic wealth and political stability. In choosing tasks, we must be careful not to focus solely on planning and implementing conservation action and, thereby, neglect the monitoring that can lead to modifications of our actions. Finally, the overriding priority is to try to deal with the root causes of biodiversity loss, rather than the symptoms.

FURTHER READING

Johnson (1995) provides a thorough and readable review of approaches for setting geographic priorities for biodiversity conservation (available on the web at www.bsponline.org). Usher (1986) and Spellerberg (1992) also cover a significant portion of the relevant literature.

TOPICS FOR DISCUSSION

What are the highest priorities for conservation action in your region? Having generated a list of 10 priorities, perhaps in collaboration with your classmates, try to identify the criteria that shaped your thinking in arriving at the list (e.g., rarity, feasibility, urgency, etc.). Try to assign numerical weights to these criteria to reflect their relative importance in your evaluation. Next, use these criteria and their weights to develop a quantitative system analogous to the one presented in Box 14.1. Finally, test this system by evaluating some conservation activities from outside your region. Do the priorities generated make sense?

Part IV

The Human Factors

Conservation biologists are constantly reminded of what our species has done to extirpate or threaten other life-forms. This awareness seems to make some conservation biologists a bit misanthropic. Perhaps it is the other way around though; maybe people who are not particularly fond of humanity are more likely to select careers in which they can interact with other species. In either case, many conservation biologists are not very comfortable in dealing with human institutions such as the social, economic, and political systems that are the subjects of our last three chapters. Yet, this does not diminish the importance of these systems to conservation biology. If people are the primary force degrading biodiversity, then people must change their behavior. If we wish to facilitate these changes, we must understand social, economic, and political systems. These three chapters are far too brief to provide a real foundation for understanding sociology, economics, and politics, but they can give you an appreciation of how important these subjects are to conservation biology.

Chapter 15
Social Factors

When I am hungry, a date palm gives me food. When my belly is full, behold, the tree is beautiful.

This statement, ascribed to a Bedouin of Jordan by Guy Mountforth, says two important things about the values held by individuals and societies. Values differ. Values change. Values differ between hungry people and well-fed people. Values change when a person who was hungry has eaten. We will begin by examining how different groups value biodiversity—focusing on different cultures, rural and urban people, and women and men.

Values Differ

Cultures and Religions

The cultural diversity that characterizes humanity is one of our greatest assets. It is a deep, rich lode of human potential that reflects our religious, ethnic, racial, and linguistic diversity. At times, however, when universal cooperation and unanimity are needed, this diversity seems like a significant liability, a source of frustration and bafflement. As Rudyard Kipling wrote, "East is east and west is west, and never the twain shall meet."

Differences in cultural values are seen on the world stage most often when human rights are being discussed, but cultural differences in attitudes about nonhuman organisms are at least as profound. Consider the various species of rodents known as rats. In most places they are loathed as the epitome of vermin, but in several parts of the world they are relished as food; in Nigerian markets grasscutter rats sell for more than beef and pork (National Research Council 1991). In India, rats are fed and protected in temples of the Hindu goddess Bhagwati Karniji (Canby 1977). In North America and Europe dogs are beloved companions; in many East Asian

Figure 15.1. Divergent attitudes about snakes are epitomized by rattlesnake roundups in the United States, where snakes are killed in large numbers, and religious ceremonies in India, where snakes are given offerings.

countries they are prized as food; among the Zoroastrians of the Middle East they are key participants in certain religious rites. Similar stories could be told for many species: bats, snakes, whales, ravens, and more (Fig. 15.1).

Often, such differences are explained with a simplistic statement; for example, Hindus consider cows to be sacred and do not eat them because that is their culture. In his book, *Cows, Pigs, Wars, and Witches: The Riddles of Culture,* anthropologist Marvin Harris (1974) argues that the statement, "It's their culture" is not really an explanation and that usually a rational, ecologically based explanation can be found. For example, he argues that the Jewish and Islamic strictures against pigs are based on the fact that these religions originated in a desert environment where raising pigs (which require a high-quality diet compared with sheep, goats, and cattle) was a waste of limited resources. Once an idea has been codified as part of a set of cultural values, it can persist, even if the original reason for it disappears, because it becomes a mechanism for maintaining group cohesion. In other words, Jews and Moslems who do not live in desert environments still do not eat pork because it is a way to demonstrate their cultural identity.

Differences in the ways that various cultures perceive their relationship with nature are sometimes linked to theology. For example, it has been argued that Verse 26 of the first chapter of *Genesis,*

Then God said, "Let us make humankind in our image, according to our likeness; and let them have dominion over the fish of the sea, and over the birds of the air, and over the cattle, and over all the wild animals of the earth, and over every creeping thing that creeps upon the earth."

compels Christians, Jews, and Moslems to think that all other species exist for the use of people (White 1967). On the other hand, many other scholars have argued that this verse should be interpreted as a mandate to be good stewards of nature (Van Dyke et al. 1996). Certainly, an environmental ethic in the Old Testament can be found in passages such as Isaiah (5:8):

Woe to you who add house to house and join field to field till no space is left and you live alone in the land.

Nevertheless, it seems clear that these religions see humans as distinct from the rest of creation in some way, and in this respect they contrast with some other religions such as Hinduism, Buddhism, and Taoism that emphasize the sameness of humans and nature (Callicott 1994). Consider this passage from the Ishopanashads, a holy scripture of Hinduism:

This universe is the creation of the supreme power meant for the benefit of all His creations. Individual species must, therefore, learn to enjoy its benefits by forming a part of the system in close relation with other species. Let not any one species encroach upon the other's rights.

Ultimately, the theological distinction between religions that emphasize "sameness" versus "separation" may or may not mean a great deal in practical terms. Certainly, one can find landscapes that are nearly barren of nature or that are lush and diverse, inhabited and managed by people in any of these religions (Moncrief 1970). Clearly, religion is only part of what distinguishes cultures, and one must consider other influences such as history, politics, economics, and technology to understand cultural differences in the way people interact with nature.

Urban-Rural

Living in a city and living in the country are profoundly different experiences (so much so, that one could argue that urban and rural people have different cultures and that this section should be folded into the preceding one). Consequently, it is not surprising that rural and urban attitudes toward nature should be quite different.

Many urban people are quite isolated from the natural world that lies outside their cities, and this can limit their understanding of it. A colleague of mine once took a group of urban school children to a farm and watched, amazed, as one boy vomited after seeing a cow being milked. The boy had not known where milk came from and was nauseated by the idea of it coming from a cow. Such isolation can also lead to apathy that may be broken only by an event with a direct impact, for example, if water rationing were imposed because deforestation had ruined the city's watershed. Of course, not all urbanites are apathetic or poorly informed about nature. Indeed, many of the world's most committed naturalists and conservationists have deep urban roots. It is tempting to speculate that for some people their day-to-day distance from nature has given them a stronger appreciation for it, a sort of "distance makes the heart grow fonder" effect. Nevertheless, in general, rural people are much more likely to understand their relationship with the natural world, in part because they will be more directly affected by any problems that arise. This understanding, known in some circles as *traditional ecological knowledge*, encompasses much more than a set of information; it shapes value systems and world views (see Berkes et al. 2000 and following articles in Ecological Applications 10[5]).

Certainly, urban isolation from nature does not mean that urban people are the villains and rural people are the heroes in the drama of conservation. Through farming, fishing, logging, and similar activities many rural people interact daily with other species and have a strong utilitarian attitude toward them. Utilitarian attitudes are not a threat as long as the species in question can sustain the usage, but, sometimes, these attitudes are extended to species that are too uncommon to be exploited. To put it in more practical terms: when an endangered species is exploited, it is usually a rural person holding the gun or the axe.

If conflict arises between the well-being of biota and the well-being of people, rural people are usually the ones on the front line. Consider three brief examples. Millions of city dwellers cherish the tiger as one of the most spectacular life-forms on earth. However, tigers inspire mostly fear among the people living in the Sunderbans, a large delta on the border of India and Bangladesh where tigers kill people each year (Saberwal 1997). Similarly, urbanites around the world write to politicians and give money to conservation groups to save the rain forests of Amazonia; however, for the poor people who live there, establishing parks impedes their ability to hunt for game or cut the forest to grow crops (Schwartzman et al. 2000). Yellowstone National Park is a remote symbol of wilderness for people throughout the United States and the world, but for the people who live around it, it is a complex array of resources that directly affects their daily lives, both positively and negatively (Jobes 1991). In these cases the isolation of urban people may make it easier for them to value biodiversity because it costs them little to do so. Rural people's values are shaped by being directly entangled in the interplay of biota and people. Often, this interplay is highly positive, but some conflicts are almost inevitable.

To make a summary generalization: rural people's attitudes toward wild life—both positive and negative—are likely to be pragmatic attitudes based on regular interactions. On the other hand, the attitudes of urban people—positive, negative, and apathetic—are likely to be more removed from direct experience.

Women-Men

The heart of feminism is equal rights and opportunities for women, and in many countries this idea is widely accepted, if not practiced. However, equality of rights and opportunities does not mean that women and men are identical physically or psychologically, and such differences could lead to both divergent interactions with nature and different attitudes toward nature (Fig. 15.2). In particular, some writers—Rachel Carson and Ariel Kay Salleh, for example—have suggested that because most women bear and care for children, they are more nurturing than men, and that this quality shapes their attitudes toward nature (Norwood 1993, Mellor 1997). In contrast, many feminist writers minimize or reject the idea that women are closer to nature than men because this stereotype perpetuates a dualism that separates men from women and men from nature (Warren 1993). From their perspective, language that makes nature seem feminine (e.g., raping Mother Earth) or that makes women seem more like a part of nature (e.g.,

Figure 15.2. Women and men often interact with the natural world in different ways that may reflect or shape their values.

slang terms for women such as chick, bunny, kitten, fox, bitch, etc.) lumps women and nature together and makes them both inferior to men.

Regardless of disagreements over whether women are closer to nature than men, all of these writers agree that both nature and women have been subjected to domination by men, and that we must work toward more harmony and balance in the future. This idea is the foundation for a growing area of philosophy called ecological feminism, or *ecofeminism* (see Norwood 1993, Mellor 1997, or Warren 1997 for further details) and may partly explain why women play such a pivotal role in the environmental movement, especially at the grassroots level.

Beyond differences in attitudes toward nature in general, men and women may also differ in how they value particular species because they interact with different suites of species (Shiva 1988). In many rural parts of southern Africa men are primarily hunters of large mammals, whereas women interact with a much broader array of wild life: gathering wild plants for food, fuel, fiber, and medicine and catching birds, reptiles, fish, insects, and small mammals for food (Hunter et al. 1990). It is interesting to speculate that the tendency of conservationists to focus heavily on large mammals may, in part, reflect a male-dominated culture and a relationship between men and large mammals that stretches back to our earliest ancestors.

Additional Perspectives

Differences among cultures, rural and urban people, and women and men are but three of many ways to view social values. We could also discuss how attitudes about nature are influenced by age, occupation, income, education, and other factors. Given all this complexity, it is often difficult to sort out why people feel the way they do. If a male banker in London does not share the same attitude toward tigers as a female farmer in the Sunderbans, to what degree are the differences based on their gender, culture, geography, wealth, education, and so on? Above, I made some broad generalizations about the values of rural people versus urban people, but would it be more accurate to ascribe these differences to the fact that most of the world's rural people live in less-developed countries and tend to be poor? Sorting out these complexities is more feasible if one uses a systematic approach to describing values; in the next section we will examine one well-known example.

Describing Values

How do you feel about crocodiles? Do they frighten you? Do they fascinate you? Do you love them? Do you love them more than I do? Discussions about human values can be rather fuzzy because they are difficult to describe systematically. Stephen Kellert, a sociologist who works on conservation issues, has spent many years developing systematic techniques for describing how people feel about animals, especially wild animals, and then using them to better understand how values differ among people of different ages, education, employment, culture, race, gender, region, and so on (Kellert and Berry 1981, Kellert 1996). Kellert's basic method is to read statements to people and ask them to strongly agree, agree, slightly agree, slightly disagree, disagree, or strongly disagree. By scoring responses to statements such as, "I have owned pets that were as dear to me as another person" or "If I were going camping I would prefer staying in a modern campground than in an isolated spot where there might be wild animals around," Kellert has identified several basic types of attitudes toward animals. See Table 15.1.

We can use these attitude types and Kellert's data to reexamine briefly some of the generalizations made above during our discussion of how values differ among cultures, between rural and urban people, and between women and men. For example, comparing United States women and men, there were significant differences in the prevalence of almost every attitude type (Kellert and Berry 1987). In particular, women were much more likely than men to have strong humanistic and moralistic attitudes, whereas men were more apt to show strong utilitarian and dominionistic attitudes. Similarly, some major differences were found when comparing people of the United States living in rural areas (defined as towns with populations less than 500) with those living in urban areas (populations greater than 1,000,000) (Kellert and Berry 1981). Rural people tended to have high scores for utilitarian and naturalistic attitudes, whereas urban people showed higher scores for moralistic and humanistic attitudes than rural people. Finally, Kellert (1991, 1993) has used his techniques to compare attitudes toward wild animals in the United States, Japan, and Germany. Overall the patterns were not strikingly different between residents of Japan and the United States. The most noticeable results were the higher scores for moralistic and ecologistic attitudes in the United States and for dominionistic attitudes in Japan. Data from Germany revealed extremely high scores for moralistic attitudes and relatively low scores for dominionistic and utilitarian attitudes. As a generalization, people in all three countries seemed to care about the welfare of individual species of wild animals, typically, species with strong aesthetic, cultural, and historic associations. They did not necessarily extend these feelings to caring about broader, more conceptual entities such as ecosystems or biodiversity. Of course, Germany, Japan, and the United States are all industrial superpowers; it would be interesting to see how attitudes differ across a wider range of cultures.

TABLE 15.1 Stephen Kellert has described several types of attitudes that people have toward animals.*

Term	Definition
Naturalistic	Showing an interest in, and affection for, wild animals and the outdoors. **10%**
Ecologistic	Concerned with ecosystems, particularly the interrelationships between species and their habitats. **7%**
Humanistic	Showing a strong affection for individual animals such as pets or large wild animals. Strong tendency for anthropomorphism. **35%**
Moralistic	Concerned with ethical treatment of animals; strongly opposed to cruelty toward animals or presumed overexploitation. **20%**
Scientific	Intellectual interest in organisms as biological entities. **1%**
Aesthetic	Interested in the physical attractiveness and symbolic characteristics of animals. **15%**
Utilitarian	Interested in the practical value of animals and their habitats. **20%**
Dominionistic	Interested in the mastery and control of animals, typically in sporting situations. **3%**
Negativistic	Preferring to actively avoid animals because of dislike or fear. **2%**
Neutralistic	Preferring to passively avoid animals because of a lack of interest. **35%**

*The types are described here using definitions slightly modified from Kellert and Berry (1981). Following each definition is the percentage of United States residents (based on a survey of 2,455 people over 18 years old) who strongly exhibited that type of attitude (Kellert and Berry 1981). Note that most people have more than one type of attitude, but usually only one type is strongly held. The percentages total 148 because of people who hold more than one attitude strongly.

These techniques can also be used to try to understand how people feel about a particular issue. For example, Reading and Kellert (1993) used a modified typology of attitudes to describe how ranchers, other local residents, and urban dwellers felt about a proposed reintroduction program for black-footed ferrets. Czech et al. (1998) used a different methodology to compare people's attitudes toward different groups of endangered species (mammals and birds versus reptiles and invertebrates) in the United States to see if these were consistent with how we allocate funds for conservation. They found that amphibians and plants were gravely underfunded relative to the positive values most people had for them.

Values Change

Here is one of those paradoxical truisms worth remembering: the only thing that never changes is the fact that everything changes. A few millennia ago, when daily life revolved around being predators and grazers and avoiding becoming prey, ecologistic and utilitarian attitudes toward wild life must have been very widespread and neutralistic attitudes virtually unknown. Even looking back just a few decades can reveal some remarkable changes in values. Not very long ago, attitudes toward whales and wolves were shaped by *Moby Dick* and *Little Red Riding Hood*. At best, these creatures were irrelevant to the lives of most people; at worst, they were the embodiment of evil. Today, attitudes toward whales and wolves seem to be much more positive (Kellert et al. 1996). Dramatic photographs and evocative recordings of songs and howls have transformed these creatures into powerful and popular symbols to people who denounce the human assault on nature (Fig. 15.3). If you belong to a conservation group you have seen countless advertisements for merchandise—jewelry, mugs, T-shirts, etcetera—with whale and wolf motifs. One could even argue that

Figure 15.3. In the eyes of many people wolves have been transformed from evil vermin to a symbol of wilderness.

whales and wolves have become sacred totems for thousands of people. Some of these patterns have been quantified. Using articles about animals published in four United States newspapers between 1900 and 1976, Kellert (1985) documented a general decrease in utilitarian and negativistic attitudes.

Marked changes can also occur within a single individual. During his early career Aldo Leopold never passed up the chance to kill a wolf, but later in life the wolf became a potent symbol of wilderness for him. Reflecting on his youth, he described the death of a wolf he had shot in *Thinking like a mountain:*

> We reached the old wolf in time to watch a fierce green fire dying in her eyes. I realized then, and have known ever since, that there was something new to me in those eyes—something known only to her and to the mountain. I was young then, and full of trigger-itch; I thought that because fewer wolves meant more deer that no wolves would mean hunters' paradise. But after seeing the green fire die I sensed that neither the wolf nor the mountain agreed with such a view. (Leopold 1949)

Changing People's Values

If we are to maintain the earth's biodiversity, values must change in the future even more than they have during the last few decades. In Kellert's terminology, attitudes toward wild life that are naturalistic, ecologistic, aesthetic, and moralistic must wax stronger; while negativistic, neutralistic, dominionistic, and utilitarian values must wane. Trying to sensitize people to the value of nature is a routine exercise for environmentalists; in particular; it is a central part of environmental education (Orr 1992). In the words of Baba Dioum: "In the end, we will conserve only what we understand and we will understand only what we are taught."

Environmental education can shift people's attitudes toward nature through two basic modes: information and experience. If we give people information about the instrumental value of biodiversity, about how important it is to the welfare of humanity and the biosphere in general, then people will probably place a higher value on it. Millions of people around the world now think of tropical rain forests as storehouses of medicinal plants and pivotal components of global climatic processes, rather than as bug-infested jungles, simply because they were given information.

Experience can shape people's values too, and environmental educators often try to get people outdoors where they can interact with the local biota. Not surprisingly, Kellert's (1980) research found that people who participated in outdoor, nature-related activities (ranging from bird-watching to fur trapping) had higher naturalistic and ecologistic scores than those who did not. Moreover, species that we enjoy during outdoor excursions automatically gain at least one type of instrumental value—recreational value. People who have encountered organisms in their natural habitat may also find it easier to accept the idea that they have intrinsic value. Even indirect exposure, through wolf and whale paraphernalia, for example, may help to shape values. Kellert (1980) found that watching nature shows on television was positively correlated with naturalistic and ecologistic attitudes. Which shapes values more, experience or information? This is probably one of those head-versus-heart, emotional-versus-rational, questions, but at least one pair of authors has argued that experience is more important in shaping attitudes toward nature (Rivas and Owens 1999).

The idea of changing people's values can be rather controversial. Environmental educators often refer to "clarifying" values rather than "changing" values to avoid the idea that they are imposing their own set of values on other people, especially children. They are confident that knowledge and experience will lead to caring without forcing one person's values onto someone else. This issue becomes even more controversial when the boundaries between cultures are crossed. For example, people in many parts of the world are flooded by a tidal wave of music, movies, television, fashion, fast food, and so on that emanate from the United States and Europe. Some people welcome this because it makes them feel modern and cosmopolitan; others resent it because it drowns their traditional culture. These conflicts become more troubling when political and economic power are used to impose foreign value systems. Consider the fact that animal-rights groups in the United States have been able to coerce some Asian nations into banning the use of dogs for food. What do you suppose the reaction in the United States would be if a group of Hindus, for whom cows are sacred animals, came to Washington, DC, to persuade Congress to ban the consumption of beef?

On the other hand, providing information seems to be an innocuous way to change values across cultural boundaries. Imagine a scenario in which a Finnish ecologist doing comparative research on Finnish and Canadian boreal forests discovers that the fruits of a particular shrub species are critical to the overwinter survival of many birds and mammals. If sharing this information results in Canadians changing their logging practices to minimize detrimental effects on that shrub species, it would be hard to argue that the Finnish ecologist's values were inappropriately imposed on the Canadians. Next, consider a survey of Costa Ricans that found a poor understanding of the relationship between overpopulation and environmental quality (Holl et al. 1995). In this case, an educational campaign on this theme might seem reasonable, but it certainly could spark a controversy if it were initiated by foreigners rather than Costa Ricans.

People living in remote areas often have no idea that a particular local species is globally significant until an outsider tells them so. The fact that this person is usually a well-educated scientist from far away gives immediate credence to the idea that this species is indeed unique. If cultivated carefully, the knowledge that a local species is unique can be the source of great pride and conservation action. Throughout the Caribbean there are nine parrots of the genus *Amazona,* most of which are endemic to single islands (Butler 1992a). In recent years, these parrots have become national treasures, celebrated with songs and plays, stamps and posters (see case study below). The initial impetus to this outpouring was often an outsider saying, "You have a special parrot."

Some conservationists would argue that all this sensitivity about the feelings of local people is missing the point. They would argue that all the earth's species belong to everyone (in other words, they are a globally shared inheritance) or that they all belong to no one (i.e., their intrinsic value is paramount). There is an attractive simplicity to this point of view, but, as we will see in the next two chapters, it is naive because it overlooks important economic and political realities about who carries the burden of conservation.

The Biggest Change Anthropocentrism Versus Biocentrism

Many environmental philosophers have argued that if we are to maintain the earth's biota, we need a major shift in human values (Naess 1989, Snyder 1990). They believe that we need to move from being *anthropocentric* (i.e., believing that people are the center of the universe) to being *biocentric* (i.e., believing that life, in all its various forms, is the center of the universe). A biocentric view (sometimes called ecocentric) recognizes that all species have intrinsic value and rejects the idea that *Homo sapiens* is more important than other species. Without such a change we may be left cataloging the instrumental value of different species and saving only those that we find useful. Biocentrism forms the philosophical foundation of what Naess has called the deep ecology movement (Devall and Sessions 1985).

As with most things in life, when ideas about biocentrism and anthropocentrism are applied to action, they are not black and white. Even the most ardent preservationists are not likely to be purely biocentric; given a choice between the survival of humanity and the survival of a small species of snail, very few people would flip a coin. Conversely, very few people would opt to eliminate a life-form simply because it is not apparently essential to human welfare. It is probably better to think of anthropocentrism and biocentrism as two poles that define a continuum. Conservationists would argue among themselves at length about exactly where we need to be on that continuum if the earth's entire biota is to thrive, but they would virtually all agree that we need to shift toward the biocentric pole from where we are now.

Another way to represent this issue is as a nested hierarchy of concern (Fig. 15.4) (Noss 1992). In this hierarchy the lowest, narrowest level is concern for one's personal well-being; the highest, broadest concern is at the level of ecosystems or the whole biosphere. Most people have some concern about the welfare of all other humans and many people care about sentient animals (i.e., those species—chiefly, mammals and birds—that they perceive to have feelings). Raising people's level of concern to embrace all species and ecosystems is an essential goal for conservation biologists. Some will argue that this can be done only if people become biocentric; others will argue that you can be anthropocentric—caring primarily about people—and still reach out to care about life in all its forms. Both of these are easier to do after basic needs are met; in other words, date trees are more beautiful to someone with a full stomach.

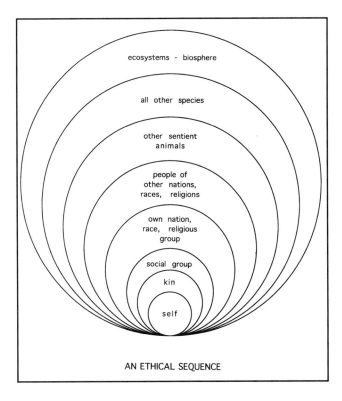

Figure 15.4. This figure conceptualizes an ethical sequence as a nested hierarchy—with concern for oneself at the lowest, narrowest level and concern for ecosystems and the whole biosphere at the highest, broadest level. The success of conservation hinges on people's expanding their level of concern to fully encompass all species and ecosystems. (Redrawn by permission from Noss 1992.)

CASE STUDY   The Bahama Parrot

On October 12, 1492, Lucayan Indians greeted Christopher Columbus on his arrival on the island they called Guanahani. They presented him with a variety of items, and Columbus seemed particularly attracted by the parrots he was given; for when he returned to Spain a few months later, he carried 40 parrots with him. Much has changed in 500 years. Guanahani is now known as San Salvador, and it is part of an island nation, the Bahamas, inhabited by 254,000 residents and visited by 3 million tourists annually. The Lucayans are gone and the parrots—so abundant that Columbus described them darkening the sky—have disappeared from Guanahani. Today, the Bahama parrot persists on only two islands, Abaco and Great Inagua, and numbers fewer than 3000 individuals. The parrot's demise can be traced

to several factors, the broadest being loss of habitat because of development, agriculture, and logging. Hunting parrots for food was a significant issue at one time, but today catching live parrots for the pet trade is a greater threat. Lastly and perhaps of most immediate importance, feral cats cause heavy losses on Abaco Island where the parrots nest in holes in the ground.

In 1990 a program to save the Bahama parrot was initiated by four organizations: the Forestry Section of the Lands and Surveys Department; the Ministry of Agriculture; the Bahamas National Trust, a private group dedicated to protecting the natural and cultural heritage of the Bahamas; and the RARE Center for Tropical Conservation, a small United States-based conservation group. A wide-reaching campaign to engender public support for the Bahama parrot followed. Here are some of the tactics employed as described by Paul Butler (1992b), RARE's director of conservation education. The visual image of the Bahama parrot and a simple conservation message were dispersed far and wide through posters, buttons, bumper stickers, billboards, puppets, grocery bags, a one dollar bill, postage stamps in five denominations, and a cancellation mark used by the post office. Over 26,000 school children were visited by Quincy (a person wearing a parrot costume) plus a counterpart who taught the children a song about the Bahama parrot, led them in a parrot dance, and told them about the plight of the Bahama parrot and other wild life. Fact sheets about the Bahama parrot were distributed widely; for example, clergy were sent these sheets along with a selection of universal prayers with an environmental theme and scriptural sources pertaining to caring for the earth. A rap song, music video, and a stream of press releases made sure the Bahama parrot program was very conspicuous in the mass media.

The campaign has worked. Questionnaires have ascertained that most Bahamians are now aware of the Bahama parrot's plight, and this has translated into direct action. In particular, a 10,700-ha national park was created on Abaco Island. It protects habitat not only for the parrot, but for many other species, at least one of which, the Kirtland's warbler, is even rarer than the parrot. It is also likely that the campaign has sensitized Bahamians to the ecological well-being and biological riches of their nation in general.

Summary

The attitudes people have toward other organisms and conservation vary enormously from person to person. While each person's values may be unique, there are patterns that can, to some extent, be explained by culture, gender, income, occupation, age, and other factors. Understanding how these factors affect someone's attitudes is easier if we use a systematic means of describing attitudes toward nature. Values change through time, and promoting positive attitudes toward the natural world is a fundamental part of environmental education. By informing people about the importance of biodiversity and encouraging them to experience nature, we can foster attitudes that move us away from an anthropocentric (human centered) view of the world toward a perspective that is more biocentric (life centered).

FURTHER READING

See Smith (1999), VanDeVeer and Pierce (1994), and Zimmerman et al. (1993) for anthologies of papers about environmental philosophy that includes work on ecofeminism, biocentrism, and other relevant topics, and see Pojman (1999) for a textbook on the same topics. Aldo Leopold's (1949) *Sand County Almanac* remains an important source. For a summary of Stephen Kellert's work, see his 1996 book. For some recent ideas about environmental education and changing values, read Orr (1992, 1994). Jacobson and McDuff (1998) make a strong argument for why conservation biologists need to understand the human dimension of conservation.

TOPICS FOR DISCUSSION

1. What factors do you feel have the most profound influence on a person's attitudes toward wild organisms? Has this changed in recent years? Will it change in the future?
2. Is trying to change another person's values generally acceptable? When is it *not* acceptable? When is it acceptable?
3. How would you encourage Chinese peasants to protect giant pandas and their habitat even though they could earn many years' wages by catching one?
4. Do you feel that you are more anthropocentric or more biocentric? Why? Has taking a conservation biology class and/or reading this text shifted your values? (See Caro et al. 1994.)

Chapter 16
Economics

Both a date palm and a tiny mite living in a crack in the palm's bark may stand equal in the eyes of people who believe that every lifeform has intrinsic value, value that is independent of the special, self-centered interests of humanity. However, intrinsic values do not eliminate or invalidate instrumental values. Instrumental values are still there, profoundly influencing the way most people view date palms and mites and thus potentially determining how they will fare in a world dominated by people. This is especially true when the instrumental values are readily translated into economic values. The phrase "money makes the world go round," may seem hyperbolic given the realities of astronomy, but it does hint at the central role of economics in most human undertakings, including conservation. Consider the World Conservation Union's definition of conservation: "the management of human use of the biosphere so that it may yield the greatest sustainable benefit to present generations while maintaining its potential to meet the needs and aspirations of future generations" (IUCN et al. 1980). Those certainly sound more like the words of an economist than the words of a biologist.

In this chapter we will first address a two-pronged question: what are the costs and benefits of maintaining biodiversity? Then we will examine how these costs and benefits are distributed among people. Perhaps you can already anticipate the take-home message of this chapter: people are compelled to overexploit species and degrade ecosystems because the costs and benefits of maintaining biodiversity are typically not distributed in a fair and sensible manner.

The Benefits

Coal, copper, and gold are fundamentally different from cod, herring, and oak in the eyes of an economist who sees the world as a collection of resources to be exploited for the benefit of people. The first three are *nonrenewable resources,* supplies of which are finite and exhaustible. The latter three are *renewable resources;* they will last indefinitely if used wisely. Most renewable resources are living, biological resources, but some nonliving resources such as clean air and water are also renewable. To exclude domestic species, which are also a renewable resource, the term *renewable natural resource* is often used. Quantifying the values of renewable resources is often more difficult than quantifying the values of nonrenewable resources. A herring is not just a commodity waiting to be sold at market. It is a living thing—dynamic, mobile, and interacting with many other species.

In this section we will describe benefits as economists do—goods and services. Briefly, goods are physical objects that you can purchase, own, and use, while services are labor that is performed for your benefit (e.g., by a professor teaching you). We will also discuss two types of benefits that are less concrete: potential values and existence values. We will continue to focus on wild species, but we will not overlook domestic species entirely. Domestic species may be a small component of biodiversity, a few hundred species among millions, but they are a significant portion of the overall economy.

Goods

Plants and animals harvested from the earth's farmlands, rangelands, forests, and waters and sold at market account for nearly $2 trillion per year (United Nations Development Programme et al. 2000), roughly 6% of the global domestic product. (All monetary figures used here will be in United States dollars.) In many less-industrialized countries such as Albania,

Burundi, Guyana, and Nepal the percentage is much higher, 30% to 50% or more. The commercial value of wild species is a significant portion of this total because two major enterprises, forestry and fisheries, rely far more heavily on wild stocks than on plantations and aquaculture. Global values for wood and fishery products have been estimated at $418 billion and $70 billion, respectively (Freese 1998). (Note: the annual production of fisheries is about $20 billion *less* than the cost because of government subsidies that we will discuss further.)

A significant portion of the goods derived from biota are not bought and sold commercially. They are used for direct subsistence by the people who collect them. For example, modern supermarkets with thousands of food products—sardines from Norway, grapes from Chile, tea from Sri Lanka—are amazing things, but far more people grow much of their own food than use supermarkets. Similarly, wild species gathered from nearby ecosystems are often very important to subsistence lifestyles, especially for fuel and building materials.

Estimating the total volume and value of wild and agricultural products that are consumed directly by the people who grow and harvest them is not easy for two reasons. First, because items move from ecosystem to home quickly, not changing hands except within a family, it is difficult for government data collectors to record these uses except by doing a house-by-house survey. Second, many wild products are seldom sold at market, making it difficult to estimate their true value. Despite these difficulties, several assessments have been made. For example, in two Amerindian villages in southern Venezuela the values of plants and animals collected in the forest were estimated at $1902 per household per year for the large village and $4696 for the small village (Melnyk and Bell 1996). It was specifically noted that these values were higher than the people could have earned if they had invested the same amount of time working for wages.

In some areas, simply determining the variety of wild species being used is a substantial exercise. For example, Phillips et al. (1994) documented uses of 57 families of woody plants in one small area of the Peruvian Amazon.

Services

The distinction between goods and services is readily applied to the benefits we receive from other living things. If you eat the dates of a date palm, it has provided you with goods; if you rest in its shade, it has provided you with a service. Of course, estimating the value of the shade provided by a date palm can be difficult unless the person who owns the date palm charges a fee for this service, and, in practice, most of the services provided by genes, species, and ecosystems are not sold.

This has not stopped creative people from devising ways to estimate the value of biological services (Westman 1977, Costanza et al. 1997a, Daily 1997, Pimentel et al. 1997). For example, horticulturalists routinely estimate the value of the aesthetic services provided by ornamental trees, usually to settle insurance claims, and values often reach thousands of dollars for a tree too large to be directly replaced (Council of Tree and Landscape Appraisers 1992). One ambitious project attempted to estimate the total value of global ecological services in two steps (Costanza et al. 1997a). First, they summed value estimates (dollars per hectare per year) for 17 types of service across 16 major types of ecosystem in a huge matrix. (Some goods were included too, but they were dwarfed by services.) For example, the value of coral reefs was estimated at $6075 per hectare per year by summing up $3008 for recreational value, $1 for cultural value, $2750 for disturbance regulation (blocking storm waves), $58 for waste treatment, and so on for eight types of service provided by coral reefs. For tropical forests, estimates for 14 different types of service totaled $2007 per hectare per year. Open oceans totaled $252 per hectare per year. Many cells in this matrix were empty because there were no estimates available; indeed, some whole ecosystems (e.g., tundra and deserts) were given no value because there were no estimates available. After they had summed estimates for each ecosystem type these were multiplied by the global area of that type. In other words, the $6075-per-hectare-per-year figure for coral reefs was multiplied by 62 million hectares to become an estimate of $375 billion per year for the ecosystem services of all the earth's coral reefs. The per-hectare figure for open ocean was much smaller, $252; however, when multiplied by 33 billion hectares, it yielded a total value over $8 trillion. Summed across all the ecosystem types, the grand estimate for global ecosystem services was $33 trillion per year with a reasonable range of $16 trillion to $54 trillion; as a point of reference, the gross national products of all the world's nations total about $18 trillion. Because of various uncertainties, notably missing estimates for many services of many ecosystem types, the estimate was considered a minimum.

Skeptics would claim that, since people do not actually pay $6075 per hectare per year for the services of a coral reef, these numbers are too speculative to be useful. Nevertheless, these numbers do help to bring home the importance of ecosystems, especially to people who view the world in economic terms. (See Box 16.1 for a more detailed example of estimating value; see Wilson and Carpenter [1999] for a comparison of three common methods.)

There are some important exceptions to the generalization that ecological services are not bought and sold. In particular, ecological services that are chiefly based on aesthetics and recreation are widely purchased. This happens whenever people pay for the privilege of visiting a natural ecosystem or encountering wild life by paying for transportation, food, lodging, equipment, guide services, licenses, entrance fees, and so on. The total value of nature-based recreational activities or ecotourism is enormous. In the United States alone, people pay about $101 billion per year just for recreation centered around wild animals: hunting, fishing, and wild animal viewing (U.S. Fish and Wildlife Service 1997). When the actual expenditures people make for recreational services are compared with the amounts they are willing to pay (see Box 16.1), it is usually discovered that people would be willing to pay more than they actually have to pay. In other words, the potential value of recreational services probably greatly exceeds the current market value.

Potential Values

Most economists are reasonably comfortable with the idea of predicting the future, but they are quite conservative about predicting the future value of all the species that are not currently of direct use. Their reticence is understandable. Who is to say which species of plant or microbe, if any, will be discovered to contain a cure for AIDS or malaria?

Given that we cannot predict where science will take us, the one thing that we can say with confidence is that all life-forms have potential value. Perhaps we can say that some life-forms have more potential value than others: for example, plants that have high levels of bioactive compounds, or wild species that are close relatives to important domesticates and that might serve as a source of genetic material. Again, however, we can never definitively say of any form of life that it has no potential value.

Biodiversity advocates usually think of potential values, also called option values, with respect to species and genes not known to be of direct value currently. However, in a broad sense, the term could be used for populations or ecosystems that are known to be of value if someone wished to initiate exploitation. For example, forests in our nature reserves or in parts of Siberia and Canada that are too remote to be logged currently have potential value because we can cut them in the future if we need to. It is not a matter of determining their potential usefulness, only of deciding whether it makes sense to use them at a particular time.

Existence Values

Your chances of ever seeing a wild snow leopard, slinking down a slope of snow and scree, are exceedingly small. Snow leopards are so rare and elusive, and their habitat in the mountain fastness of central Asia is so inaccessible, that only a handful of outsiders have ever seen one. Nevertheless, I would guess that you derive some pleasure simply from knowing that snow leopards exist in the wild. Many people feel this way about species that they will never encounter and ecosystems that they will never visit. We could argue about whether this phenomenon is based on spiritual values,

BOX 16.1	Using contingent valuation to value elements of biodiversity[1]

Kevin J. Boyle[2]

About one-half of all land in the contiguous United States is used for cropland or pasture. Biologists have expressed concern about the declining populations of grassland bird species, and loss of habitat is generally cited as the major reason for their decline. The major historical grassland area of the United States is in the plains states, where agriculture dominates the landscape. Over time, farms have been established and consolidated, leaving less undisturbed habitat for grassland birds.

Beginning in 1985 the U.S. Department of Agriculture's (USDA) Conservation Reserve Program (CRP) converted about 7% of the cropland in the 48 contiguous states to grassland. Over four-fifths of this area is concentrated in one-fifth of the counties, which are predominantly in the plains states. The Wildlife Management Institute (1994) reports that CRP grasslands cover at least twice the area of grasslands in all of the national and state wild life refuges within the continental United States. Anecdotal evidence by wild life experts (e.g., National Audubon Society) and empirical analyses suggest that the CRP has helped to reduce, stop, or reverse the declines in the populations of some grassland bird species.

Initial enrollments in the CRP were prioritized according to the erodibility of the soil. As priorities of the CRP are expanded to recognize other environmental benefits such as improved habitat for grassland birds, information is needed on the values the public places on such benefits. Revisions of the CRP in 1990 introduced the Environmental Benefits Index (EBI) as a tool to prioritize and rank landowners' offers of land for enrollment in the CRP. Contingent-valuation estimates of the values the public places on changes in the populations of grassland birds can be used to help to justify changes in grassland bird populations as a component of the EBI.

Two samples were used, one national and one of Iowa residents, and the survey was administered by mail. Individuals in the national sample were asked about changes in populations of grassland birds in the plains states, which included all of Iowa and parts of the border areas of each state that is adjacent to Iowa. Iowa is one of three areas of high concentrations of CRP lands.

Empirical analyses of grassland bird populations suggested that CRP lands may benefit populations of grassland birds whose populations have been decreasing over the last 30 years (grasshopper sparrow, Henslow's sparrow, mourning dove, eastern kingbird, northern bobwhite quail, horned lark, dickcissel, and ring-necked pheasant) and may also benefit species whose populations have been constant or increased over the last 30 years (lark sparrow, upland sandpiper, gray partridge, field sparrow, indigo bunting, killdeer, barn swallow, and house wren). We asked respondents to reveal their monetary values, through a contingent-valuation question, for restoring populations of grassland birds to their population of 30 years ago (for the species whose populations had declined), or for increasing the populations (for

BOX 16.1 *Continued*

species whose populations had been constant or increased over the same period). For declining populations the proposed increases ranged from 14% for ring-necked pheasants to 136% for grasshopper sparrows. The proposed increases for the second group of species ranged from 20% for house wrens and 84% for lark sparrows.

The survey described a proposal where changing agricultural lands to native grasses would result in the populations of each of the species specified above being increased by a certain amount. For each species, respondents were told: 1) whether it was native or introduced, 2) whether it was a permanent resident or migratory with breeding habitat in the study area, 3) the population 30 years ago, 4) the current population, and 5) the population with the habitat enhancement. Study participants were asked two questions. The Iowa sample was first asked:

"Would you vote for the proposal if passage of the proposal would increase your household's 1998 Iowa income tax?"

Those who answered YES were asked a second question:

"How would you vote on the proposal if passage of the proposal would increase your household's 1998 Iowa income tax by the following amounts?"

The dollar amounts ranged from $1 to $100. Participants in the national sample received similar questions where they were asked about increases in their federal income tax. Responses from the first question allowed us to find out if respondents held a value for changes in the populations of the specified grassland birds, and the second question allowed us to statistically estimate the value for people who hold values for the population changes.

For the national sample, 71% of respondents answered yes to the first question, and the comparable figure for the Iowa sample was 72%. The average values were $12 for the national sample and $13 for the Iowa sample.

The values for the national sample are primarily composed of existence values because these people are not expected to have any direct use of these grassland birds for viewing or hunting, while the Iowa values contain a use component. Quail, pheasant, and partridge can all be hunted in Iowa, and other species in the lists are popular for viewing.

These results indicate substantial public values for improving grassland bird habitats in the greater Iowa area from both a national and local perspective. When the averages of $12 or $13 are multiplied by the respective populations of households, the aggregate economics benefits are substantial: over $500 million at the national level. While increased populations of grassland birds are only one of the environmental benefits of the CRP, these valuation results indicate that providing habitat for grassland bird populations should be a component of USDA's EBI for prioritizing land to take out of agricultural production.

---

[1] This box is distilled from Ahearn, Boyle, and Hellerstein (in press).
[2] Department of Resource Economics and Policy, University of Maine, Orono, Maine.

ethical values, respect for intrinsic value, or other factors, but the key point is that it is not tied to goods and services. Economists prefer to speak of *existence values*, the value of simply knowing that something exists (Krutilla 1967). Sometimes, people like to know that something exists simply because it might be of use to future generations, even though it is not used now. This is called *bequest value* and one could consider it to be a special type of existence value or a special type of potential value.

Economists have estimated existence values by asking people questions such as, "How much would you pay to save blue whales from extinction, even though you will never see one?" The average amounts mentioned in response are usually rather small, typically a few dollars, although they may reach tens of dollars for well-known species such as bald eagles (Bishop and Welsh 1992). Nevertheless, if you multiply these figures by millions of people, then the total values can be quite impressive. Existence values are quite controversial because some economists object to attempts to quantify something so intangible, especially by asking people hypothetical questions about how they would spend money; see Kopp (1992), Rosenthal and Nelson (1992), Larson (1993), Stevens et al. (1994), Blomquist and Whitehead (1995), and Attfield (1998).

Consumptive Versus Nonconsumptive Uses

We need to add a coda to our list of benefits to explain some terminology often used by natural resource managers. When we use something in such a manner that it is no longer available for someone else to use (e.g., we harvest an oak tree or codfish), this is called *consumptive use*. On the other hand, if our use does not eliminate or substantially reduce its value (e.g., children climbing in the oak tree), this is *nonconsumptive use*. Generally, goods used for commerce and subsistence involve consumption, whereas services and existence values are nonconsumptive. Sometimes, using ecological services can involve a form of consumption, for example, when so many people visit an ecosystem that they degrade it simply by compacting soil, trampling vegetation, and frightening animals. Some economists also distinguish a third style of use, *indirect use*. In this sense, people who know and value the Serengeti, Amazonia, Great Barrier Reef, mountain gorillas, and blue whales through books and films, but will never encounter them directly, are making indirect use of these ecosystems and species.

The Costs

"There is no such thing as a free lunch." This truism is a favorite among environmentalists, who frequently use it when pointing out the hidden costs of environmental degradation. Consider a simple example: historically, when a businessperson weighed the costs and benefits of building a coal-fired power plant, the costs in terms of respiratory disease of people living downwind were not included in the calculations. These were external costs (externalities, in the language of economics) that did not affect business profits and losses. One consequence of the environmental movement is that many costs that were formerly external have been internalized. Today, in many countries environmental regulations mean that business people must include the costs of pollution control in their calculations. This is often called the "polluter-pays" principle. Hidden environmental costs still exist, but a precedent for internalizing them has been set. We will return to this issue later, but in the balance of this section we will turn the "no free lunch" truism upside down by asking what are the costs of maintaining biodiversity—of maintaining a healthy environment. There are basically two; we will call them explicit and implicit costs.

Explicit Costs

Human beings are doers, actors, manipulators. If we encounter a problem, we try to solve it. As it became apparent that loss of biodiversity was a problem, we began to attack the problem by using a host of technological approaches outlined in the chapters of Part III, Maintaining Biodiversity. We restore wetlands, grasslands, and other ecosystems; we translocate endangered species and provide them with food and other required resources; we redesign our existing technology to reduce environmental pollution and energy use; and so on. Collectively, we could refer to these explicit costs as the cost of environmental technology.

Our response may be grossly inadequate, but this work still requires money. Some projects have substantial budgets such as bringing California condors back from the brink of extinction (well over $1,000,000 per year) (Cohn 1993) or restoring the Everglades wetland complex of southern Florida (estimated at $7.8 billion over 20 years) (Kloor 2000). Others get by on a shoestring. For example, almost the entire world population of Robbin's cinquefoil is protected in a 1-ha area below the summit cone of Mount Washington in the White Mountain National Forest, New Hampshire. Closure of the area to the public, rerouting of the Appalachian Trail, periodic population censuses, a successful transplant program, and a hiker education program have cost less than $10,000 per year and the species is now being reviewed for removal from the U.S. Endangered Species list (K. Kimball, personal communication). In between these extremes we have estimates that $32 million to $42 million per year would pay for habitat management for 681 United States' endangered species (chiefly through the control of exotic species and the management of natural fire regimes) (Wilcove and Chen 1998).

Turning to the big picture, it has been estimated that the current global set of ecological reserves costs $6 billion per year to maintain, $453 per square kilometer, and that an additional $2.3 billion are needed to do the job adequately (James et al. 1999). Furthermore, increasing the current coverage to cover 15% of the land area would add only $19.2 billion per year for acquisition, management, and compensation for people who were displaced by the new reserves. Looking beyond reserves to the farms, forests, grasslands, and water bodies managed for commodity production, James et al. suggested that subsidies designed to make farming, logging, fishing, etcetera, ecologically friendly would bring the total global bill for conservation to $300 billion. That is, of course, a great deal of money but not when compared with the subsidies governments currently give to farmers, fishers, etcetera, usually to do things that are rather detrimental to the environment. These negative subsidies have been estimated at roughly $1.5 *trillion* dollars per year, about five times the amount needed for conservation (Myers 1998). To put it in other terms, $300 billion for conservation is only about $50 per person per year, an extraordinary bargain when you consider how many of us spend much more than this on junk food alone.

Implicit Costs

From an economist's perspective, whenever a logger is prevented from cutting a tree, a whaler from harpooning a whale, or a farmer from draining a wetland, they have suffered an implicit cost because they have lost an opportunity to use a resource to make money. These losses of opportunity will seem most acute if a specific investment has been made; for example, if the wetlands, forest land, whaling ship, logging equipment, and so on were purchased on the assumption that the trees, whales, or wetlands were available for use. If the farmer has owned the wetland for many years and only recently considered draining it, or if the whaler purchased the ship 20 years ago and long ago repaid the initial investment, then the loss may seem less severe. Similarly, the loss will seem less acute if the investment can be easily redirected somewhere else—if the logging equipment can be used to cut a different stand of trees, for example. However, if the investment is in land, it can be difficult to sell the land if it has special ecological values that restrict how it can be used.

The implicit costs imposed by environmental regulations can be significant. The timber in a single hectare of old-growth Douglas fir in the Pacific Northwest can have a stumpage value (the price paid by the logger to the landowner) of about $75,000 per hectare (Lippke and Bishop 1999). This translates into $75 million for a thousand hectares of old-growth forest (approximately the area needed by a single pair of spotted owls) and an even greater loss if you assume that the land will not be available for growing lumber in the future.

The Distribution of Benefits and Costs

Life is not always fair. Some acorns are carried by squirrels to a sunny spot with a moist, fertile soil and left there to grow; some acorns are eaten by squirrels or weevils before ever leaving the tree. Some people derive more benefits from the maintenance of biodiversity than do some other people; some people bear relatively more costs. Before going on to some ideas about how to make life a bit more fair, in this section we will briefly examine how the benefits and costs described are distributed among people.

Goods: The commercial and subsistence benefits of biodiversity flow most directly to the producers who grow domestic species or harvest wild life; to the manufacturers who generate secondary products such as paper and medicines from biotic materials; and to the merchants who distribute these items to consumers. Of course, the ultimate beneficiaries of commercial use are consumers, and we all consume other species.

Services: Everyone benefits from the ecological services of the earth's biota, even urbanites who never leave their cities, but still need clean air and water. Most people also enjoy ecological services based on aesthetics, especially if you include watching nature films on television and walking in a city park.

Potential Values: Potential values simply reflect the possible future value of goods and services, and thus one could describe their principal beneficiaries as our children and grandchildren.

Existence Values: Many people have heard of at least a few of the better-known species (e.g., ostriches, giraffes, koalas) and have a generally positive impression toward them that can be construed as an existence value. Of course, on the other hand, most people have not heard of rheas, okapis, and wombats, to say nothing of the myriad species of invertebrates and microorganisms. In other words, most people hold existence values for biota, but they are focused on a tiny subset of biodiversity.

Explicit Costs: Understanding how the costs of environmental technology are distributed is a bit complex because there are four widely overlapping groups involved: taxpayers, consumers, business owners, and volunteers. Taxpayers fund all the government agencies that undertake conservation work such as establishing and maintaining reserves, restoring ecosystems, protecting endangered species, and so forth. Consumers usually pay for environmental technology through higher prices whenever regulations mandate that businesses internalize the costs of maintaining a healthy environment. On the other hand, there are at least two circumstances under which the owners of a business will bear the cost of internalizing environmental technology through a reduction in their profits. First, this can happen if competing businesses are not subject to the same regulations. For example, imagine that government regulations require Argentinean farmers to use an expensive, low-toxicity pesticide, whereas Chilean farmers can use a cheaper but more dangerous pesticide. If farmers in both countries are competing for the same international market, the Argentinean farmers may have to lower their profits so that they can still sell their produce at the same price as the Chilean farmers. Second, some businesses (e.g., utilities companies) have their profits regulated by the government, and the government can decree that internalization of environmental costs should be borne by the company owners rather than consumers. The fourth group of people who pay the explicit cost of environmental technology are the millions of individuals who pay dues to belong to private conservation groups, thereby supporting conservation work with voluntary contributions of money and sometimes labor.

Implicit Costs: When protection of biodiversity takes precedence over someone's opportunity to use a species or ecosystem for personal gain, the costs are borne most directly by people who make their living by farming, logging, fishing, hunting, trapping, mining, developing land, and so on. Most of these people are commercial entrepreneurs, ranging from huge corporations that own millions of hectares of forest to loggers who own little more than an axe. A few are subsistence users who sell little, if any, of their harvest. Merchants and manufacturers who might have used the species or ecosystem later (e.g., converting a tree to paper and then selling it) will also experience an implicit cost through losing an opportunity to use a resource.

Problems and Solutions

Superficially, there seems to be a fairly good balance in the distribution of biodiversity maintenance costs and benefits because almost everyone sits on both sides of the equation. We all use ecological services and consume biota-based products on the one hand; we all support biodiversity maintenance through taxes and higher prices on the other hand. On closer inspection there are many fundamental imbalances. In this section we will examine four problems and outline some possible solutions. Note that while it is easy to suggest solutions, this does not mean that it is easy to implement them. As you will see, many of them require fundamental changes.

Problem 1. Many biological resources are communally owned, and thus the costs of overexploitation and degradation are shared by many people, whereas the benefits of overexploitation are taken largely by the people who are doing the overexploiting. If you are familiar with Garrett Hardin's classic essay "The Tragedy of the Commons," you will realize that this situation tends to compel overexploitation. (See Box 16.2 if you are not familiar with the "tragedy of the commons," formally known in economic circles by terms like *"open-access resource management"* [Costanza et al. 1997b] .) This phenomenon is particularly pervasive in aquatic ecosystems, especially the oceans, where private ownership is rare. Indeed, in most parts of the oceans and in Antarctica, ownership is not even claimed by nations, let alone by individuals or corporations. Ownership is also a very tenuous thing in many tropical forests where local people have only tradition, not legally binding documents, as a basis of "ownership" and are frequently at risk of being displaced by programs concocted by the government and large businesses (see Oldfield and Alcorn 1991).

The "tragedy of the commons" dilemma is also applicable to biodiversity in general if you think of genes and species as communally owned assets. From this perspective, you may own the individual sequoia trees growing on your land at this time, but the sequoia as a species is owned by everyone on earth as part of a global biological heritage. (Note that this requires an anthropocentric perspective, and it would not be accepted by a biocentrist; recall Chapter 15.) If sequoias were to become extinct, we would all lose something; however, the people who drove the species into extinction might gain more than they lost and thus would be doing something consistent with their self-interest.

Solutions: Passing and enforcing laws are the standard ways of ensuring that people do not harm society as a whole while acting in their own self-interest, and laws designed to protect biodiversity are widespread. Nevertheless, new laws and better enforcement of existing laws are needed. For example, in many countries wild animals are owned and protected by society as a whole because they can move from property to property, but wild plants are owned by whoever owns the land where they are rooted (Bean 1983). This can make it very difficult to prevent landowners from destroying plants, even if they are a highly endangered species, and thus new laws to protect endangered plants are sorely needed. Laws to protect biodiversity may simply prohibit certain actions (e.g., banning the killing of endangered species), or they may impose significant financial costs (e.g., by

BOX 16.2   Tragedy of the commons

In a classic essay entitled "Tragedy of the Commons," Garrett Hardin (1968) explained how communally owned natural resources are highly vulnerable to degradation through overuse. This is easy to understand using the metaphor of an English village commons, a tract of pasture land used by all the farmers of a village to graze their cattle. Imagine that the commons can support 100 cattle without being degraded and that currently there are 20 farmers using the commons, each of whom owns five cows. Each of these cows can produce an average of 10 kg of milk per day, so the total milk production of the commons averages 1000 kg/day, and the production for each farmer averages 50 kg/day. Now, imagine that one day one of the farmers considers buying another cow, thus increasing her herd to six cows and the commons herd to 101. She knows that, if she does this, she will push the herd above the carrying capacity of the commons and average milk production per cow will fall; let us assume it will fall 1% to 9.9 kg/day. Should she buy the extra cow? If she is acting in terms of her own immediate economic interest, she should because her cows' daily milk production will increase from 50 kg to 59.4 kg (6 cows × 9.9 kg/cow). Unfortunately, average milk production for the other 19

requiring mining companies to establish a fund that will be used to restore a mined ecosystem after the mine is closed). In small, local communities formal laws are often unnecessary because simple rules of conduct regulate sharing common property as a public trust (Ciriacy-Wantrup and Bishop 1975). Sometimes, the benefits of smaller groups can be achieved while retaining some central government control. For example, the lobster fishery along the coast of Maine is now managed in large part by seven local councils of elected lobster fishers (Acheson et al. 2000). Sharing authority, so-called comanagement (see Chap. 17), between these councils and the state government has produced one of the most successful examples of sustainable fisheries management.

BOX 16.2   *Continued*

farmers will fall to 49.5 kg/day (5 × 9.9). Now, imagine that one of these farmers decides that he needs more than 49.5 kg/day, so he buys another cow, too, even though average production per cow will fall again, to 9.8 kg/day. This will make economic sense for him because his production will be 58.8 kg/day (6 × 9.8), far better than 49.5 kg/day. This pattern can continue to snowball until all 20 farmers have six cows, total production is 960 kg/day for 120 cows, and each farmer is producing 48 kg/day. This is 2 kg less than when the cycle began, and now each farmer has to care for six cows instead of five. (This assumes that production per cow continues to drop by 0.1 kg for each cow over the carrying capacity; in reality the drop per cow may get larger as the carrying capacity is exceeded by a greater and greater number of cows.)

Although the numbers used here were contrived to make a simplified, hypothetical example, the tragedy of the commons is real. It is well illustrated in many fisheries where each fisher buys more equipment and works longer hours to catch a larger share of a fish population that is constantly dwindling because it is being overexploited (Wilson 1977, Butler et al. 1993). It remains a problem on communal grazing land in many countries (Yonzon and Hunter 1991). It also underlies one problem with ecotourism: tour guides will take their clients closer and closer to wild animals, thus assuring a good tip, until they end up molesting the animals. Although the classic image of a commons is a communal resource that everyone consumes, we can also think of the commons as a communal place where everyone deposits his or her wastes such as the earth's atmosphere and waters. In both cases, each individual, acting in accord with his or her own short-term economic interest, degrades the long-term economic well-being of everyone. For a collection of writings about the tragedy of the commons, see Hardin and Baden (1977); also see Ciriacy-Wantrup and Bishop (1975), Leal (1998), and Uphoff and Langholz (1998).

The regulatory approach to protecting our global biotic heritage becomes quite complex when international laws or treaties are required. International treaties designed to protect migratory birds have been moderately successful in terms of curbing overhunting, although they have done little to stem habitat loss. Attempts to restrict overfishing the oceans through the Law of the Sea conferences have been much less successful. A major advantage of international coordination is that it can make the regulatory approach to maintaining biodiversity more fair by compelling all the businesses that are competing for the same market to internalize the environmental costs of doing business. To return to our earlier hypothetical example, international coordination of environmental regulations can force both Chilean and Argentinean farmers to use less dangerous, but more expensive, pesticides and thus can avoid giving either group an unfair advantage. Furthermore, if collaboration fails, a government could act unilaterally by imposing "ecological tariffs" to increase the cost of goods imported from any country where environmental costs are not internalized (Costanza et al. 1997b). However, this approach runs headlong against the World Trade Organization's (WTO) General Agreement of Tariffs and Trade (GATT), which requires international agreements on any environmental constraints to free trade (Abboud 2000).

It is often argued that the solution to the "tragedy of the commons" dilemma is to privatize resources that are usually communal. Some governments have done this by giving or selling permanent ownership or long-term leases on government-controlled resources; leases of coastal waters to private aquaculture operations are a common example. Privatization of genetic information derived from natural populations of plants and animals has, to date, been resisted. In particular, germplasm of wild relatives of crop plants is traditionally considered a global heritage and is widely shared among agricultural researchers as communal property. However, things become more complicated when biotechnology is involved because of international laws protecting intellectual property rights. If plant breeders use genetic engineering to develop a new breed of rice, should they have exclusive rights to sell this breed with its unique genetic information? Should pharmaceutical researchers who develop a new medicine based on a chemical they identified in a plant have to share their profits with the people who live where the plant grows?

Considerable dissension has arisen over genetic resources, especially between developed countries, which tend to be rich in technology but relatively poor in terms of genetic diversity, and tropical developing countries, which tend to be technology poor and gene rich. The issue was particularly prominent at the 1992 Earth Summit (formally UNCED, the United Nations Conference on Environment and Development) where the Convention on Biodiversity called for a "fair and equitable" sharing of the profits obtained by biotechnological development based on biological resources. In other words, if a United States pharmaceutical company developed a new medicine from a plant obtained in Ecuador, the company would have to share a "fair and equitable" portion of its profits with the government of Ecuador. Without this provision there might be little economic incentive for Ecuador to protect biota that have potential value to biotechnology companies elsewhere. The United States has refused to sign the Convention because of ambiguity over what was "fair and equitable." See Kloppenburg (1988), Sedjo (1988), Vogel (1994), and Grajal (1999) for further insights on the issue of who owns genetic information.

Problem 2. When biodiversity maintenance generates an implicit cost by reducing opportunities to use resources, this cost often falls on relatively few people, especially poor people in rural areas. The richest parts of our planet in terms of biodiversity are often the poorest in terms of economics. The image of poor peasant farmers trying to scratch out a living by clearing garden patches in tropical forests is only the most dramatic example of a widespread discrepancy between the biotic and economic wealth of rural and urban areas. These situations are rife with unfairness. If the government of Uganda establishes a new national park to protect gorilla habitat from encroachment by local farmers, it will cost the affluent fans of gorillas in Boston and Bern virtually nothing, perhaps a few cents each if they pay an annual membership fee to an international conservation group that is supporting the project. For a young Uganda couple, living in a village on the border of the proposed park and looking for land where they can start a farm and a family, establishment of the park might severely constrain their opportunities (Eltringham 1994).

Inequities often occur within a single country too. If the people of Belgium decide that there is not enough forest remaining in the country and pass a law prohibiting the conversion of forest to farmland or housing developments, the loss of opportunity falls on the small number of Belgians who own forests.

Solutions: Simply put, maintaining biodiversity is everyone's responsibility, and therefore everyone must share part of the burden. Sometimes, this will require a net flow of funds from society as a whole to those people who experience the costs of maintaining the earth's biota most directly (Shogren et al. 1999). This is particularly important because in many parts of the world relatively poor people bear much of the cost.

Within a single government this redistribution of funds is relatively easy because governments have many mechanisms for subsidizing activities deemed to be in the public interest. For example, property taxes on lands that are managed to maintain their ecological values can be reduced or waived, or subsidized prices can be paid for carefully harvested commodities. Conservation groups and governments often purchase *conservation easements* from private landowners, a legal mechanism by which landowners give up certain property rights for conservation purposes (typically, they sell, in perpetuity, the right to develop the property). Permanent conservation easements often cost over 80% of the regular purchase price of a tract of land. Sometimes, specific ecological services are purchased on an annual basis. For example, Main et al. (1999) described resource conservation agreements designed to give private landowners a financial incentive to maintain habitat for the Florida panther—$74 to $82 per hectare per year.

Sharing the burden internationally is more difficult. Monies collected by international conservation groups in wealthy countries and spent in poor countries are one mechanism. Formerly, these monies were used almost exclusively for the cost of environmental technology, especially as salaries for foreign biologists who would travel to developing countries to try to conserve the local wild life. Now, much of this money goes to local conservationists, whose activities often include developing economic alternatives, education, medical aid, and other forms of assistance for the people whose ability to make a living is compromised by biodiversity projects. One innovative way of generating funds used by international conservation groups is debt-for-nature swaps (Webb 1994), which are explained in Box 16.3. There is also growing interest in allowing industries that produce greenhouse gases to compensate for this activity by paying to maintain forests where carbon can be sequestered, especially tropical forests in developing countries (Boscolo et al. 2000, Rotter and Danish 2000).

Another mechanism for sharing the financial burden internationally is the transferring of funds from wealthy governments to poor governments as bilateral aid either directly or channeled through an intermediary organization such as the World Bank or the United Nations. Historically, international aid has done far more harm to the environment than good, particularly through construction of dams and roads and initiation of badly designed agricultural, fishing, and logging schemes. In recent years, most international-development agencies are at least trying to ameliorate the environmental impact of their projects, and in some cases they are undertaking projects such as establishing new national parks that have a primary goal of conservation. It would be easy to digress into a long critique of large-scale projects that completely overwhelm the people they are meant to help—whether they be building a dam or a park—but that lies beyond our purview here. See Timberlake (1986), Cheru (1989), and Ayittey (1998) for examples.

Probably the best way to offset the losses experienced by people who share their land with wild life is to find ways to increase the benefits they receive from the local biota. This idea is the basis for what is often called "community-based conservation" or an "integrated conservation and development project" (Alpert 1996, Hackel 1999). Conservationists often promote ecotourism to this end because the Ugandan couple may not need a farm if they can get jobs as guides for tourists who come to see the gorillas (Paaby et al. 1991) (Fig. 16.1). Ecotourism has some problems. For example, it can lead to significant environmental degradation; some people do not relish working for demanding tourists; and much of the money it generates goes to foreign-owned airlines, hotels chains, tour companies, etcetera, rather than to local people (Bookbinder et al. 1998). However, if done properly, ecotourism can produce significant benefits for local people (Cater and Lowman 1994). Improving markets for local

BOX 16.3   Debt-for-nature swaps

Many nations have borrowed large sums of money, well over two trillion dollars in total, to invest in their economy by building roads, dams, irrigation systems, factories, and the like (United Nations Development Programme et al. 2000). Some of this debt is being steadily repaid, and it contributes to a net flow of money from the world's poorer countries to the world's richer countries of many billions of dollars per year. However, even this rate of repayment is slow compared with what is owed, and, consequently, the commercial banks that made these loans often sell the debts on the secondary market for much less than their face value, commonly at about 10% to 20%. This means, to take a hypothetical example, that if the government of a Latin American nation owes $10 million to a bank in New York City, the bank would be willing to sell the debt bond to another institution for $1 million to $2 million. Beginning in 1987 conservation groups and some wealthier nations have bought these discounted debt bonds to generate funds for conservation in what are called *debt-for-nature swaps* (Hansen 1989).

For example, in 1990 a coalition of the government of Sweden, the World Wildlife Fund, and The Nature Conservancy banded together to purchase $10,753,631 worth of Costa Rica's debt bonds, which, because they were discounted, only cost $1,953,473 (WRI 1992). They gave these bonds to Costa Rica, and in exchange, the Costa Rican government agreed to spend $9,602,904 on a series of conservation projects in Costa Rica, mutually agreed on by Costa Rica and the donors.

The advantage to the Costa Ricans is that they can repay their debt through projects that will benefit their own country, and they can pay for these projects in colones (their own national currency) rather than in a foreign currency such as United States dollars or German marks. (International banks would not accept payment in Costa Rican colones, only in so-called hard currencies, which are scarce in Costa Rica because they have to be earned through international trade.) The advantage to the donors is that they can multiply the impact of their donation dramatically, 4.9-fold in this example. Retiring international debt has another benefit because many debtor nations have tried to repay their debts through mining, logging, and ranching enterprises that are designed to generate foreign revenues quickly, even if it is at the expense of the environment and long-term, sustainable-natural-resource use. To date this mechanism has only been used a few times, but it may blossom in the future.

Figure 16.1. If carefully structured, ecotourism is one mechanism for allowing local people to obtain economic benefit from sharing their environment with tourists and the native biota.

products based on sustainable use of wild life is another way to offset costs. For example, several organizations encourage consumers to purchase tropical hardwoods that have been sustainably harvested by local people so that these people will have an incentive to use their forests judiciously (Shanley 1999).

Problem 2 has a corollary: *Environmental technology costs are often experienced by people and governments who are least able to afford them.* Why should Ugandan taxpayers pay for wardens to protect gorillas when gorillas are more highly valued by people in Boston and Bern and when the average Ugandan is far poorer than the average person in the United States or Switzerland? (One crude measure of this is the gross domestic product [sum of all economic activity] divided by the population size. In 1997 it was $331 for Ugandans, $28,651 for United States residents, and $43,120 for the Swiss [United Nations Development Programme et al. 2000].) Again, the solution involves transfer of wealth from rich nations to poor; but technical know-how needs to be shared as well. This can involve sending conservation biologists, environmental engineers, and other specialists to parts of the world where their expertise is in short supply. Preferably, it means training local people in these skills so that they can help themselves. In an evaluation of why ecologically sensitive forms of logging were not widely used in tropical forests, Putz et al. (2000) identified seven factors, including lack of expertise and appropriate equipment, but foremost among them was the cost of these techniques.

Problem 3. *Resource exploitation that yields a quick profit is more attractive to most people than harvesting programs that produce moderate profits, but are sustainable over a longer period.* The shortsightedness of "get-rich-quick" schemes has been decried since at least the days of the Brothers Grimm and their tale of the goose that laid golden eggs, and criticism of this folly formed one of the historical roots of conservation in Europe and North America. In recent years, *sustainable development* has become the catchphrase for natural resource exploitation programs designed to produce goods and services in perpetuity with little or no environmental degradation. Treating future generations equitably (i.e., leaving them a healthy, diverse planet) is the key ethical concern here. Unfortunately, "A bird in the hand is worth two in the bush," and people have a clear tendency to devalue something that they will not use until the future, especially if the future users are their descendants, not themselves (Marsh 1994).

Economists account for our tendency to devalue use by employing *discount rates* to calculate *net present value*. These are most readily explained mathematically. A simple formula for calculating net present value is NPV = V/r; where V is the current annual value for production of some commodity and r is the rate at which we discount its future value. For example, if your date palm produces a crop worth $100 per year and your discount rate is 5%, then the net present value of the date palm is 100/0.05 or $2000. This is roughly equivalent to saying that the palm tree is worth $100 for this year's crop, plus $95 for next year's crop, plus $90.25 for the following year's crop, and so on in perpetuity, for a grand total of $2000. To put it another way—if your discount rate is 5%, receiving $2000 now is equivalent to the promise of receiving $100 per year forever.

Peters et al. (1989) used net present values to argue that sustainable production of nontimber plant products from a tract of tropical rainforest was more valuable than if the forest were cut and converted to another purpose. They measured annual production of fruits and natural rubber from 1 hectare of riparian forest in Peru and estimated that it could be sold for $422 profit, after deducting the costs of collecting and transporting the products 30 km to the city of Iquitos. They used a figure of $316.50 for current annual value (having deducted 25% from $422 on the assumption that some fruit and latex should be left unharvested) and a discount rate of 5% to arrive at a net present value estimate of $6330 per hectare. They estimated that clearcutting all the commercially valuable timber on the hectare would generate an immediate net profit of $1000 on delivery to a sawmill. If the site were then converted to another use, potential NPVs might include $3184 for a tree plantation or $2960 for a cattle pasture. Either figure, when added to the $1000 for selling the timber, is still far less than $6330. (See Balick and Mendelsohn 1992 for a similar analysis based on extraction of medicinal plants.)

Are people acting irrationally when they cut tropical forests rather than harvest its fruits? Not necessarily. It may not be wise from the perspective of the biosphere, of humanity as a whole, or of their own descendants, but it still may make sense in terms of their personal economics. Phillips (1993) argued that the 5% discount rate used by Peters et al. (1989) was far too low. Many Amazonian villagers discount future value much more than this because they have little confidence that they will be able to use the forest into the future. For example, if they discount future value at 20% because they fear they will lose their access to the forest to powerful political and commercial interests, then the estimated NPV would decline from $6330 to $1582.50. Many people in less-developed countries are so impoverished—they lead lives so close to the margin of survival—that discount rates are practically 100%. Future values mean virtually nothing because, if they do not use a resource now, they will probably die.

Finally, one could also argue that uncertainty is not the major reason that people devalue future uses. Some individuals and corporations are always ready to reap a quick profit and reinvest their gains somewhere else despite the long-term consequences for the biota or the people who continue to live in the degraded area. One could commend this as an aggressive business strategy, or one could condemn it as greed. People who suffer the consequences of this approach call it greed.

Solutions: Trying to mitigate uncertainty and greed is a tall order. Let us consider greed first. Regulations can curb some of the environmental consequences of greed, but some of the most promising approaches are based on environmentally based tax reform. "Polluter-pays" taxes are the best known of these; "natural capital depletion" taxes that act as a brake on exploitation of natural resource's are another idea (Costanza et al. 1997b). Unfortunately, governments routinely find themselves caught between two goals, a desire to encourage economic activity and a desire to protect their citizens, including future generations, from the undesirable spinoffs of economic development. We could enter a long, liberal-versus-conservative debate about government that helps the rich get richer versus government that protects the disadvantaged. Suffice it to say that most environmentalists feel that too often the economic well-being of a few people takes precedence over the environmental well-being of everyone and every living thing.

Dealing with uncertainty or risk is also complex. If the problem is poor people whose land tenure is insecure, then the solution is to give them land and legal protection from those who might take the land away from them. For people whose immediate survival is in jeopardy, economic assistance is needed to allow them to see beyond finding food and fuel for tomorrow. Uncertainty is also a significant problem for large corporations. Not knowing how markets, supplies, government regulations, and other factors will affect profitability is a major catalyst for reaping short-term profits. Governments can mitigate this tendency by trying to create a stable business climate, but ironically this can conflict with the need for new environmental regulations that arises when scientists discover new problems.

Problem 4. *Not everyone agrees that the benefits of biodiversity far outweigh the costs of maintaining it.* This basic premise of biodiversity advocates is easily challenged by people who weigh only direct economic benefits against the explicit costs of environmental technology and the implicit costs of opportunity losses. These people would dismiss potential values as too speculative. They would say we should not worry about some obscure plant in the highlands of Tanzania that might have a cure for breast cancer because by the time that we figure out that it has this property, biochemists will have synthesized a cure. They would argue that while the earth's biota as a whole clearly has ecological value, it has not been proven that each species has a unique role and that disaster will ensue if we lose some species. They would dismiss existence values as too abstract, too philosophical to have meaning in the hard-headed world of business.

Solutions: The solution to this problem begins in rhetoric, with the argument that the burden of proof must lie with those who assert that a species lacks value. If some people wish to dismiss a species as without commercial, subsistence, or ecological value, let them demonstrate that it lacks value now and will probably have no value in the future. This conservative approach is the only reasonable course in the absence of deep knowledge and understanding.

Returning to economics—if the case for conserving a species relies heavily on potential values or ill-defined ecological values, then this suggests that taxpayers should pay most of the costs because one of the major responsibilities of a government is to provide for the well-being of future generations.

It may be useful to weigh the costs of environmental technology against other expensive undertakings such as medical and military endeavors. Medical costs are particularly appropriate for comparison because so many medical problems can be traced to lack of clean air, water, and food. If human overpopulation and environmental degradation continue to widen the gap between rich nations and poor, then it is also appropriate to think of efforts to maintain a healthy environment as an alternative to increasing national security through military expenditures. We could also compare the $50 million spent for a single painting by Vincent van Gogh, the product of a few weeks of his labor, with the meager sums spent to protect most endangered species, products of evolutionary processes that stretch back through the ages.

Some conservationists, notably David Ehrenfeld (1988), have argued that this problem is essentially insoluble. He believes that cost-benefit analyses will serve biodiversity badly, particularly because the species that are rarest, and thus most vulnerable to extinction, are least likely to have critical ecological roles. This argument also finds support in some economic analyses, particularly when one focuses on marginal values rather than average or total values (for instance, the amount people would be willing to pay to add one more tiger to the world's tiger population) (Bulte and Van Kooten 2000). Disavowing cost-benefit analyses would force conservationists to focus on solutions based on morality, particularly a shift from anthropocentrism toward biocentrism.

There are many additional ways to express the basic problem of an imbalance in the costs and benefits of maintaining biodiversity. Some are generic problems common to all human enterprise and are captured in truisms like "money talks" and "money corrupts." Some are much more specific, such as people who own ecosystems are seldom compensated for the ecological services they provide. We will not delve into these further because of space constraints and because the solutions are much the same as those already outlined. We can summarize the solutions by characterizing them as either economic incentives (e.g., subsidies, privatization of government-owned assets) or economic disincentives (e.g., regulations, tax penalties, tariffs). The costs of economic incentives will be borne largely by taxpayers; the costs of economic disincentives will be paid largely by the owners of biological resources and their customers. On the whole, incentives are generally preferable to disincentives because carrots work better than sticks. Moreover, it seems only fair to favor carrots when poor people are the resource owners.

CASE STUDY   Butterfly Ranching[1]

When you think of ranches, images of cattle grazing on a dusty plain under the watchful eye of a cowboy are likely to come to the fore. You probably would not envision butterflies fluttering about a small opening in a tropical forest, yet in Papua New Guinea people have created hundreds of butterfly ranches. Most of these consist of a small patch planted with some plant species that are preferred food for the caterpillars of certain butterfly species, especially the elegant birdwing butterflies. Surrounding the patch, the rancher maintains a border of plants that produce nectar and thus attract the adults. Pupae are collected from the food plants and protected from predators, chiefly ants, until they hatch. Although butterfly ranches (which rely on a wild butterfly population that will use the ranch) are the norm, there are some butterfly farms where captive butterfly populations are maintained in cages. Butterflies may seem a strange form of livestock, but not to the Papua New Guineans. They knew the basic ecology of the butterflies even before ranching was initiated and in contrast have virtually no familiarity with sheep and cattle.

There are three potential markets for ranch butterflies. First, decorative specimens are popular as tourist souvenirs and are exported to curio shops. Serious butterfly collectors are a second market; they seek to own a wide variety of butterflies, preferably of rare species. Collectors want perfect specimens, and these are provided more readily by a rancher than by someone who nets wild butterflies. Finally, there is a significant market for live specimens, which are exported, usually as pupae, to be displayed in butterfly houses, often at zoos. In 1986 over four million people visited 40 butterfly houses in Great Britain. Papua New Guinea has chosen not to enter this live market for fear that, if they export live specimens, competitors will start breeding programs with their species.

People are not getting rich through butterfly ranching—annual incomes generally range from $100 to $3000 in Papua New Guinea—but in a country where the average annual income in rural areas is $50, these are significant amounts. More importantly, butterfly ranching is the kind of development that many conservationists advocate for rural areas because it allows local people to participate in a cash economy through the sustainable use of their renewable natural resources. Indeed, for many Papua New Guineans butterfly ranching has been their first opportunity to earn cash. In other countries such as Malaysia and Costa Rica, butterfly ranching is more likely to supplement existing cash income.

Butterfly ranching is not a complete solution to the problem of overcollecting butterflies. In particular, not all marketable species lend themselves to ranching, and thus the collecting of wild specimens continues. This is generally viewed as acceptable as long as overexploitation is avoided. The key to avoiding overexploitation, black markets, and related problems is to have a reliable, well-run agency that will serve as an honest broker for the ranchers, buying their produce and marketing it overseas. When the system works, it can provide a significant incentive for people to maintain butterflies and the entire ecosystems that support both the butterflies and themselves.

[1] Primary references for this section were National Research Council (1983) and New (1991, 1994, 1997). For details on another system for using wild animals as the basis of rural development—crocodile farming—see Webb et al. (1987) and Thorbjarnarson (1999).

Summary

Recognition of the true costs and benefits associated with wise natural resource management has led to a paradigm shift in many circles. Many people are no longer asking, "Can we afford to conserve?" but, rather, "Can we afford not to conserve?" The benefits of biodiversity include the wide variety of organisms that we use as goods for commerce or subsistence, plus many ecological services such as providing us with clean air and water and recreational opportunities. Less tangible benefits include existence values (the value in simply knowing that a species or ecosystem exists) and potential values (the future, possible value of a gene, species, or ecosystem). The costs of maintaining biodiversity can be divided into two major types: explicit costs for environmental technology such as salaries for people who undertake projects that maintain and restore biodiversity, and implicit costs, the loss of potential benefits when conservation means giving up an opportunity to use a resource.

From an economic perspective, the fundamental problem with maintaining biodiversity is that there is often an imbalance between who pays for maintaining biodiversity and who enjoys the benefits. We all enjoy the goods and services provided by biodiversity, but often the costs of maintaining biodiversity fall disproportionately on rural people and developing nations. Furthermore, when biological resources are open to unrestricted access, the "tragedy of the commons" can prevail and an imbalance of costs and benefits can drive overexploitation within a single community. Solutions can be characterized as economic incentives or disincentives. Incentives can encourage people to maintain biodiversity by giving them subsidies such as tax breaks. Disincentives can encourage sound stewardship by imposing financial penalties such as fines and tariffs on activities that degrade or overexploit natural resources.

FURTHER READING

For further information on ecological economics, I recommend Barbier et al. (1995), Goodstein (1995), or Costanza et al. (1997b); Barbier et al. is strong for methods and Costanza et al. for the development of ecological economics. McNeely (1988) describes both general concepts and case studies. *Elephants, Economics and Ivory* by Barbier et al. (1990) is an interesting case study. *Ecological Economics* and *Land Economics* are key journals for this field. See www.ecologicaleconomics.org to learn more about the International Society for Ecological Economics.

TOPICS FOR DISCUSSION

1. Can you think of any creative mechanisms for sharing the benefits and costs of maintaining biodiversity more equitably?
2. Why is the environmental record of countries with a communist economic system generally worse than that of capitalist countries?
3. Some people think that we need to maximize economic growth to keep people satisfied with their lifestyle and to pay for environmental technology. What do you think?
4. What are the relative advantages and disadvantages to using financial incentives versus disincentives to effect conservation activities?
5. Many of our technological developments, notably the automobile, have had negative impacts on environmental quality. Do you think that future technological advances will have a net effect on environmental quality that is positive or negative?

Chapter 17
Politics and Action

Talk is cheap. So is hand-wringing. Conserving biodiversity requires action. Conservationists must try to shape human institutions to make them more compatible with maintaining biodiversity. Broadly speaking, politics is the art and science of governing human institutions, and thus conservationists must be political if they wish to advance their agenda.

The interface between conservation and politics is a complex landscape that can be explored in many ways. We will take a relatively short and simple route that touches on what different types of human entities—international agencies, governments, nongovernmental organizations, corporations, communities, and individuals—can do to foster biodiversity conservation. Some of the actions listed here are of an economic nature and were described in more detail in the preceding chapter; others are not based on economics. All of these actions are currently being undertaken somewhere but seldom at an adequate scope or intensity.

International Agencies

UNDP, UNEP, UNESCO, IUCN, IMF, ADB—the alphabet soup of organizations that has evolved to foster better international relationships is large and complex. (See Box 17.1 for brief descriptions of these and other organizations.) In this section we will focus on some common threads that link these diverse groups to conservation.

1. Fostering a global conservation ethic. All of these organizations have a fundamental goal of improving the well-being of humanity, but this goal cannot be achieved without careful stewardship of natural resources.

To this end it is important that the "family of nations" fosters a climate in which its members are encouraged to practice sound conservation. Various international documents have codified a global conservation ethic. Among the most important are *The World Conservation Strategy* (IUCN et al. 1980), *The World Charter for Nature* (Annex 2 in McNeely et al. 1990), and the *Rio Declaration* (Parson et al. 1992, Grubb et al. 1993). Although essentially anthropocentric, these documents suggest some movement toward biocentrism. For example, the World Charter for Nature, which was passed by the United Nations in 1982, states that "every life-form is unique, warranting respect regardless of its worth to man."

Unfortunately, the world has yet to live up to the promise of these documents. Moreover, one of the most critical elements of a global conservation ethic—limiting human population growth—is often suppressed in these documents because some people fear that population control will infringe on basic human rights, and because discussion of overpopulation turns the spotlight on less-developed countries rather than on the excessive use of resources in wealthy countries (Baltz 1999) (Fig. 17.1).

2. **Regulating globally shared resources.** Maritime law has regulated human use of the oceans for centuries—making piracy illegal, for ex-

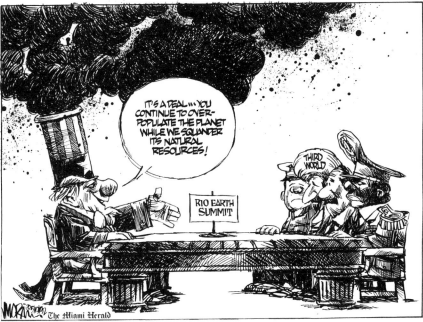

Figure 17.1. (Based on a *Miami Herald* cartoon, June 1992. Reprinted with special permission of King Features Syndicate.)

BOX 17.1   International agencies[1]

United Nations Environment Programme (UNEP, Nairobi) (www.unep.org) facilitates international cooperation on environmental issues chiefly as a catalyst and source of information. It also administers some funds for environmental projects, but this is a secondary role. It now oversees the **World Conservation Monitoring Centre** (www.unep-wcmc.org).

United Nations Development Programme (UNDP, New York) (www.undp.org) is the world's largest source of multilateral grants and funds a wide variety of projects (agriculture, transportation systems, health care, etc.) with environmental consequences. It also funds projects designed to aid conservation and is a major participant in a new program, the **Global Environmental Facility,** along with UNEP and the World Bank.

United Nations Educational Scientific and Cultural Organization (UNESCO, Paris) (www.unesco.org) facilitates international intellectual endeavors such as improving world literacy. Its mission also includes protecting the world's cultural and natural heritage, and it administers the Man and the Biosphere Programme (www.unesco.org/mab) and the list of **World Heritage Sites** (see Box 17.2).

United Nations Population Fund (UNFPA, New York) (www.unfpa.org) gathers population statistics and funds family planning services.

Several other United Nations organizations administer programs that have strong links to conservation issues, including the **Food and Agriculture Organization** (FAO, Rome), the **World Health Organization** (WHO, Geneva), the **World Food Program** (Rome), and the **United Nations International Children's Emergency Fund** (UNICEF, New York).

World Bank (Washington, DC) (www.worldbank.org) is formally known as the International Bank for Reconstruction and Development, and its goal is

---

ample—and this concept has been extended to a few treaties that help protect the marine environment, as well as Antarctica. Unfortunately, a comprehensive treaty for conserving marine resources, a major goal of the United Nations Law of the Sea Conference, has still not been completed. More recently, the atmosphere has been recognized as a collective resource in need of protection. Particular attention has focused on global warming, and the Kyoto Protocols of The Convention on Climate Change are the ongoing attempts to cope with the issue (Cameron 2000).

Recognition of species, genes, and nonmarine ecosystems as common resources has been more problematic. However, some long-standing

BOX 17.1   *Continued*

to raise the living standards of people in the developing world by distributing funds provided by wealthier nations. It does this primarily through loans and grants for developing infrastructure such as roads, dams, electrical systems, and so on. It has often been criticized for the environmental impacts of its projects, but it is trying to ameliorate these and to initiate conservation projects. There are also regional development banks: the **Asian Development Bank** (ADB), the **African Development Bank** (AFDB), and the **Inter-American Development Bank** (IDB).

International Monetary Fund (IMF, Washington, DC) (www.imf.org) was created simultaneously with the World Bank and oversees the international system for currency exchange and loans, and it negotiates loans itself.

The **World Trade Organization** (WTO) (www.wto.org) deals with the rules of trade between nations, and its goal is to help producers of goods and services, exporters, and importers conduct their business. To date its actions have been widely perceived as harmful to the environment, but it could play a role in removing harmful subsidies and in negotiating environmental treaties.

World Conservation Union (www.iucn.org) (formerly the International Union for Conservation of Nature and Natural Resources and still usually known as the IUCN) is a hybrid organization formed by over 800 member organizations: governments (chiefly, national-level, natural-resource agencies), nongovernmental conservation groups, and research institutions. Its goal is to promote the protection and sustainable use of living resources.

---

[1] Information primarily from Welsh and Butorin (1990) and websites.

treaties do accomplish the following: 1) protect natural sites of global significance, 2) conserve organisms that live outside of territorial boundaries or move among nations (e.g., whales and migratory birds), and 3) regulate international trade in endangered species. At the 1992 Earth Summit (officially UNCED, the United Nations Conference on Environment and Development) in Rio de Janeiro, 153 nations signed a biodiversity treaty, but it is still too soon to judge its efficacy. See Box 17.2 for the official titles and brief descriptions of some of the major international environmental treaties.

3. Facilitating the sharing of financial resources. Many international agencies—notably the United Nations Development Programme, World Bank,

BOX 17.2   **Environmental treaties[1]**

The **Convention on International Trade in Endangered Species of Wild Fauna and Flora** (CITES) (1973) (www.cites.org) controls international trade in endangered species of plants and animals whether they are live or dead, whole organisms, or materials derived from organisms. Species listed on their Appendix I cannot be traded internationally for commercial purposes. International trade in their Appendix-II species is regulated and monitored.

The **Convention on the Conservation of Migratory Species of Wild Animals** (1979) protects wild animals that migrate across international borders through international agreements.

The **International Convention for the Regulation of Whaling** (1946) establishes the International Whaling Commission (ourworld.compuserve.com/homepages/iwcoffice/) to regulate whaling.

The **Convention on the Conservation of Antarctic Marine Living Resources** (1980) protects the integrity of the ecosystems surrounding Antarctica and conserves marine living resources there.

The **Convention Concerning the Protection of World Cultural and Natural Heritage** (1972) establishes a system of World Heritage Sites that are protected for their natural and cultural values. Another international system of reserves called Biosphere Reserves has been established by UNESCO's **Man and the Biosphere Programme** to demonstrate the integration of rural development and environmental protection.

The **Convention on Wetlands of International Importance, Especially as Water-fowl Habitat** (1971) (often known as the Ramsar Convention because it was signed in Ramsar, Iran) promotes protection of wetland resources in general and establishes a system of Wetlands of International Importance.

The **Convention on the Prevention of Marine Pollution by Dumping of Wastes and Other Matter** (1972) prohibits the ocean dumping of some pollutants and regulates others.

The **United Nations Convention on the Law of the Sea** (1982) (www.un.org/Depts/los/index.htm) establishes a comprehensive legal framework for oceans, including regulation of marine pollution and harvesting natural resources.

The **Protocol on Substances that Deplete the Ozone Layer** (1987) requires reduction in emissions of chlorofluorocarbons and halons that deplete the ozone.

BOX 17.2   *Continued*

The Treaty Banning Nuclear Weapon Tests in the Atmosphere, in Outer Space, and Under Water (1963) prohibits tests that could distribute radioactive debris across international boundaries.

The following five documents were signed by heads of state at the United Nations Conference of Environment and Development (UNCED) (www.unep.org/unep/partners/un/unced/home.htm) in 1992; the first two are binding treaties.

The Convention on Biodiversity (www.biodiv.org) This convention's objectives are "the conservation of biological diversity, the sustainable use of its components, and the fair and equitable sharing of the benefits arising out of the utilization of genetic resources." (This last item has proven contentious, at least in the United States, because it attempts to establish a mechanism by which nations that are the site of origin for a species or gene would benefit financially if this species or gene were developed into a marketable product [e.g., a new medicine] in another country. This treaty has been signed and ratified by 179 nations but not by the United States.)

The Convention on Climate Change (www.unfcc.org) requires stabilization of the concentrations of carbon dioxide, methane, and other greenhouse gases to avoid interfering with the earth's climate. The Kyoto Protocol is the latest manifestation of this convention, but it still has not been ratified (Cameron 2000).

The Statements on Forest Principles A formal treaty on sustainable management of forests could not be negotiated, in large part because industrialized nations insisted that it apply only to tropical forests. A nonbinding statement of 17 principles was signed.

The Rio Declaration promotes general principles to guide nations in their programs for development and environmental protection.

Agenda 21 describes environmental problems and associated issues such as health and poverty and puts forth a series of action plans. These cover the legal, technical, financial, and institutional aspects of tackling a host of problems such as deforestation, desertification, atmospheric pollution, and so on. The difficult part of Agenda 21 was determining how to pay for its estimated cost of $600 billion per year.

[1] Information for this box came principally from WRI (1994), Parson et al. (1992), Grubb et al. (1993), and the listed websites.

and International Monetary Fund—were designed to allow richer nations to aid the development of poorer nations through loans or outright donations. In practice, this system has some major shortcomings (e.g., development projects that do more harm than good, aid programs that are designed to aid the donor nations more than the recipients, and exacerbation of the international debt crisis), but a critique of all that is wrong with this system is beyond our scope. Suffice it to repeat a major point of the preceding chapter: despite various problems, a system for transferring wealth from richer nations to poorer ones is an essential part of biodiversity conservation because many of the poorest nations have a vast array of biota, and it is not fair to expect them to protect this global heritage alone. The mechanisms are present to facilitate this process, but often the political will to use them seems inadequate.

4. Facilitating the sharing of information. Biodiversity conservation is a complex enterprise that requires vast amounts of information, and international agencies are uniquely positioned to facilitate this exchange of information through publications, computerized databases, and conferences. From a biodiversity perspective, the most important example of such an enterprise is the World Conservation Monitoring Centre in Cambridge, England, an effort initiated by the World Conservation Union, the World Wide Fund for Nature, and the United Nations Environment Programme and which is now run as a function of UNEP.

Governments

Governments are powerful. They strongly influence human interaction with most elements of biodiversity, as well as many key institutions: economics, education, law, and so on. Ultimate control usually lies with a sovereign nation, but in many cases proximate control is exercised at a smaller scale by state, provincial, county, or municipal governments. In some cases there is considerable overlap between national and local government (Goble et al. 1999, Ray and Ginsberg 1999); for example, having both national and state laws to protect endangered species may make the safety net of laws more thorough, or it may lead to inefficient redundancies (Press et al.1996). In this section we will review some of the most important ways that governments can shape conservation.

1. Developing and enforcing environmental regulations. Whether by setting a quota for the number of fish that can be harvested, by compelling car manufacturers to install air-pollution-control devices, or by prohibiting farmers and homeowners from wasting water, governments have an enormous, virtually unlimited, scope to protect the public interest by regulating the activities of private individuals and organizations. In theory the only limits on what democratic governments can undertake to conserve

biodiversity are constraints imposed by public opinion. In practice, environmental regulations are often constrained by powerful, special-interest groups, especially those that would prefer not to internalize the environmental costs of doing business. Furthermore, it is much easier to pass laws than to enforce them.

2. Conserving publicly owned resources. In most countries, virtually all aquatic ecosystems and many terrestrial ecosystems are publicly owned. In these areas governments have a particular responsibility to be good stewards because they are on the front line of natural-resource management, not simply looking over the shoulder of private property owners. This responsibility usually takes one of three basic forms: 1) maintaining a well-trained staff of governmental, natural-resource managers who directly manage publicly owned lands and waters; 2) issuing long-term leases to individuals and corporations (e.g., selling grazing rights to ranchers) that are designed to ensure sound conservation; or 3) working with local communities to conserve natural resources that are legally owned by the national government, but that are, practically speaking, owned by local communities that have a long tradition of using the resource. (We will return to communities below.) Additionally, in most countries wild animals are publicly owned because they can move from property to property and here, too, governments have special responsibilities.

3. Encouraging conservation through economic policy. Governments profoundly affect the economics of both individuals and corporations through many mechanisms. They can offer financial incentives (e.g., direct subsidies, or abatement of property and income taxes) for activities that contribute to conservation (Langholz et al. 2000), as well as financial disincentives (e.g., higher tax rates and fines) for activities that are harmful (James et al. 1999). Lowering property taxes for land that is used for conservation purposes is one of the best examples of an incentive.

4. Supporting environmental education and research. Most of the world's schools are public institutions; therefore governments assume a major responsibility for providing students with the education they need to be responsible citizens. Clearly, this includes education that encourages students to be careful stewards of the earth. Similarly, most environmental research is undertaken by governmental agencies and government-funded universities and research institutions; thus governments have the primary responsibility for filling the information vacuum that often hampers conservation.

Nongovernmental Organizations

"Nongovernmental organization" (NGO) is a term that covers a broad spectrum of groups ranging from the World Wide Fund for Nature, with millions of members throughout the world, to small groups of volunteers that only operate within a single community, sometimes focusing on a single topic like saving a marsh from being developed. Many NGOs have no members at all, only a professional staff supported by grants from foundations, governmental agencies, and corporations. NGOs working on conservation problems are usually easy to label as "conservation" or "environmental" groups, but some groups have their major focus on another issue (e.g., labor, health, community development, religion) and are involved in conservation because it is linked with their primary concern.

In a perfect world, there might be little need for NGOs because governments would be completely responsive to their citizens' desires and effective in meeting their needs. In practice, NGOs have diverse roles to play in the conservation movement. A complete list of their activities would be long and overlap extensively with topics already listed (e.g., fostering a global conservation ethic and supporting environmental education and research). Here, we will consider just two features that are unique to NGOs and that focus on their interactions with other organizations, especially governments.

1. Representing members to governments and other organizations. People become members of NGOs because they care, because they support the goals of the NGO, and because they wish to add their voice to the chorus calling for change. NGOs give ordinary people a vehicle for communicating with governments, and sometimes with international agencies and corporations, that are often quite inaccessible to the average citizen. Writing to elected officials and other powerful people is important, as we shall see below, but not everyone who cares can attend an official hearing and give expert testimony. However, along with like-minded people, they can join an NGO and can be represented by experts. When an NGO staff person can say: "I represent 400,000 members of the 'Save the _____ Society,'" significant clout is brought to bear.

2. Using their flexibility to undertake actions that are not open to governments. Governmental bureaucracies can be rather slow and ponderous because they are usually large and hobbled by rules designed to limit power and avoid corruption, so-called "checks and balances." In contrast, NGOs can often "strike while the iron is hot"; for example, quickly purchasing a critical ecosystem that is in imminent danger of being degraded. Moreover, because they are less encumbered by bureaucracy, NGOs can often undertake the same project at a much lower cost than a governmental agency. A fundamental part of this flexibility is the fact that NGOs can use monies obtained from their members and foundations rather than public tax dollars. Private funds usually have far fewer strings attached than public funds.

Sometimes, NGOs initiate actions that would be illegal for most governmental agencies: for example, calling for a boycott of products manufactured by an irresponsible corporation or, in extreme cases, acts of civil disobedience such as sabotaging a whaling ship or blocking a road to limit access for oil exploration or unrestrained logging.

Corporations

Corporations usually have a single primary goal—to make money—but most corporate managers believe that to achieve this goal it is necessary that they be perceived as "good corporate citizens." Traditionally, this has meant providing stable, high-salaried employment, a safe workplace, generous health and retirement benefits, donations to charitable causes—in other words, being socially responsible. Increasingly, being a good corporate citizen has come to include being environmentally responsible too (Daily and Walker 2000).

1. Internalizing the environmental costs of doing business. At a minimum, this means meeting the standards of environmental regulations; ideally, it means exceeding these standards. The greatest disincentive to this—competition in international markets with corporations that do not internalize environmental costs—can be solved through international cooperation.

2. Exceeding the standards of environmental regulations. Some corporations have learned that it can be profitable to exceed environmental standards because many consumers prefer to buy products that have been produced in an environmentally sensitive manner. This phenomenon, often known as green-labeling, first became prominent with "dolphin-free" labels on cans of tuna fish. It has long been recognized that good public relations are important to corporate success, and green labels are a mechanism for codifying the responsible behavior of corporations and conveying it to the public. Green-labeling involves an independent agency certifying that a corporation has met or exceeded high standards for responsible behavior (Sher 1997, Hayward and Vertinsky 1999, Bennett 2000). It must not be confused with the advertising many corporations use to promote themselves as good citizens; this is often based only on the corporation's image of themselves.

3. Finding innovative ways to advance conservation. Some corporations have found ways to promote conservation that are completely divorced from environmental regulations (PCEQ 1993). For example, some manufacturers have used their packaging to carry conservation messages to their consumers. Corporations that own land can take a proactive approach to conservation, ranging from planting native plants instead of exotic

species on the grounds around a corporate headquarters to restoring degraded ecosystems and extirpated species on large tracts. Others make sizable donations to conservation groups whose agendas are consistent with corporate shareholders or employees.

Communities

Groups of people who live in the same area, who share common resources, who are confronted with common problems, or who share common interests can be a very effective force for conservation (Bernhard and Young 1997). This is particularly true in the rural areas of many developing countries where access to natural resources is often based on traditional uses rather than on private property rights. In other words, in these places people's right to harvest wild plants and animals is based on the fact that their family has done so for generations, rather than on a legal document giving them exclusive ownership.

In situations such as this, effective conservation requires empowerment of the communities. From the governmental side, this begins with recognizing communities' rights. From the community side, it begins with recognizing the need for cooperation, both internally and between the community and the government. Once these hurdles are passed, the process requires sharing control between communities and governmental officials so that both community interests (e.g., continued access to natural resources such as firewood and livestock fodder) and national or global interests (e.g., maintaining biodiversity or minimizing atmospheric pollution) are met. This sharing of the authority and responsibility of management by different stakeholders has been called *comanagement*. See Berkes (1989) and West and Brechin (1991) for more information. Some of the best examples of comanagement involve fishing, where local fishers have banded together to form cooperatives, and the government has collaborated with these cooperatives to ensure sound fisheries management (Castilla and Fernandez 1998, Acheson et al. 2000).

Individuals

Last and most importantly, there are individuals. All organizations are simply assemblages of individuals, and all actions begin with one person, one catalyst. The standard advice for individual conservationists is "Think globally, act locally." Here are seven things you can do to follow this advice.

1. Be informed. Read voraciously. Listen attentively. Think critically. Learn your whole life long. Knowledge confers power. The conservation movement needs some emotion and subjectivity, but it has a greater need for people with facts and objectivity. It is easy to be a "do-gooder"; it is more

difficult to be a "good do-er," someone who has what it takes to be effective—including knowledge and credibility.

2. Become experienced. Information is not enough; you need wisdom too, and wisdom comes with experience. Experience can come with age, but there are shortcuts. Travel is one way. Immerse yourself in another culture, another biota for a few months or years. If that is not possible, seek out people from other cultures who live nearby and talk with them. Try to see the world as they do. Learn another language. Colleges and universities are wonderful places to do all these things because they are intended to be collegial and universal. Also, travel where you live; get out and explore your local environment whenever you can.

3. Communicate. Write or call your elected representatives, your local newspaper, corporations, or anyone who is in a position to make a difference, and tell them what you think. Letters and phone calls make a huge difference. The environmental movement was founded in grassroots activism, and its strength still lies there. Also, do not limit your communication to distant officials. Talk to your family, friends, and colleagues about conservation.

4. Make your lifestyle consistent with your values. In other words: do not be a hypocrite; practice what you preach. This can be difficult. I find it easy to be self-righteous around people who own big, gas-guzzling cars, have more than two children, eat lots of meat, and so on, but when I compare my lifestyle with that of my friends who have no cars, are 100% vegetarian, live in small houses heated only with wood that they grow themselves, and so on, I feel guilty. My advice is to live as frugally as you can and still be happy.

5. Support conservation groups. Many people are quite generous with their personal monies; United States citizens alone donate over $135 billion per year to charitable causes (www.independentsector.org). The lion's share of these funds goes to religion (about 60%) and education, health, and social services (about 6% to 9% each); conservation groups receive roughly 3%. If you think this distribution is unbalanced, then you should consider directing most or all of your charitable donations to conservation groups. If you have little money, give your time; most conservation groups make extensive use of volunteers.

6. Become a professional conservationist. Around the world millions of natural-resource managers, scientists, educators, and the like have dedicated their lives to conservation. The financial rewards may be modest, but the personal satisfaction offers substantial compensation. *Do it now.* Try not to think of yourself as a student who happens to be majoring in biology, natural-resource management, or whatever; think of yourself as a professional conservationist who happens to be a student. Join a professional society such as the Society for Conservation Biology and attend its meetings. Many professional societies have local and student chapters.

7. Keep your perspective. It is easy to get depressed when contem-

plating the magnitude of the biodiversity crisis and when evaluating your chances of making a measurable difference. To avoid this, keep your perspective focused on an appropriate temporal and spatial scale. Make life better where you live, in your lifetime. Also, take heart that there have been some miraculous success stories in conservation that originated through the actions of individuals like you. John Muir, Aldo Leopold, Rachel Carson, Chico Mendes, and many others had very humble origins, but accomplished great things.

Rights and Responsibilities

Who owns the giant pandas? Who has the right to use them and the responsibility to ensure their continued survival? The citizens of China? Only the people who share the giant pandas' range in the montane forests of south-central China? All the world's people? A biocentrist would argue that no human being *owns* another species; we are all equal partners in the web of life. This argument has some merit, but we will step around it because this chapter is an anthropocentric account of human institutions.

Legally speaking, the people of China—formally the national government of China—own the giant pandas with the exception of the handful owned by foreign zoos. For other species the answer is not necessarily so simple. Legal rights and responsibilities can rest with private property owners; with local, regional, or national governments; with international coalitions of governments (e.g., in the case of some marine species and migratory birds); or with no one and everyone (in the case of most marine species that live outside of territorial waters).

In an ideal world the distribution of rights and responsibilities might be shaped like a pyramid as shown in Figure 17.2. The people who live in the forests inhabited by giant pandas would have the greatest rights and responsibilities per person, but everyone, wherever they live, would have some rights and responsibilities. In other words, even though you may live halfway around the world from giant pandas and never see one, you still have the right to ask for the continued existence of giant pandas and the responsibility to do what you can to help save them, for example, by giving money to an organization that supports giant-panda conservation.

Although the rights and responsibilities of people who live far from the pandas' range are quite small on a per capita basis, collectively, they may supersede the rights and responsibilities of the people who live close by. For example, if the people who inhabited the giant pandas' range wanted to allow the panda to become extinct, their right to make this decision would be superseded by the collective rights of all the world's people who want the giant panda to survive.

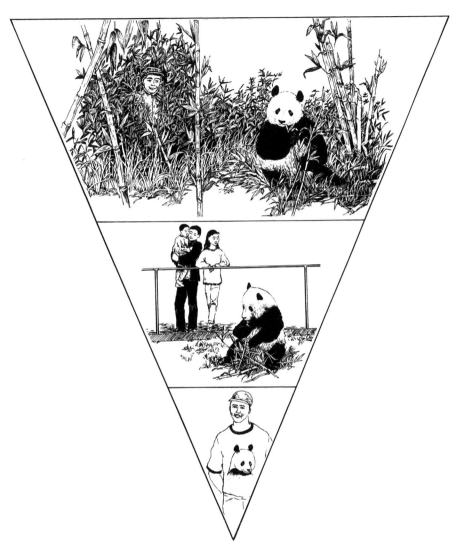

Figure 17.2. Ideally, the rights and responsibilities associated with conserving a species would be distributed such that they are greatest, on a per person basis, among the people who share the species' habitat; intermediate for people who live outside the species' habitat but nearby; and least among people who live far away.

Summary

Conservation action begins with individuals, and there is much that *you* can do to assist with efforts to maintain biodiversity. This will require working with other people in the context of various types of institutions: international and governmental agencies, conservation groups and other nongovernmental organizations, local communities, and corporations. Each of these has a special role to play in the conservation movement. Ultimately, every person has the right to enjoy the manifold benefits of biodiversity and with that right comes the responsibility to work to maintain biodiversity. This work must go forward within the fabric of social, economic, and political realities.

FURTHER READING

The *World Conservation Strategy* and its successor, *Caring for the Earth* (IUCN et al. 1980, 1991), give substantial insight on global approaches to conservation in general. For activities directly relevant to college campuses and conservation students, see Eagan and Orr (1992), Touval and Dietz (1994), and Creighton (1998). *World Resources 2000–2001* and its periodic revisions (United Nations Development Programme et al. 2000) are important compendia of information on which to base action. For a popular account of biodiversity to share with people, see Grumbine (1992) and Wilson (1992). Schaller (1993) gives a good account of the politics that have surrounded giant-panda conservation. Many journals carry articles related to conservation biology; the two most important are *Conservation Biology* and *Biological Conservation*. URLs for the websites of major international agencies and certain treaties are given in Boxes 17.1 and 17.2.

TOPICS FOR DISCUSSION

1. With the demise of the Cold War and communism versus capitalism as an organizing principle for the world, will global environmental issues become a central theme for international relationships? Why or why not?

2. Do you think corporations that undertake environmental activities are sincere or driven by public-relations concerns? Do their motivations matter?

3. How would natural-resource management in your area be different if policies were determined entirely by local communities without influence from state and national governments?

4. What species or ecosystems are threatened in your area? What can you do to help them?

5. Identify one obstacle that hinders you from taking political action. How can you overcome it?

6. How can the different entities described in this chapter work together more effectively?

Epilogue

In a universe too vast to comprehend, life is but a tiny mote, an extraordinarily rare and precious jewel. Nevertheless, life is demonstrably resilient. It has persisted for over 3 billion years and will almost certainly continue for billions of years more. If you can fully embrace a biocentric perspective, you can be heartened by the fact that the harm our species has brought to other organisms is largely insignificant in the long term. A few hundred million years from now a visitor to planet earth will probably be barely able to detect that *Homo sapiens* ever existed. For some people, it may be comforting to have this big picture firmly in view because it offers a path to freedom from despair over the state of life on earth. However, we cannot allow such intellectual ponderings to be used as a veil to disguise apathy. The earth's biota is a beautiful, incredible thing *now*. Species are being lost today because of human greed or ignorance, and each loss is a tragedy. Collectively, with thousands, probably millions, of species in jeopardy we are unraveling the tapestry of life. Although this will not lead to the eradication of life on earth, it will, unless stopped, lead to a truly dismal quality of life for each of us. Fortunately, each of us has the opportunity to make a difference. At a personal scale we can create a better environment for ourselves and the organisms that live nearby. We can choose to live in a world where a chorus of birds marks the dawn, where flowers and butterflies and myriad other creatures wait to make every day rich and full of wonderment.

Glossary

Acclimatization societies: social groups composed of European colonists during the late 19th and early 20th centuries whose sole purpose was to introduce new species; motivated largely by a love of nature and nostalgia for species left behind in Europe, e.g., songbirds to New Zealand

Adventive plants: term used by botanist for plant species living outside its native range

Agricultural seed banks: collections of seeds embraced by international agricultural community as an effective resource for preserving genetic variability in crops

Alien plants: term used by botanist for plant species living outside of its native range

Allee effect: the positive relationship between population density and the reproduction and survival of individuals (Warder Allee)

Alleles: differing configurations of DNA occupying the same locus on a chromosome; differences in the distributions of alleles are the foundation of measuring genetic diversity

Allozymes: enzymes that differ because of allelic differences

Alpha diversity: species diversity that exists within an ecosystem

Angiosperms: flowering plants

Anthropocentric: believing people are the center of the universe; people-centered

Area-sensitive species: species that do not occur in small patches of habitat

Attrition: a stage in the process of fragmentation when only very small, very isolated patches of natural vegetation remain

Augmentation program: release of individuals (wild or captive) into existing population to increase its size and genetic diversity

B

Bequest value: knowing that something exists simply because it might be of use to future generations, even though it is not used now

Beta diversity: among ecosystems, species diversity

Biocentric: believing that life, in all its various forms, is the center of the universe

Biochemical screening: testing organisms for their unique biochemical properties

Biodiversity: the variety of genes, species, and ecosystems in a given place or the world

Biological control: introduction of exotic species to control other exotic species

Biomagnification (bioamplification): a process whereby fat soluble chemicals (pesticides and PCBs) accumulate in the tissues of one species and pass from prey to predators, becoming more concentrated as they travel up the food chain

Biophilia: E.O. Wilson coinage for "love of life" which encompasses our aesthetic, spiritual, and emotional affinity for other species

Bioregion: geographic region based on ecological factors, not political boundaries

Bioregionalism: organizing conservation efforts around ecological regions

Biotic (biological) integrity: the completeness or wholeness of a biological system, especially including the presence of all the species at appropriate densities and the occurrence of all ecological processes at appropriate rates

Bleaching: unusually warm water temperatures thought to cause the massive death of coral polyps

Built ecosystems: human-made structures, e.g, cities, factories, mines, highways; urban areas and other places intensively used by people

C

Catastrophes: events such as droughts or hurricanes that occur at random intervals

Census population size: actual number of individuals in a population

Centinelan extinctions: phenomenon of species becoming extinct before they are described (E.O. Wilson)

Channelizing: making rivers and streams straighter, wider, and deeper and replacing shoreline vegetation with banks of stone and concrete

Chlorinated hydrocarbons: fat soluble chemicals (including DDT and PCBs) that pass through food chains, causing long-term and insidious damage

Climate flickers: extraordinarily rapid climate changes

Coarse-filter approach: concept of maintaining biodiversity by protecting a representative array of ecosystems

Colonization event: appearance of a subpopulation, e.g., a species of grass colonizing a forest opening after a tornado creates the opening

Connectivity: a quality of landscapes in which organisms can readily move among patches of habitat

Conservation biology: applied science of maintaining earth's biological diversity; crisis discipline focused on saving life on earth

Conservation easements: agreement to purchase certain property rights from landowners so that they can continue their traditional use of the land but cannot convert it to more intensive use such as housing, factories, or mines

Conservation forensics: identification of illegally collected species

Conservationist: advocate or practitioner of sensible and careful use of natural resources

Consumptive use: using something in such a manner that it is no longer available for someone else to use (harvesting an oak tree or cod)

Contaminant: substance which infects or makes impure by introducing foreign or undesirable material

Contingent valuation: survey method asking for the maximum values that users would pay for access to a particular activity

Controller species: major role in controlling movement of energy and nutrients

Convention on International Trade in Endangered Species of Wild Fauna and Flora (CITES): an agreement among a group of countries to ban commercial international trade in an agreed list of endangered species and to regulate and monitor trade in others that might become endangered.

Core subpopulations: subpopulations that persist for relatively long periods

Corridors: linear strips of protected ecosystems designed to maintain connectivity

Critically endangered: a taxon that is facing an extremely high risk of extinction in the wild based on several objective criteria

Cross-fostering: a method used to save species at risk whereby one species is used as a "foster parent" to the offspring of another species

Cryopreservation: *ex situ* storage of semen, embryos, or microbial organisms through storage at extremely low temperature, commonly in liquid nitrogen or its vapors

Cryptic species: genetically isolated species, not readily distinguished based on morphology

Cultivar: variety of plants a farmer selects for growing

Cultivated ecosystems: largely agricultural land; places where natural ecosystems have been replaced with a sparse assemblage of exotic and native species used in our production of food, fuel, and fiber

Cultural transmission: information transmitted through individuals and generations through a learning process, e.g., methods for exploiting novel food items

D

Damming: building barriers on rivers to impede water flow. Dams also stop or inhibit the movement of organisms; their effects on hydrology can alter ecosystems profoundly both upstream and downstream

Debt-for-nature swaps: conservation groups and wealthier nations purchasing discounted debt bonds from poorer nations to generate funds for conservation

Deforestation: conversion of forest to a nonforested ecosystem, persisting for a significantly prolonged period

Demographic population: group of interacting individuals of the same species whose structure and dynamics are relatively independent of other groups

Desertification: land degradation of grasslands and woodlands until they are dominated by sparse, relatively unproductive vegetation, resulting mainly from adverse human impact

Dispersal: movement of young plants and animals away from their parents

Diking: construction of earthen banks along edges of water bodies to prevent flooding

Dissection: a stage in the process of fragmentation when natural ecosystems are cut by roads and other human-made structures

DNA sequencing: determining the sequence of adenine, thymine, cytosine, and guanine for a given allele

Domestication: taming of wild species

Dominant species: constitutes a large portion of the biomass of an ecosystem

Double-clutching: transferring eggs from a rare mother bird to an incubator or a bird of a related species to raise, forcing mother to lay and raise a second "clutch" of eggs

Draining: lowering the water table by moving the water in a wet ecosystem elsewhere

Dredging: digging up the bottom of a water body and depositing the material elsewhere, often in a wetland that someone wants filled

E

Early successional colonizers: species adapted to disturbed ecosystems

Ecocentric: believing that life, in all its various forms, is the center of the universe

Ecological management: use of natural ecosystems as a model for resource management

Ecologist: scientist who studies relationships between organisms and their environments, often used as synonym for environmentalist

Economic incentives and disincentives: encouraging people to maintain biodiversity by giving subsidies and tax breaks, or, conversely, imposing penalties such as fines and tariffs on activities that degrade or overexploit natural resources

Economic values: utilitarian value of species, (e.g., food, medicine, clothing, shelter, fuel, tools, services, recreation)

Ecoregion: geographic region based on ecological factors, not political boundaries

Ecosystem: a group of interacting organisms (usually called a community) and the physical environment they inhabit at a given point in time

Ecosystem degradation: occurs when alterations to an ecosystem degrade or destroy habitat for many of the species that constitute the ecosystem

Ecosystem integrity: the quality of an ecosystem in which its constituent species and natural ecological processes are sustained

Ecosystem loss: occurs when the changes to an ecosystem are so profound and so many species are lost that the ecosystem is converted to another type

Ecosystem restoration: the return of an ecosystem or habitat to its original community structure and ecological functions; see reclamation, rehabilitation, and replacement

Ecotone: edge between two adjacent ecosystems

Ecotourism: travel undertaken to witness sites or regions of unique natural or ecological quality, or the provision of services to facilitate such travel

Ectothermic: dependent on environmental heat; "cold blooded"

Effective population size: number of individuals in a theoretically ideal population that would have same magnitude of random genetic drift as the actual population

Endangered species: a taxon that is facing a very high risk of extinction in the wild based on several objective criteria

Endemic: a species found only in a defined geographic area (e.g., koalas in Australia)

Endocrine disrupters: contaminants that are thought to cause problems by mimicking the action of the female sex hormone estradiol, causing sterility, delayed sexual maturity, abnormal sex organs, and an array of other problems

Endothermic: generating own body heat; "warm blooded"

Enhancement, ecosystem: any activity that improves the value of an ecosystem, even if the change is limited or the ecosystem has not been degraded (for example, installing water holes in desert reserves)

Environmental stochasticity: refers to random variation in parameters that measure habitat quality such as climate, nutrients, water, cover, pollutants, and relationships with other species

Environmentalist: someone concerned about impact of people on environmental quality

Ethnobotany: the study of the way plants are identified, classified, and used by various cultural groups

Eutrophication, cultural: an increase in the amount of nutrients, especially nitrogen and phosphorus, in a marine or aquatic ecosystem resulting from human activities

Evenness, species: component of diversity based on relative abundance of different species

Ex-situ **conservation:** maintaining organisms outside of their natural habitat

Existence values: Non-monetary value of human pleasure in fact of other species' very existence; the value of simply knowing that something exists even if one may never encounter it (e.g., snow leopard in central Asia)

Exotic species: a species living outside its native range

Explicit costs: money it costs to utilize environmental technology in restoring ecosystems, reducing environmental pollution or energy use, etc.

Exploitation: fundamental human activity to make use of wild plants and animals; including commercial, subsistence, recreational, nonconsumptive, indirect, incidental

Extinct: a taxon is extinct when there is no reasonable doubt that the last individual has died

Extinct in the wild: a taxon is extinct in the wild when it is known only to survive in cultivation or in captivity, or used to be extinct in the wild until successfully introduced

Extinction: disappearance of a species from the earth

Extirpation: small-scale disappearance of a species

Extractive reserve: allows limited extraction of resources, for example, collecting nuts and fruit, and tapping rubber trees, or subsistence hunting and fishing for native people

F

Feral: species having escaped from a state of domestication and reverted to the original wild or untamed state

Filling: material put in a wet depression until the surface of water table is well below ground, turning a wet ecosystem into a dry one

Fire suppression: removing fire from a fire-dependent ecosystem

Flagship species: charismatic species that captures the public's heart and wins support for its conservation; often a fellow mammal

Founder event: a few individuals arrive in a new area and establish a new population that is inevitably small at first, likely reducing genetic diversity

Fragmentation: process by which a natural landscape is broken into small parcels of natural ecosystems isolated from one another in a matrix of lands dominated by human activities

G

Gaia hypothesis: idea that all life on earth might constitute a giant, well-organized, self-regulating organism

Game cropping: systematic harvest of wild (neither domesticated nor captive) larger mammals, birds, and reptiles

Game ranching/farming: raising undomesticated large mammals such as bison in North America, or eland in Africa, within fenced areas

Gamma diversity: geographic-scale species diversity

Gap analysis: priority-setting technique that identifies gaps in the network of reserves designed to protect species and ecosystems

Gene: the functional unit of heredity; the part of the DNA molecule that encodes a single enzyme or structural protein on it

Genetic bottleneck: a phenomenon in which the genetic diversity of the original larger population is likely to be reduced because only a sample of the original gene pool will be retained

Genetic diversity: variation in the genetic composition of individuals within or among species; the heritable genetic variation within and among populations

Genetic population: a group possessing an allele not shared with another group, or alternatively, a group that shares less than 95% of its genetic variability with another group (Sewall Wright)

Genetic stochasticity: random variation in the gene frequencies of a population due to genetic drift, bottlenecks, inbreeding, and similar factors

Genetic swamping: when genes of one species come to dominate a common gene pool, largely excluding the genes of the second species

Geographic Information System (GIS): computer system for capturing, storing, checking, integrating, manipulating, analyzing, and displaying data related to positions on the earth's surface

Germplasm: the genetic material, especially its specific molecular and chemical constitution, that comprises the physical basis of the inherited quality of an organism

Ghost nets: lost gill nets which drift for months or years, still catching enormous quantities of fish, diving birds, seals, and other creatures

Green-ways: ecological connectivity in urban ecosystems by way of vegetated ribbons for walking and biking

Greenhouse gases: gases known to allow solar radiation to penetrate the atmosphere and warm the earth's surface, but to inhibit reradiation of energy back into space, causing the so-called "Greenhouse effect"

Growing-out: when viability of seed in seed banks deteriorates, this necessitates removing them from storage, growing new plants, then harvesting and storing the new seeds

Gymnosperms: plants, such as conifers and cycads, whose seeds are bare, the ovules not being enclosed in an ovary

H

Habitat: the physical and biological environment used by an individual, a population, a species, or perhaps a group of species

Habitat degradation: the process by which habitat quality for a given species is diminished, e.g., when contaminants reduce an area's ability to support a population

Habitat generalists: species adapted to a more varied habitat and as a result less vulnerable to extinction than habitat specialists

Habitat loss: when habitat quality is so low that the environment is no longer usable by a given species

Habitat specialists: species confined to a very specific habitat and as a result more vulnerable to environmental change

Hatchery raising ("Head Start" program): eggs collected and placed in ideal hatching conditions, cared for by humans during their vulnerable early stages, and then released into wild or raised in captivity

HCA: Habitat Conservation Areas

Heterosis: phenomenon whereby heterozygous individuals are more fit in terms of phenotypic characteristics than homozygous individuals

Heterozygosity: index of genetic diversity defined as proportion or percentage of genes at which the average individual is heterozygous

Heterozygous: possessing two different forms of a particular gene, one inherited from each parent

High-grading: forestry practice whereby the best formed trees are harvested and the worst formed or diseased trees are left behind; thought to cause an alteration of population's genetic structure

Homozygous: possessing two identical forms of a particular gene, one inherited from each parent

Hormonally active agents: contaminants that are thought to cause problems by mimicking the action of the female sex hormone estradiol, causing sterility, delayed sexual maturity, abnormal sex organs, and an array of other problems

Hot spots: areas that conservationists believe should have a high priority for establishing reserves because they are host to an unusually large number of species, e.g., tropical forests and coral reefs; home to endemic species, e.g., Madagascar, southwestern Australia; or, areas experiencing exceptional loss of habitat

Hybridization: offspring resulting from interbreeding between two apparently distinct species

I

Implicit costs: loss of opportunity to profit from natural resources when environmental regulations are imposed

In-situ **conservation:** protecting and maintaining organisms in their natural habitat

Inbreeding depression: loss of fitness in genetically uniform populations through breeding between closely related individuals

Incidental exploitation: catching species accidentally while harvesting other target species

Indicator species: health of these populations is an easy-to-monitor indication of environmental conditions or status of other species.

Indirect exploitation: human activities that indirectly kill other organisms, e.g., roads, fences, antennas, overgrazing, predation of domestic animals, or our introduction of exotic species

Indirect use: knowing and valuing an ecosystem or species only through books and films, but never actually encountering them personally

Instrumental value: the importance of a species because of its utility to people and other species

Integrated pest management (IPM): use of natural enemies of pests, specific cultivation practices (e.g., mixing crops), and limited use of pesticides to achieve pest control

Intrinsic value: the internal importance of a species without any reference to its usefulness for people or other species

Introduced species: term used to describe a species moved by humans to areas outside its native range

Invaders/Invasive species: term used for exotic populations that are expanding dramatically

International Species Inventory System (ISIS): global system for keeping track of captive animal populations

Island biogeography theory: number of species on an island represents a balance between immigration and extinction which will keep the number of species on any given island relatively constant

K

Keystone species: species that play critical ecological roles that are of greater importance than we would predict from their abundance (e.g., beavers; purple sea star)

L

Landraces: crop grown locally, often in only one small area of the world by traditional farmers

Landscape: a large-scale mosaic of ecosystems often consisting of a matrix with patches (small ecosystems) imbedded within it

Lethal recessives: alleles that are fatal when they come together in a homozygous recessive individual

Limnology: study of the chemistry, biology, and physics of freshwater

Local endemic: species found only in a small area (e.g., a small, isolated island)

Local extinction: disappearance of a species from a small area (e.g., beavers from small watershed)

M

Mangal or mangrove swamps: ecosystem dominated by woody plants found in tropical intertidal environments

Mean kinship: pedigree information that allows *ex situ* conservationists to calculate a measure of relatedness to decide who should mate with whom to maintain genetic diversity

Metapopulations: a model of population structures whereby each patch of habitat contains a different subpopulation of a species, and a group of different patch populations is collectively called a metapopulation

Minimum viable populations (MVP): The smallest viable population having a good chance of surviving for a given number of years despite the foreseeable effects of demographic, environmental, and genetic events and natural catastrophes

Mitigation of environmental impact: four major forms are: 1) impact avoided altogether 2) if impact cannot be avoided, site should be restored or rehabilitated 3) if impact is permanent, another nearby site should be restored to replace lost one 4) purchase and permanently protect natural ecosystems at a ratio of several hectares protected for every one lost

Modified ecosystem: ecosystem subject to management for commodities (e.g., wood, livestock, fish), that leaves the ecosystem in a semi-natural state

Multiple-use module (MUM): idea that a reserve is a central core buffered by concentric circles of ecosystems with decreasing degrees of naturalness

N

Natural disturbances: fires, floods, hurricanes, insect outbreaks, earthquakes, etc., that can initiate ecological succession and are often critical in maintaining natural structure and function of ecosystems

Near threatened (NT): A taxon is Near Threatened when it has been assessed against objective criteria and does not qualify for Critically Endangered, Endangered, or Vulnerable now, but is likely to qualify for a threatened category in the near future.

Nonindigenous species: term used to describe species living outside of its native range

Nonnative species: term used to describe species living outside of its native range

Nonconsumptive use: use that does not eliminate or substantially reduce value of something (e.g., naturalists viewing wildlife)

Nonconsumptive use of wildlife: non-hunting activities, including naturalists viewing or photographing, divers touching coral reefs, or the growth in ecotourism which can harm species in their natural habitats

Nongovernmental organizations (NGOs): a term covering a broad spectrum of private, not-for-profit groups

Nonpoint sources: pollutants originating from broad areas, e.g., runoff of pollutants from fields, lawns, and streets

Nurse logs: fallen trees which provide reservoirs of nutrients and moisture for seedlings

O

Orthodox seeds: seeds of certain plant species that remain viable when exposed to cold, dry conditions that reduce metabolic activity

Outbreeding depression: loss of fitness resulting from mating between individuals that are too genetically dissimilar

Overexploitation: human overuse of a population of organisms that seriously threatens its viability or radically alters the natural community in which it lives

P

Patchy distributions: species occurring in discrete patches of habitat

Pedigree: a record of the genetic history of an individual

Perforation: a stage in the process of fragmentation when natural landscapes have been broken by patches of human-altered vegetation such as agricultural fields

Phenotypic characteristics: refers to a species' adaptation to surrounding conditions, which are neither stable nor capable of being inherited

Point sources: pollutants originating from specific sites, e.g., factories

Pollutant: any substance or agent that causes pollution

Polluter-pays principle: idea that business must include pollution control in their costs

Polygynous: one male mating with multiple females

Polymorphic gene: a gene in which the frequency of the most common allele is less than some arbitrary threshold (often 95%)

Polymorphism: an index of genetic diversity based on the proportion or percentage of genes that are polymorphic

Population viability analysis (PVA models): method for organizing and enhancing our understanding of factors that shape a population's likelihood of persistence, and for comparing the effects of different management alternatives on relative probabilities of extinction

Potential value: the concept that all life-forms may have undiscovered worth

Preservationist: an advocate of allowing some land and some creatures to exist without significant human interference

Protein electrophoresis: indirect method of measuring genetic diversity by determining rate at which enzymes move through a gel when subjected to an electrical field

R

Random genetic drift: change in gene frequencies likely to occur in small populations because each generation retains just a portion of gene pool from previous generation

Rare alleles: alleles that have a frequency below some threshold; usually 0.05, 0.01, or 0.005

Rare species: species that are geographically specific, or habitat specific, or have naturally small populations

Recalcitrant seeds: species whose seeds cannot tolerate desiccation or freezing

Recessive deleterious alleles: potentially harmful alleles that are only expressed in homozygous individuals because they are recessive (i.e., not dominant)

Reclamation, ecosystem: shifting a degraded ecosystem back toward a greater value or higher use, but not all the way to its original state (e.g., reclaiming a mine site as a grazing pasture, rather than restoring it to a natural grassland)

Red tide: excess of nutrients causing explosive growth of plankton; a result of water pollution upsetting the equilibrium of marine food webs

Rehabilitation, ecosystem: see reclamation

Reintroduction program: releasing captive-bred or wild-collected individuals into an ecologically suitable site within their historic range where the species no longer occurs, with the intention of creating a new population in its original environment; terms to denote this are restorations, reestablishments, or translocation

Replacement, ecosystem: creating a completely new ecosystem out of a degraded one (e.g., creating a marsh in a mine pit that was formerly a forest, or replacing terrestrial ecosystems with wetlands)

Rescue effect: subpopulations are saved from extinction by immigration from other sub-populations

Reserve: used in text as a generic term for areas in which natural ecosystems are protected from most forms of human use

Restoration ecology: discipline that focuses on methods for restoring the structure and function of ecosystems degraded by human activities

Richness, species: number of different species in an ecosystem

Riparian: shoreline

S

Salinization: common when irrigation is used in arid environments; large volumes of water evaporate, leaving behind sodium chloride and other salts that can reach toxic concentrations

Satellite subpopulations: subpopulations that are likely to be small and a net sink, and which have rapid turnover

Secondary compounds: organic chemicals in plants that deter animals and may lend themselves to medicinal use

Seed banks: collections of seeds from the wild and from cultivated plants

Seminatural ecosystems: ecosystems modified by human activities, e.g., logging, fishing, grazing, but which are still dominated by native species

Shade intolerant trees: trees that can regenerate only in openings

Sibling species: genetically isolated species, not readily distinguished based on morphology

Sinks: subpopulations that cannot maintain themselves without a net immigration of individuals from other subpopulations

SLOSS (*single large or several small*)**:** debate regarding optimal size of reserves

Soil erosion: process whereby soil is removed especially by water and wind; it is greatly accelerated by human use of ecosystems, e.g., agriculture, overgrazing, timber harvesting, roads, construction

Sources: subpopulations which produce a substantial number of emigrants that disperse to other patches

Species: groups of actually or potentially interbreeding natural populations, which are reproductively isolated from other such groups (this is one of several alternative definitions)

Spiritual value: aesthetic, emotional, and spiritual affinity for other species

Strategic value: value of a species or ecosystem in achieving broader conservation goals; see flagship, indicator, and umbrella species for examples

Studbooks: pedigree records of captive populations for purpose of maintaining genetic diversity in *ex situ* conservation circles

Sustainability: ability to maintain something over period of time without diminishing it

Sustainable agriculture: often local and low-input style that avoids manipulating the environment with fertilizers, insecticides, herbicides, and irrigation

T

Threatened species: category of jeopardy one step below "endangered"

Tragedy of the commons: when biological resources are open to unrestricted access, and an imbalance of costs and benefits can drive overexploitation within a single community

Translocation: moving plants or animals from a location where they are about to be destroyed to another site that will provide greater protection; also sometimes called reestablishments or restorations

Trash fish: undesirable and abundant native species that compete with preferred species

Triage approach: idea that priority should be given to species with a reasonable chance of surviving if given attention

Turnover: subpopulations appearing and disappearing due to colonization and local extinction

Typology of attitudes: systematic sociological scale to determine peoples' attitudes toward animals

U

Umbrella species: species with large home ranges and broad habitat requirements; protecting habitat for their populations protects habitat for many other species across a broad set of ecosystems (e.g., tiger)

V

Vulnerable (VU): A taxon is Vulnerable if it faces a high risk of extinction in the wild based on several objective criteria

W

Waterlogging: raising of water table to the surface as a result of farmers using water to leach salts lower into soil to solve problem of salinization in irrigated croplands

Literature Cited and Author Index

Abboud, W. 2000. The WTO's Committee on Trade and Environment: Reconciling GATT 1994 with unilateral trade-related environmental measures. European Environmental Law Review 9(5):147–151. **414**

Acheson, J.M., T. Stockwell, and J.A. Wilson. 2000. Evolution of the Maine lobster co-management law. Maine Policy Review 9(2):52–62. **412, 436**

Adams, L.W., and D.L. Leedy (eds.). 1991. Wildlife conservation in metropolitan environments. National Institute for Urban Wildlife, Columbia, Maryland. 264 pp. **299**

Adkisson, C.S. 1996. Red crossbill. The birds of North America, No. 256. The Academy of Natural Sciences of Philadelphia, Philadelphia, Pennsylvania. 24 pp. **41**

Aguirre, A.A., and E.E. Starkey. 1994. Wildlife disease in U.S. national parks: Historical and coevolutionary perspectives. Conservation Biology 8:654–661. **328**

Ahearn, M.C., K.J. Boyle, and D.R. Hellerstein. (in press). Designing a contingent-valuation study to estimate the benefits of the Conservation Reserve Program on grassland bird populations. In J. Kahn and A. Alberini (eds.).Contingent valuation handbook. Edward Elgar. **405**

Aizen, M.A., and P. Feinsinger. 1994a. Forest fragmentation, pollination, and plant reproduction in a Chaco dry forest, Argentina. Ecology 75:330–351. **242**

Aizen, M.A., and P. Feinsinger. 1994b. Habitat fragmentation, native insect pollinators, and feral honey bees in Argentine "Chaco Serrano." Ecological Applications 4:378–392. **242**

Ajayi, S.S. 1979. Utilization of forest wildlife in West Africa. Food and Agriculture Organization, Rome. (FO: Misc./79/26) 76 pp. **50**

Akçakaya, H.R., S. Ferson, M.A. Burgman, et al. 2000. Making consistent IUCN classifications under uncertainty. Conservation Biology 14:1001–1013. **45**

Akerele, O., V. Heywood, and H. Synge, (eds.). 1991. The conservation of medicinal plants. Cambridge University Press, Cambridge. 362 pp. **51**

Alderson, L. (ed.). 1990. Genetic conservation of domestic livestock. C.A.B International, Wallingford, Great Britain. 242 pp. **361, 364**

Alford, R.A., and S.J. Richards. 1999. Global amphibian declines: A problem in applied ecology. Annual Review of Ecology and Systematics 30:133–165. **167**

Alison, R.M. 1981. The earliest traces of a conservation conscience. Natural History May 1981:72–77. **8, 256**

Allan, J.D., and A.S. Flecker. 1993. Biodiversity conservation in running waters. Bioscience 43:32–43. **193**

Allee, W.C. 1931. Animal aggregations. The University of Chicago Press, Chicago. 431 pp. **151**

Allegretti, M.H. 1990. Extractive reserves: An alternative for reconciling development and environmental conservation in Amazonia. Pages 252–264 in A.B. Anderson (ed.). Alternatives to deforestation: Steps toward sustainable use of the Amazon rain forest. Columbia University Press, New York. 281 pp. **290**

Allen, E.B., and L.L. Jackson. 1992. The arid west. Restoration and Management Notes 10:56–59. **306**

Allen-Wardell, G., P. Bernhardt, and R. Bitner, et al. 1998. The potential consequences of pollinator declines on the conservation of biodiversity and stability of food crop yields. Conservation Biology 12:8–17. **172**

Allendorf, F.W., D. Bayles, D.L. Bottom, et al. 1997. Prioritizing pacific salmon stocks for conservation. Conservation Biology 11:140–152. **372**

Allendorf, F.W., and R.F. Leary. 1986. Heterozygosity and fitness in natural populations of animals. Pages 57–76 in M.E. Soulé (ed.). Conservation biology: The science of scarcity and diversity. Sinauer, Sunderland, Massachusetts. 584 pp. **99**

Alley, R.B. 2000. The younger Dryas cold interval as viewed from central Greenland. Quaternary Science Reviews 19:213–226. **130**

Allison, G.W., J. Lubchenco, and M.H. Carr. 1998. Marine reserves are necessary but not sufficient for marine conservation. Ecological Applications 8(suppl):S79–S92. **260, 273**

Alonso, J.C., J.A. Alonso, and R. Munoz-Pulido. 1994. Mitigation of bird collisions with transmission lines through groundwire marking. Biological Conservation 67:129–134. **325**

Alpert, P. 1996. Integrated conservation and development projects. BioScience 46:845–855. **416**

Alvarez, L.W., W. Alvarez, F. Asaro, and H.V. Michel. 1980. Extraterrestrial cause of the Cretaceous-Tertiary extinction. Science 208:1095–1108. **122**

Alvarez, W., J. Smit, W. Lowrie, et al. 1992. Proximal impact deposits at the Cretaceous-Tertiary boundary in the Gulf of Mexico: A restudy of DSDP Leg 77 Sites 536 and 540. Geology 20:697–700. **122**

Amaral, M., A. Kozol, and T. French. 1997. Conservation status and reintroduction of the endangered American burying beetle. Northeastern Naturalist 4:121–132. **316, 331**

Ambuel, B., and S.A. Temple. 1983. Area-dependent changes in the bird communities and vegetation of southern Wisconsin forests. Ecology 64:1057–1068. **265**

Amos, B., and A.R. Hoelzel. 1992. Applications of molecular genetic techniques to the conservation of small populations. Biological Conservation 61:133–144. **92**

Anderson, A. 1989. Prodigious birds: Moas and moa-hunting in prehistoric New Zealand. Cambridge University Press, Cambridge. 238 pp. **209**

Ando, A., J. Camm, S. Polasky, and A. Solow. 1998. Species distribution, land values, and efficient conservation. Science 279:2126–2128. **263**

Andrews, C. 1990. The ornamental fish trade and fish conservation. Journal of Fish Biology 37(suppl A):53–59. **224**

Andrews, R.N.L. 1999. Managing the environment, managing ourselves: A history of American environmental policy. Yale University Press, New Haven, Connecticut. 463 pp. **18**

Angermeier, P.L. 1994. Does biodiversity include artificial diversity? Conservation Biology 8:600–602. **28, 34**

Angermeier, P.L. 2000. The natural imperative for biological conservation. Conservation Biology 14:373–381. **31, 34, 278, 291, 305**

Angermeier, P.L., and J.R. Karr. 1994. Biological integrity versus biological diversity as policy directives. Bioscience 44:690–697. **31, 32**

Antonovics, J., A.D. Bradshaw, and R.G. Turner. 1971. Heavy metal tolerance in plants. Advances in Ecological Research 7:1–85. **96**

Archibald, G.W. 1977a. Supplemental feeding and manipulation of feeding ecology of endangered birds. Pages 131–134 in S.A. Temple (ed.). Endangered birds: Management techniques for preserving threatened species. University of Wisconsin Press, Madison. 466 pp. **315**

Archibald, G.W. 1977b. Winter feeding programs for cranes. Pages 141–148 in S.A. Temple (ed.). Endangered birds: Management techniques for preserving threatened species. University of Wisconsin Press, Madison. 466 pp. **315**

Arita, H.T., J.G. Robinson, and K.H. Redford. 1990. Rarity in neotropical forest mammals and its ecological correlates. Conservation Biology 4:181–192. **46**

Armesto, J.J., R. Rozzi, C. Smith-Ramirez, and M.T.K. Arroyo. 1998. Conservation targets in South American temperate forests. Science 282:1271–1272. **260**

Armstrong, D.P., and I.G. McLean. 1995. New Zealand translocations: Theory and practice. Pacific Conservation Biology 2:39–54. **333, 334**

Armstrong, D.P., and J.K. Perrott. 2000. An experiment testing whether condition and survival are limited by food supply in a reintroduced hihi population. Conservation Biology 14:1171–1181. **316**

Arnold, M.L. 1992. Natural hybridization as an evolutionary process. Annual Review of Ecology and Systematics 23:237–261. **37**

Ashley, E.P., and J.T. Robinson. 1996. Road mortality of amphibians, reptiles and other wildlife on the Long Point Causeway, Lake Erie, Ontario. Canadian Field-Naturalist 110:403–412. **173**

Ashton, P.S. 1988. Conservation of biological diversity in botanical gardens. Pages 269–278 in E.O. Wilson and F.M. Peter (eds.). Biodiversity. National Academy Press, Washington, D.C. 521 pp. **351, 353**

Attfield, R. 1998. Existence value and intrinsic value. Ecological Economics 24:163–168. **406**

Augustine, D.J., and L.E. Frelich. 1998. Effects of white-tailed deer on populations of an understory forb in fragmented deciduous forests. Conservation Biology 12:995–1004. **327**

Avise, J.C. 1994. Molecular markers, natural history and evolution. Chapman and Hall, New York. 511 pp. **113**

Avise, J.C. 2000. Phylogeography: The history and formation of species. Harvard University Press, Cambridge, Massachusetts. 447 pp. **38**

Avise, J.C., and J.L. Hamrick (eds). 1996. Conservation genetics: Case histories from nature. Chapman and Hall, New York. 512 pp. **113**

Avise, J.C., and W.S. Nelson. 1989. Molecular genetic relationships of the extinct dusky seaside sparrow. Science 243:646–648. **350**

Ayeni, J.S.O. 1977. Waterholes in Tsavo National Park, Kenya. Journal of Applied Ecology 14:369–378. **277**

Ayittey, G.B.N. 1998. Africa in chaos. St. Martin's Press, New York. 399 pp. **416**

Bailey, R.G. 1995. Description of the ecoregions of the United States. Miscellaneous Publication No. 1391, USDA Forest Service, Washington, DC. 108 pp. (with map). **71**

Bailey, R.G. 1996. Ecosystem geography. Springer-Verlag, New York, NY. 204 pp. (with map). **71, 87**

Bailey, R.G., P.E. Avers, T. King, and W.H. McNab (eds.). 1994. Ecoregions and subregions of the United States. USDA Forest Service, Washington, DC. (with map). **71**

Baker, J.M.R., and T.R. Halliday. 1999. Amphibian colonization of new ponds in an agricultural landscape. Herpetological Journal 9:55–63. **293**

Baker, W.L. 1992. Effects of settlement and fire suppression on landscape structure. Ecology 73:1879–1887. **275**

Baldwin, A.D., Jr., J. DeLuce, and C. Pletsch (eds.). 1994. Beyond preservation: Restoring and inventing landscapes. University of Minnesota Press, Minneapolis. 280 pp. **307**

Balick, M.J., and P.A. Cox. 1996. Plants, people, and culture. Scientific American Library, New York. 228 pp. **110**

Balick, M.J., and R. Mendelsohn. 1992. Assessing the economic value of traditional medicines from tropical rain forests. Conservation Biology 6:128–130. **420**

Ballou, J.D., and R.C. Lacy. 1995. Identifying genetically important individuals for management of genetic variation in pedigreed populations. Pages 76–111 in J.D. Ballou, M. Gilpin, and T.J. Foose (eds.). Population management for survival and recovery. Columbia University Press, New York. 375 pp. **354**

Balmford, A., and K.J. Gaston. 1999. Why biodiversity surveys are good value. Nature 398:204–205. **259**

Balmford, A., K.J. Gaston, A.S.L. Rodrigues, and A. James. 2000. Integrating costs of conservation into international priority setting. Conservation Biology 14:597–605. **369, 373**

Balmford, A., G.M. Mace, and N. Leader-Williams. 1996. Designing the ark: Setting priorities for captive breeding. Conservation Biology 10:719–727. **357, 372**

Baltz, D.M., and P.B. Moyle. 1993. Invasion resistance to introduced species by a native assemblage of California stream fishes. Ecological Applications 3:246–255. **246**

Baltz, M.E. 1999. Overconsumption of resources in industrial countries: The other missing agenda. Conservation Biology 13:213–215. **427**

Bancroft, G.T., W. Hoffman, R.J. Sawicki, and J.C. Ogden. 1992. The importance of the water conservation areas in the Everglades to the endangered wood stork (*Mycteria americana*). Conservation Biology 6:392–398. **164**

Barbier, E.B., G. Brown, S. Dalmazzone, et al. 1995. The economic value of biodiversity. Pages 823–914 in V.H. Heywood and R.T. Watson (eds.). Global biodiversity assessment. Cambridge University Press, Cambridge, United Kingdom. 1140 pp. **424**

Barbier, E.B., J.C. Burgess, T.M. Swanson, and D.W. Pearce. 1990. Elephants, economics and ivory. Earthscan Publications, London. 154 pp. **424**

Barclay, J.H., and T.J. Cade. 1983. Restoration of the peregrine falcon in the eastern United States. Pages 3–40 in S.A. Temple (ed.). Bird conservation 1. University of Wisconsin

Press, Madison. 198 pp. **339**

Barker, J.R., and D.T. Tingey. 1992. Air pollution effects on biodiversity. Van Nostrand Reinhold, New York. 322 pp. **167**

Barnard, M. 1986. Sea, Salt and Sweat: A story of Nova Scotia and the vast Atlantic fisheries. Four East Publications and The Nova Scotia Department of Fisheries. Halifax, N.S. 109 pp. **225**

Barrett, S.C.H., and J.R. Kohn. 1991. Genetic and evolutionary consequences of small population size in plants: Implications for conservation. Pages 3–30 in D.A. Falk and K.E. Holsinger (eds.). Genetics and conservation of rare plants. Oxford University Press, New York. 283 pp. **99**

Bartlein, P.J., C. Whitlock, and S.L. Shafer. 1997. Future climate in the Yellowstone National Park region and its potential impact on vegetation. Conservation Biology 11:782–792. **132**

Bawa, K.S., and P.S. Ashton. 1991. Conservation of rare trees in tropical rain forests: A genetic perspective. Pages 62–74 in D.A. Falk, and K.E. Holsinger (eds.). Genetics and conservation of rare plants. Oxford University Press, New York. 283 pp. **151**

Bawa, K.S., S. Menon, and L.R. Gorman. 1997. Cloning and conservation of biological diversity: Paradox, panacea, or Pandora's box? Conservation Biology 11:829–830. **355**

Bazzaz, F.A. 1986. Life history of colonizing plants: Some demographic, genetic, and physiological features. Pages 96–110 in H.A. Mooney and J.A. Drake (eds.). Ecology of biological invasions of North America and Hawaii. Springer-Verlag, New York. 321 pp. **246**

Bean, M.J. 1983. The evolution of national wildlife law. Praeger, New York. 448 pp. **321, 411**

Beatley, J.C. 1977. Ash meadows: Nevada's unique oasis in the Mojave Desert. Mentzelia 3:20–24. **181**

Bedward, M., R.L. Pressey, and D.A. Keith. 1992. A new approach for selecting fully representative reserve networks: Addressing efficiency, reserve design and land suitability with an iterative analysis. Biological Conservation 62:115–125. **262, 368**

Beggs, J.R., and P.R. Wilson. 1991. The Kaka *Nestor meridionalis*, a New Zealand parrot endangered by introduced wasps and mammals. Biological Conservation 56:23–28. **242**

Begon, M., J.L. Harper, and C.R. Townsend. 1990. Ecology: Individuals, populations, and communities. 2nd ed. Blackwell Scientific Publications, Boston. 945 pp. **326**

Beier, P., and R.F. Noss. 1998. Do habitat corridors provide connectivity? Conservation Biology 12:1241–1252. **270, 273**

Beissinger, S.R. 1995. Modeling extinction in periodic environments: Everglade water levels and snail kite population viability. Ecological Applications 5:618–631. **164**

Beissinger, S.R., and D.R. McCullough (eds.). (in press). Population viability analysis. University of Chicago Press, Chicago, Illinois. **162**

Beissinger, S.R., and N.F.R. Snyder (eds.). 1992. New world parrots in crisis. Smithsonian Institution Press, Washington, D.C. 288 pp. **224**

Beissinger, S.R., E.C. Steadman, T. Wohlgenant, et al. 1996. Null models for assessing ecosystem conservation priorities: Threatened birds as titers of threatened ecosystems in South America. Conservation Biology 10:1343–1352. **369**

Beissinger, S.R., and M.I. Westphal. 1998. On the use of demographic models of population viability in endangered species management. Journal of Wildlife Management 62:821–841. **150, 155, 158, 162**

Ben-Shahar, R. 1993. Patterns of elephant damage to vegetation in Northern Botswana. Biological Conservation 65:249–256. **328**

Benkman, C.W. 1993. Logging, conifers, and the conservation of crossbills. Conservation Biology 7:473–479. **41**

Bennett, E.L. 2000. Timber certification: Where is the voice of the biologist? Conservation Biology 14:921–923. **435**

Bennetts, R.E., M.W. Collopy, and J.A. Rodgers, Jr. 1994. The snail kite in the Florida Everglades: A food specialist in a changing environment. Pages 507–532 in S.M. Davis and J.C. Ogden (eds.). Everglades: The ecosystem and its restoration. St. Lucie Press, Delray Beach, Florida. 826 pp. **276**

Bennitt, R., J.S. Dixon, V.H. Cahalane, et al. 1937. Statement of policy. Journal of Wildlife Management 1:1–2. **15**

Benstead, J.P., J.G. March, C.M. Pringle, and F.N. Scatena. 1999. Effects of a low-head dam and water abstraction on migratory tropical stream biota. Ecological Applications 9:656–668. **175**

Benstead, J.P., M.L.J. Stiassny, P.V. Loiselle, et al. 2000. River conservation in Madagascar. Pages 205–231 in P.J. Boon, B.R. Davies and G.E. Petts (eds.). Global perspectives on river conservation. John Wiley and Sons, New York. 548 pp. **203**

Bent, A.C. 1926. Life histories of North American marsh birds. Smithsonian Institution Bulletin 135. 392 pp. **213**

Berg, H., M. Kiibus, and N. Kautsky. 1992. DDT and other insecticides in the Lake Kariba ecosystem, Zimbabwe. Ambio 21:444–450. **171, 172**

Berger, A., J. Imbrie, J. Hays, et al. (eds.). 1984. Milankovitch and climate. Reidel, Dordrecht, The Netherlands. **129**

Berger, J. 1990. Persistence of different-sized populations: An empirical assessment of rapid extinctions in bighorn sheep. Conservation Biology 4:91–98. **139**

Berger, J. 1999. Intervention and persistence in small populations of bighorn sheep. Conservation Biology 13:432–435. **139**

Berger, J., and C. Cunningham. 1994a. Active intervention and conservation: Africa's pachyderm problem. Science 263:1241–1242. **322**

Berger, J., and C. Cunningham. 1994b. Phenotypic alterations, evolutionarily significant structures, and rhino conservation. Conservation Biology 8:833–840. **322**

Berger, J., and C. Cunningham. 1998. Natural variation in horn size and social dominance and their importance of black rhinoceros. Conservation Biology 12:708–711. **322**

Berger, J., C. Cunningham, A.A. Gawuseb, and M. Lindeque. 1993. "Costs" and short-term survivorship of hornless black rhinos. Conservation Biology 7:920–924. **322**

Berkes, F. (ed.). 1989. Common property resources: Ecology and community-based sustainable development. Belhaven Press, London. 302 pp. **436**

Berkes, F., J. Colding, and C. Folke. 2000. Rediscovery of traditional ecological knowledge as adaptive management. Ecological Applications 10:1251–1262. **386**

Berlow, E.L. 1999. Strong effects of weak interactions in ecological communities. Nature 398:330–334. **59**

Bernard, T., and J. Young. 1997. The ecology of hope: Communities collaborate for sustainability. New Society, Gabriola Island, British Columbia. 233 pp. **436**

Betancourt, J.L., W.S. Schuster, J.B. Mitton, and R.S. Anderson. 1991. Fossil and genetic history of a pinyon pine (*Pinus edulis*) isolate. Ecology 72:1685–1697. **109**

Bierregaard, R.O., Jr., T.E. Lovejoy, V. Kapos, et al. 1992. The biological dynamics of tropical rainforest fragments. Bioscience 42:859–866. **201**

Bigelow, H.B., and W.C. Schroeder. 1953. Fishes of the Gulf of Maine. Fish and Wildlife Service Fisheries Bulletin. Vol. 53. Washington, D.C. 577 pp. **226**

Bignal, E.M., and D.I. McCracken. 1996. Low-intensity farming systems in the conservation of the countryside. Journal of Applied Ecology 33:413–424. **293**

Bisby, F.A. 2000. The quiet revolution: Biodiversity informatics and the internet. Science 289:2309–2312. **66**

Bishop, R.C., and M.P. Welsh. 1992. Existence values in benefit-cost analysis and damage assessment. Land Economics 68:405–417. **406**

Bjorndal, K.A. (ed.). 1981. Biology and conservation of sea turtles. Smithsonian Institution Press, Washington, D.C. 583 pp. **337**

Bjorndal, K.A., A.B. Bolten, and C.J. Lagueux. 1994. Ingestion of marine debris by juvenilesea turtles in coastal Florida habitats. Marine Pollution Bulletin 28:154–158. **177**

Blackmore, S. 1996. Knowing the earth's biodiversity: Challenges for the infrastructure of systematic biology. Science 274:63–64. **41**

Blaustein, A.R., J.M. Kiesecker, D.P. Chivers, and R.G. Anthony. 1997. Ambient UV-b radiation causes deformities in amphibian embryos. Proceedings of the National Academy of Sciences 94:13735–13737. **167**

Blomquist, G.C., and J.C. Whitehead. 1995. Existence value, contingent valuation, and natural resources damages assessment. Growth and Change 26:573–589. **406**

Blondel, J., P.C. Dias, P. Perret, et al. 1999. Selection-based biodiversity at a small spatial scale in a low-dispersing insular bird. Science 285:1399–1402. **339**

Blouin, M.S., and E.F. Connor. 1985. Is there a best shape for nature reserves? Biological Conservation 32:277–288. **265**

Bock, C.E., and J.H. Bock. 1999. Response of winter birds to drought and short-duration grazing in southeastern Arizona. Conservation Biology 13:1117–1123. **286**

Boecklen, W.J., and N.J. Gotelli. 1984. Island biogeographic theory and conservation practice: Species-area or specious-area relationships. Biological Conservation 29:63–80. **266**

Bohnsack, J.A. 1993. Marine reserves. Oceanus 36(3):63–71. **269**

Boettner, G.H., J.S. Elkinton, and C.J. Boettner. 2000. Effects of a biological control introduction on three nontarget native species of Saturniid moths. Conservation Biology 14:1798–1806. **242**

Bolen, E.G., and W.L. Robinson. 1999. Wildlife ecology and management, 4th ed. Prentice Hall, Upper Saddle River, New Jersey. 605 pp. **234**

Bologna, P., and R. Steneck. 1993. Kelp beds as habitat for the American lobster, *Homarus americanus.* Marine Ecology Progress Series. 100:127–134. **227**

Bond, W.J., and B.W. van Wilgen. 1996. Fire and plants. Chapman and Hall, London. 263 pp. **180**

Bonnell, M.L., and R.K. Selander. 1974. Elephant seals: Genetic variation and near extinction. Science 184:908–909. **109, 156**

Bookbinder, M.P., E. Dinerstein, A. Rijal, et al. 1998. Ecotourism's support of biodiversity conservation. Conservation Biology 12:1399–1404. **416**

Boon, P.J., B.R. Davies, and G.E. Petts (eds.). 2000. Global perspectives on river conservation. John Wiley and Sons, Chichester. 548 pp. **205, 311**

Boone, R., and M.L. Hunter, Jr. 1996. Using diffusion models to simulate the effects of grizzly bear dispersal in the Rocky Mountains. Landscape Ecology 11:51–64. **150**

Boorman, S.A., and P.R. Levitt. 1973. Group selection on the boundary of a stable population. Theoretical Population Biology 4:85–128. **149**

Bormann, F.H., D. Balmori, and G.T. Geballe. 1993. Redesigning the American lawn. Yale University Press, New Haven. 166 pp. **301**

Boscolo, M., M. Powell, M. Delaney, et al. 2000. The cost of inventorying and monitoring carbon: Lessons from the Noel Kempff Climate Action Project. Journal of Forestry 98(9):24–31. **416**

Bostock, S.St.C. 1993. Zoos and animal rights. Routledge, London. 227 pp. **357**

Botkin, D.B. 1990. Discordant harmonies: A new ecology for the twenty-first century. Oxford University Press, New York. 241 pp. **87**

Bowen, B., J.C. Avise, J.I. Richardson, et al. 1993. Population structure of loggerhead turtles (*Caretta caretta*) in the northwestern Atlantic Ocean and Mediterranean sea. Conservation Biology 7:834–844. **338**

Bowen, B.W., T.A. Conant, and S.R. Hopkins-Murphy. 1994. Where are they now? The Kemp's ridley headstart project. Conservation Biology 8:583–856. **338**

Bowring, S.A., D.H. Erwin, and Y. Isozaki. 1999. The tempo of mass extinction and recovery: The end-Permian example. Proceedings of the National Academy of Sciences 96:8827–8828. **123**

Boyce, M.S. 1992. Population viability analysis. Annual Review of Ecology and Systematics 23:481–506. **156, 162**

Boyd, L., and K.A. Houpt (eds.). 1994. Przewalski's horse. State University of New York Press, Albany. 313 pp. **364**

Bradshaw, A.D. 1983. The reconstruction of ecosystems. Journal of Applied Ecology 20:1–17. **307**

Bradshaw, A.D. 1984. Ecological principles and land reclamation practice. Landscape Planning 11:35–48. **303**

Bradshaw, A.D. 1987. The reclamation of derelict land and the ecology of ecosystems. Pages 53–74 in W.R. Jordan III, M.E. Gilpin, and J.D. Aber (eds.). Restoration ecology: A synthetic approach to ecological research. Cambridge University Press, Cambridge. 342 pp. **307**

Bradshaw, A.D., and M.J. Chadwick. 1980. The restoration of land. Blackwell, Oxford. 317 pp. **306**

Bray, F. 1994. Agriculture for developing nations. Scientific American 271(1):30–37. **293**

Breymeyer, A.I. (ed.). 1990. Managed grasslands. Elsevier, Amsterdam. 387 pp. **144**

Bricklemyer, E.C., Jr., S. Iudicell, and H.J. Hartmann. 1989. Discarded catch in U.S. commercial marine fisheries. Pages 258–295 in W.J. Chandler, L. Labate, and C. Wille (eds.). Audubon Wildlife Report 1989/1990. Academic Press, San Diego. **217**

Bright, C. 1998. Life out of bounds: Bioinvasion in a borderless world. Norton, New York. 287 pp. **251**

Brisbin, I.L., Jr. 1995. Conservation of the wild ancestors of domestic animals. Conservation Biology 9:1327–1328. **362**

Brittingham, M.C. 1991. Effects of winter bird feeding on wild birds. Pages 185–190 in L.W. Adams and D.L. Leedy (eds.). Wildlife conservation in metropolitan environments. National Institute for Urban Wildlife, Columbia, Maryland. 264 pp. **315, 316**

Brittingham, M.C., and S.A. Temple. 1983. Have cowbirds caused forest songbirds to decline? Bioscience 33:31–35. **201**

Brooks, M.L. 1995. Benefits of protective fencing to plant and rodent communities of the western Mojave Desert, California. Environmental Management 19:65–74. **177**

Broomhall, S.D., W.S. Osborne, and R.B. Cunningham. 2000. Comparative effects of ambient ultraviolet-b radiation on two sympatric species of Australian frogs. Conservation Biology 14:420–427. **167**

Broseliske, G.H., J. de Jong, and H. Smit. 1991. Historical and present day management of the River Rhine. Water Science and Technology 23:111–120. **169**

Brothers, T.S. 1992. Postsettlement plant migrations in northeastern North America. American Midland Naturalist 128:72–82. **238**

Brown, J.H., and A. Kodric-Brown. 1977. Turnover rates in insular biogeography: Effect of immigration on extinction. Ecology 58:445–449. **147, 197**

Brown, J.H., and W. McDonald. 1995. Livestock grazing and conservation on southwestern rangelands. Conservation Biology 9:1644–1647. **286**

Brown, L. 1994. Ailing orchids. Nature Conservancy 44(3):8–9. **320**

Browne, R.A., C.R. Griffin, P.R. Chang, et al. 1993. Genetic divergence among populations of the Hawaiian duck, Laysan duck, and mallard. Auk 100:49–56. **244**

Bruner, A.G., R.E. Gullison, R.E. Rice, and G.A.B. da Fonseca. 2001. Effectiveness of parks in protecting tropical biodiversity. Science 291:125–128. **275**

Brush, S.B. 1995. *In situ* conservation of landraces in centers of crop diversity. Crop Science 35:346–354. **361**

Brush, S., R. Kesseli, R. Ortega, et al. 1995. Potato diversity in the Andean center of crop domestication. Conservation Biology 9:1189–1198. **361**

Buchert, G.P., O.P. Rajora, J.V. Hood, and B.P. Dancik. 1997. Effects of harvesting on genetic diversity in old-growth eastern white pine in Ontario, Canada. Conservation Biology 11:747–758. **221**

Bulte, E., and G.C. Van Kooten. 2000. Economic science, endangered species, and biodiversity loss. Conservation Biology 14:113–119. **422**

Burgess, R.L., and D.M. Sharpe (eds.). 1981. Forest island dynamics in man-dominated landscapes. Springer-Verlag, New York. 310 pp. **205**

Burgman, M.A., S. Ferson, and H.R. Akçakaya. 1993. Risk assessment in conservation biology. Chapman and Hall, London. 314 pp. **150, 162**

Burkett, D.W., and B.C. Thompson. 1994. Wildlife association with human-altered water sources in semiarid vegetation communities. Conservation Biology 8:682–690. **317**

Burney, D.A. 1993. Recent animal extinctions: Recipes for disaster. American Scientist 81:530–541. **228**

Burney, D.A., and R.D.E. MacPhee. 1988. Mysterious island: What killed Madagascar's large native animals? Natural History 97(7):46–55. **210**

Busch, B.C. 1985. The war against the seals. McGill-Queen's University Press, Kingston, Ontario. 374 pp. **224**

Bush, M.B. 1994. Amazonian speciation: A necessarily complex model. Journal of Biogeography 21:5–17. **131**

Butler, D., and D. Merton. 1992. The black robin. Oxford University Press, Auckland. 294 pp. **313, 336, 341**

Butler, J.N., J. Burnett-Herkes, J.A. Barnes, and J. Ward. 1993. The Bermuda fisheries: A tragedy of the commons averted? Environment 35(1):6–15, 25–33. **413**

Butler, P.J. 1992a. Parrots, pressure, people, and pride. Pages 25–46 in S. Beissinger and F.R.

Snyder (eds.). New World parrots in crisis. Smithsonian Institution Press, Washington, D.C. 288 pp. **394**

Butler, P. 1992b. Conservation education for the Bahamas. RARE Center for Tropical Conservation, Philadelphia, Pennsylvania. **396**

Byers, D.L., and D.M. Waller. 1999. Do plant populations purge their genetic load? Effects of population size and mating history on inbreeding depression. Annual Review of Ecology and Systematics 30:479–513. **109**

Cade, T.J., and C.G. Jones. 1993. Progress in restoration of the Mauritius kestrel. Conservation Biology 7:169–175. **336**

Cairns, J., Jr. (ed.). 1995. Rehabilitating damaged ecosystems. 2nd ed. Lewis Publishers, Boca Raton, Florida. 425 pp. **306, 311**

Caldwell, B. 1981. Islands of Maine: Where America really began. Guy Gannett Publishing Co. Portland, Maine. 241 pp. **225**

Callaway, R.M., and E.T. Aschehoug. 2000. Invasive plants versus their new and old neighbors: A mechanism for exotic invasion. Science 290:521–523. **247**

Callicott, J.B. 1990. Whither conservation ethics? Conservation Biology 4:15–20. **11, 17, 18**

Callicott, J.B. 1994. Conservation values and ethics. Pages 24–49 in G.K. Meffe and C.R. Carroll. Principles of conservation biology. Sinauer, Sunderland, Massachusetts. 600 pp. **385**

Callicott, J.B., L.B. Crowder, and K. Mumford. 1999. Current normative concepts in conservation. Conservation Biology 13:22–35. **29, 31, 311**

Calvete, C., R. Villafuerte, J. Lucientes, and J.J. Osacar. 1997. Effectiveness of traditional wild rabbit restocking in Spain. Journal of Zoology 241:271–277. **316**

Cam, E., J.D. Nichols, J.R. Sauer, et al. 2000. Relative species richness and community completeness: Birds and urbanization in the mid-Atlantic states. Ecological Applications 10:1196–1210. **31**

Cameron, P. 2000. From principle to practice: The Kyoto Protocol. Journal of Energy and Natural Resources Law 18:1–18. **428**

Canby, T.Y. 1977. The rat: Lapdog of the devil. National Geographic 152:60–87. **383**

Carlson, C.C. 1986. Maritime catchment areas: An analysis of prehistoric fishing strategies in the Boothbay region of Maine. M.S. thesis, University of Maine, Orono. 258 pp. **225**

Carlton, J.T. 1985. Transoceanic and interoceanic dispersal of coastal marine organisms: The biology of ballast water. Oceanography and Marine Biology Annual Reviews 23:313–371. **232**

Carlton, J.T. 1989. Man's role in changing the face of the ocean: Biological invasions and implications for conservation of near-shore environments. Conservation Biology 3:265–273. **236**

Carlton, J.T., and J.B. Geller. 1993. Ecological roulette: The global transport of nonindigenous marine organisms. Science 261:78–82. **233**

Carlton, J.T., G.J. Vermeij, D.R. Lindberg, et al. 1991. The first historical extinction of a marine invertebrate in an ocean basin: The demise of the eelgrass limpet *Lottia alveus*. Biological Bulletin 180:72–80. **170**

Caro, T.M., and M.K. Laurenson. 1994. Ecological and genetic factors in conservation: A cautionary tale. Science 263:485–486. **112**

Caro, T.M., and G. O'Doherty. 1999. On the use of surrogate species in conservation biology. Conservation Biology 13:805–814. **60**

Caro, T.M., N. Pelkey, and M. Grigione. 1994. Effects of conservation biology education on attitudes toward nature. Conservation Biology 8:846–852. **398**

Carroll, C.R., J.H. Vandermeer, and P. Rosset (eds.). 1990. Agroecology. McGraw-Hill, New York. 641 pp. **292, 293**

Carson, R. 1962. Silent spring. Houghton-Mifflin Co., Boston. 368 pp. **10**

Case, T.J. 1996. Global patterns in the establishment and distribution of exotic birds. Biological Conservation 78:69–96. **246**

Casey, J.M., and R.A. Myers. 1998. Near extinction of a large, widely distributed fish. Science 281:690–692. **210, 289**

Castilla, J.C., and M. Fernandez. 1998. Small-scale benthic fisheries in Chile: On co-management and sustainable use of benthic invertebrates. Ecological Applications 8 (suppl):S124–S132. **436**

Cater, E., and G. Lowman (eds.). 1994. Ecotourism: A sustainable option? John Wiley and Sons, Chichester, United Kingdom. 218 pp. **54, 74, 416**

Caughley, G. 1994. Directions in conservation biology. Journal of Animal Ecology 63:215–244. **162**

Chamberlain, D.E., R.J. Fuller, R.G.H. Bunce, et al. 2000. Changes in the abundance of farmland birds in relation to the timing of agricultural intensification in England and Wales. Journal of Applied Ecology 37:771–788. **293**

Chapin, F.S. III, M.S. Torn, and M. Tateno. 1996. Principles of ecosystem sustainability. American Naturalist 148:1016–1037. **32**

Chapin F.S. III, B.H. Walker, R.J. Hobbs, et al. 1997. Biotic control over the functioning of ecosystems. Science 277:500–504. **79**

Charlesworth, D., and B. Charlesworth. 1987. Inbreeding depression and its evolutionary consequences. Annual Review of Ecology and Systematics 18:237–268. **100**

Chen, J., J.F. Franklin, and T.A. Spies. 1992. Vegetation responses to edge environments in old-growth Douglas-fir forests. Ecological Applications 2:387–396. **201**

Chepko-Sade, B.D., and Z.T. Halpin (eds.). 1987. Mammalian dispersal patterns. The University of Chicago Press, Chicago. 342 pp. **272**

Cheru, F. 1989. The silent revolution in Africa: Debt, development and democracy. Zed Books, London. 189 pp. **416**

Christensen, B. 1983. Mangroves—what are they worth? Unasylva 35(139):2–15. **85**

Churcher, P.B., and J.H. Lawton. 1987. Predation by domestic cats in an English village. Journal of Zoology (London) 212:439–455. **218**

Cipriano, F., and S. Palumbi. 1999. Genetic tracking of a protected whale. Nature 397:307–308. **91**

Ciriacy-Wantrup, S.V., and R.C. Bishop. 1975. "Common property" as a concept in natural resources policy. Natural Resources Journal 15:713–727. **412, 413**

Claridge, M.F., H.A. Dawah, and M.R. Wilson (eds.). 1997. Species: The units of biodiversity. London, Chapman and Hall. 439 pp. **37**

Clark, J.R. 1998. Leopold's land ethic: A vision for today. Wildlife Society Bulletin 26:719–724. **14**

Clark, J.S. 1988. Effect of climate change on fire regimes in northwestern Minnesota. Nature 334:233–235. **276**

Clark, J.S., C. Fastie, and G. Hurtt, et al. 1998. Reid's paradox of rapid plant migration. BioScience 48:13–24. **134**

Clark, T.E., and M.J. Samways. 1996. Dragonflies (Odonata) as indicators of biotope quality in the Kruger National Park, South Africa. Journal of Applied Ecology 33:1001–1012. **61**

Clark, T.W., A.P. Curlee, S.C. Minta, and P.M. Kareiva. 1999. Carnivores in ecosystems: The Yellowstone experience. Yale University Press, New Haven, Connecticut. 429 pp. **68**

Clayton, L.M., E.J. Milner-Gulland, D.W. Sinaga, and A.H. Mustari. 2000. Effects of a proposed *ex situ* conservation program on *in situ* conservation of the babirusa, an endangered suid. Conservation Biology 14:382–385. **357**

Clegg, M., G.M. Brown, Jr., W.Y. Brown, et al. 1995. Species definitions and the Endangered Species Act. Pages 46–70 in Science and the Endangered Species Act. National Academy Press, Washington, D.C. 271 pp. **38**

Clevenger, A.P., and N. Waltho. 2000. Factors influencing the effectiveness of wildlife underpasses in Banff National Park, Alberta, Canada. Conservation Biology 14:47–56. **324**

Clutton-Brock, J. 1981. Domesticated animals from early times. University of Texas Press, Austin. 208 pp. **284**

Coblentz, B.E. 1978. The effects of feral goats (*Capra hircus*) on island ecosystems. Biological Conservation 13:279–286. **241**

Coblentz, B.E. 1990. Exotic organisms: A dilemma for conservation biology. Conservation Biology 4:261–265. **245**

Coblentz, B.E., and D.W. Baber. 1987. Biology and control of feral pigs on Isla Santiago, Galapagos, Ecuador. Journal of Applied Ecology 24:403–418. **241**

Cohen, A.N., and J.T. Carlton. 1998. Accelerating invasion rate in a highly invaded estuary. Science 279:555–558. **245, 246**

Cohen, A.S., R. Bills, C.Z. Cocquyt, and A.G. Caljon. 1993. The impact of sediment pollution on biodiversity in Lake Tanganyika. Conservation Biology 7:667–677. **179**

Cohn, J.P. 1993. The flight of the California condor. Bioscience 43:206–209. **407**

Colborn, T., D. Dumanoski, J.P. Myers. 1996. Our stolen future. Dutton, New York. 306 pp. **170**

Cole, F.R., L.L. Loope, A.C. Medeiros, et al. 1995. Conservation implications of introduced game birds in high-elevation Hawaiian shrubland. Conservation Biology 9:306–313. **248**

Collinge, S.K., and R.T.T. Forman. 1998. A conceptual model of land conversion processes: Predictions and evidence from a microlandscape experiment with grassland insects. Oikos 82:66–84. **199**

Collins, S.L., A.K. Knapp, J.M. Briggs, et al. 1998. Modulation of diversity by grazing and mowing in native tallgrass prairie. Science 280:745–747. **286**

Collins, W.W., and C.O. Qualset. (eds.). 1999. Biodiversity in agroecosystems. CRC Press, Boca Raton, Florida. 334 pp. **293, 295, 311**

Committee on Sea Turtle Conservation. 1990. Decline of the sea turtles: Causes and prevention. National Academy Press, Washington, D.C. 259 pp. **177**

Contreras-B., S. and M.L. Lozano-V. 1994. Water, endangered fishes, and development perspectives in arid lands of Mexico. Conservation Biology 8:379–387. **181, 187**

Conway, W.G. 1986. The practical difficulties and financial implications of endangered species breeding programmes. International Zoo Yearbook 24/25:210–219. **351**

Cook, J.H., J. Beyea, and K.H. Keeler. 1991. Potential impacts of biomass production in the United States on biological diversity. Annual Review of Energy and the Environment 16:401–431. **54**

Cooke, G.D., E.B. Welch, S.A. Peterson, and P.R. Newroth. 1993. Restoration and management of lakes and reservoirs. 2nd ed. Lewis Publishers, Boca Raton, Florida. 548 pp. **306**

Costanza, R., et al. 1997a. The value of the world's ecosystem services and natural capital. Nature 387:253–260. **74–75, 401, 402**

Costanza, R.J., J.Cumberland, C.H. Daly, R. Goodland, and R. Norgaard. 1997b. An introduction to ecological economics. St. Lucie Press, Boca Raton, Florida. 275 pp. **411, 420, 424**

Council of Tree and Landscape Appraisers. 1992. Guide for plant appraisal. (8th ed.). Savoy, Illinois. 103 pp. **402**

Cowan, P.E. 1992. The eradication of introduced Australian brushtail possums, *Trichosurus vulpecula*, from Kapiti Island, a New Zealand nature reserve. Biological Conservation 61:217–226. **278**

Cowie, R.H. 1992. Evolution and extinction of Partulidae, endemic Pacific island land snails. Philosophical Transactions of the Royal Society of London, Series B 335:167–191. **237**

Cowling, R.M. (ed). 1992. The ecology of fynbos. Oxford University Press, Cape Town, Republic of South Africa. 411 pp. **43**

Cowling, R.M., and R.L. Pressey. (in press). Rapid plant diversification: Planning for an evolutionary future. Proceedings of the National Academy of Sciences. **258**

Cowling, R.M., P.W. Rundel, B.B. Lamont, et al. 1996. Plant diversity in mediterranean-climate regions. Trends in Ecology and Evolution 11:362–366. **81**

Cowx, I.G. (ed.). 1994. Rehabilitation of freshwater fisheries. Fishing News Books, Oxford, United Kingdom. 486 pp. **289**

Cox, G.W. 1999. Alien species in North America and Hawaii: Impacts on natural ecosystems. Island Press, Washington D. C. 387 pp. **251**

Craig, J.L. 1994. Meta-populations: Is management as flexible as nature? Pages 50–66 in P.J.S. Olney, G.M. Mace, and A.T.C. Feistner (eds.). Creative conservation: Interactive management of wild and captive animals. Chapman and Hall, London. 517 pp. **99**

Creighton, S.H. 1998. Greening the ivory tower. The MIT Press, Cambridge, Massachusetts. 337 pp. **300, 311, 440**

Cronon, W. 1983. Changes in the land: Indian, colonists, and the ecology of New England. Hill and Wang, New York. 241 pp. **9, 18**

Crooks, K.R., and M.E. Soulé. 1999. Mesopredator release and avifaunal extinctions in a fragmented system. Nature 400:563–566. **199**

Crosby, A.W. 1986. Ecological imperialism: The biological expansion of Europe, 900–1900. Cambridge University Press, Cambridge. 368 pp. **231, 241, 246, 251**

Crowder, L.B., D.T. Crouse, S.S. Heppell, and T.H. Martin. 1994. Predicting the impact of turtle excluder devices on loggerhead sea turtle populations. Ecological Applications 4:437–445. **217**

Crozier, R.H. 1992. Genetic diversity and the agony of choice. Biological Conservation 61:11–15. **63, 372**

Cuddihy, L.W., and C.P. Stone. 1990. Alteration of native Hawaiian vegetation: Effects of humans, their activities and introductions. University of Hawaii Press, Honolulu. 138 pp. **59**

Culotta, E. 1994. Is marine biodiversity at risk? Science 263:918–920. **125, 170, 223**

Culver, D.C., L.L. Master, M.C. Christman, and H.H. Hobbs III. 2000. Obligate cave fauna of the 48 contiguous United States. Conservation Biology 14:386–401. **140**

Cumming, D.H.M. 1981. The management of elephants and other large mammals in Zimbabwe. Pages 91–118 in P.A. Jewell, S. Holt, and D. Hart (eds.). Problems in the management of locally abundant wild mammals. Academic Press, New York. 361 pp. **328**

Cunningham, A.A. 1996. Disease risks of wildlife translocations. Conservation Biology 10:349–353. **334**

Curatolo, J.A., and S.M. Murphy. 1986. The effects of pipelines, roads, and traffic on the movements of Caribou, *Rangifer tarandus*. Canadian Field-Naturalist 100:218–224. **176**

Curnutt, J.L., J. Comiskey, M.P. Nott, and L.J. Gross. 2000. Landscape-based spatially explicit species index models for Everglades restoration. Ecological Applications 10:1849–1860. **164, 276**

Curnutt, J., J. Lockwood, H-K Luh, et al. 1994. Hotspots and species diversity. Nature 367:326–327. **263**

Czech, B., P.R. Krausman, and R. Borkhataria. 1998. Social construction, political power, and the allocation of benefits to endangered species. Conservation Biology 12:1103–1112. **390**

Czech, B., P.R. Krausman, and P.K. Devers. 2000. Economic associations among causes of species endangerment in the United States. BioScience 50:593–601. **165, 217, 239**

D'Agrosa, C., C.E. Lennert-Cody, and O. Vidal. 2000. Vaquita bycatch in Mexico's artisanal gillnet fisheries: Driving a small population to extinction. Conservation Biology 14:1110–1119. **217**

Dahl, T.E. 1991. Wetlands, status and trends in the conterminous United States mid-1970s to mid-1980s. U.S. Fish and Wildlife Service, Washington, D.C. 22 pp. **191**

Dahlsten, D.L. 1986. Control of invaders. Pages 275–302 in H.A. Mooney, and J.A. Drake (eds.). Ecology of biological invasions of North America and Hawaii. Springer-Verlag, New York. 321 pp. **251, 328**

Daily, G.C. 1995. Restoring value to the world's degraded lands. Science 269:350–354. **302**

Daily, G.C. (ed.). 1997. Nature's services. Island Press, Washington, D.C. 392 pp. **74, 401**

Daily, G.C., and B.H. Walker. 2000. Seeking the great transition: Environmentally sustainable economies are unachievable without enhanced participation of the private sector. Scientists must facilitate this process. Nature 403:231–235. **435**

Dale, V.H., S. Brown, R.A. Haeuber, et al. 2000. Ecological principles and guidelines for managing the use of land. Ecological Applications 10:639–670. **311**

Dansgaard, W., J.W.C. White, and S.J. Johnsen. 1989. The abrupt termination of the Younger Dryas climate event. Nature 339:532–534. **130**

Davis, R.C. 1983. Encyclopedia of American forest and conservation history. 2 vols. Macmillan, New York. 871 pp. **18**

Davis, S.M., and J.C. Ogden (eds.). 1994. Everglades: The ecosystem and its restoration. St. Lucie Press, Delray Beach, Florida. 826 pp. **181, 276, 305, 306**

DeBach, P., and D. Rosen. 1990. Biological control by natural enemies (2nd ed.). Cambridge University Press, Cambridge. 440 pp. **236**

Debinski, D.M., and R.D. Holt. 2000. A survey and overview of habitat fragmentation experiments. Conservation Biology 14:342–355. **195**

De Leo, G.A., and S. Levin. 1997. The multifaceted aspects of ecosystem integrity. Conservation Ecology [online]1(1): 3. URL: *http://www.consecol.org/vol1/iss1/art3* **31, 32**

DeLong, D.C., Jr. 1996. Defining biodiversity. Wildlife Society Bulletin 24:738–749. **34**

DeMauro, M.M. 1993. Relationship of breeding system to rarity in the Lakeside daisy (*Hymenoxys acaulis var. glabra*). Conservation Biology 7:542–550. **339** deMaynadier, P.G., and M.L. Hunter, Jr. 1997. The role of keystone ecosystems in landscapes. Pages 68–76 in A. Haney and M. Boyce (eds.). Ecosystem management. Yale University Press, New Haven, Connecticut. 361 pp. **76**

Derrickson, S.R., and N.F.R. Snyder. 1992. Potentials and limits of captive breeding in parrot conservation. Pages 133–163 in S.R. Beissinger and N.F.R. Snyder (ed.). New world parrots in crisis. Smithsonian Institution Press, Washington. 288 pp. **351**

Devall, B., and G. Sessions. 1985. Deep ecology. Gibbs Smith, Salt Lake City, Utah. 267 pp. **394**

DeVeaux, J.S., and E.B. Shultz, Jr. 1985. Development of buffalo gourd (*Cucurbita foetidissima*) as a semiarid-land starch and oil crop. Economic Botany 39:454–472. **50**

Devine, R. 1998. Alien invasion: America's battle with non-native animals and plants. National Geographic Society, Washington, D.C. 280 pp. **251**

Dewar, R.E. 1984. Extinctions in Madagascar: The loss of the subfossil fauna. Pages 574–593 in P.S. Martin and R.G. Klein (eds.). Quaternary extinctions. University of Arizona Press, Tucson, Arizona. 892 pp. **203, 210**

Diamond, J.M. 1975. The island dilemma: Lessons of modern biogeographical studies for the design of natural preserves. Biological Conservation 7:129–146. **264**

Diamond, J.M. 1976. Island biogeography and conservation: Strategies and limitations. Science 193:1027–1029. **266**

Diamond, J.M. 1987. Extant unless proven extinct? Or, extinct unless proven extant? Conservation Biology 1:77–79. **125**

Diamond, J. 1989. Overview of recent extinctions. Pages 37–41 in D. Western, and M.C. Pearl (eds.). 1989. Conservation for the twenty-first century. Oxford University Press, New York. 365 pp. **79**

Diamond, J. 1997. Guns, germs, and steel. W. W. Norton and Company, New York. 480 pp. **231, 233, 241**

Dinerstein, E., and G.F. McCracken. 1990. Endangered greater one-horned rhinoceros carry high levels of genetic variation. Conservation Biology 4:417–422. **109**

Dinerstein, E., and E.D. Wikramanayake. 1993. Beyond "hotspots": How to prioritize investments to conserve biodiversity in the Indo-Pacific region. Conservation Biology 7:53–65. **374**

Doak, D., P. Karieva, and B. Klepetka. 1994. Modeling population viability for the desert tortoise in the western Mojave desert. Ecological Applications 4:446–460. **150**

Doak, D.F., and L.S. Mills. 1994. A useful role for theory in conservation. Ecology 75:615–626. **195**

Dobson, A.P, A.D. Bradshaw, and A.J.M. Baker. 1997. Hopes for the future: Restoration ecology and conservation biology. Science 277:515–522. **302**

Dobson, A., and A. Lyles. 2000. Black-footed ferret recovery. Science 288:985–988. **350**

Dobson, A.P., J.P. Rodriquez, W.M. Roberts, and D.S. Wilcove. 1997. Geographic distribution of endangered species in the United States. Science 275:550–553. **144**

Dodd, C.K., Jr. and R.A. Seigel. 1991. Relocation, repatriation, and translocation of amphibians and reptiles: Are they conservation strategies that work? Herpetologica 47:336–350. **333**

Dodd, J.L. 1994. Desertification and degradation in sub-Saharan Africa. Bioscience 44:28–34. **187**

Dodson, C.H., and A.H. Gentry. 1991. Biological extinction in western Ecuador. Annals of the Missouri Botanical Garden 78:273–295. **125**

Donovan, T.M., F.R. Thompson III, J. Faaborg, and J.R. Probst. 1995. Reproductive success of migratory birds in habitat sources and sinks. Conservation Biology 9:1380–1395. **146–147**

Dorweiler, J., A. Stec, J. Kermicle, and J. Doebley. 1993. Teosinte glume architecture 1: A genetic locus controlling a key step in maize evolution. Science 262:233–235. **50**

Dossaji, S.F., R. Wrangham, and E. Rodriguez. 1989. Selection of plants with medicinal properties by wild chimpanzees. Fitoterapia 60(4):378–380. **53**

Douthwaite, R.J. 1995. Occurrence and consequences on DDT residues in woodland birds following tsetse fly spraying operations in NW Zimbabwe. Journal of Applied Ecology 32:727–738. **172**

Dowdeswell, W.H. 1987. Hedgerows and verges. Allen and Unwin, London. 190 pp. **293, 319**

Dresser, B.L. 1988. Cryobiology, embryo transfer, and artificial insemination in *ex situ* animal conservation programs. Pages 296–308 in E.O. Wilson and F.M. Peter (eds.). Biodiversity. National Academy Press, Washington, D.C. 521 pp. **353**

Drewien, R.C., and E.G. Bizeau. 1977. Cross-fostering whooping cranes to sandhill crane foster parents. Pages 201–222 in S.A. Temple (ed.). Endangered birds: Management techniques for preserving threatened species. University of Wisconsin Press, Madison. 466 pp. **336**

Dunn, C.P., F. Stearns, G.R. Guntenspergen, and D.M. Sharpe. 1993. Ecological benefits of the Conservation Reserve Program. Conservation Biology 7:132–139. **296**

Dudgeon, D. 2000. Large-scale hydrological changes in tropical Asia: Prospects for riverine biodiversity. BioScience 50:793–806. **144, 193**

Dugan, J.E., and G.E. Davis. 1993. Applications of marine refugia to coastal fisheries management. Canadian Journal of Fisheries and Aquatic Sciences 50:2029–2042. **274**

Dugan, P. (ed.). 1993. Wetlands in danger. Oxford University Press, New York. 187 pp. **191**

Duggins, D.O. 1980. Kelp beds and sea otters: An experimental approach. Ecology 6:447–453. **223**

Duncan, R.P. 1997. The role of competition and introduction effort in the success of passeriform birds introduced to New Zealand. American Naturalist 149:903–915. **245**

Duncan, R.P., and J.R. Young. 2000. Determinants of plant extinction and rarity 145 years after European settlement of Auckland, New Zealand. Ecology 81:3048–3061. **139, 144**

Dunham, K.M. 1998. Use of artificial water supplies by captive-born mountain gazelles (*Gazella gazella*) released in central Arabia. Journal of Zoology 246:449–454. **317**

Eagan, D.J., and D.W. Orr. 1992. The campus and environmental responsibility. Jossey-Bass, San Francisco. 133 pp. **440**

Eames, J.C., and A.W. Tordoff. (in press). National planning for protected areas: Expanding the protected areas network in Vietnam for the 21st century. In: Proceedings of the conference on *in-situ* an *ex-situ* biodiversity conservation in the new millennium. 19–22 June 2000. Sabah Museum, Kota Kinabalu, Malaysia. **376**

Eames, J.C., L.T. Trai, N. Cu, and R. Eve. 1999. New species of Barwing *Actinodura* (Passeriformes: Sylviinae: Timaliini) from the Western Highlands of Vietnam. Ibis 141:1–10. **376**

Echelle, A.A., and A.F. Echelle. 1997. Genetic introgression of endemic taxa by non-natives: A case study with Leon Springs pupfish and sheepshead minnow. Conservation Biology 11:153–161. **244, 350**

Edington, J.M., and M.A. Edington. 1986. Ecology, recreation, and tourism. Cambridge University Press, Cambridge. 200 pp. **216**

Edwards, J.L., M.A. Lane, and E.S. Nielsen. 2000. Interoperability of biodiversity databases: Biodiversity information on every desktop. Science 289:2312–2314. **66**

Egler, F.E. 1977. The nature of vegetation: Its management and mismanagement. Aton Forest, Norfolk, Connecticut. 527 pp. **305**

Ehrenfeld, D. 1988. Why put a value on biodiversity? Pages 212–216 in E.O. Wilson and F.M. Peter (eds.). Biodiversity. National Academy Press, Washington, D.C. 521 pp. **421**

Ehrlich, P.R. 1986. Which animals will invade? Pages 79–95 in H.A. Mooney, and J.A. Drake (eds.). Ecology of biological invasions of North America and Hawaii. Springer-Verlag, New York. 321 pp. **246**

Ehrlich, P., and A. Ehrlich. 1981. Extinction. Random House, New York. 305 pp. **79**

Ehrlich, P.R., and H.A. Mooney. 1983. Extinction, substitution, and ecosystem services. Bioscience 33:248–254. **59, 79**

Ehrlich, P.R., and E.O. Wilson. 1991. Biodiversity studies: Science and policy. Science 253:758–762. **375**

Eisto, A., M. Kuitunen, A. Lammi, et al. 2000. Population persistence and offspring fitness in the rare bellflower *Campanula cervicaria* in relation to population size and habitat quality. Conservation Biology 14:1413–1421. **139, 316**

Eldridge, M.D.B., J.M. King, A.K. Loupis, et al. 1999. Unprecedented low levels of genetic variation and inbreeding depression in an island population of the black-footed rock-wallaby. Conservation Biology 13:531–541. **99**

Elias, T.S. (ed.). 1987. Conservation and management of rare and endangered plants. California Native Plant Society, Sacramento, California. 630 pp. **344**

Ellis, R.H., and E.H. Roberts. 1998. How to store seeds to conserve biodiversity. Nature 395:758. **355**

Ellstrand, N.C. 1992. Gene flow by pollen: Implications for plant conservation genetics. Oikos 63:77–86. **101, 244, 341**

Ellstrand, N.C., and D.R. Elam. 1993. Population genetic consequences of small population size: Implications for plant conservation. Annual Review of Ecology and Systematics 24:217–242. **99, 334, 341**

Ellstrand, N.C., H.C. Prentice, and J.F. Hancock. 1999. Gene flow and introgression from domesticated plants into their wild relatives. Annual Review of Ecology and Systematics 30:539–563. **244**

Elphick, C.S. 2000. Functional equivalency between rice fields and seminatural wetland habitats. Conservation Biology 14:181–191. **294**

Elton, C.S. 1958. The ecology of invasions by animals and plants. Methuen, London. 181 pp. **245, 246, 251**

Eltringham, S.K. 1994. Can wildlife pay its way? Oryx 28:163–168. **415**

Elzinga, C.L., D.W. Salzer, J.W. Willoughby, and J.P. Gibbs. 2001. Monitoring plant and animal populations. Blackwell Science, Malden, Massachusetts. 368 pp. **375**

Embley T.M., R.P. Hirt, and D.M. Williams. 1994. Biodiversity at the molecular level: The domains, kingdoms and phyla of life. Philosophical Transactions of the Royal Society of London B 345:21–33. **41**

Empson, R.A., and C.M. Miskelly. 1999. The risks, costs and benefits of using brodifacoum to eradicate rats from Kapiti Island, New Zealand. New Zealand Journal of Ecology 23:241–254. **278, 326**

Errington, P.L., and F.N. Hamerstrom, Jr. 1937. The evaluation of nesting losses and juvenile mortality of the ring-necked pheasant. Journal of Wildlife Management 1:3–20. **15**

Erwin, D.H. 1993. The great Paleozoic crisis: Life and death in the Permian. Columbia University Press, New York. 327 pp. **123, 124**

Erwin, D.H. 1994. The Permo-Triassic extinction. Nature 367:231–236. **122, 123, 124**

Erwin, T.L. 1991. An evolutionary basis for conservation strategies. Science 253:750–752. **63**

Estes, J.A., D.O. Duggins, and G.B. Rathbun. 1989. The ecology of extinctions in kelp forest communities. Conservation Biology 3:252–264. **79**

Estes, J.A., M.T. Tinker, T.M. Williams, and D.F. Doak. 1998. Killer whale predation on sea otters linking oceanic and nearshore ecosystems. Science 282:473–476. **223**

Evans, S. 1999. The green republic: A conservation history of Costa Rica. University of Texas Press, Austin, Texas. 317 pp. **18**

Ewel, J.J. 1991. Yes, we got some bananas. Conservation Biology 5:423–425. **296**

Fahrig, L., and G. Merriam. 1994. Conservation of fragmented populations. Conservation Biology 8:50–59. **157**

Fahrig, L., J.H. Pedlar, S.E. Pope, et al. 1995. Effect of road traffic on amphibian density. Biological Conservation 73:177–182. **173**

Faith, D.P. 1992. Conservation evaluation and phylogenetic diversity. Biological Conservation 61:1–10. **63, 372**

Falk, D.A., and K.E. Holsinger (eds.). 1991. Genetics and conservation of rare plants. Oxford University Press, New York. 283 pp. **113**

Falk, D.A., and P. Olwell. 1992. Scientific and policy considerations in restoration and reintroduction of endangered species. Rhodora 94:287–315. **335, 339**

Falk, D.A., C.I. Millar, and M. Olwell (eds.). 1996. Restoring diversity: Strategies for reintroduction of endangered plants. Island Press, Washington, D. C. 505 pp. **344**

Farnsworth, N.R. 1988. Screening plants for new medicines. Pages 83–97 in E.O. Wilson and F.M. Peter (eds.). Biodiversity. National Academy Press, Washington, D.C. 521 pp. **51**

Fauth, P. T. 2000. Reproductive success of wood thrushes in forest fragments in northern Indiana. Auk 117:194–204. **202**

Fearnside, P.M. 1989. Extractive reserves in Brazilian Amazonia. Bioscience 39:387–393. **290**

Feber, R.E., H. Smith, and D.W. Macdonald. 1996. The effects on butterfly abundance of the management of uncropped edges of arable fields. Journal of Applied Ecology 33:1191–1205. **293**

Fenster, C.B., and L.F. Galloway. 2000. Inbreeding and outbreeding depression in natural populations of *Chamaecrista fasciculata* (Fabaceae). Conservation Biology 14:1406–1412. **101**

Fernández-Juricic, E. 2000. Avifaunal use of wooded streets in an urban landscape. Conservation Biology 14:513–521. **301**

Ferreira, J., and S. Smith. 1987. Methods of increasing native populations of *Erysimum menziesii*. Pages 507–512 in T.S. Elias (ed.). Conservation and management of rare and endangered plants. California Native Plant Society, Sacramento, California. 630 pp. **338**

Ferrer, M., and F. Hiraldo. 1991. Evaluation of management techniques for the Spanish imperial eagle. Wildlife Society Bulletin 19:436–442. **325**

Ferreras, P., J.J. Aldama, J.F. Beltrán, and M. Delibes. 1992. Rates and causes of mortality in a fragmented population of Iberian lynx *Felis pardina* Temminck, 1824. Biological Conservation 61:197–202. **174**

Field, C.D. 1999. Mangrove rehabilitation: Choice and necessity. Hydrobiologia 413:47–52. **86**

Fischer, J., and D.B. Lindenmayer. 2000. An assessment of the published results of animal relocations. Biological Conservation 96:1–11. **333**

Fischer, M., and J. Stöcklin. 1997. Local extinctions of plants in remnants of extensively used calcareous grasslands 1950–1985. Conservation Biology 11:727–737. **142**

Fisher, J., and R.A. Hinde. 1949. The opening of milk bottles by birds. British Birds 42:347–357. **110**

Fisher, J., N. Simon, and J. Vincent. 1969. Wildlife in danger. Viking Press, New York. 368 pp. **125, 349**

Fitzgerald, S. 1989. International wildlife trade: Whose business is it? World Wildlife Fund, Washington, D.C. 459 pp. **211**

Flannery, T.F. 1995. The future eaters. George Braziller, New York. 423 pp. **223, 228**

Flather, C.H., M.S. Knowles, and I.A. Kendall. 1998. Threatened and endangered species geography. Bioscience 48:365–376. **165**

Fleischner, T.L. 1994. Ecological costs of livestock grazing in western North America. Conservation Biology 8:629–644. **187, 286**

Florentin, J-M., R. Maurrasse, and G. Sen. 1991. Impacts, tsunamis, and the Haitian Cretaceous-Tertiary boundary layer. Science 252:1690–1693. **122**

Floyd, T., and R.A. Nelson. 1990. Bone metabolism in black bears. International Conference on Bear Research and Management 8:135–137. **53**

Food and Agriculture Organization. 1993. Forest resources assessment 1990: Tropical countries. Food and Agriculture Organization of the United Nations, Food and Agriculture Organization Forestry Paper 112, Rome. **182**

Foose, T.J. 1983. The relevance of captive populations to the conservation of biotic diversity. Pages 374–401 in C.M. Schonewald-Cox, S.M. Chambers, B. MacBryde, and W.L. Thomas (eds.). Genetics and conservation. Benjamin/Cummings Publishing, Menlo Park, California. 722 pp. **357**

Forbush, E.H. 1929. Birds of Massachusetts and other New England states. Vol. 3. Berwick and Smith, Norwood, Massachusetts. 466 pp. **62**

Forbush, E.H., and C.H. Fernald. 1896. The gypsy moth. Wright and Potter, Boston. 495 pp. **236**

Ford-Lloyd, B., and M. Jackson. 1986. Plant genetic resources: An introduction to their conservation and use. Edward Arnold, London. 146 pp. **364**

Forest Ecosystem Management Assessment Team. 1993. Forest ecosystem management: An ecological, economic, and social assessment. U.S. Department of Agriculture Forest Service, Portland, Oregon. **310**

Forman, R.T.T. 1995. Land mosaics. Cambridge University Press, Cambridge, United Kingdom. 632 pp. **83, 87, 199**

Forman, R.T.T. 2000. Estimate of the area affected ecologically by the road system in the United States. Conservation Biology 14:31–35. **172**

Forman, R.T.T., and L.E. Alexander. 1998. Roads and their major ecological effects. Annual Review of Ecological Systematics 29:207–231. **173**

Forman, R.T.T., and M. Godron. 1986. Landscape ecology. John Wiley and Sons, New York. 619 pp. **302**

Fossey, D. 1983. Gorillas in the mist. Houghton Mifflin, Boston, Massachusetts. 326 pp. **60**

Fowler, M.E. 1993. Zoo and wild animal medicine: Current therapy. W. B. Saunders, Philadelphia. 617 pp. **344**

Fox, M.D., and B.J. Fox. 1986. The susceptibility of natural communities to invasion. Pages 57–66 in R.H. Groves and J.J. Burdon (eds.). Ecology of biological invasions. Cambridge University Press, Cambridge. 166 pp. **246**

Frankel, O.H., A.H.D. Brown, and J.J. Burdon. 1995. The conservation of plant biodiversity. Cambridge University Press, Cambridge, United Kingdom. 299 pp. **355, 361**

Frankel, O.H., and M.E. Soulé. 1981. Conservation and evolution. Cambridge University Press, Cambridge. 327 pp. **103, 104, 105, 113**

Frankham, R. 1996. Relationship of genetic variation to population size in wildlife. Conservation Biology 10:1500–1508. **339**

Frankham, R. 1998. Inbreeding and extinction: Island populations. Conservation Biology 12:665–675. **99**

Franklin, I.R. 1980. Evolutionary change in small populations. Pages 135–149 in M.E. Soulé and B.A. Wilcox (eds.). Conservation biology: An evolutionary-ecological perspective. Sinauer, Sunderland, Massachusetts. 584 pp. **158**

Franklin, J.F., D.R. Berg, D.A. Thornburgh, and J.C. Tappeiner. 1997. Alternative silvicultural approaches to timber harvesting: Variable retention harvest systems. Pages 111–139 in K.A. Kohm and J.F. Franklin. (eds.) Creating a forestry for the 21st century. Island Press, Washington, D. C. 475 pp. **283, 310**

Frazer, N.B. 1992. Sea turtle conservation and halfway technology. Conservation Biology 6:179–184. **337, 338**

Freese, C.H. 1998. Wild species as commodities. Island Press, Washington, D.C. 319 pp. **400**

Frelich, L.E., and K.J. Puettmann. 1999. Restoration ecology. Pages 498–524 in M.L. Hunter, Jr. (ed.). Maintaining biodiversity in forest ecosystems. Cambridge University Press, Cambridge, United Kingdom. 698 pp. **307**

Frenkel, R.E. 1970. Ruderal vegetation along some California roadsides. University of California Publications in Geography 20:1–163. **174**

Fritts, T.H., and G.H. Rodda. 1998. The role of introduced species in the degradation of island ecosystems: A case history of Guam. Annual Review of Ecology and Systematics 29:113–140. **240**

Frost, L.C. 1981. The study of *Ranunculus ophioglossifolius* and its successful conservation at the Badgeworth Nature Reserve, Gloucestershire. Pages 481–489 in H. Synge (ed.). The biological aspects of rare plant conservation. John Wiley and Sons, Chichester, United Kingdom. 558 pp. **331**

Fry, G.L.A. 1991. Conservation in agricultural ecosystems. Pages 415–443 in I.F. Spellerberg, F.B. Goldsmith, and M.G. Morris (eds.). The scientific management of temperate communities for conservation. Blackwell, Oxford. 566 pp. **296**

Fuccillo, D., L. Sears, and P. Stapleton (eds.). 1997. Biodiversity in trust: Conservation and use of plant genetic resources in CGIAR centres. Cambridge University Press, Cambridge, United Kingdom. 371 pp. **101, 360, 361**

Fuller, T.K., W.E. Berg, G.L. Radde, et al. 1992. A history and current estimate of wolf distribution and numbers in Minnesota. Wildlife Society Bulletin 20:42–55. **174**

Gadgil, M., and R. Guha. 1992. This fissured land: An ecological history of India. University of California Press, Berkeley. 274 pp. **18**

Gaines, S.D., and M. Bertness. 1993. The dynamics of juvenile dispersal: Why field ecologists must integrate. Ecology 74:2430–2435. **272**

Galicia, E., and G.A. Baldassarre. 1997. Effects of motorized tourboats on the behavior of nonbreeding American flamingos in Yucatan, Mexico. Conservation Biology 11:1159–1165. **216**

Garrott, R.A., P.J. White, and C.A. Vanderbilt White. 1993. Overabundance: An issue for conservation biologists? Conservation Biology 7:946–949. **321**

Gascon, C., et al. 1999. Matrix habitat and species richness in tropical forest remnants. Biological Conservation 91:223–229. **263, 269**

Gaston, K.J. 2000. Global patterns in biodiversity. Nature 405:220–227. **81, 263**

Gates, D.M. 1993. Climate change and its biological consequences. Sinauer, Sunderland, Massachusetts. 280 pp. **136**

Geist, C., and S.M. Galatowitsch. 1999. Reciprocal model for meeting ecological and human needs in restoration projects. Conservation Biology 13:970–979. **305**

Gelfand, D.H. 1989. Taq DNA polymerase. Pages 17–22 in H.A. Erlich (ed.). PCR technology. Stockton, New York. 246 pp. **62**

Gentry, A.H. 1986. Endemism in tropical versus temperate plant communities. Pages 153–181 in M.E. Soulé (ed.). Conservation biology: The science of scarcity and diversity. Sinauer, Sunderland, Massachusetts. 584 pp. **140**

Gibbs, J.P. 1998. Genetic structure of redback salamander *Plethodon cinereus* populations in continuous and fragmented forests. Biological Conservation 86:77–81. **174**

Gibbs, J.P., M.L. Hunter, Jr., and E.J. Sterling. 1998. Problem-solving in conservation biology and wildlife management: Exercises for class, field, and laboratory. Blackwell Science, Malden, Massachusetts. 215 pp. **xv, 162**

Gibbs, J.P., J.R. Longcore, D.G. McAuley, and J.K. Ringelman. 1991. Use of wetland habitats by selected nongame water birds in Maine. U.S. Fish Wildlife Service, Fish and Wildlife Research No. 9. 57 pp. **197**

Gibbs, J.P., and E.J. Stanton. 2001. Habitat fragmentation and arthropod community change: Carrion beetles, phoretic mites, and flies. Ecological Applications 11:79–85. **200**

Gilbert, O.L. 1989. The ecology of urban habitats. Chapman and Hall, London. 369 pp. **298, 301, 311**

Gilpin, M.E., and M.E. Soulé. 1986. Minimum viable populations: Processes of species extinction. Pages 19–34 in M.E. Soulé (ed.). Conservation biology: The science of scarcity and diversity. Sinauer, Sunderland, Massachusetts. 584 pp. **157**

Ginsberg, J.R., and E.J. Milner-Gulland. 1994. Sex-biased harvesting and population dynamics in ungulates: Implications for conservation and sustainable use. Conservation Biology 8:157–166. **221**

Gipps, J.H.W. 1991. Beyond captive breeding: Re-introducing endangered mammals to the wild. Clarendon Press, Oxford. 284 pp. **356**

Girondot, M., H. Fouillet, and C. Pieau. 1998. Feminizing turtle embryos as a conservation tool. Conservation Biology 12:353–362. **338**

Gittleman, J.L., and S.L. Pimm. 1991. Crying wolf in North America. Nature 351:524–525. **39**

Given, D.R. 1994. Principles and practice of plant conservation. Timber Press, Oregon. 292 pp. **344**

Givnish, T.J. 1994. Does diversity beget stability? Nature 371:113–114. **81**

Glenn, T.C., W. Stephan, and M.J. Braun. 1999. Effects of a population bottleneck on whooping crane mitochondrial DNA variation. Conservation Biology 13:1097–1107. **104**

Goble, D.D., S.M. George, K. Mazaika, et al. 1999. Local and national protection of endangered species: An assessment. Environmental Science and Policy 2:43–59. **432**

Goerck, J.M. 1997. Patterns of rarity in the birds of the Atlantic forest of Brazil. Conservation Biology 11:112–118. **46**

Goldschmidt, T., F. Witte, and J. Wanink. 1993. Cascading effects of the introduced Nile perch on the detritivorous/phytoplanktivorous species in the sublittoral areas of Lake Victoria. Conservation Biology 7:686–700. **240**

Goldsmith, F.B. (ed.). 1991. Monitoring for conservation and ecology. Chapman and Hall, London. 275 pp. **375**

Gompper, M.E., and E.S. Williams. 1998. Parasite conservation and the black-footed ferret recovery program. Conservation Biology 12:730–732. **329**

Goodman, S.M., and B.D. Patterson (eds.). 1997. Natural change and human impact in Madagascar. Smithsonian Institution Press, Washington D.C. 432 pp. **203**

Goodrich, J.M., and S.W. Buskirk. 1995. Control of abundant native vertebrates for conservation of endangered species. Conservation Biology 9:1357–1364. **326**

Goodstein, E.S. 1995. Economics and the environment. Prentice Hall, Englewood Cliffs, New Jersey. 575 pp. **424**

Goreau, T., T. McClanahan, R. Hayes, and A. Strong. 2000. Conservation of coral reefs after the 1998 global bleaching event. Conservation Biology 14:5–15. **167**

Gould, S.J. 1990. The golden rule—a proper scale of our environmental crisis. Natural History 1990(9):24–30. **124**

Gould, S.J. 1993. A special fondness for beetles. Natural History 102(1):4–12. **41**

Graham, R.W. 1986. Response of mammalian communities to environmental changes during the late Quaternary. Pages 300–313 in J. Diamond and T.J. Case (eds.), Community ecology, Harper and Row, New York. **132**

Graham, R.W., and E.L. Lundelius, Jr. 1984. Coevolutionary disequilibrium and Pleistocene extinctions. Pages 223–249 in P.S. Martin and R.G. Klein (eds.). Quaternary extinctions. The University of Arizona Press, Tucson. 892 pp. **208**

Grajal, A. 1999. Biodiversity and the nation state: Regulating access to genetic resources limits biodiversity research in developing countries. Conservation Biology 13:6–10. **414**

Grant, P.T., and B.R. Grant. 1992. Demography and the genetically effective sizes of two populations of Darwin's Finches. Ecology 73:766–784. **107**

Grant, V. 1981. Plant speciation (2nd ed.). Columbia University Press, New York. 563 pp. **37**

Grassle, J.F., and N.J. Maciolek. 1992. Deep-sea species richness: Regional and local diversity estimates from quantitative bottom samples. American Nautralist 139:313–341. **41**

Graveland, J., R. van der Wal, J.H. van Balen, and A.J. van Noordwijk. 1994. Poor reproduction in forest passerines from decline of snail abundance on acidified soils. Nature 368:446–448. **167**

Gray, A. 1996. Genetic diversity and its conservation in natural populations of plants. Biodiversity Letters 3:71–80. **97**

Green, B.H. 1989. Conservation in cultural landscapes. Pages 182–198 in D. Western and M. Pearl (eds.). Conservation for the Twenty-first Century. Oxford University Press, New York. 365 pp. **295**

Green, G.M., and R.W. Sussman. 1990. Deforestation history of the eastern rain forests of Madagascar from satellite images. Science 248:212–215. **203**

Green, R.E. 1997. The influence of numbers released on the outcome of attempts to introduce exotic bird species to New Zealand. Journal of Animal Ecology 66:25–35. **245**

Greenberg, C.H., S.H. Crownover, and D R. Gordon. 1997. Roadside soils: A corridor for invasion of xeric scrub by nonindigenous plants. Natural Areas Journal 17:99–109. **174**

Greig, J.C. 1979. Principles of genetic conservation in relation to wildlife management in Southern Africa. South African Journal of Wildlife Research 9:57–78. **100, 244, 340**

Griffin, A.S., D.T. Blumstein, and C.S. Evans. 2000. Training captive-bred or translocated animals to avoid predators. Conservation Biology 14:1317–1326. **336**

Griffin, G. J. 2000. Blight control and restoration of the American chestnut. Journal of Forestry 98(2):22–27. **241**

Griffith, B., J.M. Scott, J.W. Carpenter, and C. Reed. 1989. Translocation as a species conservation tool: Status and strategy. Science 245:477–480. **333**

Grimm, N.B., J.M. Grove, S.T.A. Pickett, and C.L. Redman. 2000. Integrated approaches to long-term studies of urban ecological systems. BioScience 50:571–584. **298**

Groom, M.J. 1994. Quantifying the loss of species due to tropical deforestation. Pages 121–122 in G.K. Meffe and C.R. Carroll. Principles of conservation biology. Sinauer, Sunderland, Massachusetts. 600 pp. **126**

Groombridge, B. (ed.). 1992. Global biodiversity: Status of the Earth's living resources. Chapman and Hall, London. 585 pp. **66, 87, 203**

Groombridge, J.J., C.G. Jones, M.W. Bruford, and R.A. Nichols. 2000. 'Ghost' alleles of the Mauritius kestrel. Nature 403: 616. **109**

Gross, M.R. 1991. Salmon breeding behavior and life history evolution in changing environments. Ecology 72:1180–1186. **221**

Grout, B.W.W., G.J. Morris, and M.R. McLellan. 1992. Cryopreservation of gametes and embryos of aquatic organisms. Pages 63–71 in H.D.M. Moore, W.V. Holt, and G.M. Mace (eds.). Biotechnology and the conservation of genetic diversity. Clarendon Press, Oxford. 240 pp. **355**

Grove, R.H. 1992. Origins of western environmentalism. Scientific American 267(1):42–47. **8**

Groves, C.P., and G.B. Schaller. 2000. The phylogeny and biogeography of the newly discovered Annamite Artiodactyls. Pages 261–282 in E.S. Vrba and G.B. Schaller (eds). Antelopes, deer, and relatives: Fossil record, behavioral ecology, systematics, and conservation. Yale University Press, New Haven. **376**

Grubb, M., M. Koch, A. Munsun, et al. 1993. The Earth Summit agreements. Earthscan, London. 180 pp. **427, 431**

Grumbine, E. 1990. Protecting biological diversity through the greater ecosystem concept. Natural Areas Journal 10:114–120. **269**

Grumbine, R.E. 1992. Ghost bears: Exploring the biodiversity crisis. Island Press, Washington, D.C. 294 pp. **440**

Grumbine, R.E. 1994. What is ecosystem management? Conservation Biology 8:27–38. **311**

Guzman, H.M. 1991. Restoration of coral reefs in Pacific Costa Rica. Conservation Biology 5:189–195. **332**

Hackel, J.D. 1999. Community conservation and the future of Africa's wildlife. Conservation Biology 13:726–734. **416**

Haig, S.M., J.D. Ballou, and S.R. Derrickson. 1990. Management options for preserving genetic diversity: Reintroduction of Guam rails to the wild. Conservation Biology 4:290–300. **354**

Haight, R.G., D.J. Mladenoff, and A.P. Wydeven. 1998. Modeling disjunct gray wolf populations in semi-wild landscapes. Conservation Biology 12:879–888. **279**

Haila, Y. 1999. Islands and fragments. Pages 234–264 in M.L. Hunter, Jr. (ed.) Maintaining biodiversity in forest ecosystems. Cambridge University Press, Cambridge, United Kingdom. 698 pp. **195, 199**

Hall, L.A. 1987. Transplantation of sensitive plants as mitigation for environmental impacts. Pages 413–420 in T.S. Elias (ed.). Conservation and management of rare and endangered plants. California Native Plant Society, Sacramento, California. 630 pp. **333**

Hall, L.S., P.R. Krausman, and M.L. Morrison. 1997. The habitat concept and a plea for standard terminology. Wildlife Society Bulletin 25:173–182. **163**

Hall, S.J.G., and J. Ruane. 1993. Livestock breeds and their conservation: A global overview. Conservation Biology 7:815–825. **361**

Hallam, A., and P.B. Wignall. 1997. Mass extinctions and their aftermath. Oxford University Press, Oxford. 320 pp. **136**

Halliday, T. 1978. Vanishing birds. Holt, Rinehart and Winston, New York. 296 pp. **138**

Hamilton, M.B. 1994. *Ex situ* conservation of wild plant species: Time to reassess the genetic assumptions and implications of seed banks. Conservation Biology 8:39–49. **355**

Hammond, P. 1992. Species inventory. Pages 17–39 in B. Groombridge (ed.). Global biodiversity: Status of the Earth's living resources. Chapman and Hall, London. 585 pp. **41**

Hammond, P. 1995. The current magnitude of biodiversity. Pages 113–138 in V.H. Heywood and R.T. Watson (eds.). Global biodiversity assessment. Cambridge University Press, Cambridge, United Kingdom. 1140 pp. **40**

Hansen, S. 1989. Debt for nature swaps—overview and discussion of key issues. Ecological Economics 1:77–93. **417**

Hanski, I. 1998. Metapopulation dynamics. Nature 396:41–49. **146, 149**

Hanski, I. 1999. Metapopulation ecology. Oxford University Press, Oxford, United Kingdom. 313 pp. **149, 162**

Hanski, I., J. Clobert, and W. Reid. 1995. Ecology of extinctions. Pages 232–245 in V.H. Heywood and R.T. Watson (eds.). Global biodiversity assessment. Cambridge University Press, Cambridge, United Kingdom. 1140 pp. **36**

Hanski, I., and M. Gilpin. 1991. Metapopulation dynamics: Brief history and conceptual domain. Biological Journal of the Linnean Society 42:3–16. **146**

Hanski, I., and M.E. Gilpin. (eds.). 1997. Metapopulation biology. Academic Press, San Diego. 512 pp. **162**

Hanski, I., and D. Simberloff. 1997. The metapopulation approach, its history, conceptual domain, and application to conservation. Pages 5–26 in I. Hanski and M.E. Gilpin (eds.) Metapopulation biology. Academic Press, San Diego. 512 pp. **146, 149**

Hansson, L., L. Soderstrom, and C. Solbreck. 1992. The ecology of dispersal in relation to conservation. Pages 162–200 in L. Hansson (ed.). Ecological principles of nature conservation. Elsevier, London. 436 pp. **272**

Harcourt, C. 1992. Wetlands. Pages 293–306 in B. Groombridge (ed.). Global biodiversity. Chapman and Hall, London. 585 pp. **191**

Hardin, G. 1968. The tragedy of the commons. Science 162:1243–1248. **213, 412**

Hardin, G., and J. Baden (eds.). 1977. Managing the commons. W. H. Freeman, New York. 294 pp. **413**

Hargrove, E.C. 1989. An overview of conservation and human values: Are conservation goals merely cultural attitudes? Pages 227–231 in D. Western, and M.C. Pearl (eds.). 1989. Conservation for the twenty-first century. Oxford University Press, New York. 365 pp. **42**

Harker, D., G. Libby, K. Harker, et al. 1999. Landscape restoration handbook. 2nd ed. Lewis Publishers, Boca Raton, Florida. 145 pp. **311**

Harrington, R., N. Owen-Smith, P. C. Viljoen, et al. 1999. Establishing the causes of the roan antelope decline in the Kruger National Park, South Africa. Biological Conservation 90:69–78. **277**

Harris, L.D. 1984. The fragmented forest: Island biogeography theory and the preservation of biotic diversity. University of Chicago Press, Chicago. 211 pp. **205, 310**

Harris, L.D., and W.P. Cropper, Jr. 1992. Between the devil and the deep blue sea: Implications of climate change for Florida's fauna. Pages 309–324 in R.L. Peters and T.E. Lovejoy (eds.). Global warming and biological diversity. Yale University Press, New Haven. 386 pp. **134**

Harris, M. 1974. Cows, pigs, wars, and witches. The riddles of culture. Random House, New York. 276 pp. **384**

Harrison, S. 1994. Metapopulations and conservation. Pages 111–128 in P.J. Edwards, R.M. May, and N.R. Webb (eds.). Large-scale ecology and conservation biology. Blackwell, Oxford. 375 pp. **146, 150**

Harrison, S., and E. Bruna. 1999. Habitat fragmentation and large-scale conservation: What do we know for sure? Ecography 22:225–232. **195**

Harrison, S., D.D. Murphy, and P.R. Ehrlich. 1988. Distribution of the bay checkerspot butterfly *Euphydryas editha bayensis:* Evidence for a metapopulation model. American Naturalist 132:360–382. **148**

Hartl, D.L. 2000. A primer of population genetics, 3rd ed. Sinauer, Sunderland, Massachusetts. 221 pp. **89, 92, 113**

Hartl, D.L., and A.G. Clark. 1997. Principles of population genetics, 3rd ed., Sinauer, Sunderland, Massachusetts. 542 pp. **89, 92, 93, 95, 104, 106, 113**

Hartl, G.B., and Z. Pucek. 1994. Genetic depletion in the European bison (*Bison bonasus*) and the significance of electrophoretic heterozygosity for conservation. Conservation Biology 8:167–174. **94**

Hartley, M.J., and M.L. Hunter, Jr. 1998. A meta-analysis of forest cover, edge effects, and artificial nest predation rates. Conservation Biology 12:465–469. **202**

Haskell, D.G. 2000. Effects of forest roads on macroinvertebrate soil fauna of the southern Appalachian Mountains. Conservation Biology 14:57–63. **174**

Hatcher, P.E., and K.N.A. Alexander. 1994. The status and conservation of the netted carpet *Eustroma reticulatum* (Denis and Schiffermuller, 1775) (*Lepidoptera: Geometridae*), a threatened moth species in Britain. Biological Conservation 67:41–47. **316**

Hawkins, J.P., C.M. Roberts, T. Van't Hof, et al. 1999. Effects of recreational scuba diving on Caribbean coral and fish communities. Conservation Biology 13:888–897. **216**

Hawksworth, D.L. 1991. The fungal dimension of biodiversity: Magnitude, significance, and conservation. Mycological Research 95:641–655. **41**

Hayward, J., and I. Vertinsky. 1999. High expectations, unexpected benefits: What managers and owners think of certification. Journal of Forestry 97(2):13–17. **435**

Hedrick, P.W., and P.S. Miller. 1992. Conservation genetics: Techniques and fundamentals. Ecological Applications 2:30–46. **89, 92**

Heeney, J.L., J.F. Evermann, A.J. McKeirnan, et al. 1990. Prevalence and implications of feline coronavirus infections of captive and free-ranging cheetahs (*Acinonyx jubatus*). Journal of Virology 64:1964–1972. **112**

Helldén, U. 1991. Desertification—time for an assessment? Ambio 20:372–383. **187**

Hemley, G. (ed.). 1994. International wildlife trade. Island Press, Washington, D.C., 166 pp. **211**

Henderson, I.G., J. Cooper, R.J. Fuller, and J. Vickery. 2000. The relative abundance of birds on set-aside and neighbouring fields in summer. Journal of Applied Ecology 37:335–347. **293**

Henderson, M.T., G. Merriam, and J. Wegner. 1985. Patchy environments and species survival: Chipmunks in an agricultural mosaic. Biological Conservation 31:95–105. **273**

Hendrickson, D.A., and J.E. Brooks. 1991. Transplanting short-lived fishes in North American Deserts: Review, assessment, and recommendations. Pages 283–298 in W.L. Minckley and J.E. Deacon (eds.). Battle against extinction: Native fish management in the American west. University of Arizona Press, Tucson. 517 pp. **333**

Hendrix, P.F. (ed.) 1995. Earthworm ecology and biogeography in North America. Lewis Publishers, Boca Raton, Florida. 244 pp. **231**

Hendry, A.P., J.K. Wenburg, P. Bentzen, et al. 2000. Rapid evolution of reproductive isolation in the wild: Evidence from introduced salmon. Science 290:516–518. **109**

Hess, G.R. 1994. Conservation corridors and contagious disease: A cautionary note. Conservation Biology 8:256–262. **273**

Hewitt, G. 2000. The genetic legacy of the Quaternary ice ages. Nature 405:907–913. **131**

Heywood, V.H., and R. T. Watson (eds.). 1995. Global biodiversity assessment. Cambridge University Press, Cambridge, United Kingdom. 1140 pp. **34, 40, 66**

Higgins, S.I., D.M. Richardson, R.M. Cowling, and T.H. Trinder-Smith. 1999. Predicting the landscape-scale distribution of alien plants and their threat to plant diversity. Conservation Biology 13:303–313. **247**

Higgs, A.J., and M.B. Usher. 1980. Should nature reserves be large or small? Nature 285:568–569. **267**

Higgs, E.S. 1997. What is good ecological restoration? Conservation Biology 11:338–348. **305**

Higuchi, R., B. Bowman, M. Freiberger, et al. 1984. DNA sequences from the quagga, an extinct member of the horse family. Nature 312:282–284. **355**

Hirsch, U. 1977. Artificial nest ledges for bald ibises. Pages 61–69 in S.A. Temple (ed.). Endangered birds: Management techniques for preserving threatened species. University of Wisconsin Press, Madison. 466 pp. **318**

Hobbs, R.J. 1989. The nature and effects of disturbance relative to invasions. Pages 389–405 in J.A. Drake, H.A. Mooney, F. di Castri, et al. (eds.). Biological invasions: A global perspective. John Wiley and Sons, Chichester. 525 pp. **246**

Hobbs, R.J. 1992. The role of corridors in conservation: Solution or bandwagon? Trends in Ecology and Evolution 7:389–392. **273**

Hobbs, R.J., and H.A. Mooney. 1998. Broadening the extinction debate: Population deletions and additions in California and western Australia. Conservation Biology 12:271–283. **159**

Hoctor, T.S., M.H. Carr, and P.D. Zwick. 2000. Identifying a linked reserve system using a regional landscape approach: The Florida Ecological Network. Conservation Biology 14:984–1000. **273**

Hoelzel, A.R. (ed.). 1998. Molecular genetic analysis of populations (second edition). Oxford University Press, Oxford, United Kingdom. 445 pp. **92**

Hoelzel, A.R., J. Halley, S.J. O'Brien, et al. 1993. Elephant seal genetic variation and the use of simulation models to investigate historic population bottlenecks. Journal of Heredity 84:443–449. **109**

Hoffman, W.A., and R.B. Jackson. 2000. Vegetation-climate feedbacks in the conversion of tropical savanna to grassland. Journal of Climate 13:1593–1602. **76**

Hogbin, P.M., and R. Peakall. 1999. Evaluation of the contribution of genetic research to the management of the endangered plant *Zieria prostrata*. Conservation Biology 13:514–522. **341**

Holdaway, R.N., and C. Jacomb. 2000. Rapid extinction of the Moas (Aves: Dinornithiformes): Model, test, and implications. Science 287:2250–2254. **210**

Holden, P.B. 1991. Ghosts of the Green River: Impacts of Green River poisoning on management of native fishes. Pages 43–54 in W.L. Minckley and J.E. Deacon (eds.). Battle against extinction: Native fish management in the American West. University of Arizona Press, Tucson. 517 pp. **234, 289**

Holechek, J.L., R.D. Pieper, and C.H. Herbel. 2001. Range Management. 4th ed. Prentice Hall, Upper Saddle River, New Jersey. 587 pp. **286, 287**

Holgate, S.T., J.M. Samet, H.S. Koren, and R.L. Maynard (eds.). 1999. Air pollution and health. Academic Press, San Diego. 1065 pp. **167**

Holl, K.D., G.C. Daily, and P.R. Ehrlich. 1995. Knowledge and perceptions in Costa Rica regarding environment, population, and biodiversity issues. Conservation Biology 9:1548–1558. **393**

Holmes, S.B. 1998. Reproduction and nest behavior of Tennessee warblers *Vermivora peregrina* in forests treated with Lepidoptera-specific insecticides. Journal of Applied Ecology 35:185–194. **172**

Holt, W.V. 1992. Advances in artificial insemination and semen freezing in mammals. Pages 19–35 in H.D.M. Moore, W.V. Holt, and G.M. Mace (eds.). Biotechnology and the conservation of genetic diversity. Clarendon Press, Oxford. 240 pp. **355**

Holt, W.V. 1994. Reproductive technologies. Pages 144–166 in P.J.S. Olney, G.M. Mace, and A.T.C. Feistner (eds.). Creative conservation: Interactive management of wild and captive animals. Chapman and Hall, London. 517 pp. **353, 355**

Holt, W.V., P.M. Bennett, and V. Volobouev. 1996. Genetic resource banks in wildlife conservation. Journal of Zoology 238:531–544. **353, 355**

Hoopes, M.F., and S. Harrison. 1998. Metapopulation, source-sink and disturbance dynamics. Pages 135–151 in W.J. Sutherland (ed.) Conservation science and action. Blackwell Science, Oxford, United Kingdom. 363 pp. **148**

Houghton, J. 1997. Global warming: The complete briefing. Cambridge University Press, Cambridge, United Kingdom. 251 pp. **133, 136**

Houghton, R.A., D.L. Skole, C.A. Nobre, et al. 2000. Annual fluxes of carbon from deforestation and regrowth in the Brazilian Amazon. Nature 403:301–304. **184**

Houlahan, J.E., C.S. Findlay, B.R. Schmidt, et al. 2000. Quantitative evidence for global amphibian population declines. Nature 404:752–755. **167, 192**

Houston D.B., and E.G. Schreiner. 1995. Alien species in national parks: Drawing lines in space and time. Conservation Biology 9:204–209. **277**

Howarth, F.G. 1991. Environmental impacts of classical biological control. Annual Review of Entomology 36:485–509. **236**

Hoyt, E. 1988. Conserving the wild relatives of crops. International Board for Plant Genetic Resources, Rome. 45 pp. **102**

Hudson, R.J., K.R. Drew, and L.M. Baskin (eds.). 1989. Wildlife production systems: Economic utilisation of wild ungulates. Cambridge University Press, Cambridge. 469 pp. **285**

Huenneke, L.F. 1991. Ecological implications of genetic variation in plant populations. Pages 31–44 in D.A. Falk, and K.E. Holsinger (eds.). 1991. Genetics and conservation of rare plants. Oxford University Press, New York. 283 pp. **339**

Hughes, J.B., G.C. Daily, and P.R. Ehrlich. 1997. Population diversity: Its extent and extinction. Science 278:689–692. **126**

Hughes, L. 2000. Biological consequences of global warming: Is the signal already apparent? Trends in Ecology and Evolution 15:56–61. **133**

Human, K.G., and D.M. Gordon. 1997. Effects of Argentine ants on invertebrate biodiversity in northern California. Conservation Biology 11:1242–1248. **242**

Hunter, M.L., Jr. 1990. Wildlife, forests, and forestry: Principles of managing forests for biological diversity. Prentice-Hall, Englewood Cliffs, New Jersey. 370 pp. **27, 84, 132, 183, 220, 222, 280, 281, 310, 311**

Hunter, M.L., Jr. 1991. Coping with ignorance: The coarse-filter strategy for maintaining biodiversity. Pages 266–281 in K. Kohm (ed.). Balancing on the edge of extinction. Island Press, Washington, D.C. 315 pp. **71, 77, 259**

Hunter, M.L., Jr. 1992. Paleoecology, landscape ecology, and the conservation of neotropical migrant passerines in boreal forests. Pages 511–523 in J. Hagan and D. Johnston (eds.). Ecology and conservation of neotropical migrant landbirds. Smithsonian Institution Press. 608 pp. **128**

Hunter, M.L., Jr. 1993. Natural fire regimes as spatial models for managing boreal forests. Biological Conservation 65:115–120. **282**

Hunter, M.L., Jr. 1996. Benchmarks for managing ecosystems: Are human activities natural? Conservation Biology 10:695–697. **31, 32, 34, 278**

Hunter, M.L., Jr. 1997. The biological landscape. Pages 57–67 in K.A. Kohm and J.F. Franklin (eds.). Creating a forestry for the 21st century. Island Press, Washington D.C. 475 pp. **200, 271, 273**

Hunter, M.L., Jr. (ed.). 1999. Maintaining biodiversity in forest ecosystems. Cambridge University Press, Cambridge, United Kingdom. 698 pp. **29, 40, 222, 280, 311**

Hunter, M.L., Jr., and A. Calhoun. 1995. A triad approach to land use allocation. Pages 447–491 in R. Szaro and D. Johnston (eds.). Biodiversity in managed landscapes. Oxford University Press, New York. 778 pp. **188, 296, 297**

Hunter, M.L., Jr., G. Jacobson, and T. Webb. 1988. Paleoecology and the coarse-filter approach to maintaining biological diversity. Conservation Biology 2:375–385. **78, 132, 259, 272**

Hunter, M.L., Jr., R.K. Hitchcock, and B. Wyckoff-Baird. 1990. Women and wildlife in southern Africa. Conservation Biology 4:448–451. **388**

Hunter, M.L., Jr., and A. Hutchinson. 1994. The virtues and shortcomings of parochialism: Conserving species that are locally rare, but globally common. Conservation Biology 8:1163–1165. **368**

Hunter, M.L., Jr., and J. Witham. 1985. Effects of a carbaryl-induced depression of arthropod abundance on the foraging behavior of Parulinae warblers. Canadian Journal of Zoology 63:2612–2616. **172**

Hunter, M.L., Jr., and P. Yonzon. 1993. Altitudinal distributions of birds, mammals, people, forests and parks in Nepal. Conservation Biology 7:420–423. **260**

Huntley, B., and T. Webb III. 1989. Migration: Species response to climatic variations caused by changes in the earth's orbit. Journal of Biogeography 16:5–19. **128**

Huston, M.A. 1994. Biological diversity: The coexistence of species on changing landscapes. Cambridge University Press, Cambridge. 681 pp. **66, 87, 140, 184, 245**

Huston, M.A. 1997. Hidden treatments in ecological experiments: Re-evaluating the ecosystem function of biodiversity. Oecologia 110:449–460. **81**

Hutchins, M., and R.J. Wiese. 1991. Beyond genetic and demographic management: The future of the Species Survival Plan and related AAZPA conservation efforts. Zoo Biology 10:285–292. **357**

Hutchins, M., K. Willis, and R.J. Wiese. 1995. Strategic collection planning: Theory and practice. Zoo Biology 14:5–25. **357**

ICBP. 1991. Spix's macaw update. World Birdwatch 13(3):4. **373**

Iltis, H.H., J.F. Doebley, R. Guzman, and B. Pazy. 1979. *Zea diploperennis* (Gramineae): A new teosinte from Mexico. Science 203:186–188. **49**

Imbrie, J., and K.P. Imbrie. 1986. Ice ages: Solving the mystery. Harvard University Press, Cambridge, Massachusetts. 224 pp. **129, 131, 136**

Inouye, D.W., B. Barr, K.B. Armitage, and B.D. Inouye. 2000. Climate change is affecting altitudinal migrants and hibernating species. Proceedings of the National Academy of Sciences 97:1630–1633. **133**

Isozaki, Y. 1997. Permo-triassic boundary superanoxia and stratified superocean: Records from lost deep sea. Science 276:235–238. **124**

IUCN/UNEP/WWF. 1980. World Conservation Strategy. IUCN, Gland, Switzerland. **399, 427, 440**

IUCN/UNEP/WWF. 1991. Caring for the earth. IUCN/UNEP/WWF, Gland, Switzerland. 228 pp. **440**

Jablonski, D. 1991. Extinctions: A paleontological perspective. Science 253:754–757. **120, 162**

Jablonski, D. 1993. The tropics as a source of evolutionary novelty through geological time. Nature 364:142–144. **81**

Jablonski, D. 1995. Extinctions in the fossil record. Pages 25–44 in J.H. Lawton and R.M. May (eds.). Extinction rates. Oxford University Press, Oxford, United Kingdom. 233 pp. **124**

Jackson, W., and J. Piper. 1989. The necessary marriage between ecology and agriculture. Ecology 70:1591–1593. **293**

Jacobson, E.R., J.M. Gaskin, M.B. Brown, et al. 1991. Chronic upper respiratory tract disease of free-ranging desert tortoises (*Xerobates agassizii* Journal of Wildlife Diseases 27:296–316. **334**

Jacobson, G.L., Jr. and A. Dieffenbacher-Krall. 1995. White pine and climate change: Lessons from the past, management alternatives for the future. Journal of Forestry. **59**

Jacobson, G.L., Jr., T. Webb III, and E.C. Grimm. 1987. Patterns and rates of vegetation change during the deglaciation of eastern North America. Pages 277–288 in W.F. Ruddiman and H.E. Wright, Jr. (eds.), North America and adjacent oceans during the last deglaciation, Geological Society of America, Boulder, Colorado. **132**

Jacobson, M. (ed.). 1988. Phytochemical pesticides. Vol. I. The neem tree. CRC Press, Boca Raton, Florida. 178 pp. **65**

Jacobson, S.K. 1990. Graduate education in conservation biology. Conservation Biology 4:431–440. **14, 15, 16, 18**

Jacobson, S.K., and M.D. McDuff. 1998. Training idiot savants: The lack of human dimensions in conservation biology. Conservation Biology 12:263–267. **397–398**

Jaenike, J. 1991. Mass extinction of European fungi. Trends in Ecology and Evolution 6:174–175. **167**

James, A.N., K.J. Gaston, and A. Balmford. 1999. Balancing the Earth's accounts. Nature 401:323–324. **408, 433**

Janick, J. (ed.). 1996. Progress in new crops. ASHS Press, Alexandria Virginia. 660 pp. **50**

Janss, G.F.E. 2000. Avian mortality from power lines: A morphologic approach of a species-specific mortality. Biological Conservation 95:353–359. **176**

Janssen, R., and J.E. Padilla. 1999. Preservation or conversion? Valuation and evaluation of a mangrove forest in the Philippines. Environmental and Resource Economics 14:297–331. **86**

Jansson, R., C. Nilsson, M. Dynesius, and E. Andersson. 2000a. Effects of river regulation on river-margin vegetation: A comparison of eight boreal rivers. Ecological Applications 10:203–224. **176, 191**

Jansson, R., C. Nilsson, and B. Renöfält. 2000b. Fragmentation of riparian floras in rivers with multiple dams. Ecology 81:899–903. **176**

Janzen, D.H. 1986. The eternal external threat. Pages 286–303 in M.E. Soulé (ed.) Conservation biology. Sinauer, Sunderland, Massachusetts. 584 pp. **202, 268**

Janzen, D.H. 1988a. Tropical dry forests: The most endangered major tropical ecosystem. Pages 130–137 in E.O. Wilson and F.M. Peter (eds.). Biodiversity. National Academy Press, Washington, D.C. 521 pp. **184**

Janzen, D.H. 1988b. Guanacaste National Park: Tropical ecological and biocultural restoration. Pages 143–192 in J. Cairns, Jr. (ed.). Rehabilitating damaged ecosystems. CRC Press, Boca Raton, Florida. 222 pp. **306**

Janzen, D.H. 1992. The neotropics. Restoration and Management Notes 10:8–13. **306**

Järvinen, O. 1982. Conservation of endangered plant populations: Single large or several small reserves? Oikos 38:301–307. **267**

Järvinen, O., and E. Ranta. 1987. Patterns and processes in species assemblages on Northern Baltic islands. Annales Zoologica Fennici 24:249–266. **196**

Jefferson, R.G., and M.B. Usher. 1986. Ecological succession and the evaluation of non-climax communities. Pages 70–91 in M.B. Usher (ed.). Wildlife conservation evaluation. Chapman and Hall, London. 394 pp. **299**

Jeffreys, A.J., V. Wilson, and S.L. Thein. 1985. Hypervariable "minisatellite" regions in human DNA. Nature 314:67–73. **109**

Jenkins, M.D. (ed.). 1987. Madagascar: An environmental profile. IUCN, Gland, Switzerland. 374 pp. **203**

Jenkins, M. 1992. Species extinction. Pages 192–233 in B. Groombridge (ed.). Global biodiversity: Status of the Earth's living resources. Chapman and Hall, London. 585 pp. **119, 142**

Jennings, M.D. 2000. Gap analysis: Concepts, methods, and recent results. Landscape Ecology 15:5–20. **262**

Jewell, P. 1985. Rare breeds of domestic livestock as a gene bank. Ark 12:158–168. **361**

Jewell, S.D. 1998. Fixing all the parts. Endangered Species Bulletin 23(6):23. **164**

Jin, Y.G., Y. Wang, W. Wang, et al. 2000. Pattern of marine mass extinction near the Permian-Triassic boundary in South China. Science 289:432–436. **123**

Jobes, P.C. 1991. The Greater Yellowstone social system. Conservation Biology 5:387–394. **386**

Johnsen, S.J., H.B. Clausen, W. Dansgaard, et al. 1992. Irregular glacial interstadials recorded in a new Greenland ice core. Nature 359:311–313. **130**

Johnson, J.E., and B.L. Jensen. 1991. Hatcheries for endangered freshwater fishes. Pages 199–217 in W.L. Minckley and J.E. Deacon (eds.). Battle against extinction: Native fish management in the American west. University of Arizona Press, Tucson. 517 pp. **338**

Johnson, K.H., K.A. Vogt, H.J. Clark, et al. 1996. Biodiversity and the productivity and stability of ecosystems. Trends in Ecology and Evolution 11:372–377. **79, 81**

Johnson, L.E., and D.K. Padilla. 1996. Geographic spread of exotic species: Ecological lessons and opportunities from the invasion of the zebra mussel *Dreissena polymorpha*. Biological Conservation 78:23–33. **232**

Johnson, N.C. 1995. Biodiversity in the balance: Approaches to setting geographic conservation priorities. Biodiversity Support Program, World Wildlife Fund, Washington, DC. 116 pp. **379**

Johst, K., R. Brandl, and R. Pfeifer. 2001. Foraging in a patchy and dynamic landscape: Human land use and the white stork. Ecological Applications 11:60–69. **294**

Jolly, A. 1980. A world like our own: Man and nature in Madagascar. Yale University Press, New Haven, Connecticut. 272 pp. **203**

Jolly, A., P. Oberlé, and R. Albignac (eds.). 1984. Madagascar. Pergamon Press, Oxford. 239 pp. **203**

Jones, C.G., W. Heck, R.E. Lewis, et al. 1991. A summary of the conservation management of the Mauritius kestrel *Falco punctatus* 1973–1991. Dodo 27:81–99. **336**

Jones, J.A., F.J. Swanson, B.C. Wemple, and K.U. Snyder. 2000. Effects of roads on hydrology, geomorphology, and disturbance patches in stream networks. Conservation Biology 14:76–85. **174**

Joyce, C. 1993. Taxol: Search for a cancer drug. Bioscience 43:133–136. **52**

Juniper, T., and C. Yamashita. 1990. The conservation of Spix's macaw. Oryx 24:224–228. **373**

Kaiser, J. 2000. Rift over biodiversity divides ecologists. Science 289:1282–1283. **81**

Kaiser, M.J., F.E. Spence, and P.J.B. Hart. 2000. Fishing-gear restrictions and conservation of benthic habitat complexity. Conservation Biology 14:1512–1525. **323**

Kalinowski, S.T., P.W. Hedrick, and P.S. Miller. 2000. Inbreeding depression in the Speke's gazelle captive breeding program. Conservation Biology 14:1375–1384. **109**

Kallman, H., C.P. Agee, W.R. Goforth, and J.P. Linduska (eds.). 1987. Restoring America's wildlife: 1937–1987. U.S. Fish and Wildlife Service, Washington, D.C. 394 pp. **215–216**

Karieva, P., M. Marvier, and M. McClure. 2000. Recovery and management options for spring/summer Chinook salmon in the Columbia River basin. Science 290:977–979. **175**

Karl, S.A., and B.W. Bowen. 1999. Evolutionary significant units versus geopolitical taxonomy: Molecular systematics of an endangered sea turtle (genus Chelonia). Conservation Biology 13:990–999. **38**

Karow, A.M., and J.K. Critser (eds.). 1997. Reproductive tissue banking. Academic Press, San Diego. 472 pp. **355**

Katahira, L.K., P. Finnegan, and C.P. Stone. 1993. Eradicating feral pigs in montane mesic habitat at Hawaii Volcanoes National Park. Wildlife Society Bulletin 21:269–274. **328**

Kaufman, J.H., D. Brodbeck, and O.R. Melroy. 1998. Critical biodiversity. Conservation Biology 12:521–532. **81**

Keenleyside, M.H.A. 1991. Cichlid fishes. Chapman and Hall, London. 378 pp. **140**

Keller, A.E., and S.G. Zam. 1990. Simplification of in vitro culture techniques for freshwater mussels. Environmental Toxicology and Chemistry 9:1291–1296. **338**

Keller, L.F., P. Arces, J.N.M. Smith, et al. 1994. Selection against inbred song sparrows during a natural population bottleneck. Nature 372:356–357. **99**

Kellert, S.R. 1980. Activities of the American public relating to animals. U.S. Government Printing Office, Washington, D.C. 178 pp. **392, 393**

Kellert, S.R. 1985. Historical trends in perceptions and uses of animals in 20th century America. Environmental Review 9:34–53. **392**

Kellert, S.R. 1991. Japanese perceptions of wildlife. Conservation Biology 5:297–308. **389**

Kellert, S.R. 1993. Attitudes, knowledge, and behavior toward wildlife among the industrial superpowers: United States, Japan, and Germany. Journal of Social Issues 49:53–69. **389**

Kellert, S.R. 1996. The value of life: Biological diversity and human society. Island Press, Washington, DC. 263 pp. **389**

Kellert, S.R., and J.K. Berry. 1981. Knowledge, affection and basic attitudes toward animals in American society. U.S. Government Printing Office, Washington, D.C. 162 pp. **389, 390**

Kellert, S.R., and J.K. Berry. 1987. Attitudes, knowledge, and behaviors toward wildlife as affected by gender. Wildlife Society Bulletin 15:363–371. **389**

Kellert, S.R., M. Black, C.R. Rush, and A.J. Bath. 1996. Human culture and large carnivore conservation in North America. Conservation Biology 10:977–990. **391**

Kelly, M.J., and S.M. Durant. 2000. Viability of the Serengeti cheetah population. Conservation Biology 14:786–797. **112, 326**

Kemf, E. (ed.). 1993. The law of the mother: Protecting indigenous peoples in protected areas. Sierra Club Books, San Francisco. 296 pp. **290**

Kerley, G.I.H., M.H. Knight, and M. De Kock. 1995. Desertification of subtropical thicket in the Eastern Cape, South Africa: Are there alternatives? Environmental Monitoring and Assessment 37:211–230. **285**

Kerley, G.I.H., and W.G. Whitford. 2000. Impact of grazing and desertification in the Chihuahuan Desert: Plant communities, granivores and granivory. American Midland Naturalist 144:78–91. **186**

Kettlewell, B. 1973. The evolution of melanism. Clarendon Press, Oxford. 423 pp. **96**

Khogali, M.M. 1991. Famine, desertification and vulnerable populations: The case of Umm Ruwaba District, Kordofan Region, Sudan. Ambio 20:204–206. **187**

Kie, J.G., V.C. Bleich, A.L. Medina, et al. 1994. Managing rangelands for wildlife. Pages 663–688 in T.A. Bookhout (ed.). Research and management techniques for wildlife and habitats (5th ed.). The Wildlife Society, Bethesda, Maryland. 740 pp. **176**

Kim, K.C. 1997. Preserving biodiversity in Korea's demilitarized zone. Science. 278:242–243. **270**

Kimura, M., and T. Ohta. 1971. Theoretical aspects of population genetics. Princeton University Press, Princeton, New Jersey. 219 pp. **145**

King, C. 1984. Immigrant killers. Oxford University Press, Aukland, New Zealand. 224 pp. **180, 236, 246, 251**

King, C.M. (ed.). 1990. The handbook of New Zealand mammals. Oxford University Press, Auckland. 600 pp. **251**

Kingdon, J. 1989. Island Africa. Princeton University Press, Princeton, New Jersey. 287 pp. **74**

Kinnaird, M.F. 1992. Competition for a forest palm: Use of *Phoenix reclinata* by human and nonhuman primates. Conservation Biology 6:101–107. **223**

Kinzig, A.P., and J. Harte. 2000. Implications of endemics-area relationships for estimates of species extinctions. Ecology 81:3305–3311. **127**

Kirsop, B.E., and J.J.S. Snell (eds.). 1984. Maintenance of microorganisms. Academic Press, London. 205 pp. **354**

Kleijn, D., and L.A.C. van der Voort. 1997. Conservation headlands for rare arable weeds: The effects of fertilizer application and light penetration on plant growth. Biological Conservation 81:57–67. **293**

Kleiman, D.G. 1989. Reintroduction of captive mammals for conservation. Bioscience 39:152–161. **110**

Kloor, K. 2000. Everglades restoration plan hits rough waters. Science 288:1166–1167. **407**

Kloppenburg, J.R. 1988. First the seed: The political economy of plant biotechnology, 1492–2000. Cambridge University Press, Cambridge. 349 pp. **414**

Knapp, R.A., and K.R. Matthews. 2000. Non-native fish introductions and the decline of the mountain yellow-legged frog from within protected areas. Conservation Biology 14:428–438. **240**

Knapp, E.E., and K.J. Rice. 1998. Comparison of isozymes and quantitative traits for evaluating patterns of genetic variation in purple needlegrass (*Nassella pulchra*). Conservation Biology 12:1031–1041. **307**

Knight, R.L., and D.P. Anderson. 1990. Effects of supplemental feeding on an avian scavenging guild. Wildlife Society Bulletin 18:388–394. **315**

Knight, R.L., G.N. Wallace, and W.E. Riebsame. 1995. Ranching the view: Subdivisions versus agriculture. Conservation Biology 9:459–461. **298**

Knoll, A.H. 1984. Patterns of extinction in the fossil record of vascular plants. Pages 21–68 in M.H. Nitecki (ed.). Extinctions. University of Chicago Press, Chicago. 354 pp. **122, 123**

Knoll, A.H., R.K. Bambach, D.E. Canfield, and J.P. Grotzinger. 1996. Comparative earth history and late Permian mass extinction. Science 273:452–457. **124**

Knopf, F.L. 1992. Faunal mixing, faunal integrity, and the biopolitical template for diversity conservation. Transactions of the North American Wildlife and Natural Resources Conference 57:330–342. **273**

Knopf, F.L., and F.B. Samson. 1994. Scale perspectives on avian diversity in western riparian ecosystem. Conservation Biology 8:669–676. **273**

Kopp, R.J. 1992. Why existence value *should* be used in cost-benefit analysis. Journal of Policy Analysis and Management 11:123–130. **406**

Koshland, D.E., Jr. 1994. The case for diversity. Science 264:639. **329**

Kremen, C., R.K. Colwell, T.L. Erwin, et al. 1993. Terrestrial arthropod assemblages: Their use in conservation planning. Conservation Biology 7:796–808. **61**

Kremen, C., A.M. Merenlender, and D.D. Murphy. 1994. Ecological monitoring: A vital need for integrated conservation and development programs in the tropics. Conservation Biology 8:388–397. **375**

Kremen, C., I. Raymond, and K. Lance. 1998. An interdisciplinary tool for monitoring conservation impacts in Madagascar. Conservation Biology 12:549–563. **375**

Kremen, C., V. Razafimahatratra, R.P. Guillery, et al. 1999. Designing the Masoala National Park in Madagascar based on biological and socioeconomic data. Conservation Biology 13:1055–1068. **269**

Kress, S.W. 1983. The use of decoys, sound recordings, and gull control for re-establishing a tern colony in Maine. Colonial Waterbirds 6:185–196. **320, 326**

Kress, S.W. 1985. The Audubon Society guide to attracting birds. Charles Scribner's Sons, New York. 377 pp. **301, 325**

Kress, S.W., and D.N. Nettleship. 1988. Re-establishment of Atlantic puffins (*Fratercula arctica*) at a former breeding site in the Gulf of Maine. Journal of Field Ornithology 59:161–170. **320, 368**

Kricher, J. 1997. A neotropical companion, 2nd ed. Princeton University Press, Princeton, New Jersey. 451 pp. **184**

Krutilla, J.V. 1967. Conservation reconsidered. American Economic Review 57:777–786. **406**

Kuchler, A.W. 1964. Potential natural vegetation of the conterminous United States. American Geographical Society, New York, New York. **71**

Kurki, S., A. Nikula, P. Helle, and H. Lindén. 2000. Landscape fragmentation and forest composition effects on grouse breeding success in boreal forests. Ecology 81:1985–1997. **202**

Labandeira, C.C., and J.J. Sepkoski, Jr. 1993. Insect diversity in the fossil record. Science 261:310–315. **123**

Lackey, R.T. 1995. Ecosystem health, biological diversity, and sustainable development: Research that makes a difference. Renewable Resources Journal 13(2):8–13. **32**

Lacy, R.C., J.D. Ballou, F. Princée, et al. 1995. Pedigree analysis for population management. Pages 57–75 in J.D. Ballou, M. Gilpin, and T.J. Foose (eds.). Population management for survival and recovery. Columbia University Press, New York. 375 pp. **354**

Lacy, R.C., and T.W. Clark. 1990. Population viability assessment of the eastern barred bandicoot in Victoria. Pages 131–146 in T.W. Clark and J.H. Seebeck (eds.). The management and conservation of small populations. Chicago zoological society, Chicago. **160**

Laikre, L., and N. Ryman. 1996. Effects on intraspecific biodiversity from harvesting and enhancing natural populations. Ambio 25:504–509. **221**

Laist, D.W. 1987. Overview of the biological effects of lost and discarded plastic debris in the marine environment. Marine Pollution Bulletin 18:319–326. **177**

Lal, R. 1987. Tropical ecology and physical edaphology. John Wiley and Sons, Chichester. 732 pp. **182**

Lamberson, R.H., B.R. Noon, C. Voss, and K.S. McKelvey. 1994. Reserve design for territorial species: The effects of patch size and spacing on the viability of the northern spotted owl. Conservation Biology 8:185–195. **148**

Lande, R. 1988. Genetics and demography in biological conservation. Science 241:1455–1460. **157**

Lande, R., and G.F. Barrowclough. 1987. Effective population size, genetic variation, and their use in population management. Pages 87–123 in M.E. Soulé (ed.). Viable populations for conservation. Cambridge University Press, Cambridge. 189 pp. **106**

Langholz, J., J. Lassoie, and J. Schelhas. 2000. Incentives for Biological conservation: Costa Rica's private wildlife refuge program. Conservation Biology 14:1735–1743. **433**

Langton T.E.S. (ed.) 1989. Amphibians and roads. ACO Polymer Products Ltd., Sheford, Bedfordshire. **324**

Larson, D.M. 1993. On measuring existence value. Land Economics 69:377–388. **406**

Laurenson, M.K., N. Wielebnowski, and T.M. Caro. 1995. Extrinsic factors and juvenile mortality in cheetahs. Conservation Biology 9:1329–1331. **112**

Lawrence, E.A. 1993. The sacred bee, the filthy pig, and the bat out of hell: Animal symbolism as cognitive biophilia. Pages 301–341 in S.R. Kellert and E.O. Wilson (eds.). The biophilia hypothesis. Island Press, Washington, D.C. 484 pp. **56**

Laws, R.M., I.S.C. Parker, and R.C.B. Johnstone. 1975. Elephants and their habitats. Clarendon Press, Oxford. 376 pp. **64**

Lawton, J.H., and R.M. May. (eds.). 1995. Extinction rates. Oxford University Press, Oxford. 233 pp. **136**

Lawton, J.H., D.E. Bignell, B. Bolton, et al. 1998. Biodiversity inventories, indicator taxa and effects of habitat modification in tropical forest. Nature 391:72–76. **184**

Laycock, G. 1966. The alien animals. Natural History Press, Garden City, New York. 240 pp. **235**

Leach, M.K., and T.J. Givnish. 1996. Ecological determinants of species loss in remnant prairies. Science 273:1555–1558. **179, 200**

Leal, D.R. 1998. Community-run fisheries: Avoiding the "tragedy of the commons." Population and the Environment 19:225–245. **413**

Ledig, F.T. 1986. Heterozygosity, heterosis, and fitness in outbreeding plants. Pages 77–104 in M.E. Soulé (ed.). Conservation biology: The science of scarcity and diversity. Sinauer, Sunderland, Massachusetts. 584 pp. **99**

Lehman, S.J., and L.D. Keigwin. 1992. Sudden changes in North Atlantic circulation during the last deglaciation. Nature 356:757–762. **130**

Lehmkuhl, J.F., R.K. Upreti, and U.R. Sharma. 1988. National parks and local development: Grasses and people in Royal Chitwan National Park, Nepal. Environmental Conservation 15:143–148. **274**

Lélé, S., and R.B. Norgaard. 1996. Sustainability and the scientist's burden. Conservation Biology 10:354–365. **32**

Leopold, A. 1939. A biotic view of land. Journal of Forestry 37:113–116. **13**

Leopold, A. 1949. A Sand County almanac and sketches here and there. Oxford University Press, New York. 226 pp. **9, 392, 397**

Lesica, P., and F.W. Allendorf. 1995. When are peripheral populations valuable for conservation? Conservation Biology 9:753–760. **339, 368**

Lesica, P., and H.E. Atthowe. 2000. Should we use pesticides to conserve rare plants? Conservation Biology 14:1549–1550. **328**

Lesica, P., and S.V. Cooper. 1999. Succession and disturbance in sandhills vegetation: Constructing models for managing biological diversity. Conservation Biology 13:293–302. **307**

Levin, D.A., J. Francisco-Ortega, and R.K. Jansen. 1996. Hybridization and the extinction of rare plant species. Conservation Biology 10:10–16. **101, 244, 341**

Levine, J.M. 2000. Species diversity and biological invasions: Relating local process to community pattern. Science 288:852–854. **246**

Levine, J.M., and C.M. D'Antonio. 1999. Elton revisited: A review of evidence linking diversity and invasibility. Oikos 87:15–26. **245**

Lewis, D.M., and P. Alpert. 1997. Trophy hunting and wildlife conservation in Zambia. Conservation Biology 11:59–68. **216**

Lewis, D., G.B. Kaweche, and A. Mwenya. 1990. Wildlife conservation outside protected areas—lessons from an experiment in Zambia. Conservation Biology 4:171–180. **285**

Lewis, R.R. III. 1983. Impact of oil spills on mangrove forests. Pages 171–183 in H.J. Teas (ed.). Biology and ecology of mangroves. Junk, The Hague. 188 pp. **86**

Liddle M. 1997. Recreation ecology. Chapman and Hall, London. 639 pp. **215, 216, 229**

Lindenmayer, D.B., M.A. Burgman, H.R. Akcakaya, et al. 1995. A review of the generic computer programs ALEX, RAMAS/space and VORTEX for modelling the viability of wildlife metapopulations. Ecological Modelling 82:161–174. **155**

Lindenmayer, D.B., T.W. Clark, R.C. Lacy, and V.C. Thomas. 1993. Population viability analysis as a tool in wildlife conservation policy: With reference to Australia. Environmental Management 17:745–758. **155, 158, 162**

Lindenmayer, D.B., and R.C. Lacy. 1995. Metapopulation viability of Leadbeater's possum, *Gymnobelides leadbeateri,* in fragmented old-growth forest. Ecological Applications 5:164–182. **150**

Lindenmayer, D.B., R.C. Lacy, and M.L. Pope. 2000. Testing a simulation model for population viability analysis. Ecological Applications 10:580–597. **150**

Lippke, B.R., and J.T. Bishop. 1999. The economic perspective. Pages 597–638 in M.L. Hunter, Jr. (ed.) Maintaining biodiversity in forest ecosystems. Cambridge University Press, Cambridge, United Kingdom. 698 pp. **409**

Lloyd, B.D., and R.G. Powlesland. 1994. The decline of kakapo *Strigops habroptilus* and attempts at conservation by translocation. Biological Conservation 69:75–85. **327**

Loeschcke, V., J. Tomiuk, and S.K. Jain (eds.). 1994. Conservation genetics. Birkhäuser Verlag, Basel, Switzerland, 440 pp. **113**

Long, J.L. 1981. Introduced birds of the world. Universe Books, New York. 528 pp. **231, 234, 245**

Lopez, B.H. 1986. Arctic dreams. Scribner, New York. 464 pp. **213**

Lorimer, C.G., and L.E. Frelich. 1994. Natural disturbance regimes in old-growth northern hardwoods: Implications for restoration efforts. Journal of Forestry 92(1):33–38. **280**

Louda, S.M., and R.L. Bevill. 2000. Exclusion of natural enemies as a tool in managing rare plant species. Conservation Biology 14:1551–1552. **328**

Lövel, G.L. 1997. Global change through invasion. Nature 388:627–628. **245**

Lovelock, J.E. 1979. Gaia. Oxford University Press, Oxford. 157 pp. **72, 87**

Lovett, G.M. 1994. Atmospheric deposition of nutrients and pollutants in North America: An ecological perspective. Ecological Applications 4:629–650. **167**

Lowe, D.W., J.R. Matthews, C.J. Moseley (eds.). 1990. The official World Wildlife Fund guide to endangered species of North America. 3 vols. Beacham, Washington, D.C. **258**

Ludwig, D., R. Hilborn, and C. Walters. 1993. Uncertainty, resource exploitation, and conservation: Lessons from history. Science 260:17,36. **324**

Luoma, J.R. 1987. A crowded ark. Houghton Mifflin, Boston. 209 pp. **356, 364**

MacArthur, R.H., and E.O. Wilson. 1967. The theory of island biogeography. Princeton University Press, Princeton, New Jersey. 203 pp. **126, 194**

MacDonald, I.A.W., L.L. Loope, M.B. Usher, and O. Hamann. 1989. Wildlife conservation and the invasion of nature reserves by introduced species: A global perspective. Pages 215–255 in J.A. Drake, H.A. Mooney, F. di Castri, R.H. Groves, F.J. Kruger, M. Rejmanek, and M. Williamson (eds.). Biological invasions: A global perspective. John Wiley and Sons, Chichester. 525pp. **244**

Mace, G.M., and R. Lande. 1991. Assessing extinction threats: Towards a reevaluation of IUCN threatened species categories. Conservation Biology 5:148–157. **43**

Mace, R.D., and J.S. Waller. 1998. Demography and population trend of grizzly bears in the Swan Mountains, Montana. Conservation Biology 12:1005–1016. **279**

MacEwen, W.M. 1987. Ecological regions and districts of New Zealand. (3rd ed.). Department of Conservation, Wellington, New Zealand. **71**

Mack, R.N., D. Simberloff, W.M. Lonsdale, et al. 2000. Biotic invasions: Causes, epidemiology, global consequences, and control. Ecological Applications 10:689–710. **244, 245, 247, 251**

MacPhee, R.D.E. (ed.). 1999. Extinction in near time: Causes, contexts, and consequences. Kluwer, New York. 394 pp. **228**

Mader, H.J. 1984. Animal habitat isolation by roads and agricultural fields. Biological Conservation 29:81–96. **174**

Madigan, M.T., J.M. Martinko, and J. Parker (eds.). 1997. Brock biology of microorganisms, 8th ed. Prentice Hall, Upper Saddle River, New Jersey. 986 pp. **52**

Madsen, T., R. Shine, M. Olsson, and H. Wittzell. 1999. Restoration of an inbred adder population. Nature 402:34–35. **334, 335, 340**

Madsen, T., B. Stille, and R. Shine. 1996. Inbreeding depression in an isolated population of adders *Vipera berus*. Biological Conservation 75:113–118. **99**

Maguire, L.A., R.C. Lacy, R.J. Begg, and T.W. Clark. 1990. An analysis of alternative strategies for recovering the eastern barred bandicoot in Victoria. Pages 147–164 in T.W. Clark and J.H. Seebeck (eds.). Management and conservation of small populations. Chicago Zoological Society, Brookfield, Illinois. 295 pp. **160**

Magurran, A.E. 1988. Ecological diversity and its measurement. Princeton University Press, Princeton, New Jersey. 179 pp. **23, 70**

Main, M.B., F.M. Roka, and R.F. Noss. 1999. Evaluating costs of conservation. Conservation Biology 13:1262–1272. **298, 415**

Maina, G.G., and H.F. Howe. 2000. Inherent rarity in community restoration. Conservation Biology 14:1335–1340. **307**

Mainguet, M. 1994. Desertification: Natural background and human mismanagement, 2nd ed. Springer-Verlag, Berlin. 314 pp. **187, 205**

Mainguet, M. 1999. Aridity: Droughts and human development. Springer-Verlag, Berlin. 302 pp. **185**

Marble, A.D. 1992. A guide to wetland functional design. Lewis Publishers, Boca Raton, Florida. 222 pp. **304**

Mares, M.A. 1992. Neotropical mammals and the myth of Amazonian biodiversity. Science 255:976–979. **184**

Margules, C.R. 1989. Introduction to some Australian developments in conservation evaluation. Biological Conservation 50:1–11. **262**

Margules, C.R., and R L. Pressey. 2000. Systematic conservation planning. Nature 405:243–253. **259, 263, 368**

Marmontel, M., S.R. Humphrey, and T.J. O'Shea. 1997. Population viability analysis of the

Florida manatee (*Trichechus manatus latirostris*), 1976–1991. Conservation Biology 11:467–481. **177**

Marsh, J.S. 1994. The ecological component of economic policy. Pages 331–344 in P.J. Edwards, R.M. May, and N.R. Webb (eds.). Large-scale ecology and conservation biology. Blackwell, Oxford. 375 pp. **419**

Marshall, C.R., and P.D. Ward. 1996. Sudden and gradual molluscan extinctions in the latest Cretaceous of western European tethys. Science 274:1360–1363. **121**

Marshall, L.G. 1984. Who killed Cock Robin? An investigation of the extinction controversy. Pages 785–806 in P.S. Martin and R.G. Klein (eds.). Quaternary extinctions. The University of Arizona Press, Tucson. 892 pp. **209**

Martin, P.S. 1984. Prehistoric overkill: The global model. Pages 354–403 in P.S. Martin and R.G. Klein (eds.). Quaternary extinctions. The University of Arizona Press, Tucson. 892 pp. **208**

Martin, P.S. 1986. Refuting late Pleistocene extinction models. Pages 107–130 in D.K. Elliott, (ed.). Dynamics of extinction. John Wiley and Sons, New York. 294 pp. **208**

Martin, P.S., and D.A. Burney. 1999. Bring back the elephants! Wild Earth 9(1):57–64. **248, 307**

Martin, P.S., and RG. Klein (eds.). 1984. Quaternary extinctions. The University of Arizona Press, Tucson. 892 pp. **228**

Martin, P.S., and D.W. Steadman. 1999. Prehistoric extinctions on islands and continents. Pages 17–35 in R.D.E. MacPhee (ed.). Extinctions in near time: Causes, contexts, and consequences. Kluwer, New York. 394 pp. **208**

Maschinski, J., R. Frye, and S. Rutman. 1997b. Demography and population viability of an endangered plant species before and after protection from trampling. Conservation Biology 11:990–999. **325**

Maschinksi, J., T.E. Kolb, E. Smith, and B. Phillips. 1997a. Potential impacts of timber harvesting on a rare understory plant, *Clematis hirsutissima* var. *arizonica*. Biological conservation 80:49–61. **316**

Maslin, M.A., and S.J. Burns. 2000. Reconstruction of the Amazon Basin effective moisture availability over the past 14,000 years. Science 290:2285–2287. **131**

Mast, J.N., P.Z. Fulé, M.M. Moore, et al. 1999. Restoration of presettlement age structure of an Arizona ponderosa pine forest. Ecological Applications 9:228–239. **307**

Master, L. 1990. The imperiled status of North American aquatic animals. Biodiversity Network News 3(3):1–2, 7. **193**

Matlack, G.R. 1993. Microenvironment variation within and among forest edge sites in the eastern United States. Biological Conservation 66:185–194. **201**

Matlack, G.R., and J.A. Litvaitis. 1999. Forest edges. Pages 210–233 in M.L. Hunter, Jr. (ed.). Maintaining biodiversity in forest ecosystems. Cambridge University Press, Cambridge, United Kingdom. 698 pp. **201**

Matson, P.A., W.J. Parton, A.G. Power, and M.J. Swift. 1997. Agricultural intensification and ecosystem properties. Science 277:504–509. **295**

Matthiessen, P. 1987. Wildlife in America. 2nd ed. Viking Press, New York. 332 pp. **210, 229**

Mattick, J.S., E.M. Ablett, and D.L. Edmonson. 1992. The gene library—preservation and analysis of genetic diversity in Australasia. Pages 15–35 in R.P. Adams and J.E. Adams (eds.). Conservation of plant genes: DNA banking and *in vitro* biotechnology. Academic Press, San Diego. 345 pp. **335**

Mattson, D.J., and M.M. Reid. 1991. Conservation of the Yellowstone grizzly bear. Conservation Biology 5:364–372. **340**

Maugh, T.H. 1982. Leprosy vaccine trials to begin soon. Science 215:1083–1086. **53**

Maunder, M. 1992. Plant reintroduction: An overview. Biodiversity and Conservation 1:51–61. **333**

Maunder, M., A. Culham, B. Alden, et al. 2000. Conservation of the Toromiro tree: Case study in the management of a plant extinct in the wild. Conservation Biology 14:1341–1350. **358**

Maxted, N., B.V. Ford-Lloyd, and J.G. Hawkes (eds.). 1997. Plant genetic conservation: The *in situ* approach. Chapman and Hall, London, United Kingdom. 446 pp. **101, 360**

May, R.M. 1988. Conservation and disease. Conservation Biology 2:28–30. **328**

May, R.M. 1995. The cheetah controversy. Nature 374:309–310. **112**

Mayfield, H.F. 1977. Brood parasitism: Reducing interactions between Kirtland's warblers and brown-headed cowbirds. Pages 85–91 in S.A. Temple (ed.). Endangered birds: Management techniques for preserving threatened species. University of Wisconsin Press, Madison. 466 pp. **331**

Mayle, F.E., R. Burbridge, and T.J. Killeen. 2000. Millennial-scale dynamics of southern Amazonian rain forests. Science 290:2291–2294. **131**

Mayr, E. 1942. Systematics and the origin of species. Columbia University Press, New York. 334 pp. **37**

McCann, K.S. 2000. The diversity-stability debate. Nature 405:228–233. **80, 81**

McClanahan, T.R., and S. Mangi. 2000. Spillover of exploitable fishes from a marine park and its effect on the adjacent fishery. Ecological Applications 10:1792–1805. **269**

McClay, W. 2000. Rotenone use in North America (1988–1997). Fisheries 25(5)15–21. **331**

McClenaghan, L.R., Jr., J. Berger, and H.D. Truesdale. 1990. Founding lineages and genic variability in plains bison (*Bison bison*) from Badlands National Park, South Dakota. Conservation Biology 4:285–289. **92, 93**

McComb, W., and D. Lindenmayer. 1999. Dying, dead, and down trees. Pages 335–372 in M.L. Hunter, Jr. (ed.) Maintaining biodiversity in forest ecosystems. Cambridge University Press, Cambridge, United Kingdom. 698 pp. **222, 274, 280**

McCullough, D.R. (ed.) 1996. Metapopulations and wildlife conservation. Island Press, Washington, D.C. 429 pp. **162**

McIntosh, R.P. 1980. The relationship between succession and the recovery process in ecosystems. Pages 11–62 in J. Cairns, Jr. (ed.). The recovery process in damaged ecosystems. Ann Arbor Science Publ., Ann Arbor, Michigan. 167 pp. **72, 330**

McIntyre, S., G.W. Barrett, R.L. Kitching, H.F. Recher. 1992. Species triage—seeing beyond wounded rhinos. Conservation Biology 6:604. **372**

McLellan, B.N., and D.M. Shackleton. 1988. Grizzly bears and resource-extraction industries: Effects of roads on behaviour, habitat use and demography. Journal of Applied Ecology 25:451–460. **174**

McNaughton, S.J. 1993. Grasses and grazers, science and management. Ecological Applications 3:17–20. **286**

McNeely, J.A. 1988. Economics and biological diversity: Developing and using economic incentives to conserve biological resources. International Union for Conservation of Nature and Natural Resources, Gland, Switzerland. 236 pp. **424**

McNeely, J.A., K.R. Miller, W.V. Reid, et al. 1990. Conserving the world's biological diversity. IUCN, Gland, Switzerland. 193 pp. **427**

McShea, W.J., and J.H. Rappole. 2000. Managing the abundance and diversity of breeding bird populations through manipulation of deer populations. Conservation Biology 14:1161–1170. **277, 295**

Meesters, E.H., R.P.M. Bak, S. Westmacott, et al. 1998. A fuzzy logic model to predict coral reef development under nutrient and sediment stress. Conservation Biology 12:957–965. **170**

Meffe, G.K. 1992. Techno-arrogance and halfway technologies: Salmon hatcheries on the Pacific coast of North America. Conservation Biology 6:350–354. **176, 338**

Meffe, G.K., and R.C. Vrijenhoek. 1988. Conservation genetics in the management of desert fishes. Conservation Biology 2:157–169. **94, 341**

Meffe, G.K., and C.R. Carroll. 1994. Principles of conservation biology. Sinauer, Sunderland, Massachusetts. 600 pp. **xiii**

Melnyk, M., and N. Bell. 1996. The direct-use values of tropical moist forest foods: The Houttuja (Piaroa) Amerindians of Venezuela. Ambio 25:468–472. **401**

Mellor, M. 1997. Feminism and ecology. New York University Press, New York. 221 pp. **387, 388**

Melvin, E.F., J.K. Parrish, and L.L. Conquest. 1999. Novel tools to reduce seabird bycatch in coastal gillnet fisheries. Conservation Biology 13:1386–1397. **323**

Melvin, S.M., L.H. MacIvor, and C.R. Griffin. 1992. Predator exclosures: A technique to reduce predation at piping plover nests. Wildlife Society Bulletin 20:143–148. **326**

Menge, B.A., and J.P. Sutherland. 1987. Community regulation: Variations in disturbance competition, and predation in relation to environmental stress and recruitment. American Naturalist 130:730–757. **226**

Menges, E.S. 1990. Population viability analysis for an endangered plant. Conservation Biology 4:52–63. **148**

Menges, E.S. 2000. Population viability analyses in plants: Challenges and opportunities. Trends in Ecology and Evolution 15:51–57. **162**

Menotti-Raymond, M., and S.J. O'Brien. 1993. Dating the genetic bottleneck of the African cheetah. Proceedings of the National Academy of Science 90:3172–3176. **111**

Meretsky, V.J., N.F.R. Snyder, S.R. Beissinger, et al. 2000. Demography of the California condor: Implications for reestablishment. Conservation Biology 14:957–967. **315, 336**

Merola, M. 1994. A reassessment of homozygosity and the case for inbreeding depression in the cheetah, *Acinonyx jubatus:* Implications for conservation. Conservation Biology 8:961–971. **112**

Mighetto, L. 1991. Wild animals and American environmental ethics. University of Arizona Press, Tucson. 177 pp. **215**

Milchunas, D.G., W.K. Lauenroth, and I.C. Burke. 1998. Livestock grazing: Animal and plant biodiversity of shortgrass steppe and the relationship to ecosystem function. Oikos 83:65–74. **285**

Miller, B., R.Reading, J. Hoogland, et al. 2000. The role of prairie dogs as a keystone species: Response to Stapp. Conservation Biology 14:318–321. **287**

Miller, G.T., Jr. 1992. Living in the environment 7th ed. Wadsworth, Belmont, California. 706 pp. **219**

Miller, G.T., Jr. 1999. Living in the environment 11th ed. Wadsworth, Belmont, California. 911 pp. **205**

Miller, J.R., and P. Cale. 2000. Behavioral mechanisms and habitat use by birds in a fragmented agricultural landscape. Ecological Applications 10:1732–1748. **293**

Miller, K., and L. Tangley. 1991. Trees of life: Saving tropical forests and their biological wealth. Beacon Press, Boston. 217 pp. **181, 182**

Miller, M., and G. Aplet. 1993. Biological control: A little knowledge is a dangerous thing. Rutgers Law Review 45:285–334. **236**

Miller, S.G., S.P. Bratton, and J. Hadidian. 1992. Impacts of white-tailed deer on endangered and threatened vascular plants. Natural Areas Journal 12:67–74. **28**

Mills, L.S., and F.W. Allendorf. 1996. The one-migrant-per-generation rule in conservation and management. Conservation Biology 10:1509–1518. **146, 340**

Minckley, W.L., and J.E. Deacon (eds.). 1991. Battle against extinction: Native fish management in the American West. University of Arizona Press, Tucson. 517 pp. **181, 344**

Mitchell, J.G. 1982. The hunt. Alfred Knopf, New York, 243 pp. **215**

Mitsch, W.J., and J.G. Gosselink. 2000. Wetlands, 3rd ed. John Wiles and Sons, New York. 920 pp. **191, 192, 205, 293, 308**

Mladenoff, D.J., T.A. Sickley, R.G. Haight, and A.P. Wydeven. 1995. A regional landscape analysis and prediction of favorable gray wolf habitat in the northern Great Lakes region. Conservation Biology 9:279–294. **174**

Mladenoff, D.J., M.A. White, T.R. Crow, and J. Pastor. 1994. Applying principles of landscape design and management to integrate old-growth forest enhancement and commodity use. Conservation Biology 8:752–762. **269**

Monaghan, P. 1996. Relevance of the behaviour of seabirds to the conservation of marine environments. Oikos 77:227–237. **61**

Moncrief, L.W. 1970. The cultural basis for our environmental crisis. Science 170:508–512. **385**

Montalvo, A.M., and N.C. Ellstrand. 2000. Transplantation of the subshrub *Lotus scoparius*: Testing the home-site advantage hypothesis. Conservation Biology 14:1034–1045. **339**

Moore, J.W. 1972. Composition and structure of algal communities in a tributary stream of Lake Ontario. Canadian Journal of Botany 50:1663–1674. **193**

Morgan, J.W. 2000. Reproductive success in reestablished versus natural populations of a threatened grassland daisy (*Rutidosis leptorrhynchoides*). Conservation Biology 14:780–785. **333**

Moritz, C.C. 1994. Defining "evolutionary significant units" for conservation. Trends in Ecology and Evolution 9:373–375. **38**

Moritz, C.C., S. Lavery, and R. Slade. 1995. Using allele frequency and phylogeny to define units for conservation and management. American Fisheries Society Symposium. 17:49–262. **38**

Moss, C. 1988. Elephant memories. William Morrow, New York. 336 pp. **110**

Mowat, F. 1984. Sea of slaughter. McClelland and Stewart, Toronto. 438 pp. **210, 229**

Moyle, P.B. 1976a. Inland fishes of California. University of California Press, Berkeley. **32**

Moyle, P.B. 1976b. Fish introductions in California: History and impact on native fishes. Biological Conservation 9:101–118. **234, 244, 246**

Moyle, P.B., H.W. Li, and B.A. Barton. 1986. The Frankenstein effect: Impact of introduced fishes on native fishes in North America. Pages 415–426 in R.H. Stroud, (ed.). Fish culture in fisheries management. American Fisheries Society, Bethesda, Maryland. 481 pp. **234**

Moyle, P.B., and T. Light. 1996. Biological invasions of fresh water: Empirical rules and assemble theory. Biological Conservation 78:149–161. **245, 246**

Moyle, P.B., and J.E. Williams. 1990. Biodiversity loss in the temperate zone: Decline of the native fish fauna of California. Conservation Biology 4:275–284. **181**

Muir, J. 1916. A thousand-mile walk to the gulf. Houghton Mifflin, Boston, Massachusetts. 219 pp. **13**

Mumme, R.L., S.J. Schoech, G.E. Woolfenden, and J.W. Fitzpatrick. 2000. Life and death in the fast lane: Demographic consequences of road mortality in the Florida scrub-jay. Conservation Biology 14:501–512. **174**

Murphy, D.D., and S.B. Weiss. 1992. Effects of climate change on biological diversity in western North America: Species losses and mechanisms. Pages 355–368 in R.L. Peters and T.E. Lovejoy (eds.). Global warming and biological diversity. Yale University Press, New Haven. 386 pp. **134**

Murray, J., E. Murray, M.S. Johnson, and B. Clarke. 1988. The extinction of *Partula* on Moorea. Pacific Science 42:150–153. **237**

Musters, C.J.M., H.J. de Graaf, and W.J. ter Keurs. 2000. Can protected areas be expanded in Africa? Science 287:1759–1760. **263**

Mwalyosi, R.B.B. 1991. Ecological evaluation for wildlife corridors and buffer zones for Lake Manyara National Park, Tanzania, and its immediate environment. Biological Conservation 57:171–186. **272**

Myers, J.H., D. Simberloff, A.M. Kuris, and J.R. Carey. 2000. Eradication revisited: Dealing with exotic species. Trends in Ecology and Evolution 15:316–320. **236**

Myers, N. 1988. Threatened biotas: 'Hot-spots' in tropical forests. Environmentalist 8:187–208. **262**

Myers, N. 1989a. A major extinction spasm: Predictable and inevitable? Pages 42–49 in D. Western, and M.C. Pearl (eds.). 1989. Conservation for the twenty-first century. Oxford University Press, New York. 365 pp. **79**

Myers, N. 1989b. Deforestation rates in tropical forests and their climatic implications. Friends of the Earth, London. 78 pp. **127**

Myers, N. 1990. The biodiversity challenge: Expanded hot-spots analysis. Environmentalist 10:243–256. **262**

Myers, N. 1992. Synergisms: Joint effects of climate change and other forms of habitat destruction. Pages 344–354 in R.L. Peters and T.E. Lovejoy (eds.). Global warming and biological diversity. Yale University Press, New Haven. 386 pp. **133**

Myers, N. 1998. Lifting the veil on perverse subsidies. Nature 392:327–328. **408**

Myers, N., R.A. Mittermeier, C.G. Mittermeier, G.A.B. da Fonseca, and J. Kent. 2000. Biodiversity hotspots for conservation priorities. Nature 403:853–858. **236, 242, 262, 263, 369**

Nabhan, G.P. 1982. The desert smells like rain: A naturalist in Papago country. North Point Press, San Francisco, California. 148 pp. **28**

Nabhan, G.P. 1987. Nurse plant ecology of threatened desert plants. Pages 377–383 in T.S. Elias (ed.). Conservation and management of rare and endangered plants. California Native Plant Society, Sacramento, California. 630 pp. **318**

Nabhan, G.P. 1989. Enduring seeds: Native American agriculture and wild plant conservation. North Point Press, San Francisco. 225 pp. **355**

Naeem, S. 1998. Species redundancy and ecosystem reliability. Conservation Biology 12:39–45. **81**

Naess, A. 1989. Ecology, community, and lifestyle: Outline of an ecosophy. Cambridge University Press, Cambridge. 223 pp. **394**

Nash, R.F. 1988. The rights of nature: A history of environmental ethics. University of Wisconsin Press, Madison. 290 pp. **18**

Nash, R. 1990. American environmentalism: Readings in conservation history, 3rd ed. McGraw-Hill, New York. 364 pp. **18**

Nason, J.D., E.A. Herre, and J.L. Hamrick. 1998. The breeding structure of a tropical keystone plant resource. Nature 394:685–687. **293**

National Research Council. 1983. Butterfly farming in Papua New Guinea. National Academy Press, Washington, D.C. 35 pp. **423**

National Research Council. 1991. Microlivestock: Little-known small animals with a promising economic future. National Academy Press, Washington, D.C. 446 pp. **50, 383**

National Research Council. 1999. Hormonally active agents in the environment. National Academy Press, Washington, DC. 430 pp. **170**

Nazarea, V.D. 1998. Cultural memory and biodiversity. The University of Arizona Press, Tucson, Arizona. 189 pp. **110**

Nei, M., and S. Kumar. 2000. Molecular evolution and phylogenetics. Oxford University Press, Oxford, United Kingdom. 333 pp. **94**

Nelson, A.J. 1990. Going wild. American Demographics 12(2):34–37, 50. **347**

Nelson, R., and N. Horning. 1993. AVHRR-LAC estimates of forest area in Madagascar, 1990. International Journal of Remote Sensing 14:1463–1475. **203**

Nepstad, D.C., A. Verissimo, A. Alencar, et al. 1999. Large-scale impoverishment of Amazonian forests by logging and fire. Nature 398:505–508. **182**

Neves, R.J., L.R. Weaver, and A.V. Zale. 1985. An evaluation of host fish suitability for glochidia of *Villosa vanuxemi* and *V. Nebulosa* (Pelecypopda: Unionidae). American Midland Naturalist 113:13–19. **320**

Nevo, E., and A. Beiles. 1991. Genetic diversity and ecological heterogeneity in amphibian evolution. Copeia 1991:565–592. **97**

New, T.R. 1991. Butterfly conservation. Oxford University Press, Melbourne, Australia. 224 pp. **320, 423**

New, T.R. 1994. Butterfly ranching: Sustainable use of insects and sustainable benefit to habitats. Oryx 28:169–172. **423**

New, T.R. 1997. Butterfly conservation, 2nd ed. Oxford University Press, Melbourne, Australia. 248 pp. **172, 216, 423**

Newman, A., M. Bush, D.E. Wildt, et al. 1985. Biochemical genetic variation in eight endangered or threatened felid species. Journal of Mammalogy 66:256–267. **111**

Newman, E.I. 1993. Applied ecology. Blackwell, Oxford. 328 pp. **288**

Newmark, W.D. 1985. Legal and biotic boundaries of western North American national parks: A problem of congruence. Biological Conservation 33:197–208. **269**

Newmark, W.D. 1987. A land-bridge island perspective on mammalian extinctions in western North American parks. Nature 325:430–432. **270**

Newmark, W.D. 1995. Extinction of mammal populations in western North American national parks. Conservation Biology 9:512–526. **270**

Newmark, W.D. 1996. Insularization of Tanzanian parks and the local extinction of large mammals. Conservation Biology 10:1549–1556. **270**

Newmark, W.D., D.N. Manyanza, D-GM. Gamassa, and H.I. Sariko. 1994. The conflict between wildlife and local people living adjacent to protected areas in Tanzania: Human density as a predictor. Conservation Biology 8:249–255. **294**

Newton, P. 1991. The use of medicinal plants by primates: A missing link? Trends in Ecology and Evolution 6:297–299. **53**

Nichols, W.F., K.T. Killingbeck, and P.V. August. 1998. The influence of geomorphological heterogeneity on biodiversity. II. A landscape perspective. Conservation Biology 12:371–379. **259**

Niemelä, P., and W.J. Mattson. 1996. Invasion of North American forests by European phytophagous insects. BioScience 46:741–753. **246–247**

Niklasson, M., and A. Granström. 2000. Numbers and sizes of fires: Long-term spatially explicit fire history in a Swedish boreal landscape. Ecology 81:1484–1499. **275**

Nilsson, C., and K. Berggren. 2000. Alterations of riparian ecosystems caused by river regulation. BioScience 50:783–792. **190**

Nilsson, S.G., J. Bengtsson, and S. As. 1988. Habitat diversity or area *per se*? Species richness of woody plants, carabid beetles, and land snails on islands. Journal of Animal Ecology 57:685–704. **267**

Nilsson, C., R. Jansson, and U. Zinko. 1997. Long-term responses of river-margin vegetation to water-level regulation. Science 271:798–800. **191**

Noon, B.R., and R.H. Lamberson, M.S. Boyce, and L.L. Irwin 1999. Population viability analysis: A primer on its principal technical components. Pages 87–134 in R.C. Szaro, N.C. Johnson, W.T. Sexton, and A.J. Malk (eds.). Ecological stewardship: A common reference for ecosystem management. Vol. II. Elsevier Science, Oxford, United Kingdom. 1788 pp. **162**

Norse, E.A. (ed.). 1993. Global marine biological diversity. Island Press, Washington, D.C. 383 pp. **170, 205**

North, S.G., D.J. Bullock, and M.E. Dulloo. 1994. Changes in the vegetation and reptile populations on Round Island, Mauritius, following eradication of rabbits. Biological Conservation 67:21–28. **179, 241**

Norton, B.G., M. Hutchins, E.F. Stevens, and T.L. Maple (eds.). 1995. Ethics on the ark. Smithsonian Institution, Washington. 330 pp. **357**

Norton, D.A. 1992. Disruption of natural ecosystems by biological invasion. Pages 309–319 in J.H.M. Willison, S. Bondrup-Nielsen, C. Drysdale, et al. (eds.). Science and the management of protected areas. Elsevier, Amsterdam. **244**

Norwood, V. 1993. Made from this earth: American women and nature. University of North Carolina Press, Chapel Hill. 368 pp. **387, 388**

Noss, A.J. 1998. The impacts of cable snare hunting on wildlife populations in the forests of the Central African Republic. Conservation Biology 12:390–398. **217**

Noss, R.F. 1983. A regional landscape approach to maintain diversity. Bioscience 33:700–706. **84**

Noss, R.F. 1987. From plant communities to landscapes in conservative inventories: A look at the Nature Conservancy (USA). Biological Conservation 41:11–37. **77, 259**

Noss, R.F. 1990. Indicators for monitoring biodiversity: A hierarchical approach. Conservation Biology 4:355–364. **366**

Noss, R.F. 1992. Issues of scale in conservation biology. Pages 242–250 in P.L. Fiedler and S.K. Jain (eds.). Conservation biology: The theory and practice of nature conservation, preservation, and management. Chapman and Hall, New York. 507 pp. **395**

Noss, R.F. 1993. A conservation plan for the Oregon Coast Range: Some preliminary suggestions. Natural Areas Journal 13:276–290. **260**

Noss, R.F., and A.Y. Cooperrider. 1994. Saving nature's legacy. Island Press, Washington, D.C. 416 pp. **186, 263, 284, 285, 286, 311, 368, 375**

Noss, R.F., and L.D. Harris. 1986. Nodes, networks, and MUM's: Preserving diversity at all scales. Environmental Management 10:299–309. **269**

Noss, R.F, E.T. LaRoe III, and J.M. Scott. 1995. Endangered ecosystems of the United States: A preliminary assessment of loss and degradation. U.S. Dept. of the Interior, National Biological Service, Washington, D.C. 58 pp. **74**

Nowak, R.M., and N.E. Federoff. 1998. Validity of the red wolf: Response to Roy et al. Conservation Biology 12:722–725. **37**

Nunney, L., and D.R. Elam. 1994. Estimating the effective population size of conserved populations. Conservation Biology 8:175–184. **106**

O'Brien, S.J. 1987. The ancestry of the giant panda. Scientific American 257(5):102–107. **63**

O'Brien, S.J. 1994. The cheetah's conservation controversy. Conservation Biology 8:1153–1155. **112**

O'Brien, S.J., and E. Mayr. 1991. Bureaucratic mischief: Recognizing endangered species and subspecies. Science 251:1187–1188. **39**

O'Brien, S.J., M.E. Roelke, L. Marker, et al. 1985. Genetic basis for species vulnerability in the cheetah. Science 227:1428–1434. **111, 112**

O'Brien, S.J., D.E. Wildt, M. Bush, et al. 1987. East African cheetahs: Evidence for two population bottlenecks? Proceedings of the National Academy of Science 84:508–511. **111**

O'Brien, S.J., D.E. Wildt, D. Goldman, et al. 1983. The cheetah is depauperate in genetic variation. Science 221:459–462. **111**

O'Connor, R.J., and M.J. Shrubb. 1986. Farming and birds. Cambridge University Press, Cambridge, Great Britain. **172, 294, 311**

O'Hara, K.J. 1988. Plastic debris and its effects on marine wildlife. Pages 395–434 in W.J. Chandler, L. Labate, and C. Wille (eds.). Audubon Wildlife Report: 1988/1989. Academic Press, San Diego. 817 pp. **177**

O'Leary, C.H., and D.W. Nyberg. 2000. Treelines between fields reduce the density of grassland birds. Natural Areas Journal 20:243–249. **296**

O'Shea, T.J., C.A. Beck, R.K. Bonde, et al. 1985. An analysis of manatee mortality patterns in Florida, 1976–81. Journal of Wildlife Management 49:1–11. **325**

Office of Technology Assessment. 1993. Harmful non-indigenous species in the United States. U.S. Congress, Office of Technology Assessment. 391 pp. **233, 251**

Ojeda, F.P., and J.H. Dearborn. 1989. Diversity, abundance, and spatial distribution of fishes and crustaceans in the rocky subtidal zone of the Gulf of Maine. Fishery Bulletin 88:403–410. **226**

Oldfield, M.L. 1984. The value of conserving genetic resources. U.S. Department of the Interior, National Park Service, Washington, D.C. 360 pp. (Republished by Sinauer in 1989.) **52, 66, 74**

Oldfield, M.L., and J.B. Alcorn. 1987. Conservation of traditional agroecosystems. Bioscience 37:199–208. **361**

Oldfield, M.L., and J.B. Alcorn (eds.). 1991. Biodiversity: Culture, conservation, and ecodevelopment. Westview Press, Boulder, Colorado. 349 pp. **411**

Oliver, I., and A.J. Beattie. 1996. Invertebrate morphospecies as surrogates for species: A case study. Conservation Biology 10:99–109. **41**

Olney P.J.S., G.M. Mace, and A.T.C. Feistner (eds.). 1994. Creative conservation: Interactive management of wild and captive animals. Chapman and Hall, London. 517 pp. **364**

Olson, D.M., and E. Dinerstein. 1998. The Global 200: A representation approach to conserving the earth's most biologically valuable ecoregions. Conservation Biology 12:502–515. **260, 261, 373**

Olson, S.L., and H.F. James. 1984. The role of Polynesians in the extinction of the avifauna of the Hawaiian Islands. Pages 768–780 in P.S. Martin and R.G. Klein (eds.). Quaternary extinctions. The University of Arizona Press, Tucson. 892 pp. **210**

Oostermeijer, J.G.B., S.H. Luijten, Z.V. Křenová, and HC.M. Den Nijs. 1998. Relationships between population and habitat characteristics and reproduction of the rare *Gentiana pneumonanthe* L. Conservation Biology 12:1042–1053. **317**

Orr, D.W. 1992. Ecological literacy. State University of New York Press, Albany. 210 pp. **392, 397**

Orr, D.W. 1994. Earth in mind. Island Press, Washington, D.C. **397**

Osborn, J.G., and J.F. Polsenberg. 1996. Meeting of the mangrovellers: The interface of biodiversity and ecosystem function. Trends in Ecology and Evolution 11:354–356. **85**

Ostrowski, S., E. Bedin, D.M. Lenain, and A.H. Abuzinada. 1998. Ten years of Arabian oryx conservation breeding in Saudi Arabia—achievements and regional perspectives. Oryx 32:209–222. **362**

Otte, D., and J.A. Endler (eds.). 1989. Speciation and its consequences. Sinauer, Sunderland, Massachusetts. 679 pp. **37**

Owen-Smith, N. 1989. Megafaunal extinctions: The conservation message from 11,00 years B.P. Conservation Biology 3:405–412. **223**

Paaby, P., D.B. Clark, and H. Gonzalez. 1991. Training rural residents as naturalist guides: Evaluation of a pilot project in Costa Rica. Conservation Biology 5:542–547. **416**

Packer, C. 1979. Inter-troop transfer and inbreeding avoidance in papio anubis. Animal Behaviour 27:1–36. **98**

Packer, C., A.E. Pusey, H. Rowley, et al. 1991. Case study of a population bottleneck: Lions of the Ngorongoro Crater. Conservation Biology 5:219–230. **99**

Paddack, M.J., and J.A. Estes. 2000. Kelp forest fish populations in marine reserves and adjacent exploited areas of central California. Ecological Applications 10:855–870. **269**

Paetkau, D. 1999. Using genetics to identify intraspecific conservation units: A critique of current methods. Conservation Biology 13:1507–1509. **38**

Paine, R.T. 1966. Food web complexity and species diversity. American Naturalist 100:65–75. **59**

Palik, B., and R.T. Engstrom. 1999. Species composition. Pages 65–94 in M.L. Hunter, Jr. (ed.) Maintaining biodiversity in forest ecosystems. Cambridge University Press, Cambridge, United Kingdom. 698 pp. **223, 284**

Palik, B.J., P.C. Goebel, L.K. Kirkman, and L. West. 2000. Using landscape hierarchies to guide restoration of disturbed ecosystems. Ecological Applications 10:189–202. **305**

Palmer, M.A., N.G. Hodgetts, M.J. Wigginton, et al. 1997. The application to the British flora of the World Conservation Union's revised Red List criteria and the significance of Red Lists for species conservation. Biological Conservation 82:219–226. **42–43**

Panzer, R., and M. Schwartz. 2000. Effects of management burning on prairie insect species richness within a system of small, highly fragmented reserves. Biological Conservation 96:363–369. **286**

Paoletti, M.G., D. Pimentel, B.R. Stinner, and D. Stinner. 1992. Agroecosystem biodiversity: Matching production and conservation biology. Agriculture, Ecosystems and Environment 40:3–23. **293**

Parendes, L.A., and J.A. Jones. 2000. Role of light availability and dispersal in exotic plant invasion along roads and streams in the H. J. Andrews Experimental Forest, Oregon. Conservation Biology 14:64–75. **174, 246**

Parker, I.M., and P. Karieva. 1996. Assessing the risks of invasion for genetically engineered plants: Acceptable evidence and reasonable doubt. Biological conservation 78:193–203. **236**

Parker, P.G., and T.A. Waite. 1997. Mating systems, effective population size, and conservation of natural populations. Pages 243–261 in J.R. Clemmons and R. Buchholz (eds.). Behavioral approaches to conservation in the wild. Cambridge University Press, Cambridge, United Kingdom. 382 pp. **107**

Parson, E.A., P.M. Haas, and M.A. Levy. 1992. A summary of the major documents signed at the Earth Summit and the Global Forum. Environment 34(8):12–15, 34–36. **427, 431**

Paton, D.C. 1993. Honeybees in the Australian environment. Bioscience 43:95–103. **242, 243**

Paton, P.W.C. 1994. The effect of edge on avian nest success: How strong is the evidence. Conservation Biology 8:17–26. **201**

Pauly, D., V. Christensen, J. Dalsgaard, et al. 1998. Fishing down marine food webs. Science 279:860–863. **210, 227, 228, 289**

Pavlick, B.M., D.L. Nickrent, and A.M. Howald. 1993. The recovery of an endangered plant: I. Creating a new population of *Amsinckia grandiflora*. Conservation Biology 7:510–526. **330**

Payne, N.F. 1992. Techniques for wildlife habitat management of wetlands. McGraw-Hill, New York. 549 pp. **276**

PCEQ (President's Commission on Environmental Quality). 1993. Biodiversity on private lands. Executive Office of the President, Washington, D.C. 20 pp. **435**

Pell, A.S., and C.R. Tidemann. 1997. The impact of two exotic hollow-nesting birds on two native parrots in savannah and woodland in eastern Australia. Biological Conservation 79:145–153. **242**

Pendergrass, K.L., P.M. Miller, J.B. Kauffman, and T.N. Kaye. 1999. The role of prescribed burning in maintenance of an endangered plant species, *Lomatium bradshawii*. Ecological Applications 9:1420–1429. **330**

Peters, C.M., A.H. Gentry, and R.O. Mendelsohn. 1989. Valuation of an Amazonian rainforest. Nature 339:655–656. **419, 420**

Peters, R.L., and T.E. Lovejoy (eds.). 1992. Global warming and biological diversity. Yale University Press, New Haven. 386 pp. **136, 200**

Peterson, A.T., and A.G. Navarro-Sigüenza. 1999. Alternate species concepts as bases for determining priority conservation areas. Conservation Biology 13:427–431. **37, 39**

Philander, S.G. 1998. Is the temperature rising? The uncertain science of global warming. Princeton University Press, Princeton, New Jersey. 262 pp. **136**

Phillips, O. 1993. Using and conserving the rainforest. Conservation Biology 7:6–7. **420**

Phillips, O., A.H. Gentry, C. Reynel, et al. 1994. Quantitative ethnobotany and Amazonian conservation. Conservation Biology 8:225–248. **401**

Phillips, T., and M.A. Maun. 1996. Population ecology of *Cirsium pitcheri* on Lake Huron sand dunes I. Impact of white-tailed deer. Canadian Journal of Botany 74:1439–1444. **327**

Pickett, S.T A., and J.N. Thompson. 1978. Patch dynamics and the design of nature reserves. Biological Conservation 13:27–37. **265**

Pimentel, D. 1992. Land degradation and environmental resources. Pages 330–332 in Miller, G.T., Jr. 1992. Living in the environment.7th ed. Wadsworth, Belmont, California. 706 pp. **179**

Pimentel, D., L. Lach, R. Zuniga, and D. Morrison. 2000. Environmental and economic costs of nonindigenous species in the United States. BioScience 50:53–65. **239, 241**

Pimentel, D.S., and P.H. Raven. 2000. Bt corn pollen impacts on nontarget Lepidoptera: Assessments of effects in nature. Proceedings of the National Academy of Sciences 97:8198–8199. **172**

Pimentel, D., C. Wilson, C. McCullum, et al. 1997. Economic and environmental benefits of biodiversity. Bioscience 47:747–757. **401**

Pimm, S.L. 1991. The balance of nature? University of Chicago Press, Chicago. 434 pp. **87, 182**

Pimm, S.L. 1998. Extinction. Pages 20–38 in W. J. Sutherland (ed.). Conservation science and action. Blackwell Science, Oxford, United Kingdom. 363 pp. **136**

Pimm, S.L., J.L. Gittleman, G.F. McCracken, and M. Gilpin. 1989. Plausible alternatives to bottlenecks to explain reduced genetic diversity. Trends in Ecology and Evolution 4:176–178. **111**

Pimm, S.L., H.L. Jones, and J. Diamond. 1988. On the risk of extinction. American Naturalist 132:757–785. **139, 162**

Pimm, S.L., and P. Raven. 2000. Extinction by numbers. Nature 403:843–845. **184**

Pimm, S.L., G.J. Russell, J.L. Gittleman, and T.M. Brooks. 1995. The future of biodiversity. Science 269:347–350. **127**

Pinchot, G. 1947. Breaking new ground. Harcourt Brace, New York. 522 pp. **13**

Pitman, N.C.A., J. Terborgh, M.R. Silman, and P. Nuñez.V. 1999. Tree species distributions in an upper Amazonian forest. Ecology 80:2651–2661. **141**

Podolsky, R., and S.W. Kress. 1992. Attraction of the endangered dark-rumped petrel to recorded vocalizations in the Galápagos Islands. Condor 94:448–453. **320**

Pojman, L.P. 1999. Global environmental ethics. Mayfield, Mountain View, California. 393 pp. **397**

Policansky, D., and J.J. Magnuson. 1998. Genetics, metapopulations, and ecosystem management of fisheries. Ecological Applications 8(suppl):S119–S123. **339**

Ponting, C. 1991. A green history of the world: The environment and the collapse of great civilizations. St. Martin's Press, New York. 432 pp. **18**

Pope, K.O., S.L. D'Hondt, and C.R. Marshall. 1998. Meteorite impact and the mass extinction of species at the Cretaceous/Tertiary boundary. Proceedings of the National Academy of Sciences 95:11028–11029. **122**

Possingham, H.P., D.B. Lindenmayer, and T.W. Norton. 1993. A framework for the improved management of threatened species based on Population Viability Analysis (PVA). Pacific Conservation Biology 1:39–45. **158**

Post, D.M., J.P. Taylor, J.F. Kitchell, et al. 1998. The role of migratory waterfowl as nutrient vectors in a managed wetland. Conservation Biology 12:910–920. **277**

Postel, S.L., G.C. Daily, and P.R. Ehrlich. 1996. Human appropriation of renewable fresh water. Science 271:785–788. **180**

Pounds, J.A., and M.L. Crump. 1987. Harlequin frogs along a tropical montane stream: Aggregation and the risk of predation by frog-eating flies. Biotropica 19:306–309. **134**

Pounds, J.A., and M.L. Crump. 1994. Amphibian declines and climate disturbance: The case of the golden toad and the harlequin frog. Conservation Biology 8:72–85. **156**

Pounds, J.A., M.P.L. Fogden, and J.H. Campbell. 1999. Biological response to climate change on a tropical mountain. Nature 398:611–615. **134, 167**

Povilitis, T. 1994. Common ground for conservation biologists? Conservation Biology 8:598–599. **34**

Power, M.E., D. Tilman, J.A. Estes, et al. 1996. Challenges in the quest for keystones. Bioscience 46:609–620. **58**

Powell, G.V.N., and R. Bjork. 1995. Implications of intratropical migration on reserve design: A case study using *Pharomachrus mocinno*. Conservation Biology 9:354–362. **272**

Prance, G., D.J. Chadwick, and J. Marsh. (eds.). 1994. Ethnobotany and the search for new drugs. John Wiley and Sons, Chichester, United Kingdom. 280 pp. **53**

Prendergast, J.R., R.M. Quinn, J.H. Lawton, et al. 1993. Rare species, the coincidence of diversity hotspots and conservation strategies. Nature 365:335–337. **263**

Prescott-Allen, C., and R. Prescott-Allen. 1986. The first resource. Yale University Press, New Haven, Connecticut 529 pp. **66**

Prescott-Allen, R., and C. Prescott-Allen. 1982. What's wildlife worth? Earthscan, London. 92 pp. **50, 66, 214**

Prescott-Allen, R., and C. Prescott-Allen. 1990. How many plants feed the world? Conservation Biology 4:365–374. **49, 359**

Press, D., D.F. Doak, and P. Steinberg. 1996. The role of local government in the conservation of rare species. Conservation Biology 10:1538–1548. **432**

Pressey, R.L. 1994. *Ad hoc* reservations: Forward or backward steps in developing representative reserve systems? Conservation Biology 8:662–668. **260**

Pressey, R.L., S. Ferrier, T.C. Hager, et al. 1996. How well protected are the forests of northeastern New South Wales? Analyses of forest environments in relation to formal protection measures, land tenure, and vulnerability to clearing. Forest Ecology and Management 85:311–333. **260**

Preston, D.J. 1976. The rediscovery of *Betula uber*. American Forests 82(8):16–20. **142**

Price, M.R.S. 1989. Animal reintroductions: The Arabian oryx in Oman. Cambridge University Press, Cambridge. 291pp. **362**

Primack, R. 1993. Essentials of conservation biology. Sinauer, Sunderland, Massachusetts. 564pp. **xiii**

Primack, R., H. Kobori, and S. Mori. 2000. Dragonfly pond restoration promotes conservation awareness in Japan. Conservation Biology 14:1553–1554. **299**

Principe, P.P. 1996. Monetizing the pharmacological benefits of plants. Pages 191–218 in M.J. Balick, E. Elisabetsky, and S.A. Laird (eds.). Medicinal resources of the tropical forest. Columbia University Press, New York. 440 pp. **52**

Probst, J.R., and J. Weinrich. 1993. Relating Kirtland's warbler population to changing landscape composition and structure. Landscape Ecology 8:257–271. **180**

Purvis, A., P. Agapow, J.L. Gittleman, and G.M. Mace. 2000. Nonrandom extinction and the loss of evolutionary history. Science 288:328–330. **63, 372**

Putman, R.J. (ed). 1989. Mammals as pests. Chapman and Hall, London. 271 pp. **234**

Putz, F.E., D.P. Dykstra, and R. Heinrich. 2000. Why poor logging practices persist in the tropics. Conservation Biology 14:951–956. **418**

Pykälä, J. 2000. Mitigating human effects on European biodiversity through traditional animal husbandry. Conservation Biology 14:705–712. **286**

Quammen, D. 1996. The song of the dodo. Simon and Schuster, New York. 702 pp. **203, 205, 311**

Quinn, J.F., and S.P. Harrison. 1988. Effects of habitat fragmentation and isolation on species richness: Evidence from biogeographic patterns. Oecologia 75:132–140. **267**

Rabenold, K.N., P.T. Fauth, B.W. Goodner, et al. 1998. Response of avian communities to disturbance by an exotic insect in spruce-fir forests of the southern Appalachians. Conservation Biology 12:177–189. **244**

Rabinowitz, A. 1995. Helping a species go extinct: The Sumatran rhino in Borneo. Conservation Biology 9:482–488. **358**

Rabinowitz, D. 1981. Seven forms of rarity. Pages 205–217 in H. Synge (ed.) The biological aspects of rare plant conservation. John Wiley and Sons, Chichester, England. 558 pp. **46, 140**

Rabinowitz, D., S. Cairns, and T. Dillon. 1986. Seven forms of rarity and their frequency in the flora of the British Isles. Pages 182–204 in M.E. Soulé (ed.). Conservation biology: The science of scarcity and diversity. Sinauer, Sunderland, Massachusetts. 584 pp. **140**

Rabinowitz, D., J.K. Rapp, and P.M. Dixon. 1984. Competitive abilities of sparse grass species: Means of persistence or cause of abundance. Ecology 65:1144–1154. **140**

Rachlow, J.L., and J. Berger. 1997. Conservation implications of patterns of horn regeneration in dehorned white rhinos. Conservation Biology 11:84–91. **322**

Rahel, F.J. 2000. Homogenization of fish faunas across the United States. Science 288:854–856. **234**

Ralls, K., and J. Ballou. 1983. Extinction: Lessons from zoos. Pages 164–184 in C.M. Schonewald-Cox, S.M. Chambers, B. MacBryde, and W.L. Thomas (eds.). Genetics and conservation. Benjamin/Cummings, Menlo Park, California. 722 pp. **98, 348**

Ralls, K., J.D. Ballou, and A. Templeton. 1988. Estimates of lethal equivalents and the cost of inbreeding in mammals. Conservation Biology 2:185–193. **98**

Ralls, K., P.H. Harvey, and A.M. Lyles. 1986. Inbreeding in natural populations of birds and mammals. Pages 35–56 in M.E. Soulé (ed.). Conservation biology: The science of scarcity and diversity. Sinauer, Sunderland, Massachusetts. 584 pp. **99**

Ramakrishnan, P.S., and P.M. Vitousek. 1989. Ecosystem-level processes and the consequences of biological invasions. Pages 281–300 in J.A. Drake, H.A. Mooney, F. di Castri, et al. (eds.). Biological invasions: A global perspective. John Wiley and Sons, Chichester. 525 pp. **244**

Rambo, J.L., and S.H. Faeth. 1999. Effect of vertebrate grazing on plant and insect community structure. Conservation Biology 13:1047–1054. **286**

Ranker, T.A., and A.M. Arft. 1994. Allopolyploid species and the U.S. Endangered Species Act. Conservation Biology 8:895–897. **39**

Rapport, D. 1998. Ecosystem health: An integrative science. Pages 3–50 in D. Rapport, R. Costanza, P.R. Epstein, et al. (eds.) Ecosystem health. Blackwell Science, Malden, Massachusetts. 372 pp. **31**

Raup, D.M. 1979. Size of the Permo-Triassic bottleneck and its evolutionary implications. Science 206:217–218. **123**

Raup, D.M. 1991. Extinction: Bad genes or bad luck? Norton, New York. 210 pp. **119, 120, 121, 136**

Ray, J.C., and J.R. Ginsberg. 1999. Endangered species legislation beyond the borders of the United States. Conservation Biology 13:956–958. **432**

Raymond, H.L. 1979. Effects of dams and impoundments on migrations of juvenile chinook salmon and steelhead from the Snake River, 1966 to 1975. Transactions of the American Fisheries Society 108:505–529. **175**

Reading, R.P., T.W. Clark, and B. Griffith. 1997. The influence of valuational and organizational considerations on the success of rare species translocations. Biological Conservation 79:217–225. **334**

Reading, R.P., and S.R. Kellert. 1993. Attitudes toward a proposed reintroduction of black-footed ferrets (*Mustela nigripes*). Conservation Biology 7:569–580. **390**

Redford, K.H. 1992. The empty forest. Bioscience 42:412–422. **223**

Redman, C.L. 1999. Human impact on ancient environments. University of Arizona Press. Tucson, Arizona. 239 pp. **205**

Reed, J.M., D.D. Murphy, and P.F. Brussard. 1998. Efficacy of population viability analysis. Wildlife Society bulletin 26:244–251. **158**

Reed, J.M., J.R. Walters, T.E. Emigh, and D.E. Seaman. 1993. Effective population size in red-cockaded woodpeckers: Population and model differences. Conservation Biology 7:302–308. **107**

Regal, P.J. 1993. The true meaning of 'exotic species' as a model for genetically engineered organisms. Experientia 49:225–234. **236**

Reichard, S.H., and C.W. Hamilton. 1997. Predicting invasions of woody plants introduced into North America. Conservation Biology 11:193–203. **247**

Reid, W.V. 1992. How many species will there be? Pages 55–73 in T.C. Whitmore, and J.A. Sayer (eds.). Tropical deforestation and species extinction. Chapman and Hall, London. 153 pp. **126**

Reijnen, R., R. Foppen, C.T. Braak, and J. Thissen. 1995. The effects of car traffic on breeding bird populations in woodland. III. Reduction of density in relation to the proximity of main roads. Journal of Applied Ecology 32:187–202. **174**

Reinthal, P.N., and M.L.J. Stiassny. 1991. The freshwater fishes of Madagascar: A study of an endangered fauna with recommendations for a conservation strategy. Conservation Biology 5:231–243. **203**

Reisner, M. 1991. Game wars. Viking, New York. 294 pp. **321**

Rejmánek, M. 1989. Invasibility of plant communities. Pages 369–388 in J.A. Drake, H.A. Mooney, F. di Castri, et al. (eds.). Biological invasions: A global perspective. John Wiley and Sons, Chichester. 525 pp. **246**

Rejmánek, M. 1996. A theory of seed plant invasiveness: The first sketch. Biological Conservation 78:171–181. **246**

Rettalack, G.J. 1995. Permian-Triassic life crisis on land. Science 267:77–80. **123**

Rhymer, J.M., and D. Simberloff. 1996. Extinction by hybridization and introgression. Annual Review of Ecology and Systematics 27:83–109. **244, 334, 341**

Ricciardi, A., and H.J. MacIsaac. 2000. Recent mass invasion of the North American Great Lakes by Ponto-Caspian species. Trends in Ecology and Evolution 15:62–65. **232**

Ricciardi, A., R.J. Neves, and J.B. Rasmussen. 1998. Impending extinctions of North American freshwater mussels (Unionoida) following the zebra mussel (*Dreissena polymorphia*) invasion. Journal of Animal Ecology 67:613–619. **233**

Ricciardi, A., and J. Rasmussen. 1999. Extinction rates of North American freshwater fauna. Conservation Biology 13:1220–1222. **193**

Rich, W.H. 1930. Fishing grounds of the Gulf of Maine. Report of the United Sates Commissioner of Fisheries 1929:51–117. **226**

Richardson, D.M. 1998. Forestry trees as invasive aliens. Conservation Biology 12:18–26. **233**

Richter, A.R., S.R. Humphrey, J.B. Cope, and V. Brack, Jr. 1993. Modified cave entrances: Thermal effect on body mass and resulting decline of endangered Indiana bats (*Myotis sodalis*). Conservation Biology 7:407–415. **324, 325**

Richter, B.D., J.V. Baumgartner, R. Wigington, and D.P. Braun. 1997. How much water does a river need? Freshwater Biology 37:231–249. **144, 307**

Richter, B.D., D.P. Braun, M.A. Mendelson, and L.L. Master. 1997. Threats to imperiled freshwater fauna. Conservation Biology 11:1081–1093. **168**

Ricklefs, R.E. 1995. The distribution of biodiversity. Pages 139–173 in V.H. Heywood and R.T. Watson (eds.). Global biodiversity assessment. Cambridge University Press, Cambridge, United Kingdom. 1140 pp. **81**

Ricklefs, R.E. 2000. Rarity and diversity in Amazonian forest trees. Trends in Ecology and Evolution 15:83–84. **141**

Ricklefs, R.E., and D. Schluter, (eds.). 1993. Species diversity in ecological communities. University of Chicago Press, Chicago. 414 pp. **81, 87**

Rinne, J.N., and P.R. Turner. 1991. Reclamation and alteration as management techniques, and a review of methodology in stream renovation. Pages 219–244 in W.L. Minckley and J.E. Deacon (eds.). Battle against extinction: Native fish management in the American west. University of Arizona Press, Tucson. 517 pp. **331**

Ritchie, M.E., and H. Olff. 1999. Spatial scaling laws yield a synthetic theory of biodiversity. Nature 400:557–560. **83**

Rivas, J.A., and R.Y. Owens. 1999. Teaching conservation effectively: A lesson from life-history strategies. Conservation Biology 13:453–454. **393**

Robbins, C.S., D.K. Dawson, and B.A. Dowell. 1989. Habitat area requirements of breeding forest birds of the Middle Atlantic states. Wildlife Monographs 103:1–34. **195, 200**

Robbins, C.T. 1993. Wildlife feeding and nutrition. 2nd ed. Academic Press, San Diego. 352 pp. **316**

Roberts, C.M. 1997. Connectivity and management of Caribbean coral reefs. Science 278:1454–1457. **273**

Roberts, C.M., and N.V.C. Polunin. 1993. Marine reserves: Simple solutions to managing complex fisheries? Ambio 22:363–368. **269**

Roberts, D.L., R.J. Cooper, and L.J. Petit. 2000. Use of premontane moist forest and shade coffee agroecosystems by army ants in western Panama. Conservation Biology 14:192–199. **293**

Robinson, J.G., and E.L. Bennett (eds.). 2000. Hunting for sustainability in tropical forests. Columbia University Press, New York. 582 pp. **214, 291**

Robinson, J.G., and K.H. Redford (eds.). 1991. Neotropical wildlife use and conservation. The University of Chicago Press, Chicago. 520 pp. **214, 285**

Robinson, M.A. 1984. Trends and prospects in world fisheries. Food and Agriculture Organization of the United Nations, Rome. Food and Agriculture Organization Fisheries Circular No. 772. **289**

Robinson, M.H. 1988. Bioscience education through bioparks. Bioscience 38:630–634. **347**

Robinson, S.K., F.R. Thompson III, T.M. Donovan, et al. 1995. Regional forest fragmentation and the nesting success of migratory birds. Science 267:1987–1990. **28**

Rochelle, J.A., L.A. Lehmann, and J. Wisniewski. 1999. Forest fragmentation. Brill, Leiden, Netherlands. 301 pp. **199, 205**

Rojas, M. 1992. The species problem and conservation: What are we protecting? Conservation Biology 6:170–178. **39**

Romme, W.H., and D.G. Despain. 1989. Historical perspective on the Yellowstone fires of 1988. Bioscience 39:695–699. **276**

Roosenburg, W.M., and J.P Green. 2000. Impact of a bycatch reduction device on diamondback terrapin and blue crab capture in crab pots. Ecological Applications 10:882–889. **323**

Rosenthal, D.H., and R.H. Nelson. 1992. Why existence value should *not* be used in cost-benefit analysis. Journal of Policy Analysis and Management 11:116–122. **406**

Rosenzweig, M.L. 1995. Species diversity in space and time. Cambridge University Press, Cambridge, United Kingdom. 436 pp. **81, 87**

Rotter, J., and K. Danish. 2000. Forest carbon and the Kyoto Protocol's clean development mechanism. Journal of Forestry 98(5):38–47. **416**

Rowe, J.S. 1996. Land classification and ecosystem classification. Environmental Monitoring and Assessment 39:11–20. **71**

Roy, K., J.W. Valetine, D. Jablonski, and S.M. Kidwell. 1996. Scales of climatic variability and time averaging in Pleistocene biotas: Implications for ecology and evolution. Trends in Ecology and Evolution 11:458–463. **132**

Roy, M.S., E. Geffen, D. Smith, and R.K. Wayne. 1996. Molecular genetics of pre-1940 red wolves. Conservation Biology 10:1413–1424. **37**

Ruane, J. 2000. A framework for prioritizing domestic animal breeds for conservation purposes at the national level: A Norwegian case study. Conservation Biology 14:1385–1393. **361, 372**

Rubes, J., L. Borkovec, Z. Horinova, et al. 1992. Cytogenetic monitoring of farm animals under conditions of environmental pollution. Mutation Research 283:199–210. **167**

Rudnicky, T.C., and M.L. Hunter, Jr. 1993. Reversing the fragmentation perspective: Effects of clearcut size on bird species richness in Maine. Ecological Applications 3:357–366. **196**

Ruggiero, L.F., K.B. Aubry, A.B. Carey, and M.H. Huff (eds.). 1991. Wildlife and vegetation of unmanaged Douglas-fir forests. USDA Forest Service General Technical Report PNW-GTR-285. 533 pp. **310**

Ruiz, G.M., J.T. Carlton, E.D. Grosholz, and A.H. Hines. 1997. Global invasions of marine and estuarine habitats by non-indigenous species: Mechanisms, extent, and consequences. American Zoologist 37:621–632. **232**

Ruiz, G.M., T.K. Rawlings, F.C. Dobbs, et al. 2000. Global spread of microorganisms by ships. Nature 408:49–50. **241**

Ryder, O.A. 1986. Species conservation and the dilemma of subspecies. Trends in Ecology and Evolution 1:9–10. **38**

Ryder, O.A. 1993. Przewalski's horse: Prospects for reintroduction into the wild. Conservation Biology 7:13–15. **353**

Ryder, O.A., A. McLaren, S. Brenner, et al. 2000. DNA banks for endangered animal species. Science 288:275–277. **355**

Saberwal, V.K. 1997. Saving the tiger: More money or less power? Conservation Biology 11:815–817. **386**

Saccheri, I., M. Kuussaari, M. Kankare, et al. 1998. Inbreeding and extinction in a butterfly metapopulation. Nature 392:491–494. **99**

Saenger, P., E.J. Hegerl, and J.D.S. Davie (eds.). 1983. Global status of mangrove ecosystems. IUCN, Gland, Switzerland. 88 pp. **86**

Safina, C. 1998. Song for the blue ocean. Henry Holt, New York. 458 pp. **229**

Sala, O.E., F.S. Chapin, III, J.J. Armesto, et al. 2000. Global biodiversity scenarios for the year 2100. Science 287:1770–1774. **369, 373**

Salafsky, N., B.L. Dugelby, and J.W. Terborgh. 1993. Can extractive reserves save the rain forest? An ecological and socioeconomic comparison of nontimber forest product extraction systems in Peten, Guatemala, and West Kalimantan, Indonesia. Conservation Biology 7:39–52. **290**

Samson, F.B., and F.L. Knopf. (eds.). 1996. Prairie Conservation. Island Press, Washington, D. C. 339 pp. **311**

Sanderson, G.C., and F.C. Bellrose. 1986. A review of the problem of lead poisoning in waterfowl. Illinois Natural History Survey Special Publication 4:1–34. **177**

Sandlund, O.T., P.J. Schei, and Å. Viken (eds). 1999. Invasive species and biodiversity management. Kluver Academic, Dordrecht, Netherlands. 431 pp. **251**

Sankaran, M., and S.J. McNaughton. 1999. Determinants of biodiversity regulate compositional stability of communities. Nature 401:691–693. **81**

Sauer, L.J. 1998. The once and future forest. Island Press, Washington, D.C. 381 pp. **306**

Saunders, D.A., and R.J. Hobbs (eds.). 1991. Nature Conservation: The role of corridors. Surrey Beatty and Sons, Chipping Norton, Australia. **273**

Saunders, D.A., R.J. Hobbs, and C.R. Margules. 1991. Biological consequences of ecosystem fragmentation: A review. Conservation Biology 5:18–32. **193**

Savory, A., and J. Butterfield. 1999. Holistic management. Island Press, Washington, D.C. 616 pp. **180, 186**

Schaller, G.B. 1993. The last panda. University of Chicago Press, Chicago. 291 pp. **217, 364, 440**

Schelhas, J., and R. Greenberg. 1996. Forest patches in tropical landscapes. Island Press, Washington, D.C. 426 pp. **205**

Scher, A. 1997. Green labels: Can they build a new marketplace? Dollars and Sense 211:22–25, 46–47. **435**

Schindler, D.E., J.F. Kitchell, and R. Ogutu-Ohwayo. 1998. Ecological consequences of alternative gill net fisheries for Nile Perch in Lake Victoria. Conservation Biology 12:56–64. **234**

Schlesinger, W.H., J.F. Reynolds, G.L. Cunningham, et al. 1990. Biological feedbacks in global desertification. Science 247:1043–1048. **186**

Schmidt, W. 1989. Plant dispersal by motor cars. Vegetatio 80:147–152. **232**

Schmutterer, H. (ed.) 1995. The neem tree, *Azadirachta indica* A. Juss. and other meliaceous plants. VCH, Weinheim, Germany. 696 pp. **65**

Schofield, E.K. 1989. Effects of introduced plants and animals on island vegetation: Examples from the Galapagos Archipelago. Conservation Biology 3:227–238. **241**

Schonewald-Cox, C.M., S.M. Chambers, B. MacBryde, and W.L. Thomas (eds.). 1983. Genetics and conservation. Benjamin/Cummings, Menlo Park, California. 722pp. **113**

Schopf, J.W., and C. Klein (eds.). 1992. The proterozoic biosphere. Cambridge University Press, Cambridge. 1348 pp. **120**

Schorger, A.W. 1973. The passenger pigeon: Its natural history and extinction. University of Oklahoma Press, Norman. 424 pp. **142, 151, 350**

Schulze, E-D., and H.A. Mooney (eds.). 1993. Biodiversity and ecosystem function. Springer-Verlag, Berlin. 525 pp. **87**

Schwartzman, S., A. Moreira, and D. Nepstad. 2000. Rethinking tropical forest conservation: Perils in parks. Conservation Biology 14:1351–1357. **275, 386**

Scott, J.M., F. Davis, B. Csuti, et al. 1993. Gap analysis: A geographic approach to protection of biological diversity. Wildlife Monograph 123. 41 pp. **262**

Scott, J.M., and M.D. Jennings. 1998. Large-area mapping of biodiversity. Annals of the Missouri Botanical Garden 85:34–47. **377**

Scott, M.E. 1988. The impact of infection and disease on animal populations: Implications for conservation biology. Conservation Biology 2:40–56. **328**

Scudder, G.G.E. 1989. The adaptive significance of marginal populations: A general perspective. Pages 180–185 in C.D. Levings, L.B. Holtby, and M.A. Henderson (eds.). Proceedings of the National Workshop on Effects of Habitat Alteration on Salmonid Stocks. Canadian Special Publication Fisheries and Aquatic Science 105. 199 pp. **339, 368**

Seddon, P.J., and P.S. Soorae. 1999. Guidelines for subspecific substitutions in wildlife restoration projects. Conservation Biology 13:177–184. **307, 339**

Sedjo, R.A. 1988. Property rights and the protection of plant genetic resources. Pages 293–314 in J.R. Kloppenburg (ed.). Seeds and sovereignty: The use and control of plant genetic resources. Duke University Press, Durham, North Carolina. 368 pp. **414**

Sedjo, R.A., and D. Botkin. 1997. Using forest plantations to spare natural forests. Environment 39(10):14–20, 30. **296**

Seebeck, J.H. 1990. Recovery management of the eastern barred bandicoot in Victoria: Statewide strategy. Pages 165–177 in T.W. Clark and J.H. Seebeck (eds.). Management and conservation of small populations. Chicago Zoological Society, Brookfield, Illinois. 295 pp. **161**

Seehausen, O., J.J.M. van Alphen, and F. Witte. 1997. Cichlid fish diversity threatened by eutrophication that curbs sexual selection. Science 277:1808–1811. **240**

Seehausen, O., F. Witte, E.F. Katunzi, et al. 1997. Patterns of the remnant cichlid fauna in southern Lake Victoria. Conservation Biology 11:890–904. **169**

Sepkoski, J.J., Jr. 1982. Mass extinctions in the Phanerozoic oceans: A review. Pages 283–289 in L.T. Silver, and P.H. Schultz (eds.). Geological implications of impacts of large asteroids and comets on the Earth. The Geological Society of America Special Paper 190, Boulder, Colorado. 528 pp. **121**

Sepkoski, J.J., Jr. 1984. A kinetic model of Phanerozoic taxonomic diversity. III. Post-Paleozoic families and mass extinctions. Paleobiology 10:246–267. **122**

Sepkoski, J.J. 1995. Large-scale history of biodiversity. Pages 202–212 in V.H. Heywood and R.T. Watson (eds.). Global biodiversity assessment. Cambridge University Press, Cambridge, United Kingdom. 1140 pp. **120**

Seymour, R.S., and M.L. Hunter, Jr. 1992. New Forestry in eastern spruce-fir forests: Principles and applications to Maine. Maine Agricultural and Forest Experiment Station Miscellaneous Publication 716. 36 pp. **297**

Seymour, R.S., and M.L. Hunter, Jr. 1999. Principles of ecological forestry. Pages 22–64 in M.L. Hunter, Jr. (ed.) Maintaining biodiversity in forest ecosystems. Cambridge University Press, Cambridge, United Kingdom. 698 pp. **280**

Shafer, C.L. 1990. Nature Reserves: Island theory and conservation practice. Smithsonian Institution Press, Washington, D.C. 189 pp. **311**

Shaffer, M.L. 1981. Minimum population sizes for species conservation. Bioscience 31:131–134. **151, 158**

Shanley, P. 1999. To market, to market. Natural History 108(8):44–51. **418**

Sheail, J. 1985. Pesticides and nature conservation. Clarendon Press, Oxford. 276 pp. **172**

Shiva, V. 1988. Staying alive: Women, ecology, and development. Zed Books, London. 224 pp. **388**

Shogren, J.F., J. Tschirhart, T. Anderson, et al. 1999. Why economics matters for endangered species protection. Conservation Biology 13:1257–1261. **415**

Sigler, W.F. 1972. Wildlife law enforcement. 2nd ed. Wm. C. Brown, Dubuque, Iowa. 360 pp. **321**

Silva, J.L., and S.D. Strahl. 1991. Human impact on populations of chachalacas, guans, and curassows (Galliformes: Cracidae) in Venezuela. Pages 37–52 in J.G. Robinson, and K.H. Redford (eds.). Neotropical wildlife use and conservation. The University of Chicago Press, Chicago. 520 pp. **61**

Silvertown, J., M. Franco, and E. Menges. 1996. Interpretation of elasticity matrices as an aid to the management of plant populations for conservation. Conservation Biology 10:591–597. **150**

Simberloff, D.S. 1986. Design of nature reserves. Pages 315–337 in M.B. Usher (ed.) Wildlife conservation evaluation. Chapman and Hall, London. 394 pp. **266**

Simberloff, D.S. 1988. The contribution of population and community biology to conservation science. Annual Review of Ecology and Systematics 19:473–511. **356**

Simberloff, D. 1998. Flagships, umbrellas, and keystones: Is single-species management passé in the landscape era? Biological Conservation 83:247–257. **60**

Simberloff, D.S., and L.G. Abele. 1976a. Island biogeography theory and conservation practice. Science 191:285–286. **266, 267**

Simberloff, D.S., and L.G. Abele. 1976b. Island biogeography and conservation: Strategy and limitations. Science 193:1032. **266**

Simberloff, D.S., and L.G. Abele. 1982. Refuge design and island biogeographic theory: Effects of fragmentation. American Naturalist 120:41–50. **266, 267**

Simberloff, D.S., and L.G. Abele. 1984. Conservation and obfuscation: Subdivision of reserves. Oikos 42:399–401. **267**

Simberloff, D.S., and J. Cox. 1987. Consequences and costs of conservation corridors. Conservation Biology 1:63–71. **273**

Simberloff, D.S., J.A. Farr, J. Cox, and D.W. Mehlman. 1992. Movement corridors: Conservation bargains or poor investments? Conservation Biology 6:493–504. **273**

Simberloff, D.S., and Gotelli, N. 1984. Effects of insularization on plant species richness in the prairie forest ecotone. Biological Conservation 29:27–46. **267**

Simberloff, D., and P. Stiling. 1996. Risks of species introduced for biological control. Biological Conservation 78:185–192. **236**

Skinner, J.A., K.A. Lewis, K.S. Bardon, et al. 1997. An overview of the environmental impact of agriculture in the U. K. Journal of Environmental Management 50:111–128. **295**

Skole, D., and C. Tucker. 1993. Tropical deforestation and habitat fragmentation in the Amazon: Satellite data from 1978 to 1988. Science 260:1905–1910. **182**

Smith, D.M., B.C. Larson, M.J. Kelty, and P.M.S. Ashton. 1997. The practice of silviculture: Applied forest ecology. 9th ed. John Wiley and Sons, New York. 537 pp. **323**

Smith, M.A. 1988. Reclamation and treatment of contaminated land. Pages 61–89 in J. Cairns (ed.). Rehabilitating damaged ecosystems. Vol. I. CRC Press, Boca Raton, Florida. 192 pp. **306**

Smith, M.J. (ed.). 1999. Thinking through the environment. Routledge, London. 435 pp. **397**

Smith, M., J. Whitelegg, and N. Williams. 1998. Greening the built environment. Earthscan Publications, London. 248 pp. **300**

Smith, S.H. 1968. Species succession and fishery exploitation in the Great Lakes. Journal of the Fisheries Research Board of Canada 25:667–693. **238**

Smith, T.B., R.K. Wayne, D.J. Girman, and M.W. Bruford. 1997. A role for ecotones in generating rainforest biodiversity. Science 276:1855–1857. **260**

Snyder, G. 1990. The practice of the wild: Essays. North Point Press, San Francisco. 190 pp. **394**

Snyder, N.F.R., S.R. Derrickson, S.R. Beissinger, et al. 1996. Limitations of captive breeding in endangered species recovery. Conservation Biology 10:338–348. **358**

Snyder, N.F.R., J.W. Wiley, and C.B. Kepler. 1987. The parrots of Luquillo: Natural history and conservation of the Puerto Rican parrot. Western Foundation of Vertebrate Zoology, Los Angeles, California. 384 pp. **318**

Solberg, E.J., A. Loison, B. Sæther, and O. Strand. 2000. Age-specific harvest mortality in a Norwegian moose *Alces alces* population. Wildlife Biology 6:41–52. **220**

Solomon, B.D. 1998. Impending recovery of Kirtland's warbler: Case study in the effectiveness of the Endangered Species Act. Environmental Management 22:9–17. **331**

Soule, J.D., and J.K. Piper. 1992. Farming in nature's image. Island Press, Washington, D.C. 286 pp. **293, 295**

Soulé, M.E. 1985. What is conservation biology? Bioscience 35:727–734. **14, 17, 18**

Soulé, M.E. (ed.). 1986. Conservation biology: The science of scarcity and diversity. Sinauer, Sunderland, Massachusetts. 584 pp. **113**

Soulé, M.E. (ed.). 1987. Viable populations for conservation. Cambridge University Press, Cambridge. 184 pp. **150, 158, 162**

Soulé, M.E. 1991. Conservation: Tactics for a constant crisis. Science 253:744–750. **366, 375**

Soulé, M.E., and M.A. Sanjayan. 1998. Conservation targets: Do they help? Science 279:2060–2061. **263**

Soulé, M.E., and D. Simberloff. 1986. What do genetics and ecology tell us about the design of nature reserves? Biological Conservation 35:19–40. **267**

Soulé, M.E., and B.A. Wilcox (eds.) 1980. Conservation biology: An evolutionary-ecological perspective. Sinauer, Sunderland, Massachusetts. 395 pp. **15**

Spalton, J.A., M.W. Lawrence, and S.A. Brend. 1999. Arabian oryx reintroduction in Oman: Successes and setbacks. Oryx 33:168–175. **362**

Spellerberg, I.F. 1992. Evaluation and assessment for conservation. Chapman and Hall, London. 260 pp. **368, 379**

Spies, T.A., and M.G. Turner. 1999. Dynamic forest mosaics. Pages 95–160 in M.L. Hunter, Jr. (ed.) Maintaining biodiversity in forest ecosystems. Cambridge University Press, Cambridge, United Kingdom. 698 pp. **280, 282, 310**

Spirn, AW. 1984. The granite garden: Urban nature and human design. Basic Books, New York. 334 pp. **311**

St. John, T.V. 1987. Mineral acquisition in native plants. Pages 529–535 in T.S. Elias (ed.). Conservation and management of rare and endangered plants. California Native Plant Society, Sacramento, California. 630 pp. **316**

Stachowicz, J.J., R.B. Whitlatch, and R.W. Osman. 1999. Species diversity and invasion resistance in a marine ecosystem. Science 286:1577–1579. **245**

Stamps, J.A. 1991. The effect of conspecifics on habitat selection in territorial species. Behavioral Ecology and Sociobiology 28:29–36. **200**

Stanley, S.M. 1987. Extinction. Scientific American Books, New York. 227 pp. **123, 136**

Steadman, D.W. 1995. Prehistoric extinctions of Pacific Island birds: Biodiversity meets zooarchaeology. Science 267:1123–1131. **210**

Steenkamp, H.E., and S.L. Chown. 1996. Influence of dense stands of an exotic tree, *Prosopis glandulosa* Benson, on a savanna dung beetle (Coleoptera: Scarabaeinae) assemblage in southern Africa. Biological Conservation 78:305–311. **245**

Steidl, R.J., and R.G. Anthony. 2000. Experimental effects of human activity on breeding bald eagles. Ecological Applications 10:258–268. **216**

Stein, S. 1993. Noah's Garden: Restoring the ecology of our own back yards. Houghton Mifflin Company, Boston. 294 pp. **301**

Steneck, R.S. 1997. Fisheries-induced biological changes to the structure and function of the Gulf of Maine Ecosystem. Plenary Paper. Pages 151–165 in G.T. Wallace and E.F. Braasch (eds). Proceedings of the Gulf of Maine ecosystem dynamics scientific symposium and workshop. RARGOM Report, 91-1. Regional Association for Research on the Gulf of Maine. Hanover, NH. **226, 227**

Steneck, R.S. 1998 Human influences on coastal ecosystems: Does overfishing create trophic cascades? Trends in Ecology and Evolution. 13:429–430. **226, 227**

Steneck, R.S., and J.T. Carlton. 2000. Human alterations of marine communities: Students beware! Pages 445–468 in M.D. Bertness, S.D. Gaines, and M.E. Hay (eds). Marine community ecology. Sinauer, Sunderland, Massachusetts. 550 pp. **226, 227**

Stevens, P.F. 1989. New Guinea. Pages 120–132 in D.G. Campbell, and H.D. Hammond (eds.). Floristic inventory of tropical countries. The New York Botanical Garden, New York. 545 pp. **37**

Stevens, P.F. 1990. Nomenclatural stability, taxonomic instinct, and flora writing—a recipe for disaster? Pages 387–410 in P. Baas, K. Kalkman, and R. Geesink (eds.). The plant diversity of Malesia. Kluwer Academic Publishers, Dordrecht, Netherlands. 420 pp. **39**

Stevens, T.H., T.A. More, and R.J. Glass. 1994. Interpretation and temporal stability of cv bids for wildlife existence: A panel study. Land Economics 70:355–363. **406**

Stewart, W.N., and G.W. Rothwell. 1993. Paleobotany and the evolution of plants. 2nd ed. Cambridge University Press, Cambridge. 521 pp. **122**

Stohlgren, T.J., D.Binkley, G.W. Chong, et al. 1999. Exotic plant species invade hot spots of native plant diversity. Ecological Monographs 69:25–46. **245**

Strickland, M.D., H.J. Harju, K.R. McCaffery, et al. 1994. Harvest management. Pages 445–473 in T.A. Bookhout (ed.). Research and management techniques for wildlife and habitats. The Wildlife Society, Bethesda, Maryland. 740 pp. **323**

Strong, D.R., and R.W. Pemberton. 2000. Biological control of invading species—risk and reform. Science 288:1969–1970. **236, 242**

Stuckey, I.H. 1967. Environmental factors and the growth of native orchids. American Journal of Botany 54:232–241. **316**

Sunquist, F. 1995. End of the ark? International wildlife 25(6):22–29. **347, 354**

Surridge, A.K., R.J. Timmins, G.M. Hewitt, and D.J. Bell. 1999. Striped rabbits in Southeast Asia. Nature 400:726. **376**

Sutcliffe, O.L., and C.D. Thomas. 1996. Open corridors appear to facilitate dispersal by ringlet butterflies (*Aphantopus hyperantus*) between woodland clearings. Conservation Biology 10:1359–1365. **273**

Suter, G.W. II. 1993. A critique of ecosystem health concepts and indexes. Environmental Toxicology and Chemistry 12:1533–1539. **31**

Sutherland, W.J. 2000. The conservation handbook: Research, management and policy. Blackwell Science, Oxford, United Kingdom. 278 pp. **344**

Sutherland, W.J., and D.A. Hill (eds.). 1995. Managing habitats for conservation. Cambridge University Press, Cambridge, United Kingdom. 399 pp. **278, 311**

Suzan, H., G.P. Nabhan, and D.T. Patten. 1994. Nurse plant and floral biology of a rare night-blooming Cereus, *Peniocereus striatus* (Brandegee) F. Buxbaum. Conservation Biology 8:461–470. **318**

Swengel, A.B. 1998. Effects of management on butterfly abundance in tallgrass prairie and pine barrens. Biological Conservation 83:77–89. **286**

Symstad, A.J. 2000. A test of the effects of functional group richness and composition on grassland invasibility. Ecology 81:99–109. **245**

Szafer, W. 1968. The Ure-ox, extinct in Europe since the seventeenth century: An early attempt at conservation that failed. Biological Conservation 1:45–47. **49**

Tainton, N.M. (ed.). 1999. Veld management in South Africa. University of Natal Press, Pietermaritzburg, South Africa. 472 pp. **284**

Teer, J.G., L.A. Renecker, and R.J. Hudson. 1993. Overview of wildlife farming and ranching in North America. Transactions of the North American Wildlife and Natural Resources Conference 58:448–459. **285**

Temple, S.A. (ed.). 1977. Endangered birds: Management techniques for preserving threatened species. University of Wisconsin Press, Madison, Wisconsin. 466 pp. **344**

Templeton, A.R. 1986. Coadaptation and outbreeding depression. Pages 105–116 in M.E. Soulé (ed.). Conservation biology: The science of scarcity and diversity. Sinauer, Sunderland, Massachusetts. 584 pp. **100**

Templeton, A.R., and B. Read. 1983. The elimination of inbreeding depression in a captive herd of Speke's gazelle. Pages 241–261 in C.M. Schonewald-Cox, S.M. Chambers, B. MacBryde, and W.L. Thomas (eds.). Genetics and conservation. Benjamin/Cummings, Menlo Park, California. 722 pp. **109**

Templeton, A.R., and B. Read. 1994. Inbreeding: One word, several meanings, much confusion. Pages 91–105 in V. Loeschcke, J. Tomiuk and S.K. Jain (eds.). Conservation genetics. Birkhäuser Verlag, Basel, Switzerland. 440 pp. **107**

Templeton, A.R., K. Shaw, E. Routman, and S.K. Davis. 1990. The genetic consequences of habitat fragmentation. Annals of the Missouri Botanical Garden 77:13–27. **149**

Terborgh, J. 1976. Island biogeography and conservation: Strategy and limitations. Science 193:1029–1030. **266**

Terborgh, J. 1986. Keystone plant resources in the tropical forest. Pages 330–344 in M.E. Soulé (ed.) Conservation biology. Sinauer, Sunderland, Massachusetts. 584 pp. **59**

Terborgh, J. 1992. Diversity and the tropical rain forest. Scientific American Library, New York. 242 pp. **184**

Terborgh, J., and B. Winter. 1980. Some causes of extinction. Pages 119–133 in M.E. Soulé, and B.A. Wilcox (eds.). Conservation biology: An evolutionary-ecological perspective. Sinauer, Sunderland, Massachusetts. 584 pp. **162**

Tewksbury, J.J., G.P. Nabhan, D. Norman, et al. 1999. *In situ* conservation of wild chiles and their biotic associates. Conservation Biology 13:98–107. **362**

The Nature Conservancy. 1982. Natural heritage program operations manual. The Nature Conservancy, Arlington, Virginia. **77, 78**

The Wildlife Society. 1993. Conserving biological diversity. The Wildlifer 256:3. **21**

Thiel, R.P. 1985. Relationship between road densities and wolf habitat suitability in Wisconsin. American Midland Naturalist 133:404–407. **174**

Thomas, D.S.G., and N.J. Middleton. 1993. Salinization: New perspectives on a major desertification issue. Journal of Arid Environments 24:95–105. **187**

Thomas, D.S.G., and N.J. Middleton. 1994. Desertification: Exploding the myth. John Wiley and Sons, Chichester, United Kingdom. 194 pp. **187**

Thomas, L. 1974. The lives of a cell. Viking Press, New York. 153 pp. **375**

Thompson, K., and A. Jones. 1999. Human population density and prediction of local plant extinction in Britain. Conservation Biology 13:185–189. **165**

Thorbjarnarson, J. 1999. Crocodile tears and skins: International trade, economic constraints, and limits to the sustainable use of crocodilians. Conservation Biology 13:465–470. **423**

Thorne, E.T., and E.S. Williams. 1988. Disease and endangered species: The black-footed ferret as a recent example. Conservation Biology 2:66–74. **329, 349**

Thornhill, N.W. (ed.). 1993. The natural history of inbreeding and outbreeding. University of Chicago Press, Chicago. 575 pp. **99, 101**

Thrash, I. 1998. Impact of large herbivores at artificial watering points compared to that at natural watering points in Kruger National Park, South Africa. Journal of Arid Environments 38:315–324. **277**

Tilman, D. 1996. Biodiversity: Population versus ecosystem stability. Ecology 77:350–363. **79**

Tilman, D. 1999. The ecological consequences of changes in biodiversity: A search for general principles. Ecology 80:1455–1474. **79, 81**

Tilman, D. 2000. Causes, consequences and ethics of biodiversity. Nature 405:208–211. **80**

Tilman, D., and J.A. Downing. 1994. Biodiversity and stability in grasslands. Nature 367:363–365. **79, 80**

Tilson, R.L., and U. Seal. 1987. Tigers of the world. Noyes, Park Ridge, New Jersey. 510 pp. **60**

Timberlake, L. 1986. Africa in crisis: The causes, the cures of environmental bankruptcy. New Society, Philadelphia. 232 pp. **416**

Timmins, R.J., and V.C. Trinh Viet Cuong. 2000. An assessment of the conservation importance of the Huong Son (Annamite) Forest, Ha Tinh Province, Viet Nam, based on the results of a field survey for large mammals and birds. Center for Biodiversity and Conservation at the American Museum of Natural History, New York, USA. 102 pp. **378**

Tisdell, C., and J.M. Broadus. 1989. Policy issues related to the establishment and management of marine reserves. Coastal Management 17:37–53. **260**

Tomlinson, P.B. 1986. The botany of mangroves. Cambridge University Press, Cambridge. 413 pp. **85**

Torsvik, V., J. Goksoyr, and F.L. Daae. 1990b. High diversity in DNA of soil bacteria. Applied and Environmental Microbiology 56:782–787. **41**

Torsvik, V., K. Salte, R. Sorheim, and J. Goksoyr. 1990a. Comparison of phenotypic diversity and DNA heterogeneity in a population of soil bacteria. Applied and Environmental Microbiology 56:776–781. **41**

Touval, J.L., and J.M. Dietz. 1994. The problem of teaching conservation problem solving. Conservation Biology 8:902–904. **440**

Tovar, D.C. 1989. Air pollution and forest decline near Mexico City. Environmental Monitoring and Assessment 12:49–58. **167**

Towns, D.R., C.H. Daugherty, and I.A.E. Atkinson (eds.). 1990. Ecological restoration of New Zealand islands. Department of Conservation, Wellington, New Zealand. 320 pp. **251, 278, 306**

Towns, D.R., D. Simberloff, and I.A.E. Atkinson. 1997. Restoration of New Zealand islands: Redressing the effects of introduced species. Pacific Conservation Biology 3:99–124. **251, 306, 326, 332**

Townsend, C.R. 1996. Invasion biology and ecological impacts of brown trout *Salmo trutta* in New Zealand. Biological Conservation 78:13–22. **244**

Tracy, C.R., and T.L. George. 1992. On the determinants of extinction. American Naturalist 139:102–122. **139**

Trefethen, J.B. 1964. Wildlife management and conservation. Heath, Boston. 120 pp. **9, 210**

Trimble, S.W., and P. Crosson. 2000. U.S. soil erosion rates—myth and reality. Science 289:248–250. **179**

Trippel, E.A., M.B. Strong, J.M. Terhune, and J.D. Conway. 1999. Mitigation of harbour porpoise (*Phocoena phocoena*) by-catch in the gillnet fishery in the lower Bay of Fundy. Canadian Journal of Fisheries and Aquatic Sciences 56:113–123. **323**

Trombulak, S.C., and C.A. Frissell. 2000. Review of ecological effects of roads on terrestrial and aquatic communities. Conservation Biology 14:18–30. **174**

Turner, A.M., J.C. Trexler, C.F. Jordan, et al. 1999. Targeting ecosystem features for conservation: Standing crops in the Florida Everglades. Conservation Biology 13:898–911. **21**

Tudge, C. 1992. Last animals at the zoo. Island Press, Washington, D.C. 266 pp. **364**

Turcek, F.J. 1951. Effect of introductions on two game populations in Czechoslovakia. Journal of Wildlife Management 15:113–114. **100, 340**

Tyler, V.E. 1986. Plant drugs in the twenty-first century. Economic Botany 40:279–288. **51**

U.S. Fish and Wildlife Service. 1993. National survey of fishing, hunting, and wildlife-associated recreation. U.S. Government Printing Office, Washington, D.C. 124pp. **54**

U.S. Fish and Wildlife Service. 1997. 1996 National survey of fishing, hunting, and wildlife-associated recreation. U.S. Government Printing Office, Washington, DC. 176 pp. **215, 216, 403**

U.S. National Park Service. 1999. National Park Service statistical abstract 1998. National Park Service, Denver, Colorado. 52 pp. **274**

Udvardy, M.D.L. 1975. A classification of the biogeographical provinces of the world. IUCN Occasional Paper No. 18. IUCN, Gland, Switzerland. **71**

Uhl, C. 1998. Conservation biology in your own front yard. Conservation Biology 12: 1175–1177. **299**

United Nations Development Programme, United Nations Environment Programme, World Bank, and World Resources Institute. 2000. World Resources 2000–2001. World Resources Institute, Washington, DC. 389 pp. **53, 182, 205, 263, 289, 311, 338, 400, 417, 418, 440**

Uphoff, N., and J. Langholz. 1998. Incentives for avoiding the tragedy of the commons. Environmental Conservation 25:251–261. **413**

Usher, M.B. (ed.) 1986. Wildlife conservation evaluation. Chapman and Hall, London. 394 pp. **368, 369, 379**

Usher, M.B., T.J. Crawford, and J.L. Banwell. 1992. An American invasion of Great Britain: The case of the native and alien squirrel (*Sciurus*) species. Conservation Biology 6:108–117. **242**

Valutis, L.L., and J.M. Marzluff. 1999. The appropriateness of puppet-rearing birds for reintroduction. Conservation Biology 13:584–591. **336**

VanDeVeer, D., and C. Pierce. 1994. The environmental ethics and policy book. Wadsworth, Belmont California. 649 pp. **397**

Van Dyke, F. , D.C. Mahan, J.K. Sheldon, and R.H. Brand. 1996. Redeeming creation: The biblical basis for environmental stewardship. InterVarsity Press, Downers Grove, Illinois. 213 pp. **385**

Vane-Wright, R.I., C.J. Humphries, and P.H. Williams. 1991. What to protect? Systematics and the agony of choice. Biological Conservation 55:235–254. **63, 372**

Van Horne, B. 1983. Density as a misleading indicator of habitat quality. Journal of Wildlife Management 47:893–901. **164**

Van Riper, C. III, S.G. van Riper, M.L. Goff, and M. Laird. 1986. The epizootiology and ecological significance of malaria in Hawaiian land birds. Ecological Monographs 56:327–344. **241**

Vaughan, D.A., and T-T Chang. 1992. *In situ* conservation of rice genetic resources. Economic Botany 46:368–383. **361**

Vaughn, C.C., and C.M. Taylor. 1999. Impoundments and the decline of freshwater mussels: A case study of an extinction gradient. Conservation Biology 13:912–920. **191**

Veblen, T.T., T. Kitzberger, and J. Donnegan. 2000. Climatic and human influences on fire regimes in ponderosa pine forests in the Colorado Front Range. Ecological Applications 10:1178–1195. **276**

Vermeij, G.J. 1996. An agenda for invasion biology. Biological Conservation 78:3–9. **247**

Verney, P. 1979. Animals in peril: Man's war against wildlife. Brigham Young University Press, Provo, Utah. 187 pp. **229**

Vierling, K. 2000. Source and sink habitats of red-winged blackbirds in a rural/suburban landscape. Ecological Applications 10:1211–1218. **147**

Vietmeyer, N.D. 1992. Neem: A tree for solving global problems. National Academy Press, Washington, D.C. 141 pp. **65**

Viggers, K.L., D.B. Lindenmayer, and D.M. Spratt. 1993. The importance of disease in reintroduction programmes. Wildlife Research 20:687–698. **334**

Virchow, D. 1999. Conservation of genetic resources. Springer-Verlag, Berlin, Germany. 243 pp. **101, 360**

Vitousek, P.M. 1986. Biological invasions and ecosystem properties: Can species make a difference? Pages 163–176 in H.A. Mooney, and J.A. Drake (eds.). Ecology of biological invasions of North America and Hawaii. Springer-Verlag, New York. 321 pp. **244**

Vitousek, P.M. 1990. Biological invasions and ecosystem processes: Towards an integration of population biology and ecosystem studies. Oikos 57:7–13. **244**

Vitousek, P.M., P.R. Ehrlich, A.H. Ehrlich, and P.A. Matson. 1986. Human appropriation of the products of photosynthesis. Bioscience 36:368–373. **163, 172**

Vitousek, P.M., H.A. Mooney, J. Lubchenco, and J.M. Melillo. 1997. Human domination of earth's ecosystems. Science 227:494–499. **163**

Vogel, J.H. 1994. Genes for sale. Oxford University Press, New York. 155 pp. **414**

Vogler, A.P., and R. Desalle. 1994. Diagnosing units of conservation management. Conservation Biology 8:354–363. **39**

Vucetich, J.A., and S. Creel. 1999. Ecological interactions, social organization, and extinction risk in African wild dogs. Conservation Biology 13:1172–1182. **326**

Vucetich, J.A., T.A. Waite, and L. Nunney. 1997. Fluctuating population size and the ratio of effective to census population size. Evolution 51:2017–2021. **106, 107, 158**

Vucetich, J.A., T.A. Waite, L. Qvarnemark, and S. Ibargüen. 2000. Population variability and extinction risk. Conservation Biology 14:1704–1714. **139**

Vyas, N.B., J.W. Spann, G.H. Heinz, et al. 2000. Lead poisoning of passerines at a trap and skeet range. Environmental Pollution 107:159–166. **177**

Wahle, R.A., and R.S. Steneck. 1992. Habitat restrictions in early benthic life: Experiments on habitat selection and *in situ* predation with the American lobster. Journal of Experimental Marine Biology and Ecology. 157:91–114. **226**

Walhberg, N., A. Moilanen, and I. Hanski. 1996. Predicting the occurrence of endangered species in fragmented landscapes. Science 273:1536–1538. **149**

Walsh, G.E. 1977. Exploitation of mangal. Pages 347–362 in V.J. Chapman (ed.). Wet coastal ecosystems. Elsevier, Amsterdam. 428 pp. **86**

Walsh, V., and J. Goodman. 1999. Cancer chemotherapy, biodiversity, public and private property: The case of the anti-cancer drug Taxol. Social Science and Medicine 49:1215–1225. **52**

Ward, D., and B.T. Ngairorue. 2000. Are Namibia's grasslands desertifying? Journal of Range Management 53:138–144. **187**

Wardle, P. 1991. Vegetation of New Zealand. Cambridge University Press, Cambridge. 672 pp. **251**

Warren, K.J. 1993. Ecofeminism: Introduction. Pages 253–267 in M.E. Zimmerman, J.B. Callicott, G. Sessions, et al. (eds.). Environmental philosophy: From animal rights to radical ecology. Prentice Hall, Englewood Cliffs, New Jersey. 437 pp. **387**

Warren, K.J. (ed.). 1997. Ecofeminism: Women, culture, nature. Indiana University Press, Bloomington. 459 pp. **388**

Watling, L., and E.A. Norse. 1998. Disturbance of the seabed by mobile fishing gear: A comparison to forest clearcutting. Conservation Biology 12:1180–1197. **217**

Wayne, R.K., and J.L. Gittleman. 1995. The problematic red wolf. Scientific American 273:36–39. **37**

Wayne, R.K., W.S. Modi, and S.J. O'Brien. 1986. Morphological variability and asymmetry in the cheetah (*Acinonyx jubatus*), a genetically uniform species. Evolution 40:78–85. **111**

Webb, E. 1994. Debt for nature swaps: The past, the present and some possibilities for the future. Environmental Planning and Law Journal 11:222–244. **416**

Webb, G.J.W., S.C. Manolis, and P.J. Whitehead (eds.). 1987. Wildlife management: Crocodiles and alligators. Surrey Beatty and Sons, Chipping Norton, New South Wales, Australia. 552 pp. **423**

Webb, S.D. 1984. Ten million years of mammal extinctions. Pages 189–210 in P.S. Martin and R.G. Klein (eds.). Quaternary extinctions. The University of Arizona Press, Tucson. 892 pp. **208**

Webb, T. III. 1992. Past changes in vegetation and climate: Lessons for the future. Pages 59–75 in R.L. Peters and T.E. Lovejoy (eds.). Global warming and biological diversity. Yale University Press, New Haven. 386 pp. **131**

Wege, D.C., A.J. Long, M.K. Vinh, et al. 1999. Expanding the protected areas network in Vietnam for the 21st century: An analysis of the current system with recommendations for equitable expansion. BirdLife International Vietnam Programme, Hanoi. **376**

Wehausen, J.D. 1999. Rapid extinction of mountain sheep populations revisited. Conservation Biology 13:378–384. **139**

Welsh, B.W.W., and P. Butorin (eds.). 1990. Dictionary of development. Garland, New York. 2 vols. 1194 pp. **429**

West, P.C., and S.R. Brechin (eds.). 1991. Resident peoples and national parks: Social dilemmas and strategies in international conservation. University of Arizona Press, Tucson. 443 pp. **436**

Westemeier, R.L., J.D. Brawn, S.A. Simpson, et al. 1998. Tracking the long-tern decline and recovery of an isolated population. Science 282:1695–1698. **334**

Wester, L. 1994. Weed management and the habitat protection of rare species: A case study of the endemic Hawaiian fern *Marsilea villosa*. Biological Conservation 68:1–9. **330**

Westing, A.H. (ed.). 1984. Herbicides in war: The long-term ecological and human consequences. Taylor and Francis, London. 210 pp. **166**

Westing, A.H. 1980. Warfare in a fragile world. Taylor and Francis, London. 249 pp. **166**

Westman, W.E. 1977. How much are nature's services worth? Science 197:960–964. **401**

Westman, W.E. 1990. Park management of exotic plant species: Problems and issues. Conservation Biology 4:251–260. **247**

Whelan, R.J. 1995. The ecology of fire. Cambridge University Press, Cambridge, United Kingdom. 346 pp. **180**

Whelan, T. (ed.). 1991. Nature tourism. Island Press, Washington, D.C. 223 pp. **54, 74**

Whisenant, S G. 1999. Repairing damaged wildlands. Cambridge University Press, Cambridge, United Kingdom. 312 pp. **306**

Whitcomb, R.F., J.F. Lynch, P.A. Opler, and C.S. Robbins. 1976. Island biogeography and conservation: Strategy and limitations. Science 193:1030–1032. **266**

White, L., Jr. 1967. The historical roots of our ecologic crisis. Science 155:1203–1207. **385**

Whitehead, H., J. Christal, and S. Dufault. 1997. Past and distant whaling and the rapid decline of sperm whales off the Galápagos Islands. Conservation Biology 11:1387–1396. **221**

Whitford, W.G. 1997. Desertification and animal biodiversity in the desert grasslands of North America. Journal of Arid Environments 37:709–720. **188**

Whitham, T.G., P.A. Morrow, and B.M. Potts. 1991. Conservation of hybrid plants. Science 254:779–780. **37, 39**

Whitmore, T.C., and G.T. Prance (eds.). 1987. Biogeography and quaternary history in tropical America. Oxford University Press, New York. 214 pp. **131**

Whittaker, R.H. (ed.) 1973. Ordination and classification of communities part V. Handbook of vegetation science. Junk, The Hague. 737 pp. **69**

Whittaker, R.H. 1960. Vegetation of the Siskiyou Mountains, Oregon and California. Ecological Monographs 30:279–338. **26**

Wiese, R.J., and M. Hutchins. 1994. Species survival plans. American Zoo and Aquarium Association, Bethesda, Maryland. 64 pp. **353, 354, 364**

Wilcove, D.S. 1999. The condor's shadow. W. H. Freeman and Company, New York. 339 pp. **210, 229**

Wilcove, D.S., and M.J. Bean (eds.). 1994. The big kill: Declining biodiversity in America's lakes and rivers. Environmental Defense Fund, Washington, D.C. 275 pp. **144, 193, 287, 311**

Wilcove, D.S., M. Bean, and P.C. Lee. 1992. Fisheries management and biological diversity: Problems and opportunities. Transactions of the North American Wildlife and Natural Resources Conference 57:373–383. **289**

Wilcove, D.S., and L.Y. Chen. 1998. Management costs for endangered species. Conservation Biology 12:1405–1407. **408**

Wilcove, D.S., and R.M. May. 1986. National park boundaries and ecological realities. Nature 324:206–207. **269**

Wilcove, D.S., D. Rothstein, J. Dubow, et al. 1998. Quantifying threats to imperiled species in the United States. BioScience 48:607–615. **144**

Wildlife Management Institute. 1994. The Conservation Reserve Program: A wildlife conservation legacy. Washington, D.C. **404**

Williams, G.C. 1966. Adaptation and natural selection. Princeton University Press, Princeton, New Jersey. 307 pp. **99**

Williams, P., D. Gibbons, C. Margules, et al. 1996. A comparison of richness hotspots, rarity hotspots, and complementary areas for conserving the diversity of British birds. Conservation Biology 10:155–174. **263**

Williamson, D., J. Williamson, and K.T. Ngwamotsoko. 1988. Wildebeest migration in the Kalahari. African Journal of Ecology 26:269–280. **176**

Williamson, M.H. 1981. Island populations. Oxford University Press, Oxford. 286 pp. **126**

Williamson, M., and A. Fitter. 1996a. The varying success of invaders. Ecology 77:1661–1666. **245**

Williamson, M.H., and A. Fitter. 1996b. The characters of successful invaders. Biological Conservation 78:163–170. **246**

Willis, E.O. 1984. Conservation, subdivision of reserves, and the anti-dismemberment hypothesis. Oikos 42:396–398. **267**

Willis, K., and R.J. Wiese. 1993. Genetic variation maintenance strategy. CBSG News 4(2):11. **354**

Willis, K.J., and R.J. Whittaker. 2000. The refugial debate. Science 287:1406–1407. **131**

Wilson, E.O. 1984. Biophilia. Harvard University Press, Cambridge, Massachusetts. 157 pp. **56**

Wilson, E.O. 1988. The current state of biological diversity. Pages 3–18 in E.O. Wilson and F.M. Peter (eds.). National Academy Press, Washington, D.C. 521 pp. **95**

Wilson, E.O. 1992. The diversity of life. Harvard University Press, Cambridge, Massachusetts. 424 pp. **34, 66, 125, 126, 127, 136, 183, 184, 440**

Wilson, J.A. 1977. A test of the tragedy of the commons. Pages 96–111 in G. Hardin and J. Baden (eds.). Managing the commons. W. H. Freeman, New York. 294 pp. **413**

Wilson, M.A., and S.R. Carpenter. 1999. Economic valuation of freshwater ecosystem services in the United States: 1971–1997. Ecological Applications 9:772–783. **402**

Witherington, B.E., and K.A. Bjorndal. 1991. Influences of artificial lighting on the seaward orientation of hatchling loggerhead turtles *Caretta caretta*. Biological Conservation 55:139–149. **177**

Witman, J.D., and K.P. Sebens. 1992. Regional variation in fish predation intensity: A historical perspective in the Gulf of Maine. Oecologia 90:305–315. **226**

Witte, F., T. Goldschmidt, P.C. Goudswaard, et al. 1992b. Species extinction and concomitant ecological changes in Lake Victoria. Netherlands Journal of Zoology 42:214–232. **240**

Witte, F., T. Goldschmidt, J. Wanink, et al. 1992a. The destruction of an endemic species flock: Quantitative data on the decline of the haplochromine cichlids of Lake Victoria. Environmental Biology of Fishes 34:1–28. **234, 240**

Wolf, A.T., S.P. Harrison, and J.L. Hamrick. 2000. Influence of habitat patchiness on genetic diversity and spatial structure of a serpentine endemic plant. Conservation Biology 14:454–463. **341**

Wolf, C.M., B. Griffith, C. Reed, and S.A. Temple. 1996. Avian and mammalian translocations: Update and reanalysis of 1987 survey data. Conservation Biology 10:1142–1154. **333**

Woodroffe, R. 1999. Managing disease threats to wild mammals. Animal Conservation 2:185–193. **328, 329**

World Commission on Environment and Development. 1987. Our common future. Oxford University Press, Oxford, United Kingdom. 400 pp. **263**

World Conservation Monitoring Centre. 1990. 1990 IUCN red list of threatened animals. IUCN, Gland, Switzerland. 192 pp. **44**

WRI (World Resources Institute). 1992. World resources 1992–93. Oxford University Press, New York. 385 pp. **417**

WRI. 1994. World resources 1994–95. Oxford University Press, New York. 385pp. 400pp. **431**

World Resources Institute, United Nations Environment Programme, United Nations Development Programme, and The World Bank. 1998. World Resources 1998–1999. Oxford University Press, New York. 369 pp. **170**

Worster, D. 1977. Nature's economy: The roots of ecology. Sierra Club books, San Francisco. 404 pp. **18**

Wright, M.G., and M.J. Samways. 1998. Insect species richness tracking plant species richness in a diverse flora: Gall-insects in the Cape Floristic Region, South Africa. Oecologia 115:427–433. **82**

Wright, R.G. (ed.). 1996. Nation parks and protected areas: Their role in environmental protection. Blackwell Science, Cambridge Massachusetts. 470 pp. **311**

Wright, S. 1978. Evolution and the genetics of populations: Vol. 4: Variability within and among natural populations. University of Chicago Press, Chicago. 580 pp. **145**

Wunderle, J.M., Jr., and S.C. Latta. 2000. Winter site fidelity of Nearctic migrants in shade coffee plantations of different sizes in the Dominican Republic. Auk 117:596–614. **293**

Xerces Society. 1990. Butterfly gardening. Sierra Club Books, San Francisco. 192 pp. **301**

Yaeger, C.P. 1997. Orangutan rehabilitation in Tanjung Puting National Park, Indonesia. Conservation Biology 11:802–805. **335**

Yaffee, S.L. 1999. Three faces of ecosystem management. Conservation Biology 13:713–725. **311**

Yerli, S., A.F. Canbolat, L.J. Brown, and DW. Macdonald. 1997. Mesh grids protect loggerhead turtle *Caretta caretta* nests from red fox *Vulpes vulpes* predation. Biological Conservation 82:109–111. **326**

Yoakum, J., W.P. Dasmann, H.R. Sanderson, et al. 1980. Habitat improvement techniques. Pages 329–403 in S.D. Schemnitz (ed.). Wildlife management techniques manual. 4th ed. The Wildlife Society, Washington, D.C. 686 pp. **317, 324**

Yonzon, P., and M.L. Hunter, Jr. 1991. Cheese, tourists, and red pandas in the Nepal Himalayas. Conservation Biology 5:196–202. **413**

Zedler, J.B. 2000. Progress in wetland restoration ecology. Trends in Ecology and Evolution 15:402–407. **305**

Zimmerman, B.L., and R.O. Bierregaard. 1986. Relevance of the equilibrium theory of island biogeography and species-area relations to conservation with a case from Amazonia. Journal of Biogeography 13:133–143. **195**

Zimmerman, M.E., J.B. Callicott, G. Sessions, et al. (eds.). 1993. Environmental philosophy: From animal rights to radical ecology. Prentice Hall, Englewood Cliffs, New Jersey. 437 pp. **397**

Species Index[1]

---

[1] Taxonomy follows original references or Parker, S.P. (ed.). 1982. Synopsis and classification of living organisms. McGraw -Hill, New York.

Subject Index